Golden sedge (*Carex aurea*), see p. 190

WETLAND *and* AQUATIC PLANTS *of the* NORTHERN GREAT PLAINS

A field guide for North and South Dakota, Nebraska, eastern Montana and eastern Wyoming

STEVE W. CHADDE

WETLAND AND AQUATIC PLANTS OF THE NORTHERN GREAT PLAINS

A field guide for North and South Dakota, Nebraska, eastern Montana and eastern Wyoming

Steve W. Chadde

Printed in the United States of America.

ISBN: 978-1-951682-15-6

Portions of this work were originally published as:
Larson, Gary E. 1993. *Aquatic and wetland vascular plants of the northern Great Plains.* Gen. Tech. Rep. RM-238. Fort Collins, CO: U.S. Department of Agriculture, Forest Service, Rocky Mountain Forest and Range Experiment Station. 681 p. (public domain).

Grateful acknowledgment is given to the Biota of North America Project (*www.bonap.org*) for permission to use their data to generate the distribution maps. Photographs were taken by the author, obtained from public domain sources, or used under Creative Commons commercial use licences. Special thanks go to photographers Matt Lavin, Joshua Mayer, Doug McGrady, Andreas Rockstein, Forest and Kim Starr, Len Worthington, and Andrey Zharkikh; examples of their excellent work can be seen at *www.flickr.com.*

The author can be contacted via email: steve@chadde.net

VERSION 1.1 (4/10/2020)

PREFACE

WETLAND AND AQUATIC PLANTS OF THE NORTHERN GREAT PLAINS is an updated and expanded version of *Aquatic and Wetland Vascular Plants of the Northern Great Plains,* originally prepared by Dr. Gary E. Larson, South Dakota State University, and published in 1993 as General Technical Report RM-238 by the Rocky Mountain Forest and Range Experiment Station (U.S. Department of Agriculture, Forest Service). While I have made many changes to the text of the original work, I have retained many of Dr. Larson's keys and descriptions, and I extend my sincere appreciation to him for his exhaustive research in preparing the original publication, a first for the vascular plants of the region's wetlands.

Included in this work are keys, descriptions, illustrations, and distribution maps for approx. 500 species found in the wetland and aquatic habitats of North and South Dakota, Nebraska, and the non-mountainous regions of eastern Montana and eastern Wyoming.

After the passage of nearly 30 years, changes have occurred in our understanding of the region's wetland flora. For example, botanical surveys and new collections have expanded or refined the range of many species within the region, and the maps presented here provide useful insights into each species' currently known distribution. Taxonomic research has redefined taxa and adjusted their nomenclature and is incorporated into this guide. In 2016, the National Wetland Plant List published new wetland status indicator ratings, and also based the new ratings on broad eco-regions rather than state boundaries as in previous lists (Lichvar et al., 2016). These ratings are mandated for use in wetland delineation studies and are also of use when conducting wetland studies.

Overall objectives in publishing *Wetland and Aquatic Plants of the Northern Great Plains* were to:

- Update the botanical nomenclature for the region's wetland flora. In most cases nomenclature follows that of the published volumes of The Flora of North America series (1993+), the Synthesis of the North American Flora (Kartesz 2014), and the Integrated Taxonomic Information System (ITIS, *www.itis.gov*).
- Incorporate the 2016 wetland indicator classification ratings of the National Wetland Plant List for the three regions included within the Northern Great Plains.
- Illustrate the known county distribution of each species using maps adapted from data of the Biota of North America Project (BONAP, *www.bonap.org*).
- Identify species which are state or federally listed as threatened or endangered.
- Provide a practically sized field guide for land managers, botanists, educators, students, and others interested in the wetland flora of the region. While the work is semi-technical in nature, an extensive glossary is provided. The distribution maps, in addition to the keys, descriptions, and illustrations, can be of great help in identifying unknown specimens.

STEVE W. CHADDE
November 2019

Prairie pothole region in northeastern South Dakota (aerial view). The numerous wetlands and small ponds are critical habitat for waterfowl, and home to many of the plants described in this book.

CONTENTS

INTRODUCTION

Vegetation is usually the most conspicuous aspect of natural habitats, and in the case of wetlands this is especially true. The vascular plants growing in wetlands are crucial to the overall functioning of wetland and adjacent terrestrial ecosystems. Wetland vegetation provides food (directly or indirectly), cover, and shelter needed by wildlife and waterfowl populations. Some wetland plants are excellent forages for livestock while others may be undesirable weeds. Marsh plants stabilize fine-textured, erodible soils of basins and waterways with their extensive root and rhizome systems, while they also absorb and recycle some of the excess nutrients contained in agricultural and municipal runoff. Wetland plants also add organic matter to the sediments that gradually accumulate as peat in wetland basins. Some hydrophytes are reliable indicators of surface water quality and permanence, and for this reason plant species are often used as indicators in defining and classifying wetlands (Stewart and Kantrud 1971,1972; Cowardin et al. 1979).

From whatever standpoint wetland ecosystems are studied or managed, the botanical component must be taken into account. Some wetland plants are more valuable than others in fulfilling various ecological roles. Management techniques such as mowing, prescribed burning, periodic flooding and drainage, or controlled grazing can be used to alter the composition of wetland vegetation. Species desirable for wildlife or livestock can thus be favored or diminished depending upon the influence of management practices. Before instituting management procedures, however, the habitat manager must know the identities of the plant species comprising the vegetation. Once a species is identified, information on the plant's reproductive ecology, habitat preferences, and relative value to wildlife and grazing animals is needed so that management procedures can be implemented to achieve desired results.

For many parts of North America, aquatic plant manuals have been produced to provide biologists and wetland managers with a means of identifying the plants encountered in wetland habitats. Noteworthy examples are the manuals by Muenscher (1944) and Fassett (1957) treating the aquatic plants for the predominantly wooded region of eastern North America, Correll and Correll (1975) covering the wetland flora of the southwestern USA, and Godfrey and Wooten (1979, 1981) providing a treatment for the southeastern USA. Manuals such as these are valuable tools for wetland managers and researchers, as well as botanists, for they acquaint these people with the botanical resources of regional wetlands.

Given the vastness and ecological importance of the wetland resource in the northern Great Plains, there is much justification for a manual treating the regional wetland flora. Manuals produced for other parts of North America have limited application in the so-called Prairie Pothole Region and the Nebraska Sand Hills because the assemblages of plant species inhabiting these wetlands are a unique combination of eastern, western, and boreal North American elements, as well as some more cosmopolitan and introduced species. Since no existing manuals handle the combination of species encountered in northern prairie wetlands, their utility in this region is limited.

Another shortcoming of using existing manuals in the northern Great Plains is that important species found in marginal habitats (such as wet meadows and drawdown zones) are typically excluded. Outside of the northern Great Plains, many of these species characteristically occupy more upland habitats. Wet meadows, drawdown zones, and the plants that inhabit them are integral parts of prairie wetland ecosystems, so to be of greatest utility to the region's wetland researchers and managers, a manual should include these wetland species as well as the strictly aquatic species.

The aim of this publication is to provide a comprehensive manual of the vascular plants growing in wetland habitats of the northern Great Plains, and thereby to assist students and professionals in the identification of aquatic and wetland plants that occur here. The taxonomic keys, plant descriptions, distribution maps, and statements of range and habitat preferences are offered to fulfill this primary objective.

A few secondary objectives are also met in providing a treatment of the regional wetland flora. Considerable new information on the vascular flora has been generated by recent investigations, and one purpose of this publication is to incorporate new discoveries from these floristic explorations of regional wetlands. Another purpose is to provide updated nomenclature for groups affected by name changes. Some of the taxa covered by this treatment pose difficult taxonomic problems deserving of further study. Therefore, defining problematic groups and suggesting areas of taxonomic research is another objective of this study. The references cited were consulted for information pertaining to the taxonomy of specific groups, but they are also provided to acquaint the user of this manual with previous taxonomic literature. These references may be studied to gain further insight into the groups treated.

Defining the Wetland Flora

Authors inevitably differ in their concept of which species should be included in a taxonomic treatment of aquatic plants. Each has his or her own definition for the term "aquatic." Virtually all authors recognize those plants with a submersed or floating growth habit as aquatic. Most also apply the term to the common emergent species. The definition is harder to apply consistently for (1) plants growing in marginal zones of wetlands, e.g., shores, wet meadows, streambanks, etc., and (2) plants displaying a wide-ranging ecological amplitude which enables them to grow in either wet or dry situations. Whether a plant is to be designated as aquatic or not is thus based upon the plant's growth habit and the types of habitats in which it is found. How much weight is assigned to either of these criteria is a subjective decision which largely accounts for differences in the taxa treated by various aquatic plant manuals.

Much confusion can be avoided at the outset by defining the terms used to delimit the group of plants included herein. The term "aquatic" is used to describe those plants growing in water or in soils that are saturated during most of the growing season. Four categories of aquatic plants may be recognized on the basis of growth form and zone of habitation:

1. ***Free-floating*** is the term used for plants which float at or beneath the water surface without attachment to the substrate. Free-floating aquatics are transported freely by wind and currents, so they are normally found in abundance only in calm, sheltered waters. Duckweeds (*Lemna* spp.), bladderworts (*Utricularia* spp.), and coontail (*Ceratophyllum demersum*) are common examples of free-floating aquatics.

2. ***Submergent*** describes plants anchored to the bottom by roots or rhizomes. Their foliage is either entirely submersed or some floating leaves may also be present. Reproductive structures may be submersed, floating, or borne above the water surface. Submergent plants occur in very shallow to deep water, depending upon water clarity, substrate, and growth form. Some common examples include pondweed (*Potamogeton* spp.), water milfoil (*Myriophyllum* spp.), waterweed (*Elodea* spp.), and widgeon-grass (*Ruppia cirrhosa*).

3. ***Emergent*** refers to those species which occur on saturated soils or on soils covered with water for most of the growing season. The foliage of emergent aquatics is partly or entirely borne above the water surface. Examples of emergent aquatics are many, including arrowhead (*Sagittaria* spp.), cattail (*Typha* spp.), common reed (*Phragmites australis*), and bulrushes (*Schoenoplectus, Scirpus*).

4. ***Amphibious*** applies to aquatic species which are capable of growing as either submergent or emergent aquatics. These species commonly assume a semi-terrestrial growth form when stranded by a receding water level. The semi-terrestrial growth form usually differs markedly in appearance from the submersed growth form. Amphibious aquatics include yellow water-crowfoot (*Ranunculus*

flabellaris, R. gmelinii), pepperwort (*Marsilea vestita*), and water smartweed (*Persicaria amphibia*).

Many of the species included by this treatment fit none of the categories given above. These are plants that ordinarily inhabit wet meadows, shores, streambanks, exposed mud flats, and other marginal habitats where the soil is saturated for only part of the growing season. Since these habitats are inherent to prairie wetland ecosystems, the plants which inhabit them are logically part of the wetland flora. They are, therefore, included in this treatment as wetland plants.

Not all wetland plants are restricted to wetland habitats. A considerable number of them are also found in upland situations. Many are opportunistic weedy species that rapidly invade soil left bare by receding water. Plants like barnyardgrass (*Echinochloa muricata*), foxtail barley (*Hordeum jubatum*), and common plantain (*Plantago major*) are as likely to be found on disturbed upland sites as on mud flats or shorelines. Some of the agronomic weeds, e.g., sowthistle (*Sonchus arvensis*), Canada thistle (*Cirsium arvense*), and cocklebur (*Xanthium strumarium*), are well-suited to the disturbed conditions provided by shorelines and streambanks. Still other plants found in both wetland and upland habitats are not weedy at all, but are simply capable of growing in a variety of moisture regimes. The wild lily (*Lilium philadelphicum*) and blue-eyed grass (*Sisyrinchium montanum*) are nonweedy species encountered in wet meadows and boggy places, as well as in prairie habitats. Even though these plants are by no means restricted to wetlands, they are encountered in wetland habitats with considerable frequency and are included in this treatment to ensure adequate coverage of the wetland flora.

Using this Guide

This treatment is composed of taxonomic keys and plant descriptions to facilitate identification of about 500 vascular plants encountered in wetland habitats of the northern prairie region. Nearly all taxa are illustrated with a color photograph or line drawing to help the user conceptualize what is being described. When used in combination with the keys, descriptions, and maps, correct identification of unknown plant specimens becomes much more likely.

The terminology used in keys and descriptions is necessarily botanical, because to distinguish between the various kinds of plants requires observing traits that are technical by nature. Those unfamiliar with descriptive terminology of plants will find it helpful to consult the **Glossary** (p. 317), relevant illustrations, and the lead-in descriptions of genera and families. A hand lens or binocular microscope, a 15-cm ruler with mm scale (provided on last page of the Index), and dissecting implements will also help, because discerning between closely related plant taxa almost always entails close inspection and careful measurement of plant structures.

Included with the species descriptions in the guide are:

(1) common name;

(2) flowering/fruiting periods (or periods when spore-producing structures are present in ferns and fern allies);

(3) habitat descriptions;

(4) regional distribution map within the coverage area of North Dakota, South Dakota, Nebraska, eastern Montana, and eastern Wyoming (wetland species of mountainous portions of Montana and Wyoming are not specifically included in this guide, however, species of the Black Hills region are treated);

(5) nomenclatural synonyms for those species treated under other names in recent works;

(6) wetland indicator status ratings for the species' occurrence in one or more of the three regions present in northern Great Plains area (Midwest; Great Plains; Western Mountains, Valleys and Coast, see page 316);

(7) conservation status for those species having legal protection (endangered or threatened) under state or federal law.

The original nomenclature and taxonomic concepts were largely those of the Flora of the Great Plains (Great Plains Flora Association 1986). However, this updated work largely follows the nomenclature of the published volumes of The Flora of North America series (1993+), the Synthesis of the North American Flora (Kartesz 2014), and the Integrated Taxonomic Information System (ITIS, *www.itis.gov*).

The ordering of plant families is alphabetical after first dividing the flora into three, often easily recognizable groups: **ferns and fern allies, dicotyledons** ("dicots"), and **monocotyledons** ("monocots"). Notes on related taxa less commonly encountered in regional wetlands, or other items of interest, are sometimes included in a paragraph following the appropriate species description.

The regional range distribution map for each species shows the plant's verified occurrence on a county by county basis as documented by the Biota of North America Program (BONAP, *www.bonap.org*).

The diagnostic keys to taxa in the manual are strictly dichotomous, meaning that the user must choose between two leads at each step in the key. The two leads (called a couplet) always share the same number and describe contrasting characteristics or conditions. The user must determine which one of the two leads applies to the plant being identified. Once this determination is made, the next couplet directly below the chosen lead is considered in the same manner, and a choice is made between the two leads of that couplet. The process continues until the taxon (the particular family, genus, and ultimately, the species) to which the plant belongs is determined.

The traits used as key characters are those considered most readily observable, yet least likely to cause errors in identification. The use of technical terms and reference to minute traits are unavoidable in many instances, especially in the treatments of the more complex and specialized taxa like the grasses (Poaceae), sedges (Cyperaceae), and composites (Asteraceae). Specialized structures are described and often illustrated in the treatments of families and genera where they are found.

The key beginning on page 309 is a key to the plant families represented in this regional wetland flora. This key will direct the user to the appropriate family treatment, and often (in parentheses) to a particular genus or species within the manual. Keys to genera and species are provided within each family treatment wherever there are two or more different possibilities. With practice and a growing familiarity with plant characters, the keying process becomes easier and faster. The family and genus keys can be bypassed once one is able to recognize families and genera on sight.

Wetland Indicator Status

For each species described in the guide, the wetland indicator status is provided (codes and definitions below). The indicator status ratings are based on the latest National Wetland Plant List (NWPL, Lichvar et al., 2016) and are widely used in wetland delineation studies. for a map three wetland delineation regions present in the northern Great Plains (see map, page 316):

- **GP** (Great Plains)
- **MIDW** (Midwest)
- **WMTN** (Western Mountains, Valleys and Coast).

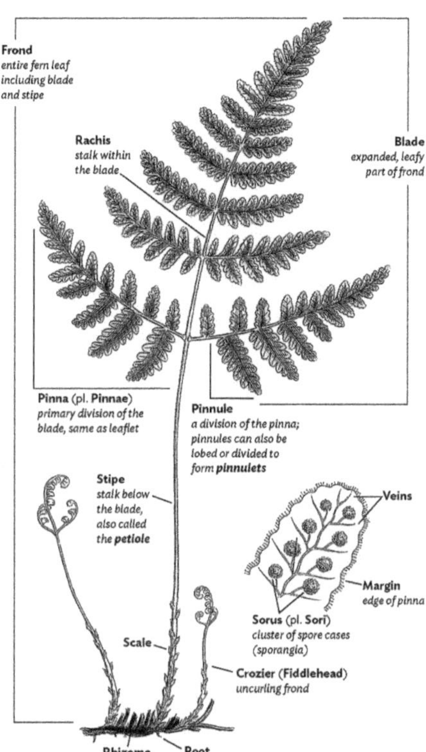

FERN TERMINOLOGY

FERNS AND FERN ALLIES

Included here are vascular plants reproducing by spores rather than by flowers and seeds. In wetlands of the northern Great Plains region, this includes the **Polypodiopsida** (true ferns) and the **Lycopodiopsida** (clubmosses, quillworts). Horsetails (Equisetaceae), formerly separated from "true ferns," are now typically included within Polypodiopsida. The Polypodiopsida are represented by nine families in wetlands of the northern Great Plains (Athyriaceae, Dryopteridaceae, Equisetaceae, Marsileaceae, Onocleaceae, Ophioglossaceae, Osmundaceae, Salviniaceae, Thelypteridaceae). Isoetaceae is the sole family within Lycopodiopsida in wetlands of the region.

All ferns and fern-allies reproduce by spores rather than seeds. The spore germinates and produces a tiny flattened thallus termed the *gametophyte*. The larger, leafy, spore-bearing plant (*sporophyte*) results from the fertilized egg produced by the pistillate part of the gametophyte. The uncurling of the young leaf or frond is characteristic. The Lycopodiopsida bear their spores on specialized leaves (sporophylls). The Equisetaceae (horsetails) bear spores in terminal cone-like structures, either on typical green stems, or on specialized fertile stems lacking chlorophyll.

Athyriaceae *lady fern family*

Athyrium filix-femina (L.) Roth
LADY FERN

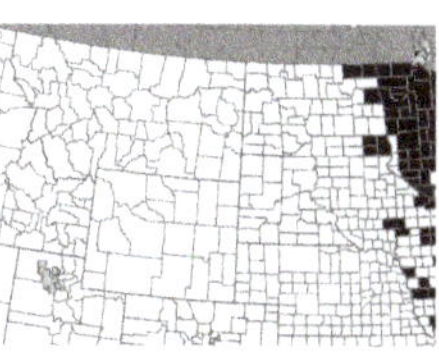

DESCRIPTION Rhizome stout, short-creeping, scaly. **Fronds** annual, alike, clustered, 4–10 dm tall, the blades bipinnate to bipinnate-pinnatifid, with the pinnules toothed to deeply dissected, lanceolate, 2.5–6 dm long, 1–3 dm wide, glabrous, acuminate, narrowed at the base; pinnae numbering mostly 15–25(30) pairs, spreading, mostly alternate, linear-lanceolate, 1–4 cm wide; veins simple to dichotomizing a few times; petioles ca. 1/2 the length of the blade, with chaffy scales mostly near the base. **Sori** round to oblong, often hooked or horseshoe-shaped, in 2 regular rows on the backs of the pinnules; **indusium** attached along its margin to one side of the sorus. June–Sept.

SYNONYM *Athyrium angustum* (Willd.) Presl.

HABITAT Swamp margins, wooded banks and alluvial woods.

WETLAND STATUS
GP FAC | MIDW FAC | WMTN FAC

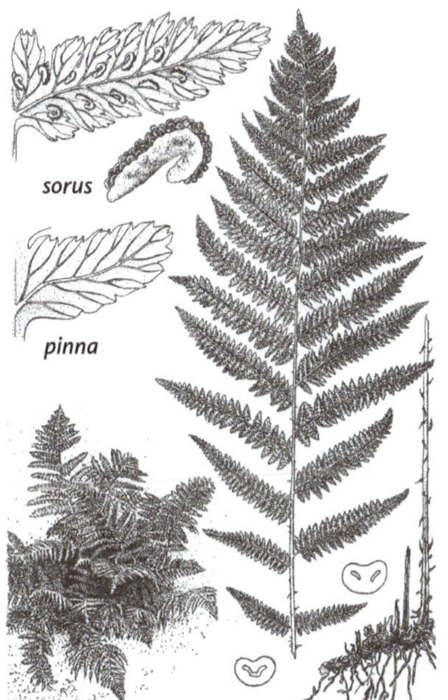

Athyrium filix-femina
LADY FERN

Dryopteridaceae
wood fern family

Dryopteris WOOD FERN

Rather large ferns of moist to wet, wooded habitats; rhizomes stout, scaly. **Fronds** alike or somewhat dimorphic, the sterile ones sometimes persisting over winter, glabrous, the blades pinnate-pinnatifid to tripinnate, ultimate segments commonly toothed or lobed; veins free, simple to once or twice dichotomous; petioles shorter than the blades, stramineous, bearing membranous scales. **Sori** round, occurring at regular intervals on the backs of the ultimate segments; **indusium** reniform, attached at its center in the middle of the sorus.

---

1 Blades bipinnate-pinnatifid to nearly tripinnate ***D. carthusiana***

1 Blades pinnate-pinnatifid to nearly bipinnate ***D. cristata***

---

fertile pinna

Dryopteris carthusiana (Vill.) H. P. Fuchs
SPINULOSE WOODFERN

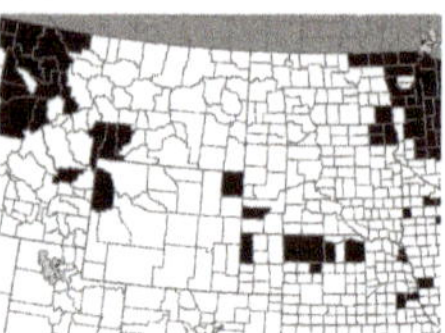

DESCRIPTION **Fronds** alike, clustered, to 10 dm tall, the blades bipinnate-pinnatifid to nearly tripinnate, lanceolate, 2–6 dm long, 6–30 cm wide, acuminate, slightly narrowed at the base; pinnae usually numbering 10–15 pairs, ascending to spreading, alternate to subopposite, lanceolate to linear-lanceolate; pinnules toothed to deeply pinnately lobed, mostly 5–40 mm long, 3- 10 mm wide, the teeth or lobes spinulose-tipped. June–Sept.

SYNONYM *Dryopteris spinulosa* (Mull.) Watt.

HABITAT Wet alluvial woods or swamps.

WETLAND STATUS
GP FACW | MIDW FACW | WMTN FAC

Dryopteris carthusiana
SPINULOSE WOODFERN

Dryopteris cristata (L.) A. Gray
CRESTED WOODFERN

DESCRIPTION **Fronds** somewhat dimorphic, clumped, the outer sterile fronds smaller than the inner fertile ones and persisting longer during winter, the fertile fronds 3–8 dm tall; blades pinnate-pinnatifid to nearly bipinnate, narrowly lanceolate, 2–6 dm long, 7–15 cm wide, acuminate, narrowed at the base; pinnae as in the preceding, except ovate-lanceolate to lanceolate, only once-pinnately lobed; pinnules toothed, mostly 4–15 mm long, 4–7 mm wide, usually spinulose. June–Sept.

HABITAT Wet alluvial woods or swamps.

WETLAND STATUS
GP OBL | MIDW OBL

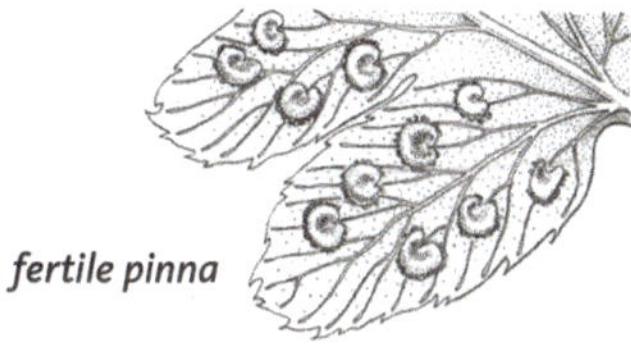

fertile pinna

Dryopteris cristata
CRESTED WOODFERN

Equisetaceae *horsetail family*

Equisetum HORSETAIL, SCOURING RUSH

Erect or decumbent perennial herbs; **stems** jointed and solid at the nodes, hollow through the internodes, longitudinally ridged, silicaceous, sometimes evergreen, some species dimorphic with colorless fertile stems which appear before the green sterile ones, the fertile stems eventually turning greenish in some; branches, if present, whorled in a regular or irregular fashion, arising from inside the sheath bases, sometimes rebranched; ridges of the stem sometimes roughened with spicules of silica deposit; rhizomes deeply buried, with adventitious roots. **Leaves** small and scalelike, often lacking chlorophyll, whorled and united at the base to form sheaths around the stem, the teeth of the sheaths (corresponding to the leaves) free or coherent by the margins in pairs or groups. **Sporangia** homosporous, borne in a terminal cone made up of whorled, peltate sporangiophores; spores spherical, with 4 spirally wound elaters. Gametophyte photosynthetic.

NOTE Perhaps only *Equisetum fluviatile* and *E. palustre* are truly wetland plants, since they are more or less restricted to permanently saturated substrates. The remaining 5 species occurring in our region are either tolerant of wide-ranging soil moisture conditions or require consistently damp substrates, and so all have been included to ensure the proper identification of any *Equiseta* encountered in wet places.

---

1 Stems fertile, brownish except that occasionally the sheaths may be green, or very small green branches may be present **2**
1 Stems fertile or sterile, green **4**

2 Teeth of the sheaths with a white margin; stem becoming branched with green branches ***E. pratense***
2 Teeth of the sheaths brown to blackish; stem simple or branched **3**

3 Stem becoming branched with green branches; teeth of the sheaths coherent in a few groups, reddish-brown, the groups with rounded to broadly acute tips ***E. sylvaticum***

3 Stem not branched, soon withering; teeth of the lower sheaths dark brown to blackish, mostly separate, with long-acuminate tips **_E. arvense_**

4 Central cavity very large in relation to the diameter of the stem (ca. 4/5 the stem diameter), the stem wall thin; small outermost cavities opposite the ridges of the stem, i.e., on the same radius* **_E. fluviatile_**

4 Central cavity large or small, stem wall usually relatively thick; outermost cavities alternate with the stem ridges, i.e., on a different radius (a third set of smaller cavities often between the central one and the outermost) 5

5 Stem not branched above the base or only irregularly so; branches erect or strongly ascending 6

5 Stem regularly branched above the base, the branches often spreading 7

6 Sheaths longer than broad, with a dark apical band **_E. laevigatum_**

6 Sheaths shorter or not much longer than broad, with both an apical and a basal dark band **_E. hyemale_**

7 Teeth of the main stem sheaths reddish-brown, coherent in a few groups, especially toward the base of the stem; branches themselves branching **_E. sylvaticum_**

7 Teeth of the main stem sheaths whitish to black, sometimes brown but not reddish, separate or coherent in pairs; branches rarely branched 8

8 Central cavity of the main stem about the same size as the outermost cavities; sheaths on the first internode of the branches with 5 or more teeth **_E. palustre_**

8 Central cavity of the main stem usually definitely larger than the outermost cavities; sheaths on the first internodes of the branches with 3 or 4 teeth 9

9 Ridges of the main stem with spicules of silica deposit, especially on the upper part of the upper internodes **_E. pratense_**

9 Ridges of the main stem usually roughened but lacking spicules of silica deposit **_E. arvense_**

---

*To observe internal stem structure a cross-section should be made near the middle of an internode of the main stem and examined with a lens. If the specimen is pressed and dried, the section should be soaked in a wetting solution (e.g., soap-water solution) before examination.

Equisetum arvense L.

COMMON HORSETAIL

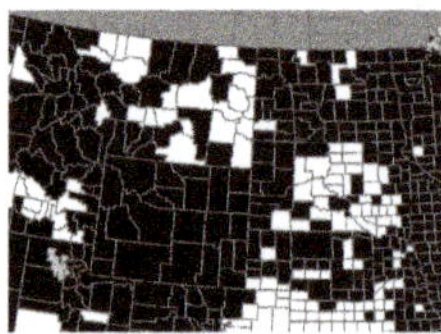

DESCRIPTION **Stems** dimorphic, annual, erect or decumbent. **Sterile stem** green, regularly branched, 1–6 dm tall, mostly 10- or 12-ridged, the ridges usually roughened but lacking spicules of silica deposit; central cavity ca. 1/4 the stem diameter; main stem sheaths 5–10 mm long, the teeth free or partly fused, with long-acuminate tips, brown to blackish, 1.5–2 mm long. **Branches** numerous, simple, ascending to occasionally recurved, 3- or 4-angled; sheaths on the first internodes of the branches with 3 or 4 teeth. **Fertile stem** unbranched, brownish, to 2 dm tall, appearing in early spring (late April-early May) and soon withering; sheaths 10–20 mm long, the teeth of the lower sheaths dark brown to blackish, mostly separate with long, acuminate tips, 4–9 mm long. **Cone** terminal, blunt, long-stalked.

Equisetum arvense (fertile stems)
COMMON HORSETAIL

HABITAT Streambanks, meadows, ditches, moist woods.

WETLAND STATUS
GP FAC | MIDW FAC | WMTN FAC

NOTE *Equisetum arvense* has many slender branches arising from the main stem in whorls. The green vegetative stems are conspicuous throughout summer after the non-green cone-bearing stems have withered.

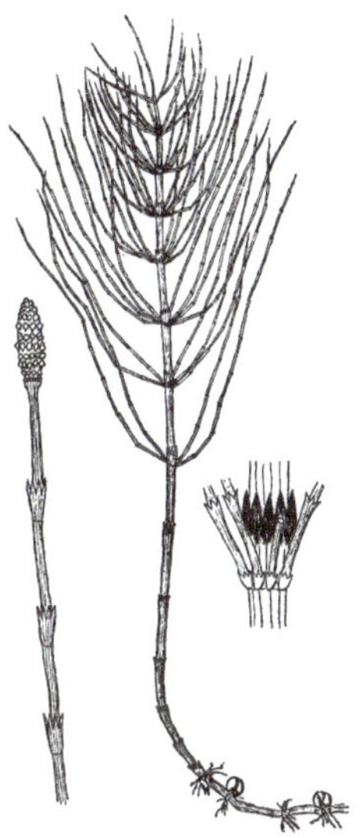

Equisetum arvense
COMMON HORSETAIL

Equisetum fluviatile L.

WATER HORSETAIL

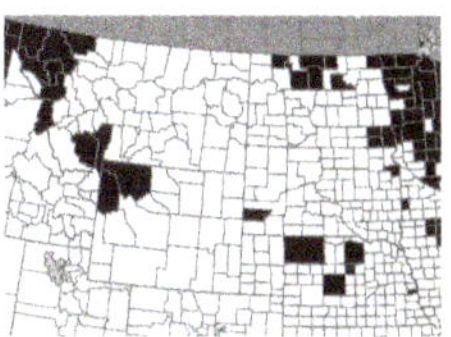

DESCRIPTION **Stems** alike, annual, erect, to 1 m tall, shallowly 9- to 25-ridged, the ridges smooth; central cavity ca. 4/5 the stem diameter, the stem wall thin; small outermost cavities opposite the ridges of the stem, i.e., on the same radius; main stem sheaths 6–10 mm long, the teeth free, dark brown to black, 2–2.5 mm long. **Branches** none to many, rather regular when present, simple, ascending, 4- to 6-angled. **Cones** terminal on the main stem or occasionally on upper branches, blunt, long-stalked, deciduous, appearing June–Aug.

HABITAT Marginal or emergent in swamps, bogs and spring-fed marshes where water is fresh.

WETLAND STATUS
GP OBL | MIDW OBL | WMTN OBL

Equisetum fluviatile
WATER HORSETAIL

Equisetum hyemale L.
COMMON SCOURING RUSH

DESCRIPTION **Stems** alike, erect, evergreen, viable for probably 2 or 3 years, simple or producing few, short, erect branches from the upper nodes, 2–12(15) dm tall, mostly 18- to 40-ridged, the ridges roughened with silica deposits; central cavity at least 3/4 the stem diameter; main stem sheaths shorter or not much longer than broad, 5–15 mm long, with both an apical and basal dark band, the teeth dark brown to black, connate by the scarious margins toward the base, 2–4 mm long, deciduous or persistent. **Cones** terminal on the main stem or produced on the short branches after the first year, sharp-pointed, sessile to short-stalked, eventually deciduous, appearing June–Sept.

SYNONYM *Equisetum praealtum* Raf.

HABITAT Often in dense colonies in seepage areas, wet meadows, ditches and along shores and streambanks, usually where sandy or gravelly.

WETLAND STATUS
GP FACW | MIDW FACW | WMTN FACW

NOTE This species was used by the early settlers to scour and clean pots and pans.

Equisetum hyemale
COMMON SCOURING RUSH

Equisetum laevigatum A. Br.
SMOOTH SCOURING RUSH

DESCRIPTION **Stems** alike, mostly annual in our range, simple or with few irregular, erect branches, commonly branching from the base if the main stem is cut, erect, 3–9 dm tall, mostly 16- to 30-ridged, central cavity usually 2/3 to 3/4 the stem diameter; main stem sheaths longer than broad, 6–15 mm long, with a single dark apical band or the lower ones sometimes with a dark basal band as well, the teeth dark brown with scarious margins, free or partly connate in pairs, 1–4 mm long, deciduous and usually missing from most of the sheaths. **Cones** terminal on the main stem or on the main branches if branched from the base, pointed or blunt, sessile to short-stalked, eventually deciduous, appearing June–July.

SYNONYM *Equisetum kansanum* J.H. Schaffner

HABITAT Wet meadows, seepage areas, streambanks, floodplains and ditches, also common in prairies and on embankments, often where sandy or gravelly; common.

WETLAND STATUS
GP FAC | MIDW FACW | WMTN FACW

NOTE A hybrid between *E. hyemale* and *E. laevigatum* is *E.x ferrissii* Clute. The cones are sterile; reproduction is entirely vegetative by rhizomes and fragmentation of the stems (although the absence of spores is most diagnostic). The range of *E. x ferrissii* corresponds with that of *E. hyemale* in the northern plains.

Equisetum laevigatum
SMOOTH SCOUING RUSH

Equisetum palustre L.

MARSH HORSETAIL

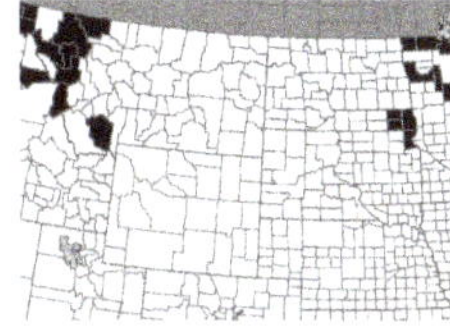

DESCRIPTION **Stems** alike, annual, regularly branched, erect, 2–6 dm tall, deeply 7- to 10-ridged, the ridges smooth to rough but without spicules; central cavity ca. 1/6 the stem diameter, about the same size as the outermost cavities; main stem sheaths 7–17 mm long, the teeth free or partly connate, brown to dark brown, 3–4 mm long. **Branches** few to many, simple, ascending, 5- or 6-angled; sheaths on the first internode of the branches with 5 or more teeth. **Cone** terminal on the main stem, blunt, long-stalked, deciduous, appearing late June–early Aug.

HABITAT Emergent or marginal in oxbow swamps and fresh spring-fed streams

WETLAND STATUS
GP FACW | MIDW FACW | WMTN FACW

Equisetum palustre
MARSH HORSETAIL

Equisetum pratense Ehrh.

MEADOW HORSETAIL

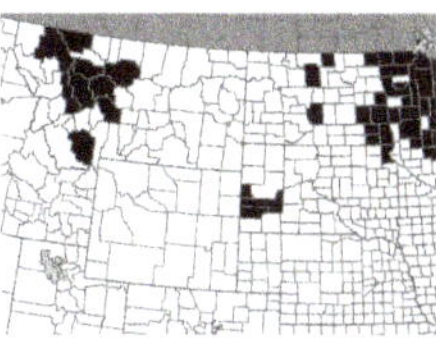

DESCRIPTION **Stems** dimorphic, annual, erect. Sterile stem green, branching regularly, 2–5 dm tall, 10- to 18-ridged, the ridges with spicules of silica deposit, especially on the upper portions of the upper internodes; central cavity 1/3 to 1/2 the stem diameter; main stem sheaths 2–6(8) mm long, the teeth free or partly connate in pairs, dark brown with white-hyaline margins, 1–3 mm long. **Branches** many, simple, spreading, mostly 3-angled; sheaths on the first internodes of the branches with 3 or 4 teeth. **Fertile stem** initially simple and brownish, eventually greening at the nodes and producing many small green branches, appearing in May before the sterile stems and persisting, mostly 1–3 dm tall; sheaths and their teeth like those of the sterile stems but somewhat longer. **Cone** terminal, blunt, long-stalked, deciduous.

HABITAT Moist woods and streambanks.

WETLAND STATUS
GP FACW | MIDW FACW | WMTN FACW

Equisetum pratense
MEADOW HORSETAIL

Equisetum sylvaticum L.
WOOD HORSETAIL

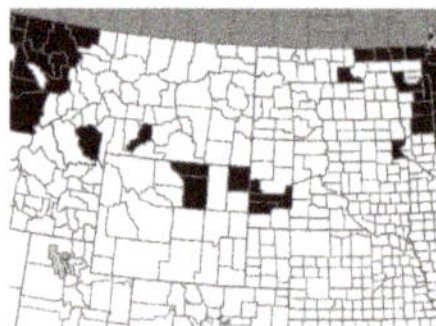

DESCRIPTION **Stems** dimorphic, annual, erect. Sterile stem green, densely branched from the nodes, 3–7 dm tall, mostly 10- to 18-ridged, the ridges usually with silicaceous spicules; central cavity usually more than 1/2 the stem diameter; main stem sheaths 5–10 mm long, the teeth coherent in usually 3–5 broad lobes, reddish-brown, 3–5 mm long. **Branches** rebranched, spreading to recurved, 4- or 5-angled. **Fertile stems** as in *E. pratense*, except usually having rebranched rather than simple branches; sheaths 10–25 mm long, the teeth fused into reddish-brown lobes 4–15 mm long. **Cones** as in *E. pratense*.

HABITAT Wet or swampy woods.

WETLAND STATUS
GP FACW | MIDW FACW | WMTN FAC

Equisetum sylvaticum
WOOD HORSETAIL

Isoetaceae *quillwort family*

Isoetes melanopoda Gay & Durieu
BLACKFOOT QUILLWORT

DESCRIPTION Perennial herb with a rosette of long, terete leaves arising from a lobed, cormlike base; roots numerous and branched, arising from the groove between lobes of the corm. **Leaves** long, slender and terete, 10–50 cm long, 0.5–3.5 mm wide, mostly hollow with 4 longitudinal air chambers surrounding the central vascular bundle, broadened at the base where the sporangia are contained; the outer leaves producing megasporangia, the inner leaves microsporangia; **sporangia** oblong, 5–30 mm long, usually brown-spotted, the megasporangia containing many rather large (0.25–0.5 mm in diameter) megaspores, the microsporangia containing numerous, tiny microspores. June–Aug.

HABITAT Submerged or in wet soil of swales and temporary ponds.

NOTE The swollen bases of the quill-like leaves contain the sporangia.

WETLAND STATUS
GP OBL | MIDW OBL | WMTN OBL

Isoetes melanopoda
BLACKFOOT QUILLWORT

Marsiliaceae *water-clover family*

Marsilea vestita Hook. & Grev.

HAIRY WATER CLOVER

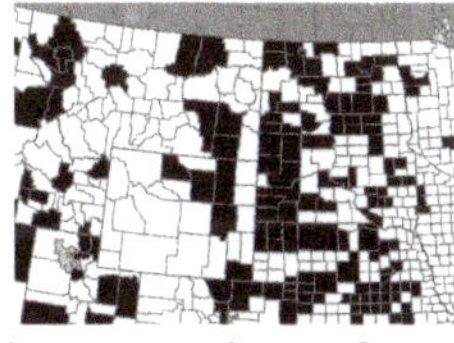

DESCRIPTION Creeping perennial (or acting as an annual when the habitat dries) in mud or shallow water. **Stem** a superficial, sprawling rhizome, rooting at the nodes. **Leaves** 4-foliate, with 4 wedge-shaped leaflets, giving the appearance of a 4-leaved clover, often floating, the leaflets obdeltate, 2–15 mm long and about as wide; petioles slender, lax when submersed, 2–20(30) cm long, greatly lengthened to accommodate the blade when submersed. **Sporangia** heterosporous, enclosed in solitary oval sporocarps borne near the base at the nodes; sporocarps brown, 4–7 mm long, strigose; sori in 2 rows inside the sporocarp, with apical megasporangia and lateral microsporangia; spores tetrahedral, the microspores numerous, the megaspores solitary in the sporangia; gametophyte aquatic, non-photosynthetic. July–Sept.

SYNONYM *Marsilea mucronata* A. Br.

HABITAT Mud or shallow water of temporary ponds, streams or ditches.

WETLAND STATUS

GP OBL | MIDW OBL | WMTN OBL

NOTE When stranded by receding water, the 4-lobed leaves are borne on erect petioles and the plant produces bean-shaped sporocarps near the petiole bases.

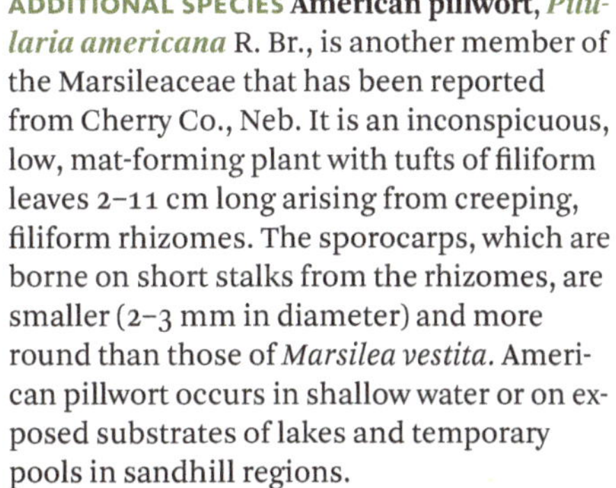

ADDITIONAL SPECIES **American pillwort,** ***Pilularia americana*** R. Br., is another member of the Marsileaceae that has been reported from Cherry Co., Neb. It is an inconspicuous, low, mat-forming plant with tufts of filiform leaves 2–11 cm long arising from creeping, filiform rhizomes. The sporocarps, which are borne on short stalks from the rhizomes, are smaller (2–3 mm in diameter) and more round than those of *Marsilea vestita*. American pillwort occurs in shallow water or on exposed substrates of lakes and temporary pools in sandhill regions.

Pilularia americana
AMERICAN PILLWORT

Marsilea vestita
HAIRY WATER CLOVER

Marsilea vestita
HAIRY WATER CLOVER

Onocleaceae *sensitive fern family*

Large, coarse ferns; sterile and fertile **fronds** strongly different, the sterile fronds deciduous, pinnatifid to 1-pinnate-pinnatifid; fertile fronds persistent. **Sori** enclosed under recurved margin of pinna segment (outer false indusium) and a tiny true inner indusium (membranous or of hairs).

1 Sterile fronds pinnate at the base, pinnatifid upward; veins of the pinnae anastomosing ***Onoclea***

1 Sterile fronds pinnate-pinnatifid; veins of the pinnae free ***Matteuccia***

Matteuccia struthiopteris (L.) Todaro
OSTRICH FERN

DESCRIPTION Rhizomes stout, branching and scaly, giving rise to short upright stems which bear the fronds. **Fronds** annual (in the northern plains), dimorphic, the fertile frond stiffly erect in the center of a circle of sterile fronds, blackish to brown, structurally unlike the green sterile fronds. **Sterile fronds** ascending, pinnate-pinnatifid, to 17 dm tall; blades much longer than the petioles, mostly 15–35 cm wide, sparsely to obviously pubescent on the rachis, abruptly narrowed to the tip, gradually tapered to the base; pinnae 20 to many pairs, ascending, mostly alternate, 7–22 cm wide, deeply divided into 20 or more pairs of pinnules, these 3–6 mm wide at the base; veins pinnate, free, not anastomosing. **Fertile frond** shorter than the sterile ones, to 5 dm tall; blade pinnate-pinnatifid, 12–25 cm long; pinnae ascending to appressed, 2–6 cm long, the margins revolute to enclose the sori, eventually spreading and the pinnules separating to expose the sporangia; **sori** several on each pinnule, the pinnules 1–2 mm wide; **indusium** hyaline, hoodlike, lacerate. Fertile fronds produced June–July, often persistent into the next year.

SYNONYM *Pteretis pensylvanica* (Willd.) Fern.

HABITAT Wet woods and stream margins.

WETLAND STATUS

GP FACW | MIDW FACW | WMTN FACW

NOTE Ostrich fern is commonly planted as an ornamental in shaded places around homes. It spreads readily by rhizomes.

Matteuccia struthiopteris
OSTRICH FERN

Onoclea sensibilis L.
SENSITIVE FERN

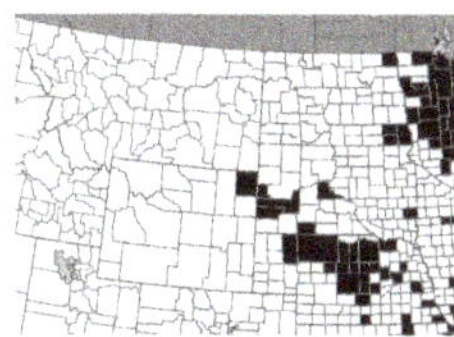

DESCRIPTION Rhizomes spreading and branching, 4–7 mm thick, the scales caducous. **Fronds** annual, erect, dimorphic, the fertile fronds brown, structurally unlike the green sterile fronds. **Sterile fronds** pinnate at the base, pinnatifid upward, the rachis broader-winged toward the tip; blades mostly 10–30 cm long, 10–35 cm wide, with 8–12 pairs of opposite pinnae, these sinuate to pinnatifid, 1–3 cm wide, sparsely white-hairy on the veins beneath, the veins anastomosing; petioles shorter to about as long as the blade. **Fertile frond** surpassed by the sterile ones; blade pinnate-pinnatifid, mostly 5–15 cm long; pinnae strongly ascending, 2–5 cm long, divided into bead-like pinnules with inrolled margins enclosing the sori, the pinnules 3–4 mm wide, becoming dry, hard, eventually separating to release the spores; **sori** globose, covered by a delicate, hoodlike indusium. Fertile fronds produced Aug–Sept, often persistent into the next year.

HABITAT Swampy woods, wet meadows and marshes, where water is fresh.

WETLAND STATUS
GP FACW | MIDW FACW | WMTN FACW

Onoclea sensibilis
SENSITIVE FERN

Ophioglossaceae
adder's-tongue family

Ophioglossum pusillum Raf.
ADDER'S TONGUE

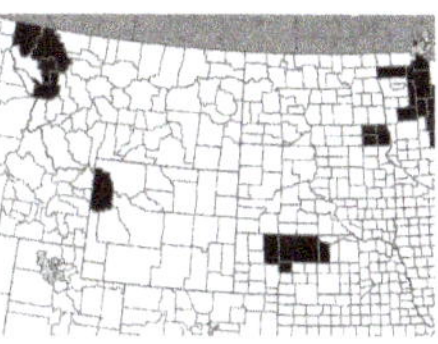

DESCRIPTION Small, erect, herbaceous perennial, 1–3 dm tall, from slender rhizomes. **Leaves** mostly 1 or paired, simple, entire, each blade borne on a stipe 3–15 cm long, the blades erect to ascending, elliptic to ovate, rounded to acute at the tip, 2.5–10 cm long, 1–4 cm wide, conspicuously net-veined. **Sporangia** homosporous, continuous in 2 rows in a terminal, unbranched spike, the spike 1–4 cm long, subtended by the leaf blade and borne above it on a stalk 3–15 cm long; spores single, with triradiate markings. Gametophyte small, subterranean, mycorhizal. June–Aug.

SYNONYM *Ophioglossum vulgatum* var. *pseudopodum* (Blake) Farw.

HABITAT Sandy wet meadows.

WETLAND STATUS
GP FACW | MIDW FACW

ADDITIONAL SPECIES **Limestone adder's-tongue** (*Ophioglossum engelmannii* Prantl), rare in c and se Neb; this species is more typical of drier meadows and more common in Missouri, Arkansas, and Okla.

Ophioglossum pusillum
ADDER'S TONGUE

Salviniaceae *water fern family*

Azolla cristata Kaulf.

MOSQUITO FERN

DESCRIPTION Small, annual, free-floating plants up to 3 cm across, often forming mats, sometimes stranded on mud; plant body comprised of dichotomously branched stems clothed with tiny, alternate, overlapping leaves in 2 rows, green or often strongly red; roots few, unbranched. **Leaves** 2-lobed, the upper lobe emersed, rhombic-ovate to obovate, 0.7–1.3 mm long, the lower lobe submersed and larger than the upper, mostly achlorophyllous. **Sporangia** heterosporous, the micro- and megasporangia contained in separate sporocarps which are usually paired on the submersed lobes of some of the leaves. July–Oct.

SYNONYM *Azolla mexicana* Schltdl. & Cham. ex C. Presl

HABITAT Quiet water of marshes, ponds, streams and ditches.

WETLAND STATUS
GP OBL | MIDW OBL

Azolla cristata
MOSQUITO FERN

Thelypteridaceae *marsh fern family*

Thelypteris palustris Schott

MARSH FERN

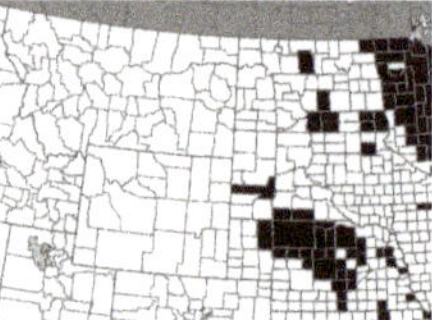

DESCRIPTION Rhizomes spreading and branching, slender, with a few appressed scales. **Fronds** annual, only slightly dimorphic, mostly erect, the blades pinnate-pinnatifid, lanceolate or oblong- lanceolate, 15–40 cm long, 8–20 cm wide, puberulent on the rachis and midveins, acuminate, only slightly narrowed at the base; pinnae usually 10–25 pairs, spreading, mostly alternate, 8–15(20) mm wide; petioles scaleless. **Sterile fronds** thin, delicate, pinnules broadly oval, obtuse, 3–5 mm wide, veins once-dichotomous; petioles shorter to longer than the blade. **Fertile fronds** surpassing the sterile ones, more firm than the sterile fronds, the petiole longer than the blade; pinnules oblong, 2–4 mm wide, the margins revolute; veins simple or once-dichotomous; **sori** round, usually confluent on the back of the pinnules; **indusium** attached at its center in the middle of the sorus, irregular in shape, usually ciliate. June–Sept.

HABITAT Boggy or marshy places, where water is fresh.

WETLAND STATUS
GP OBL | MIDW OBL | WMTN OBL

Thelypteris palustris
MARSH FERN

Flowering Seed Plants (Angiosperms)

Division Magnoliophyta

Herbs, shrubs, trees and vines, exhibiting a great diversity of growth forms and habits, featuring a life cycle in which the unisexual gametophytes are much reduced, very short-lived and nutritionally dependent upon the sporophyte. The sporophyte is typically differentiated into vascularized roots, stems and leaves, or rarely thalloid (members of former Lemnaceae, now combined into the Araceae).

Leaves simple to compound, opposite, alternate, whorled or basal, sometimes much reduced.

Sexual structures grouped into specialized short shoots termed **flowers**, the major parts of a flower typically comprised of highly modified sterile and fertile leaves in a cyclic or spiral arrangement on a receptacle; the sterile leaves (collectively termed the **perianth**) surrounding or enveloping the inner fertile ones; members of the perianth usually of 2 types, the outermost members (the **sepals**) commonly herbaceous, collectively making up the **calyx**, the innermost members (the **petals**) typically thin-textured, often brightly colored, collectively termed the **corolla**, the sepals and petals sometimes similar in appearance and referred to as **tepals**, either the calyx or the corolla or both occasionally lacking. Fertile leaves (sporophylls) making up the central portion of the flower, usually 2 kinds present, the outermost of these the stamens (microsporophylls), collectively termed the androecium, each **stamen** comprised of a slender stalk (the **filament**) with a distally attached pair of connected sacs (the **anther**); the anther sacs containing numerous pollen grains (male gametophytes), eventually rupturing to release the **pollen**; the innermost of the fertile leaves termed **carpels** (megasporophylls), collectively termed the **gynoecium**, each carpel representing an ovule-bearing leaf with the margins inrolled and fused to enclose the ovules. **Carpels** 1–many, separate so that each carpel comprises a simple pistil, or the carpels fused together to form one compound pistil; **pistil** typically composed of 3 parts, the swollen lower portion containing the ovules referred to as the **ovary**, the slender upper portion called the **style**, terminated by the sticky or hairy **stigma** which serves as the pollen-receptive surface of the pistil, the style sometimes obsolete so that the stigma is sessile on the ovary. Maturation of the ovary after fertilization results in the development of the fruit. The ovules contained within the ovary mature as seeds. **Fruits** are of many different types, of which the most common are the capsule, achene, caryopsis (or grain), follicle, drupe and berry.

Many terms are used to describe various conditions encountered in flowers. **Perfect (or bisexual) flowers** are those possessing both fertile stamens and carpels. Flowers with either stamens (**staminate or male flowers**) or carpels (**pistillate or female flowers**), but not both, are termed **imperfect** (or unisexual) flowers. The terms **regular** (or actinomorphic) and **irregular** (or zygomorphic) are used to describe flower symmetry. **Regular flowers** are radially symmetric, i.e., they may be bisected in many planes to give equal halves, whereas **irregular flowers** are bilaterally symmetric, i.e., they can be bisected in only one plane to give equal halves. The terms hypogynous, perigynous and epigynous pertain to the position of the perianth relative to the gynoecium of the flower. Flowers with the perianth lobes attached to the receptacle below the gynoecium (ovary superior) are described as **hypogynous**; those with the perianth lobes coming off above the ovary and with the ovary apparently embedded in the receptacle (ovary inferior) are termed **epigynous**; and those with the perianth lobes and stamens all attached around the margin of a disk, cup or tube, with the ovary (or ovaries) sitting free inside the base of the disk, cup or tube are considered **perigynous**. In the latter case the ovary position is considered superior because the one or more ovaries are not embedded in other tissues. In the case of perigynous and epigynous flowers, the portion of the flower surrounding (perigynous) or enclosing (epigynous) the ovary (or ovaries) is termed the **hypanthium** or floral tube.

Class MAGNOLIOPSIDA
Dicotyledons

Plants with vascular bundles of stems typically arranged in a ring in herbaceous forms, forming a cylinder which encloses a central pith, usually with a fascicular cambium; woody forms with secondary growth added on in layers by a vascular cambium; leaves mostly pinnately or palmately net-veined; flower parts (mainly the sepals, petals and stamens) usually in multiples of 4 or 5, seldom 3; plant embryo usually with 2 cotyledons.

Amaranthaceae *pigweed family*

Annual or perennial herbs, often in alkaline soil. Stems often angled or jointed, succulent in *Salicornia.* Leaves simple, alternate, or occasionally opposite (*Salicornia*), sometimes covered with thin, flaky scales giving a mealy appearance. Flowers 1 to many, small, green or red-tinged, clustered in leaf axils or at ends of stems; perfect, or staminate and pistillate flowers separate; sepals usually 5; petals absent; ovary superior, 1-chambered. Fruit a 1-seeded utricle.

Some of the common dryland weeds of this family will invade dry wetland basins during periods of drought. Notable among these are **kochia**, *Bassia scoparia* (L.) A.J. Scott, and **Russian thistle**, both *Kali tragus* (L.) Scop. and *K. collina* (Pallas) Akhani & Roalson. These plants are especially opportunistic around brackish or saline wetlands where high salt concentrations discourage other plants from pioneering exposed substrates. Since kochia and Russian thistle are well known as upland weeds, and because they are uncharacteristic of wetlands except during periods of drought, they are not further described here.

---

1 Leaves reduced to scales, opposite; stems succulent ***Salicornia***

1 Leaves not scalelike, mostly alternate; stems not succulent **2**

2 Leaves without petioles, linear, round in cross-section ***Suaeda***

2 Leaves mostly with petioles, blades broader **3**

3 Flowers perfect (with both staminate and pistillate parts), in spikes which have small leafy bracts throughout; fruit surrounded by the persistent sepals and petals ***Chenopodium***

3 Flowers unisexual (plants monoecious or dioecious); tepals and bracts acute, scarious or fruit in most if not all flowers enveloped by a pair of bracteoles (perianth absent) **4**

4 Bracts and tepals all acute, scarious (staminate flowers), absent in pistillate flowers ***Amaranthus***

4 Bracts beneath pistillate flowers broad and usually tuberculate and toothed, herbaceous in texture ***Atriplex***

Amaranthus PIGWEED

Dioecious (in those treated here) and monoecious annual herbs with taproots, usually branched, erect to spreading, often weedy. **Leaves** simple, alternate, entire or sinuate-margined, petiolate, exstipulate. Inflorescence of dense terminal and often axillary spikes (actually spikelike thyrses) or clusters. **Flowers** individually small, subtended by a few bracts, the flowers and bracts greenish or sometimes strongly purple; sepals 1–5 (none in female flowers of *A. tuberculatus*), scarious or membranous, often resembling the subtending bracts, usually aristate or mucronate; stamens 5, free; stigmas (2)3 or 4, style short or obsolete, ovary superior, 1-celled, short and broad. **Fruit** a utricle, circumscissile or irregularly splitting; seed round, lenticular, smooth and shiny.

The best known members of the genus are common weeds of fields and disturbed areas, e.g., ***Amaranthus retroflexus*** L., rough or redroot pigweed, and ***A. albus*** L., tumbleweed. Unlike the water hemps treated here, these pigweeds are monoecious. They are sometimes encountered on shores or drawdown zones. Hybrids in *Amaranthus,* even between monoecious and dioecious species, are apparently frequent.

Amaranthus tuberculatus (Moq.) Sauer
WATER HEMP

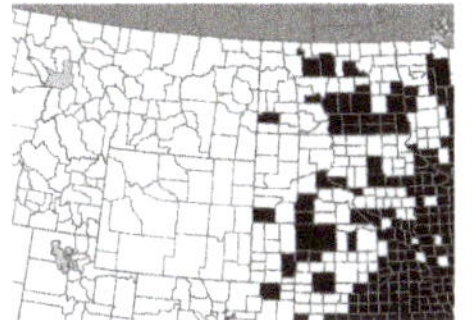

DESCRIPTION Plants erect with ascending branches, 0.5–2 m tall, or often low and spreading with branches 1–5 dm long, glabrous or nearly so, green or sometimes the stems and inflorescences strongly purple. **Leaves** oblong-lanceolate to ovate-lanceolate or rhombic-oblong, 3–10 cm long, obtuse to rounded or sometimes notched at the tip, attenuate at the base, reduced to bracts upward in the inflorescence; petioles mostly 0.5–4 cm long. **Inflorescence** of usually many terminal and axillary spikelike branches, or some of the lateral branches merely short clusters. Bracts 1–2 mm long, with a shortly excurrent midrib in the male, a conspicuously excurrent midrib in the female; **male flowers** with 5 sepals 2.2–3 mm long, the outer longer than the inner and with midveins excurrent into a mucronate or aristate tip; stamens 5; **female flowers** with 1 or 2 sepals, when 2, one of the sepals rudimentary or less than 1 mm long, the longer (or single) one ca. 2 mm long. **Utricle** 1.2–2 mm long, circumscissile, with a distinct line of dehiscence around the middle, the top coming off like a lid, often with faint ridges of tubercles on the top portion; style branches 3–4; **seed** reddish-brown to black, 0.9–1.1 mm in diameter. July–Oct.

SYNONYMS *Acnida altissima* Riddell, *Acnida tamariscina* (Nutt.) Wood, *Amaranthus rudis* Sauer, *Amaranthus tamariscinus* Nutt.

HABITAT Shores, streambanks, mud and sand bars and low places in fields where sometimes weedy.

WETLAND STATUS
GP OBL | MIDW OBL | WMTN OBL

Atriplex dioica Raf. (Nutt.) Rydb.
SALINE SALTBUSH

DESCRIPTION Monoecious to semidioecious, taprooted annual (1)2–10 dm tall; stems erect to decumbent, simple to freely branched. **Leaves** alternate, or the lowest ones often opposite, petioled or becoming sessile upward in the inflorescence; blades lanceolate to trullate or deltate, often hastate, mostly 2–8 cm long, 0.5–6 cm wide, entire to sinuate-dentate, somewhat grayish-farinose when young, becoming dull green and glabrate with age; petioles mostly 5–30 mm long. **Flowers** imperfect, minute, greenish, the male and female flowers usually mixed in glomerules which are borne in terminal and axillary spikes, the spikes simple or branched, ebracteate or only sparsely bracteate in the lower portion; **male flowers** with a deeply 5-lobed perianth, ca. 1 mm wide, stamens 5; **female flowers** lacking a perianth, subtended by 2 sepaloid bracteoles, these expanding as the fruit matures and enclosing it, ovate-deltate, 2–5 mm long and about as wide, sometimes hastate, connate toward the base, smooth or tuberculate on

Amaranthus tuberculatus
WATER HEMP

Atriplex dioica
SALINE SALTBUSH

the back. **Fruit** with a transparent membranous pericarp adherent to the seed, the seed oriented vertically between the bracteoles, dark brown to black, lenticular, 1.5–2.5 mm in diameter. Aug–early Oct.

SYNONYM *Atriplex subspicata* (Nutt.) Rydb.

HABITAT Shores, streambanks and flats, usually where alkaline or saline; also in disturbed places.

WETLAND STATUS
GP FACW | MIDW FACW | WMTN FACW

NOTE *A. dioica* is highly variable throughout its range, and many species and varieties have been described on the basis of variations in growth habit and leaf shape.

Chenopodium GOOSEFOOT

Erect to spreading, taprooted annuals with simple to freely branched **stems** and alternate, petiolate **leaves**, the blades somewhat fleshy, lanceolate to ovate-oblong or trullate to deltate, sinuate-dentate to sinuate-lobed, seldom entire, occasionally hastate, sometimes farinose, especially on the lower surface. **Flowers** perfect, tiny and numerous, ca. 0.5 mm across at anthesis, greenish or sometimes reddish-tinged, densely clustered in glomerules which are borne in short terminal and axillary spikes, the spikes bracteate throughout with the bracts reducing in size upward; perianth 2- to 5-lobed, the segments rather fleshy, obtuse, often incurved over the fruit; stamens 1–5; styles 2(3), short. **Fruit** oriented horizontally or vertically to the persistent perianth, sometimes both horizontal and vertical, the thin membranous perianth loosely adherent to the lenticular seed.

The two chenopods treated here are consistently found on alkaline to saline shores and mudflats. Other weedy species such as *Chenopodium berlandieri* Moq. and *C. strictum* Roth are occasional on dry shores but more typical of disturbed upland habitats.

1 Leaves persistently whitish-farinose on the lower surface, dull green above, mostly 0.7–3 cm long, 0.2–1.2 cm wide ***C. glaucum***

1 Leaves not whitish-farinose at maturity, green on both surfaces, often reddish-tinged, drying dark, mostly 2–8 cm long, 0.7–5 cm wide. ***C. rubrum***

Chenopodium glaucum L.
OAK-LEAVED GOOSEFOOT

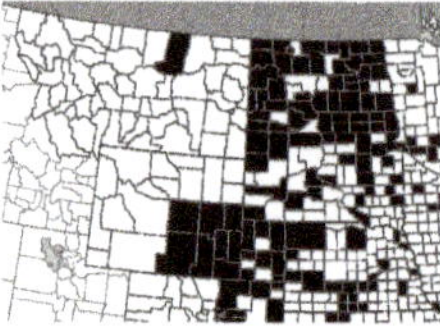

DESCRIPTION Plants erect to low and spreading, usually widely branching from the base; stems sometimes reddish toward the base, 1–7 dm long. **Leaf blades** dull green above, persistently whitish-farinose beneath, lanceolate to ovate-oblong or ovatedeltate, 0.7–3(4) cm long, 0.2–1.2(2) cm wide, sinuate-dentate; petioles about as long as or much shorter than the blades. **Perianth** 3- to 5-lobed; hit varying from horizontal to vertical, the pericarp greenish; **seed** dark brown, shiny, orbicular, 0.8–1.2 mm wide. Aug–Oct.

SYNONYM *Oxybasis glauca* (L.) S. Fuentes, Uotila & Borsch

HABITAT Shores, streambanks, flats and disturbed areas, usually where alkaline or saline; common; naturalized from Eurasia.

WETLAND STATUS
GP FAC | MIDW FACW | WMTN FAC

Chenopodium glaucum
OAK-LEAVED GOOSEFOOT

Chenopodium rubrum L.

RED GOOSEFOOT

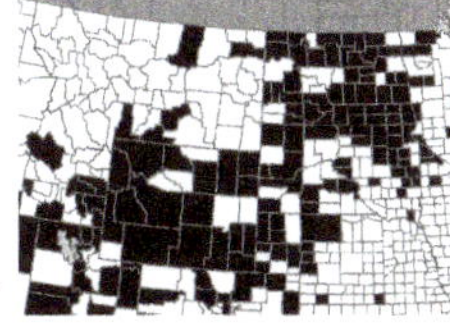

DESCRIPTION Plants erect to decumbent, sometimes low and spreading, 1–10 dm tall; stems simple to freely branched. **Leaf blades** green on both surfaces, often reddish-tinged, drying dark, only weakly farinose when young, soon glabrate, lanceolate to trullate or deltate, sometimes weakly hastate, 2–8(10) cm long, 0.7–5(8) cm wide, irregularly sinuate-dentate to sinuate-lobed, seldom entire; petioles mostly 0.5–3(5) cm long. **Perianth** 2- to 4-lobed; fruit vertical, the pericarp brown; **seed** dark brown, shiny, orbicular to oval, 0.5–0.8 mm in diameter. Aug–Oct.

SYNONYM *Oxybasis rubra* (L.) S. Fuentes, Uotila & Borsch

HABITAT Same habitats as *Chenopodium glaucum;* common, often locally abundant.

WETLAND STATUS
GP OBL | MIDW OBL | WMTN FACW

Chenopodium rubrum
RED GOOSEFOOT

Salicornia rubra A. Nels.

GLASSWORT, SALTWORT

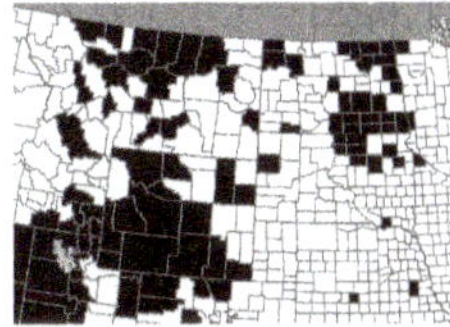

DESCRIPTION Low, erect to ascending, taprooted annual, succulent, green to strongly red throughout, 0.5–2 dm tall; **stems** oppositely branched, fleshy, jointed at the nodes, often brittle and breaking with a crackling noise when the plants are walked upon. **Leaves** opposite, small and scalelike, mostly 1–2 mm long, obtuse to acute, connate at the base to form a short sheath at each node, scarious on the margins, the internodes shortened in the spikes, with the leaves serving as scalelike bracts. **Flowers** perfect or partly female, embedded in the fleshy, terminal spikes, arranged in groups of 3 above each bract, the central flower above the lateral 2, about reaching the next node upward, the joints of the spike ca. 2 mm long and about as wide in dried condition; calyx essentially unlobed, fleshy, completely enclosing the flower except for the slitlike opening through which the stamens and style branches barely protrude at anthesis; stamens 2(1); style branches 2. **Fruit** olive, ellipsoid, 1–1.2 mm long. Aug–early Oct.

HABITAT Saline or alkaline soil of flats, shores, seepage areas and ditches

STATUS Neb threatened

NOTE The succulent stems usually turn a distinctive red in fall.

WETLAND STATUS
GP OBL | MIDW OBL | WMTN OBL

Salicornia rubra
GLASSWORT, SALTWORT

Suaeda calceoliformis (Hook.) Moq.
SEA BLITE

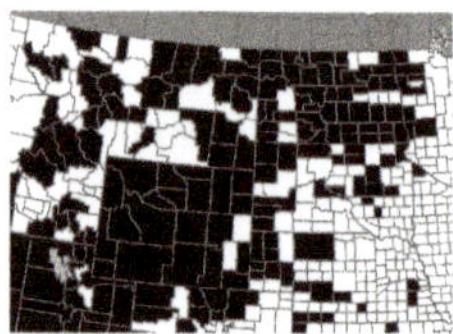

DESCRIPTION Erect to low and spreading, taprooted annual, sometimes reddish-tinged; **stems** simple to more often freely branched, (0.5)1–6 dm long. **Leaves** numerous, alternate, sessile, linear, semiterete, rather succulent, 5–30 mm long, ca. 1 mm wide, acute, reduced to short bracts in the inflorescence. **Flowers** perfect to imperfect, greenish or sometimes reddish, very small, 1–2 mm across in fruit, smaller in flower, tightly clustered in sessile glomerules of 3–7 flowers in the ails of the bracts, the glomerules forming elongate spikes which usually comprise the bulk of the mature plant; bracts mostly 1–6 mm long, somewhat broader than the leaves; perianth deeply 5-parted, the lobes very unequal, fleshy, corniculate on the back, 1 or 2 much more corniculate on the back than the others, cucullate, ca. 1 mm long at fruit maturity; stamens 5 or fewer; styles usually 2(3–5), very short. **Fruit** enclosed by the perianth, horizontal, the membranous pericarp very loose on the seed; **seed** black, shiny, oval, ca. 1 mm in diameter. Aug–Sept.

SYNONYM *Suaeda depressa* (Pursh) S. Wats.

HABITAT Alkaline or saline flats, shores, streambanks and seepage areas; common.

WETLAND STATUS
GP FACW | MIDW FACW | WMTN FACW

Suaeda calceoliformis
SEA BLITE

Apiaceae
parsley family

Glabrous, biennial to perennial herbs, fibrous or tuberous-rooted, some poisonous; **stems** hollow, sometimes thickened at the base. **Leaves** compound, 1–3x pinnate, alternate, petioled, the leaflets subentire to serrate or incised; petioles broadened at the base, with wings sheathing the stem. **Inflorescence** of few to many compound umbels, usually with an involucre of bracts subtending the primary branches (rays) of the umbels and an involucel of bractlets subtending the pedicels of the umbellets. **Flowers** tiny, perfect, regular, 5-merous; calyx minutely toothed or obsolete; petals white, separate, inflexed at the tip; stamens 5; styles 2, swollen at the base to form a stylopodium, ovary inferior, 2-celled, maturing into a flattened, corky-ribbed schizocarp which ultimately splits along a central commissure into 2, 1-seeded mericarps, the mericarps separating from the base upward to reveal a slender carpophore between them, both mericarps attached to the carpophore at their tips, the carpophore eventually splitting lengthwise as the mericarps separate.

1 Leaves 2–3x pinnate; roots usually partly tuberous; primary veins of the leaflets directed toward the sinuses between the teeth of the margin — ***Cicuta***

1 Leaves once-pinnate; roots fibrous; primary veins of the leaflets directed toward the teeth of the margin — **2**

2 Leaflets of upper leaves irregularly incised; fruits with obscure ribs — ***Berula***

2 Leaflets of upper leaves regularly serrate; fruits with conspicuous ribs — ***Sium***

Sium suave
WATER PARSNIP

Berula erecta (Huds.) Cov.

CUT-LEAF WATER-PARSNIP

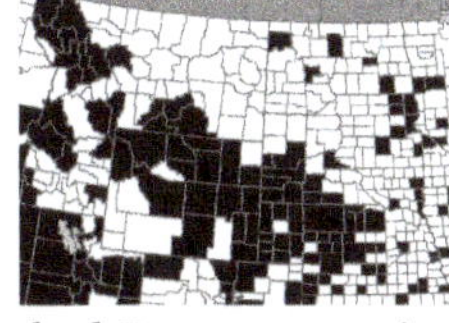

DESCRIPTION Fibrous-rooted perennial; **stems** erect to decumbent, 3–8 dm long, often rooting along the prostrate portion, sparsely branched. **Leaves** once-pinnate, basal leaves often present, larger than the cauline leaves, with the leaflets broader and less dissected than those of the cauline leaves; blades oblong, 3–20(30) cm long, 2–10 cm wide; leaflets lanceolate to ovate or oblong, subentire to toothed or incised. **Umbels** few to many, terminal and lateral, the lateral ones often overtopping the terminal one, 2.5–6 cm across in fruit; rays 5–15, 1–3 cm long; bracts narrow, entire or toothed; umbellets 1–2 cm wide; pedicels 2–5 mm long; bractlets several, narrow. **Flowers** white, 1–2 mm across; sepals minute or obsolete. **Fruit** elliptic or orbicular, somewhat flattened laterally, obscurely ribbed, 1.5–2 mm long, seldom maturing. July–Sept.

HABITAT Springs and spring-fed streams; locally common at scattered locations.

WETLAND STATUS
GP OBL | MIDW OBL | WMTN OBL

Berula erecta
CUT-LEAF WATER-PARSNIP

Cicuta WATER HEMLOCK

Poisonous biennials or perennials, usually with few to many tuberous-thickened roots, the stem often thickened and chambered at the base. **Leaves** 2–3x pinnate, the leaflets linear to linear-lanceolate or ovate-lanceolate, irregularly incised or regularly serrate, the primary veins directed toward the sinuses rather than the teeth. **Umbels** few to many, terminal and lateral, on stout peduncles; rays many, spreading-ascending; involucre usually lacking; pedicels spreading; involucel comprised of several bractlets or rarely absent. **Flowers** white or greenish, 1–2 mm across; sepals evident, triangular. Mature **fruits** oval or orbicular, constricted or not constricted at the commissure, the ribs alternating with darker intervals.

NOTE These plants are highly poisonous to humans and livestock. The tuberous roots, stem base and young shoots are especially toxic. Livestock poisoning from *Cicuta* is more common in the arid w and sw USA, where grazing animals are attracted to low areas for green forage during dry spells.

---

1 Leaflets linear, seldom to 5 mm wide; axils of at least the reduced upper leaves bearing bulbils ***C. bulbifera***

1 Leaflets linear-lanceolate to ovate-lanceolate, mostly 5–35 mm wide; axils not bearing bulbils ***C. maculata***

Cicuta bulbifera L.

BULBOUS WATER HEMLOCK

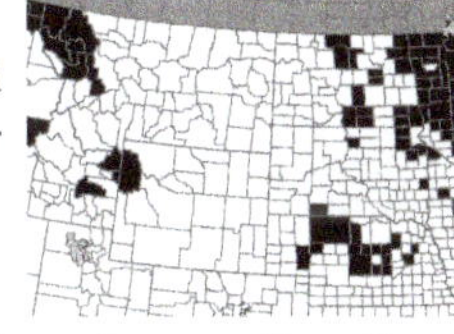

DESCRIPTION Erect biennial or perennial 3–10 dm tall, fibrous-rooted or with a few tuberousthickened roots; **stems** slender, not thickened at the base. **Leaves** all cauline, the blades to 15 cm long, to 10 cm wide, with linear or very narrowly lanceolate leaflets mostly 2–9 cm long, 0.5–5 mm wide, sparsely and irregularly dentate to incised or subentire, the upper leaves much reduced, with few segments or undivided, many bearing 1-few axillary, ovoid bulbils 1–3 mm long. **Umbels** often not produced, seldom to 5 cm wide; rays 1–2.5 cm long; bracts none or few, narrow and inconspicuous; umbellets 1–1.5 cm wide at maturity, the pedicels 2–5 mm long; bractlets several or rarely none, narrow. **Fruits** seldom maturing, orbicular, 1.5–2 mm in diameter, the lateral ribs separated by a constriction on the commissure. Aug–Sept.

HABITAT Pond margins, springs and streambanks.

WETLAND STATUS
GP OBL | MIDW OBL | WMTN OBL

Cicuta bulbifera
BULBOUS WATER-HEMLOCK

Cicuta maculata L.
COMMON WATER HEMLOCK

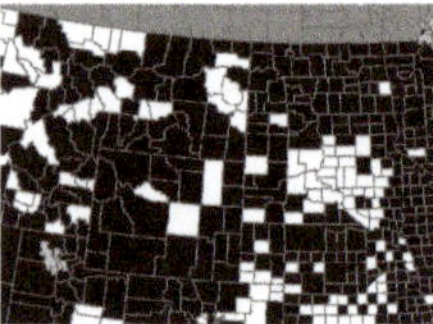

DESCRIPTION Stout biennial or perennial from sterile offshoots produced from the dying main plant, 5–20 dm tall; **stems** solitary or few together from a tuberous-thickened and chambered base, conspicuously hollow above the base, the roots partly tuberous also. **Leaves** basal and cauline, the basal leaves larger and longer-petioled than cauline leaves, the blades oblong to ovate in general outline, mostly 8–25(35) cm long, 5–15(20) cm wide; **leaflets** linear-lanceolate to ovate-lanceolate, mostly 3–10 cm long, 5–35 mm wide, serrate. **Umbels** several to many, 6–12 cm wide in fruit, on stout peduncles 5–15 cm long; rays 2–6 cm long; bracts none or few to several, narrow; umbellets 1–2 cm wide at maturity, the pedicels 3–10 mm long; bractlets ovate-lanceolate to linear, scarious-margined, 2–15 mm long. **Fruits** rotund-ovate to orbicular, 2–4.5 mm long, glabrous, constricted or not constricted at the commissure, with prominent dorsal and lateral corky ribs. July–Sept.

HABITAT Wet meadows, marshes, shores, streambanks, springs and other wet places.

WETLAND STATUS
GP OBL | MIDW OBL | WMTN OBL

Cicuta maculata
COMMON WATER HEMLOCK

Sium suave Walt.

WATER PARSNIP

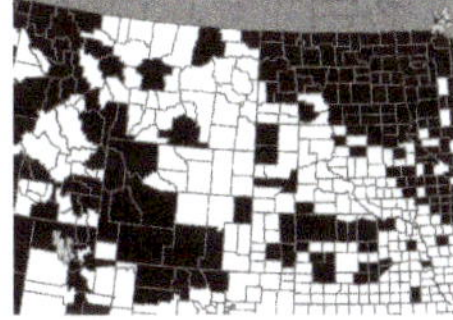

DESCRIPTION Stout, fibrous-rooted perennial 4–20 dm tall, from a short erect crown; **stems** solitary, strongly ribbed upward, thickened and hollow with cross-partitions at the base, partitioned at the nodes above. **Leaves** once-pinnate, the blades oblong to ovate in outline, 5–25(35) cm long, 6–20(30) cm wide; **leaflets** usually 7–17 per leaf, linear to lanceolate, 3–15 cm long, 3–15 mm wide, sharply serrate; petioles long below, shorter upward, hollow, septate-nodulose; **submersed leaves** often present early in the season, finely bipinnately dissected. **Umbels** terminal and lateral, few to many, 4–12 cm wide in fruit, on peduncles 3–10 cm long; rays 1.5–4 cm long; bracts 6–10, narrowly lanceolate, entire or incised, reflexed; umbellets 7–12 mm wide, the pedicels 3–5 mm long; bractlets 4–8, smaller than the bracts, entire. **Flowers** white or greenish-white, 1.5–2 mm across; sepals minute or none. **Fruits** oval to orbicular, 2–3 mm long, slightly constricted on the commissure, the ribs prominent and corky, subequal. July–Sept.

HABITAT Marshes, swamps, springs, ditches and margins of streams, ponds and lakes, often emergent.

WETLAND STATUS
GP OBL | MIDW OBL | WMTN OBL

Sium suave
WATER PARSNIP

Apocynaceae *dogbane family*

Apocynaceae now includes former members of the Milkweed Family (Asclepiadaceae).

Apocynum cannabinum L.

PRAIRIE DOGBANE, INDIAN HEMP

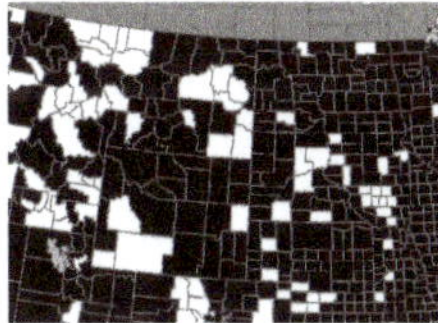

DESCRIPTION Erect, milky-juiced herb 3–10 dm tall, perennial from spreading rhizomes, weedy and often forming large patches, glabrous or nearly so; **stem** simple below, dichotomously branched mainly in the upper half. **Leaves** simple, opposite, sessile and clasping the stem to short-petiolate and not clasping, the lower leaves tending more to be sessile or shorter-petioled than the upper ones; blades ovate to oblong or lanceolate, 3–14 cm long, 0.7–4.5(7) cm wide, acute to rounded and apiculate at the tip, acute to rounded or cordate at the base. **Inflorescence** of 1 or more dense, terminal cymes; bracts linear to lanceolate, mostly inconspicuous, the lower ones sometimes rather leaflike and conspicuous. **Flowers** small, white or greenish-white, erect to drooping, perfect, regular; calyx 5-parted to near the base, the lobes linear to lanceolate, 1.2–3(3.5) mm long; corolla 5-lobed, narrowly campanulate to urceolate or short-cylindric, 2.6–4.7 mm long, the lobes triangular, 1/2 or less the length of the corolla tube; stamens 5, inserted near the base of the corolla tube, an-

Apocynum cannabinum
PRAIRIE DOGBANE, INDIAN HEMP

thers triangular, slightly adherent to the stigma and converging to form a cone above it; carpels 2, separate below, each with its own ovary, united above and sharing a sessile stigma, ovaries superior, subtended by 5 nectaries that alternate with the stamens. **Fruit** of 2 or (by abortion) 1, many-seeded follicles, these divergent to pendulous, linear-cylindric, 7–20 cm long; seeds narrowly fusiform, 3–6 mm long, with a coma of white to tawny, silky hairs at the tip. Flowering June–Aug, fruiting Aug–Oct.

SYNONYM *Apocynum sibiricum* Jacq.

HABITAT Borders of marshes, lakes and streams and other wet to moist places, often where disturbed.

WETLAND STATUS
GP FAC | MIDW FAC | WMTN FAC

Asclepias incarnata L.

SWAMP MILKWEED

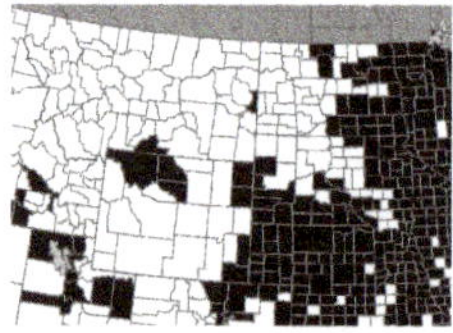

DESCRIPTION Tall, stout, milky-juiced perennial 6–15 dm tall, from a thick rootstock, glabrous except for short, appressed or curved pubescence in the uppermost part. **Leaves** simple, opposite, short-petioled, the blades linear-lanceolate to lanceolate or seldom ovate-lanceolate, 6–15 cm long, 1–5 cm wide, acute-tipped, entire, cuneate to rounded at the base. **Flowers** usually numerous in terminal and axillary umbels, deep pink to purplish-red, perfect, regular, 5-merous, 4–6 mm wide; calyx spreading to eventually reflexed, the sepals 1.5–2.5 mm long, often hidden beneath the reflexed petals which are 3–6 mm long; corolla with a petaloid corona above the petals, the corona comprised of 5 hoods which are 2–3 mm long, each with a subulate, incurved horn projecting from the orifice; stamens 5, the anthers united around the gynoecium, adherent to the stigma to form a structure called the gynostegium, the pollen of each anther sac contained in a waxy mass termed the pollinium, the pollinium of each anther sac connected to the pollinium in the sac of the adjacent anther by a translator, these double pollinia released intact at pollination; carpels 2, enclosed by the gynostegium, the ovaries separate but the styles and stigmas fused. **Fruit** produced by only a few of the flowers, usually 1 of the 2 carpels maturing, the fruit a large fusiform follicle containing many seeds which bear a fluffy white coma. July–Aug.

HABITAT Swamps, marshes, ditches, streambanks, springs and fens, where water is fairly fresh.

WETLAND STATUS
GP FACW | MIDW OBL | WMTN OBL

Asclepias incarnata
SWAMP MILKWEED

Asclepias incarnata
SWAMP MILKWEED

Asteraceae *aster family*

Annual, biennial or perennial herbs (those included here) of various growth habits. **Leaves** simple or compound, exstipulate, opposite, alternate or both, sometimes whorled, occasionally the principal leaves basal.

Flowers reduced in size, crowded into involucrate heads and sharing a common receptacle, the **disk**, the heads each resembling a single flower and arranged in various kinds of inflorescences; flowers of the head usually of 2 types, **ray** (or *ligulate*) florets simulating petals around the outside of the disk, and the usually less conspicuous **disk** (or *tubular*) florets occupying the central portion of the disk (to form a **radiate head**, or the head comprised entirely of ray florets (**ligulate head**), or **disk florets** (discoid head), the disk florets of a radiate or discoid head sometimes intermixed with bracts so that the receptacle is chaffy, otherwise the receptacle naked or pitted; involucral bracts sepaloid or foliaceous, surrounding and subtending the disk in 1–several series, often imbricate.

Individual flowers perfect or imperfect,* regular (disk florets) or irregular (ray florets), lacking a definite calyx; with a pappus of capillary bristles, scales or awns often encircling the summit of the ovary outside the corolla, commonly accrescent and persistent in fruit. Corollas of 2 types, those of the **ray florets** tubular only at the base, expanded into the flat, petaloid ray or ligule, entire or toothed at the tip; the **disk corollas** tubular, with 5 equal lobes or teeth at the summit. Stamens 5, epipetalous, usually with the elongate anthers united into a tube around the style; style branches usually 2, ovary inferior, 2-carpellary, 1-celled and 1-ovuled, ripening into an achene.

*Imperfect flowers in composite flower heads are commonplace. The ray florets of radiate heads are pistillate or sterile, whereas the disk florets are perfect or functionally staminate. In ligulate and discoid heads, usually all the florets are perfect, but in some discoid types, e.g., *Xanthium,* staminate and pistillate florets are borne in 2 very different types of discoid heads.

1 Heads ligulate, comprised entirely of yellow ray florets; plants with milky juice 2
1 Heads radiate or discoid; plants with watery juice 3

2 Leaves in a basal rosette like those of a dandelion; mature achenes terete ***Crepis***
2 Leaves cauline; mature achenes flattened ***Sonchus***

3 Leaves with sharp spines ***Cirsium***
3 Leaves spineless 4

4 Leaves all opposite or whorled 5
4 Leaves alternate or partly so, or the leaves chiefly basal 9

5 Receptacle naked 6
5 Receptacle chaffy 7

6 Leaves whorled, short-petiolate ***Eutrochium***
6 Leaves opposite, connate-perfoliate ***Eupatorium***

7 Pappus none; leaves, or at least the upper ones, connate-perfoliate, forming a cup around the stem ***Silphium***

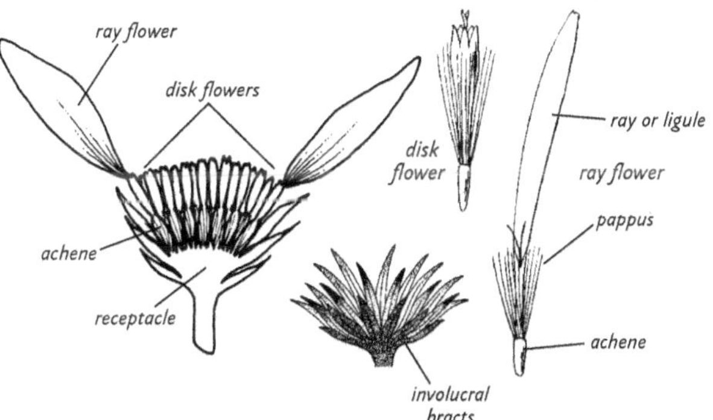

Typical Aster Family inflorescence, with both disk and ray flowers.

7 Pappus of awns or awn-tipped scales; leaves sessile or petiolate, sometimes slightly connate, but not forming a cup around the stem 8

8 Pappus of 2–4 retrorsely barbed awns **Bidens**

8 Pappus of 2 awn-tipped scales **Helianthus**

9 Heads radiate with yellow rays 10

9 Heads discoid or radiate with rays colored other than yellow 15

10 Pappus of 2-several awns or scales 11

10 Pappus of numerous capillary bristles 12

11 Receptacle chaffy; leaves short-petioled, not decurrent on the stem **Helianthus**

11 Receptacle naked; leaves tapered to the base, decurrent as wings on the stem **Helenium**

12 Involucral bracts in one series, not imbricate, sometimes with a few reduced bracts below 13

12 Involucral bracts in few to several series, imbricate 14

13 Plants annual or biennial, hollow-stemmed; leaves basically all similar in shape **Tephroseris**

13 Plants perennial, solid-stemmed; basal leaves differing in shape from cauline leaves **Packera**

14 Inflorescence corymbiform; leaves linear to linear-lanceolate or linear-elliptic, 2–10 mm wide, entire, glandular-punctate **Euthamia**

14 Inflorescence paniculiform; leaves lanceolate to elliptic, 10–40 mm wide, serrate, not glandular **Solidago**

15 Heads unisexual and dimorphic, the male florets in small heads above the larger female heads; involucres of the female heads spiny, completely enclosing the pistillate florets to form a bur **Xanthium**

15 Heads bisexual or rarely unisexual, all alike; involucres not spiny 16

16 Principal leaves basal, sagittate or palmate, white-woolly at least on the lower surface; flowering in spring or early summer **Petasites**

16 Principal leaves cauline, shaped other than sagittate or palmate, not white-woolly; flowering late summer or autumn (except *Erigeron philadelphicus*) 17

17 Heads discoid 18

17 Heads radiate (the rays very narrow and only slightly, if at all, exceeding the involucre in *Erigeron lonchophyllus* and *E. canadensis,* therefore inconspicuous) 20

18 Leaves pinnately dissected; pappus none **Artemisia**

18 Leaves simple, entire or toothed; pappus of numerous capillary bristles 19

19 Perennial with lanceolate to ovate-lanceolate leaves; involucral bracts purple-tipped **Vernonia**

19 Annual with linear leaves; involucral bracts green **Symphyotrichum**

20 Pappus of 2 awns and several minute bristles **Boltonia**

20 Pappus of numerous capillary bristles 21

21 Plants taprooted annuals blooming in late summer or fall; involucre 3–4(5) mm high **Erigeron canadensis**

21 Plants fibrous-rooted perennials, often with rhizomes, blooming in early or late summer or fall; involucre usually more than 4 mm high 22

22 Involucral bracts green, often chartaceous at the base; rays wider than 0.5 mm **Symphyotrichum**

22 Involucral bracts hyaline at the tip and on the margins above, green in the middle and at the base; rays 0.1–0.6 mm wide **Erigeron**

Almutaster pauciflorus (Nutt.) Á. & D. Löve

MARSH-ASTER

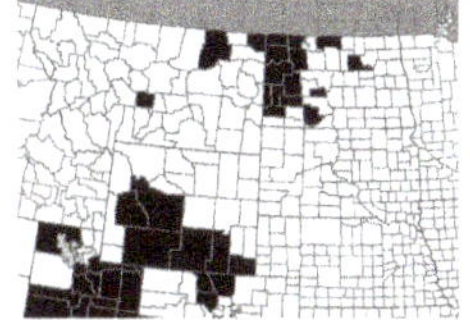

DESCRIPTION Perennial 1.5–4 dm tall, from a slender rootstock; **stems** slender, single or clumped, erect to decumbent, glandular-pubescent above, especially in the inflorescence. **Leaves** much reduced upward, the basal leaves sometimes petiolate with a broadened blade, otherwise the leaves sessile, linear to linear-oblanceolate, the lower ones 4–15(20) cm long, 1–6 cm wide, smaller above, rather thick, acute, entire. **Heads** rather few, uncrowded, solitary at the ends of branches in a corymbiform arrangement, 1–1.5 cm across; involucre 3–8 mm high, the bracts linear-lanceolate, acute, loosely imbricate, glandular-puberulent; **rays** mostly 15–25, white to light blue or lavender, 4–8 mm long. Late July–Sept.

SYNONYM *Aster pauciflorus* Nutt.

HABITAT Wet alkaline or saline ground of shores, marshes, streambanks and seepage areas.

WETLAND STATUS

GP FACW | WMTN FACW

NOTE See *Symphyotrichum* key, page xxx.

Almutaster pauciflorus
MARSH-ASTER

Artemisia biennis Willd.

BIENNIAL WORMWOOD

DESCRIPTION Erect, taprooted annual or biennial, 3- 12 dm tall, glabrous, weakly aromatic, plants strict to pyramidal in shape. **Leaves** alternate, pinnately dissected nearly to the middle, 4–12 cm long, 2–5 cm wide, the segments toothed or themselves pinnatisect on lower leaves. **Heads** discoid, small and numerous, in a dense, spikelike inflorescence or in spikelike branches; involucres campanulate, 2–3 mm high, the bracts imbricate, obovate with broad scarious margins; receptacle naked. **Achenes** ellipsoid, slightly flattened, 4- to 5- nerved, ca. 0.5 mm long; pappus none. Aug–Sept.

HABITAT A weedy species of shores, streambanks, ditches, mud flats and other places where water stands temporarily; native to nw USA, but now widespread in N America as a weed.

WETLAND STATUS

GP FACU | MIDW FACW | WMTN FACW

Artemisia biennis
BIENNIAL WORMWOOD

Bidens BEGGARTICKS

Weedy annuals varying greatly in size and degree of branching. **Leaves** simple or ternately to pinnately compound, opposite, obscurely to coarsely serrate, sessile or petiolate. **Heads** discoid or radiate, the ray florets, when present, rather few, sterile or rarely pistillate, the ligules yellow; involucral bracts biseriate and dimorphic, the outer row usually foliaceous and spreading, the inner row short and erect, membranous, striate; receptacle flat to convex, chaffy, the chaffy bracts narrow, membranous, conspicuously few-nerved; disk florets perfect, numerous or sometimes few, the corollas yellow; style branches flattened, bearded above. **Achenes** flattened parallel to the involucral bracts, narrowed at the base and widening upward, more or less truncate at the apex, with a pappus of 2–4 retrorsely barbed awns which persist atop the achene, the achene body also barbed or stiffly appressed-hairy at least on the angles, the achenes commonly functioning as stick-tights in animal fur and clothing by means of the barbs.

1 Leaves simple and toothed, some rarely incised or ternate 2

1 Leaves all ternately or pinnately compound 4

2 Achene margins (and often the faces) barbed upward at least toward the base ***B. tripartita***

2 Achene margins barbed downward for their entire length 3

3 Achenes with 4 awns; heads nodding after flowering (except on depauperate specimens) ***B. cernua***

3 Achenes with 3 awns; heads erect ***B. tripartita***

4 Rays well-developed and conspicuous, 1 cm or more long ***B. trichosperma***

4 Rays absent or poorly developed and inconspicuous, less than 5 mm long 5

5 Leaves pinnate-pinnatifid to bipinnate, the leaflets lobed; achenes with (2)3–4 awns ***B. bipinnata***

5 Leaves pinnate or 3-foliate, the leaflets serrate; achenes with 2 awns 6

6 Outer involucral bracts 5–10, typically 8; achenes dark brown to blackish ***B. frondosa***

6 Outer involucral bracts 10–16 or more, typically 13; achenes olivaceous to yellowish, or occasionally brown ***B. vulgata***

Bidens bipinnata L.
SPANISH NEEDLES

DESCRIPTION Erect, branching annual 3–15 dm tall, glabrous or seldom hispidulous. **Leaves** pinnate-pinnatifid to bipinnate, 4–20 cm long including the petiole which is 2–5 cm long, the leaflets ovate to oblong and deeply lobed, lobes acute to nearly rounded. **Heads** inconspicuously radiate, appearing discoid, 4–6 mm across; outer involucral bracts 7–10, linear, acute, shorter than the inner ones; **rays** yellowish-white and dark striate, less than 5 mm long, not exceeding the disk florets. **Achenes** with (2)3–4 retrorsely barbed, yellowish awns, the body brown to blackish, linear, mostly 10–18 mm long and much surpassing the involucre or the outermost achenes shorter. Aug–Oct.

HABITAT Moist to wet disturbed areas and shores; e Neb, where probably a sporadic introduction from farther south.

WETLAND STATUS
GP FACU | MIDW FAC

Bidens bipinnata
SPANISH NEEDLES

Bidens cernua L.

NODDING BEGGARTICKS

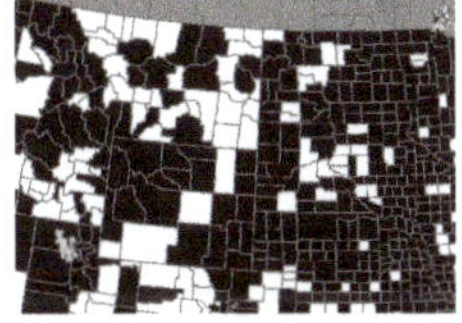

DESCRIPTION Small to large, often bushy annual 1–12 dm tall; **stem** glabrous or with spreading hairs. **Leaves** simple, linear-lanceolate to ovate-lanceolate, 3–18 cm long, 0.5–4.5 cm wide, acuminate, shallowly to coarsely toothed and usually scabrous on the margin, sessile and usually clasping at the base. **Heads** discoid or radiate, hemispheric, 1.5–3 cm across, usually nodding after flowering; outer involucral bracts 4–8, linear-lanceolate, unequal, often exceeding the disk; **rays**, when present, 6–8, yellow, to 1.5 cm long. **Achenes** with 4 retrorsely barbed awns, the body often purplish with pale, thickened margins, 5–7 mm long, retrorsely barbed on the margins. Aug–Oct.

HABITAT Shores, streambanks, marshes, ditches, wet meadows and other wet places.

WETLAND STATUS
GP OBL | MIDW OBL | WMTN OBL

Bidens frondosa L.

DEVIL'S-PITCHFORK

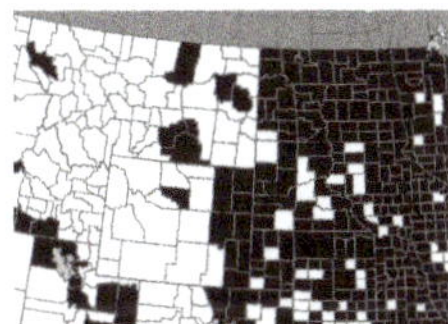

DESCRIPTION Erect annual 1.5–8 dm tall, usually branched; **stems** rather slender, often purplish, glabrous except slightly villous at upper nodes. **Leaves** all or mostly ternate, some larger ones sometimes pinnately divided into 5 leaflets, the leaflets ovate to lanceolate, glabrous to puberulent above, sparingly pubescent beneath, acuminate, serrate, the terminal leaflet largest, 3–9 cm long, 1–3.5 cm wide; petioles 1–4 cm long. **Heads** discoid, campanulate to hemispheric, 1–2 cm across; outer involucral bracts 5–10, typically 8, often sparingly ciliate, usually surpassing the disk. **Achenes** with 2 retrorsely barbed awns, the body dark brown to blackish, 4–9 mm long, antrorsely appressed-hairy. Late Jul–Oct.

HABITAT Shores, streambanks, marshes, ditches, wet meadows and other wet places.

WETLAND STATUS
GP FACW | MIDW FACW | WMTN FACW

Bidens cernua
NODDING BEGGARTICKS

Bidens frondosa
DEVIL'S-PITCHFORK

Bidens trichosperma (Michx.) Britt.
TICKSEED SUNFLOWER

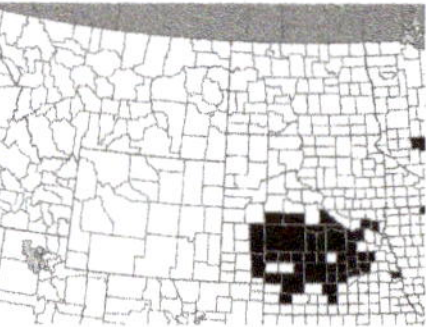

DESCRIPTION Glabrous annual (possibly partly biennial) 3–15 dm tall, rather strict to widely branched, the **stems** often purplish. **Leaves** pinnately divided into 3–7(9) narrow leaflets, short-petioled, to 15 cm long in overall length; leaflets linear to linear-lanceolate, entire to coarsely toothed or pinnately incised. **Heads** radiate, showy, the disk 8–15 mm across; outer involucral bracts 6–8(11), linear to linear-spatulate, to 10 mm long, short-hairy on the margins; inner involucral bracts shorter; **rays** bright golden-yellow, 1–2.5 cm long. **Achenes** with a pappus of 2 apical teeth or short awns, these antrorsely hispid or barbed, the body brown, cuneate-oblong or the inner ones longer, cuneate-linear, up to 9 mm long, antrorsely hispid on the margins. Sept–Oct.

SYNONYM *Bidens coronata* (L.) Britt.

HABITAT Shores, streambanks, marshes, floodplains and sand bars.

WETLAND STATUS
GP OBL | MIDW OBL

Bidens tripartita L.
THREE-LOBE BEGGARTICKS

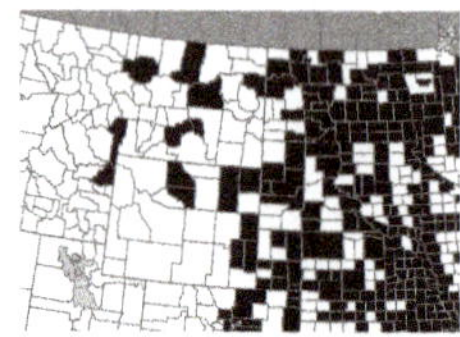

DESCRIPTION Simple and erect to branched and spreading annual 1–12 dm tall, glabrous. **Leaves** simple, oblong-lanceolate to ovate-lanceolate, 3.5–15 cm long, 0.5–5 cm wide, acuminate, shallowly to sharply toothed and scabrous on the margin, sessile or tapered to a winged, subpetiolate base. **Heads** discoid, broadly campanulate to hemispheric, 1–2.5 cm across, remaining erect after flowering; outer involucral bracts 5–10 or more, often greatly enlarged and much surpassing the disk. **Achenes** with 3 retrorsely barbed awns, the body reddish-brown to dark brown at maturity, 3.5–7 mm long, retrorsely barbed on the margins. Aug–Oct.

SYNONYMS *Bidens comosa* (A. Gray) Wieg., *Bidens connata* Muhl. ex Willd., *Bidens acuta* (Wieg.) Britt.

HABITAT Shores, streambanks, marshes, ditches, wet meadows and other wet places.

WETLAND STATUS
GP FACW | MIDW OBL | WMTN FACW

Bidens trichosperma
TICKSEED SUNFLOWER

Bidens tripartita
THREE-LOBE BEGGARTICKS

Bidens vulgata Greene
TALL BEGGARTICKS

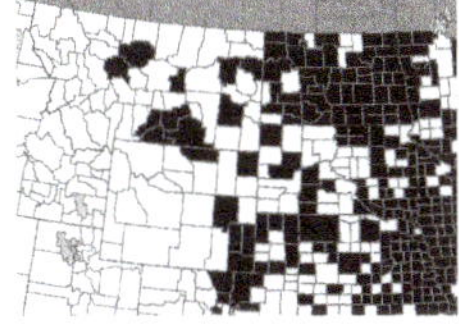

DESCRIPTION Similar in habit to *Bidens frondosa;* stems pubescent at the nodes and in the upper part. **Leaves** commonly divided into 3–5, or occasionally 7 leaflets, the leaflets ovate-lanceolate to lanceolate, sometimes incised, puberulent or nearly glabrous above, sparingly pubescent beneath, acuminate, serrate, the terminal leaflet 3–10 cm long, 1–3 cm wide; petioles 1–5 cm long. **Heads** discoid, averaging larger than in *B. frondosa,* 1–3 cm across, outer involucral bracts 10–16 or more, typically 13, ciliate. **Achenes** with 2 retrorsely barbed awns, the body olivaceous to yellowish or occasionally brown, antrorsely appressed-hairy. Aug–Oct.

HABITAT Shores, streambanks, marshes, ditches, wet meadows and other wet places.

WETLAND STATUS
GP FAC | MIDW FACW | WMTN OBL

Bidens vulgata
TALL BEGGARTICKS

Boltonia asteroides (L.) L'Her
BOLTONIA

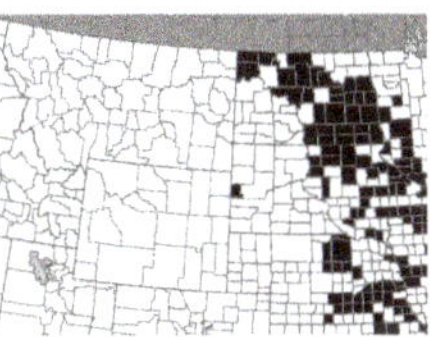

DESCRIPTION Stout, erect, glabrous, short-lived perennial 3–11 dm tall, fibrous-rooted, sometimes stoloniferous. **Leaves** simple, entire, linear to lanceolate or elliptic-lanceolate, 5–18 cm long, 0.5–2(4) cm wide, much smaller in the inflorescence, acute, scabrous on the margin, narrowed to a sessile or weakly clasping base. **Heads** several to usually numerous in a corymbiform or paniculiform inflorescence, radiate, 1.5–2.5 cm across; **ray florets** many, pistillate, the ligules white, pink, light blue or lavender, 5–15 mm long; involucres 2.5–5 mm high, the bracts imbricate, scarious-margined with a green midvein, the bracts linear and acute to spatulate and obtuse; receptacle hemispheric or conic, naked; **disk florets** perfect, their corollas yellow, drying brownish; style branches flattened, with short, lanceolate, hairy appendages. **Achenes** flattened, obovate, wing-margined, 1.5–2 mm long; pappus of 2 awns and several shorter bristles, reduced or absent in ray achenes. July–Aug.

SYNONYM *Boltonia latisquama* A. Gray.

HABITAT Wet meadows, marshes, shores, low prairie, occasionally in wet woods.

WETLAND STATUS
GP FACW | MIDW OBL | WMTN OBL

Boltonia asteroides
BOLTONIA

Cirsium arvense (L.) Scop.

CANADA THISTLE

DESCRIPTION Weedy stout perennial from deep rhizomes, dioecious or mostly so, 5–15 dm tall, often forming large patches. **Leaves** alternate, oblong, toothed to irregularly pinnatifid, spiny on the margins, 4–14 cm long, 0.5–4(5) cm wide, glabrous or occasionally white-tomentose beneath, acute to obtuse at the apex, tapered to a sessile base. **Heads** several to many in a terminal, usually branched inflorescence, discoid, 1.5–2 cm across; involucres campanulate, 1–2 cm high, the bracts in several series and strongly imbricate, sharply acute; receptacle flat, bristly; disk corollas pinkish-purple to occasionally white, elongate with 5 narrow lobes; pappus of numerous, plumose bristles, surpassing the corollas in pistillate heads, shorter than the corollas in staminate heads. **Achenes** tan, 3–4 mm long, the pappus deciduous as a ring of bristles. Late June–Aug.

HABITAT Wet meadows, shores, streambanks, ditches and drier places as well, especially where disturbed; introduced from Eurasia and well established as a noxious weed throughout n USA and s Canada.

WETLAND STATUS

GP FACU | MIDW FACU | WMTN FAC

ADDITIONAL SPECIES **Flodman's thistle** (*Cirsium flodmanii* (Rydb.) Arthur), is a native perennial thistle often encountered in wet prairies and sometimes in wet meadow zones. It is easily distinguished from Canada thistle by its fewer and larger heads borne singly at the tips of branches, with involucres 2–3 cm high. The plant is not weedy nor does it form dense patches.

Crepis runcinata (James) Torr. & Gray

HAWK'S-BEARD

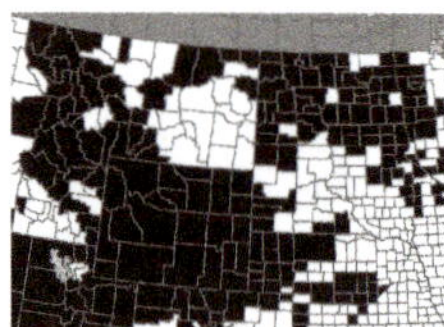

DESCRIPTION Milky-juiced perennial from 1–several strong roots, the principal leaves in a basal rosette (appearing like those of a dandelion); **stems** 1–3, scapose or nearly so, (1.5)2.5–5(7) dm tall, glabrous or sparsely hairy, the cauline leaves much reduced and bractlike. **Basal leaves** oblanceolate to elliptic or obovate, narrowed to a petiolate base, 4–15(25) cm long, 1.2–4(8) cm wide, obtuse to rounded, entire to toothed or runcinate. **Heads** ligulate, usually 1–several, rarely to 20, 1–2(3) cm across, mostly 20- to 50-flowered, sometimes with white tomentum on the outside of the receptacle, often glandular-pubescent on the involucre and onto the peduncle; involucre cylindrical to campanulate, 8–13 mm high, the bracts not numerous, linear, in 2 series, the outer ones few and reduced; receptacle naked; ligules yellow, 9–18 mm long. **Achenes** reddish-brown, terete, 4–5.5 mm long; pappus of numerous white capillary bristles. June–July.

Cirsium arvense
CANADA THISTLE

Crepis runcinata
HAWK'S-BEARD

HABITAT Wet meadows, shores and prairie swales, especially where alkaline.

WETLAND STATUS
GP FAC | MIDW FACW | WMTN FACU

Doellingeria umbellata (P. Mill.) Nees
TALL FLAT-TOPPED WHITE ASTER

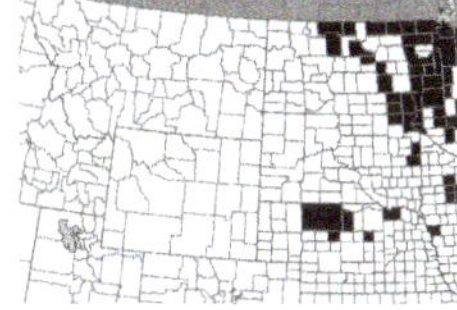

DESCRIPTION Stout, erect perennial 5–15 dm tall, from a thick rootstock; **stem** appressed-puberulent mainly above. **Leaves** subsessile to short-petiolate, narrowly to broadly elliptic or elliptic-ovate to elliptic-lanceolate, 4–14 cm long, 0.8–3(4) cm wide, scabrous above, densely puberulent beneath, acute to acuminate, entire, tapered to the subsessile or short-petiolate base. **Heads** usually many (few in depauperate specimens), 1–1.5 cm across, in a terminal, often flat-topped, corymbiform inflorescence; involucre 3–5 mm high, the bracts long-triangular and acute, puberulent; **rays** 4–7, white, 5–8 mm long. Late July–Aug.

SYNONYMS *Aster pubentior* Cronq., *Aster umbellatus* Mill.

HABITAT Wet meadows, springs and swampy or boggy places.

WETLAND STATUS
GP OBL | MIDW FACW

Doellingeria umbellata
TALL FLAT-TOPPED WHITE ASTER

Erigeron FLEABANE

Biennial to perennial herbs with simple, alternate leaves. **Heads** radiate, hemispheric, few to many in a terminal inflorescence; involucral bracts in 1 or 2 series, linear, equal or the outer ones a little shorter, hyaline at the tip and on the margins above, green in the middle and at the base; receptacle flat to slightly convex; outer 1–few rows of florets pistillate, with narrow, white to pink ligules 0.1–0.6 mm wide; disk florets perfect, their corollas yellow; style branches flattened, the appendages short, obtuse or acute. **Achenes** 2- to 4-nerved; pappus of 20–30 capillary bristles.

1 Rays tiny, shorter than the corolla tube and barely longer than the pappus; heads small, involucres less than 4 mm long, disks no more than 4 mm wide; disk florets numbering no more than 21 ***E. canadensis***

1 Rays conspicuous, larger; heads larger 2

2 Leaves not clasping; rays inconspicuous ***E. lonchophyllus***

2 Leaves auriculate-clasping; rays conspicuous ***E. philadelphicus***

Erigeron canadensis L.
HORSEWEED

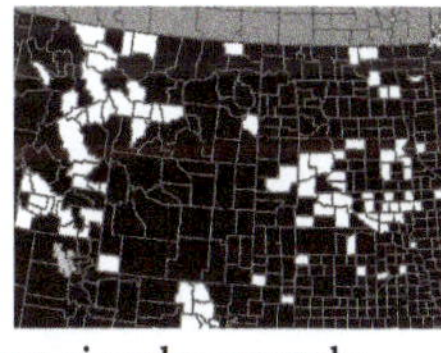

DESCRIPTION Slender, erect, weedy annual (1)2–15(25) dm tall, with a taproot, usually strongly hirsute with spreading to ascending hairs; **stem** simple or rarely branched below the inflorescence. **Leaves** numerous, alternate, ascending, mostly linear to linear-oblanceolate or lower ones elliptic-oblong to oblanceolate and often coarsely toothed, the latter often early deciduous, 2–10 cm long, 2–10 mm wide, acute, tapered to a sessile or short-petiolate base, usually hispid on the margins and at least on the midrib beneath. **Inflorescence** a terminal, elongate cluster with ascending racemiform branches; heads usually numerous (few on depauperate specimens), inconspicuously radiate; involucre 3–4 mm high, the bracts in ca. 3 series, the inner linear and much longer than the outer; receptacle smooth, slightly convex; **rays** minute, white or pink-tinged, about equal to or slightly exceeding the involucre; disk corollas yellowish to light pinkish. **Achenes** flattened, 2-ribbed, ivory to light brown, 1–1.2 mm long, appressed-puberulent; pappus of numerous, dull white capillary bristles. Late July–Oct.

SYNONYM *Conyza canadensis* (L.) Cronq.

HABITAT A weed of disturbed ground that often invades shorelines and drawdown zones; throughout s Canada and USA and widely introduced elsewhere.

WETLAND STATUS
GP FACU | MIDW FACU | WMTN FACU

Erigeron lonchophyllus Hook.
SHORT-RAY FLEABANE

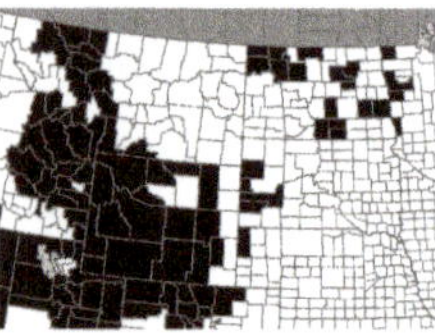

DESCRIPTION Shallowly fibrous-rooted biennial or short-lived perennial 1–4 dm tall; stems 1- few, spreading-hirsute. **Leaves** oblanceolate or linear-oblanceolate and petiolate at the base to linear and sessile above, not clasping, 4–14 cm long, 1–5 mm wide, acute, entire, hirsute on the margins. **Heads** few to many in a rather strict inflorescence, 1–1.5 cm across; involucres 6–9 mm high, the bracts hirsute, outer ones shorter than the inner; **rays** numerous, inconspicuous, white, turning brown at the tips with age, very narrow, 0.1–0.2 mm wide, about equaling to slightly exceeding the involucre; disk corollas exceeded by the pappus bristles. July–Sept.

HABITAT Wet meadows, seepage areas and boggy places.

WETLAND STATUS
GP FACW | MIDW FACW | WMTN FACW

Erigeron canadensis
HORSEWEED

Erigeron lonchophyllus
SHORT-RAY FLEABANE

Erigeron philadelphicus L.
PHILADELPHIA FLEABANE

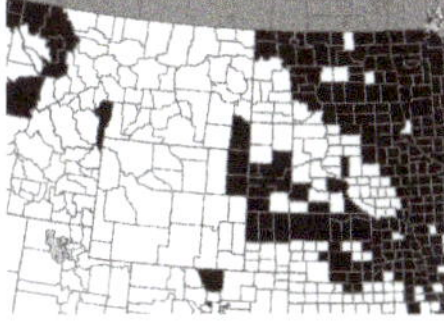

DESCRIPTION Fibrous-rooted biennial or short-lived perennial 2–7 dm tall; **stems** 1-few, hirsute. **Leaves** oblanceolate to spatulate below, oblong to lanceolate, auriculate-clasping and reduced upward, 2–15(27) cm long, 0.5–4(5) cm wide, pubescent to nearly glabrous, obtuse to rounded, entire to crenate or sinuate. **Heads** few to many in a narrow to widely branched inflorescence, 1.5–2.5 cm across; involucres 3–6 mm high, the bracts hirsute, equal or nearly so; **rays** numerous, conspicuous, white to deep pink, 5–10 mm long, 0.2–0.6 mm wide; disk corollas exceeding the pappus bristles. June–July, occasionally flowering in Aug–Sept.

HABITAT Wet meadows, shores, streambanks, wet woods, floodplains, springs and boggy places.

WETLAND STATUS
GP FAC | MIDW FACW | WMTN FACU

Erigeron philadelphicus
PHILADELPHIA FLEABANE

Eupatorium perfoliatum L.
BONESET

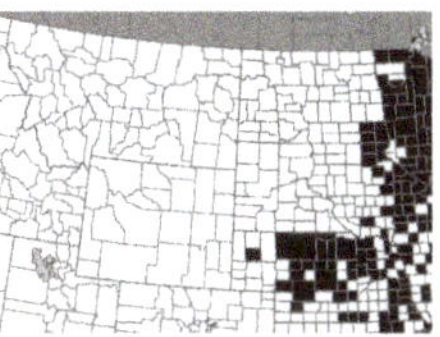

DESCRIPTION Stout, erect perennial from a thick rootstock; **stems** 3–12 dm tall, villous, especially above. **Leaves** opposite, broad-based and mostly connate-perfoliate, lanceolate, 6–15(20) cm long, 1.5–5 cm wide, sparingly pubescent to nearly glabrous above, more densely pubescent beneath, punctate with yellowish glands on both surfaces, acuminate, finely crenate-serrate and scabrous on the margins. **Heads** discoid, usually numerous and clustered in a terminal corymbiform, flat-topped inflorescence; involucres 3–5 mm high, the bracts green with white margins, glandular, weakly imbricate, acute to acuminate; corollas white. **Achenes** blackish, glandular-spotted, 1.5–2 mm long. Late July–Sept.

HABITAT Marshes, wet meadows, shores and swampy places.

WETLAND STATUS
GP FACW | MIDW OBL

Eupatorium perfoliatum
BONESET

Euthamia graminifolia (L.) Nutt.
NARROW-LEAVED GOLDENROD

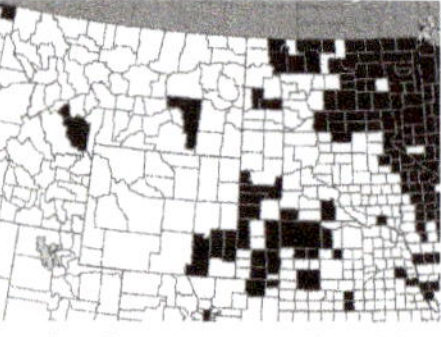

DESCRIPTION Erect rhizomatous perennial 3–9 dm tall; **stem** usually simple below and branched above, leafy except in the lower portion where the leaves are deciduous, glabrous or with scabrous lines decurrent from the leaf bases. **Leaves** simple, alternate, sessile, linear to linear-lanceolate or linear-elliptic, 2–10 cm long, 2–10 mm wide, 3- to 5-nerved, glabrous or scabrous on the margins and midrib, glandular-punctate, acute to acuminate, entire. **Heads** small, radiate, 20- to 35-flowered, usually numerous in flat-topped clusters comprising a corymbiform inflorescence; involucres campanulate to turbinate, 3–5 mm high, glutinous, the bracts in few to several series, imbricate, yellowish or green-tipped with a chartaceous base, outer ones ovate and obtuse, inner ones oblong to linear-oblong, obtuse to acute or acuminate; receptacle small, flat or convex, naked; **ray florets** pistillate, more numerous than the disk florets, the ligules yellow, 1–3 mm long; **disk florets** perfect, yellow; style branches flattened, puberulent. **Achenes** several-nerved, ca. 1 mm long, short-hairy; pappus of numerous white capillary bristles. Late July–Sept.

SYNONYM *Solidago graminifolia* (L.) Salisb.

HABITAT Fresh wet meadows, springs, fens, seepage areas, shores and streambanks, often where sandy or gravelly.

WETLAND STATUS
GP FACW | MIDW FACW | WMTN FAC

ADDITIONAL SPECIES **Texas goldentop** (*Euthamia gymnospermoides* Greene) is a similar species scattered from se ND to Neb and ne Wyo. It tends to favor drier sites than the above and differs in having heads with 12–20 florets and narrower, conspicuously punctate leaves with only 1–3 nerves.

Eutrochium maculatum (L.) E. Lamont
SPOTTED JOE-PYE WEED

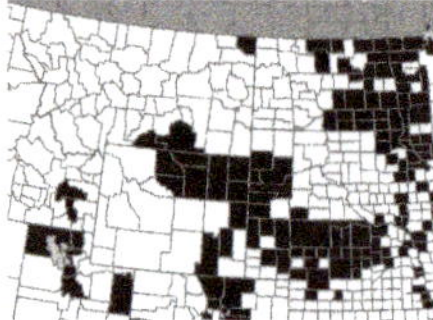

DESCRIPTION Perennial herb from a thick rhizome; **stems** stout, erect, 4–15 dm tall, short-pubescent above, densely so on branches of the inflorescence. **Leaves** in whorls of 3–6, lanceolate to ovate-lanceolate, 6–20 cm long, 2–6 cm wide, sparingly puberulent above, densely soft-puberulent beneath, acuminate, serrate, cuneate to nearly rounded at the base; petioles 0.5–2 cm long. **Inflorescence** or parts of it usually flat-topped; involucres 5–9 mm high, the bracts pinkish-purple or seldom white, imbricate, obtuse; corollas light pink to purple, seldom white. **Achenes** blackish, glandular-spotted, 2.5–4 mm long. Late July–Sept.

SYNONYM *Eupatorium maculatum* L.

HABITAT Wet meadows, marshes, shores, streambanks, ditches, springs and swampy or boggy places, where water is fresh.

WETLAND STATUS
GP OBL | MIDW OBL | WMTN OBL

Euthamia graminifolia
NARROW-LEAVED GOLDENROD

Eutrochium maculatum
SPOTTED JOE-PYE-WEED

Helenium autumnale L.

SNEEZEWEED

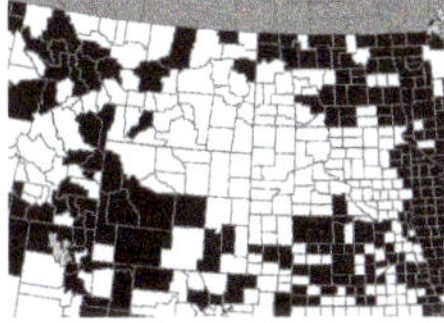

DESCRIPTION Erect, fibrous-rooted perennial 3–11 dm tall; **stems** single or clustered, puberulent or strigulose, especially above. **Leaves** alternate, ovate-lanceolate to elliptic-lanceolate or oblanceolate, 4–10 cm long, 0.8–3.5 cm wide, puberulent, glandular-punctate, acute to acuminate, entire to shallowly toothed, tapered to a narrow base which is decurrent as wings on the stem. **Heads** radiate, (1) few to many in a leafy inflorescence, hemispheric to subglobose, 1.5–4 cm across; involucral bracts in 2–3 series, not imbricate, linear and ultimately deflexed, puberulent; receptacle convex, naked; **ray florets** 10–20, pistillate or sterile, the ligules yellow, (2)3(4)-lobed, 1.5–2.5 cm long; **disk florets** perfect, the corollas yellow to brownish, glandular; style branches flat with broadened tips. **Achenes** 4- to 5-angled, 1.5–2 mm long, appressed-hairy with white to coppery hairs; pappus of several hyaline, awn-tipped scales. Late July–Sept.

HABITAT Wet meadows, shores, streambanks, seepage areas and swales.

WETLAND STATUS

GP FACW | MIDW FACW | WMTN FACW

Helenium autumnale
SNEEZEWEED

Helianthus SUNFLOWER

Stout perennials (those included here) with a stout rootstock and thick tuberous roots; **stem** simple or branched above. **Leaves** simple, petiolate, usually opposite below and alternate above, but sometimes all opposite, or opposite and alternate mixed over the length of the stem, the blades narrowly lanceolate to ovate-lanceolate, entire or serrate. **Heads** 1–several (seldom many), terminal on the main stem and lateral branches, conspicuously radiate; involucre of several series of narrow, overlapping bracts; receptacle flat to convex, chaffy, the chaffy bracts clasping the achenes (ovaries) of the disk florets; **ray florets** usually 10–20, sterile, the ligules yellow, spreading; disk florets numerous, perfect, their corollas yellow. **Achenes** flattened, narrowly obovate; pappus of 2 deciduous, awn-tipped scales.

1 Leaves coarsely serrate, often more than 4 cm wide; stem glabrous in middle and lower portions ***H. grosseserratus***

1 Leaves entire to shallowly dentate, less than 4 cm wide; stem glabrous to hispid in middle and lower portions ***H. nuttallii***

Helianthus grosseserratus Martens

SAWTOOTH SUNFLOWER

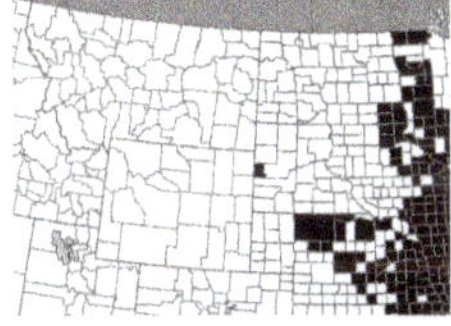

DESCRIPTION Colonial perennial 1–3 m tall; stem strigose in the inflorescence, otherwise glabrous, often purplish or

Helianthus grosseserratus
SAWTOOTH SUNFLOWER

glaucous. **Leaf blades** lanceolate to oblonglanceolate, 10–23 cm long, 2–5 cm wide, acuminate, coarsely serrate with widely spaced teeth, cuneate to abruptly tapered at the base, scabrous on both surfaces but often more densely strigose on the paler lower surface; petioles 1–4 cm long. Involucral bracts linear-lanceolate, acuminate, strigose, sometimes ciliolate; disk 1.5–2.5 cm across; **rays** 2.5–4 cm long. Achenes 3–4 mm long. Aug–Oct.

HABITAT Wet meadows, moist prairies and stream margins.

WETLAND STATUS
GP FACW | MIDW FACW | WMTN FAC

NOTE See discussion under *H. nuttallii.*

Helianthus nuttallii Torr. & Gray
NUTTALL'S SUNFLOWER

DESCRIPTION Similar to *Helianthus grosseserratus,* and grading into it in the east part of our region, differing mainly as follows: smaller in stature, 0.4–2 m tall; **stem** glabrous or sometimes hispid in the middle and lower portions, strigose in the inflorescence. **Leaf blades** narrowly lanceolate to lanceolate or ovate-lanceolate to elliptic, 4–15(20) cm long, 0.8–4.5 cm wide, entire to shallowly toothed, cuneate to rounded at the base; petioles 0.5–2(4) cm long. July–Sept.

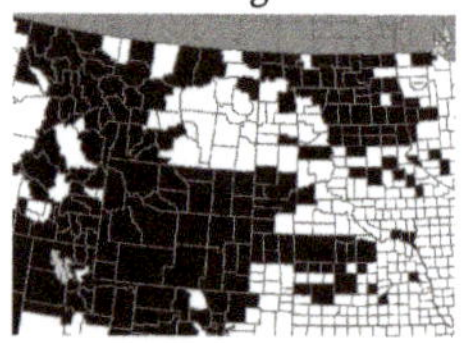

SYNONYM *Helianthus rydbergii* Britt.

HABITAT Wet meadows, marshes, shores, streambanks, ditches and other wet places.

WETLAND STATUS
GP FACW | MIDW FACW | WMTN FACW

NOTE Two phases of *H. nuttallii* are encountered in our region: *H. nuttallii* subsp. *rydbergii* (Britt.) Long is most common, with mostly opposite, ovate-lanceolate to lanceolate or elliptic leaves, acute to obtuse at the tip. The other phase is subsp. *nuttallii,* with mostly alternate, linear-lanceolate to narrowly lanceolate leaves, acute to acuminate at the tip. Also, in the e part of our region, where *H. nuttallii* and *H. grosseserratus* overlap in range, intermediate populations are found which cannot be assigned to either species with confidence.

Helianthus grosseserratus
SAWTOOTH SUNFLOWER

Helianthus nuttallii
NUTTALL'S SUNFLOWER

Packera pseudaurea (Rydb.) W.A. Weber & Á. Löve

WESTERN HEART-LEAVED GROUNDSEL

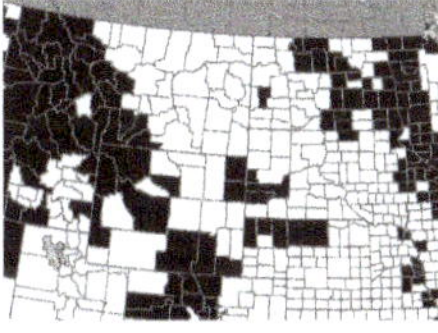

DESCRIPTION Perennial 2–5 dm tall, from a very short rhizome or caudex; stems solid, single or few clustered, glabrous or with only bits of tomentum in the leaf axils when young. **Leaves** changing in shape from the base upward on the stem; **basal leaves** with long slender petioles, the blades ovate, oval or suborbicular, 1–5 cm long, 0.8–4 cm wide, often purplish beneath, crenate, rounded to truncate or cordate at the base, transitional to the **cauline leaves** which are sessile, oblong to lanceolate or oblanceolate, 2–6 cm long, 0.4–2 cm wide, laciniate-pinnatifid at least toward the base, often clasping. **Heads** few to many, in a single terminal cluster, 1–1.5 cm across; involucre 4–7 mm high, the bracts herbaceous; **rays** pale yellow, 6–10 mm long. **Achenes** 1.5–2 mm long, glabrous; pappus slightly exceeding the disk corollas with age. Late May–June.

SYNONYM *Senecio pseudaureus* Rydb.

HABITAT Fresh wet meadows, fens and low prairie.

WETLAND STATUS

GP FACW | MIDW FACW | WMTN FACW

Petasites SWEET COLTSFOOT

Rhizomatous perennials; **principal leaves** basal, long-petioled, the blades sagittate or palmate, white-woolly at least on the lower surface; **cauline leaves** reduced and bractlike, alternate. **Heads** radiate or discoid, subdioecious, several to many in a racemiform or corymbiform inflorescence; involucral bracts in a single series, sometimes with a few reduced bracts below, broadly linear, herbaceous with scarious margins; receptacle flat, naked; female heads inconspicuously radiate, the florets all or nearly all pistillate, with a filiform corolla, some with whitish ligules; style shallowly 2-lobed, puberulent; male heads discoid, the florets with style and ovary but not developing achenes, their corollas tubular, 5-lobed. **Achenes** linear, 5- to 10-ribbed; pappus of numerous white capillary bristles, that of the sterile florets reduced.

1 Leaf blades palmately lobed *P. frigidus*

1 Leaf blades sagittate, merely toothed and not lobed *P. sagittatus*

Petasites frigidus (L.) Fries

NORTHERN SWEET COLTSFOOT

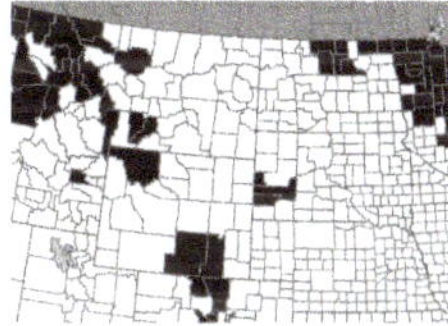

DESCRIPTION Flowering stem 3–5 dm tall, glabrous or glandular-puberulent above, especially in the inflorescence, sometimes thinly white-tomentose. **Basal leaves** expanding with or slightly after flower-

Packera pseudaurea
WESTERN HEART-LEAVED GROUNDSEL

Petasites frigidus
NORTHERN SWEET COLTSFOOT

ing, broadly triangular to reniform in outline, palmately lobed, coarsely toothed as well, 5–25 cm across, green and glabrous above, white-tomentose beneath, occasionally glabrous with age; bractlike **cauline leaves** mostly 2–6 cm long, reduced upward; petioles of basal leaves 1–3 dm long. **Heads** whitish, campanulate; involucres 4–9 mm high. May–June.

HABITAT Aspen woods and swampy places.

WETLAND STATUS
GP FAC | MIDW FACW | WMTN FACW

Petasites sagittatus (Pursh) A. Gray

ARROW-LEAF SWEET COLTSFOOT

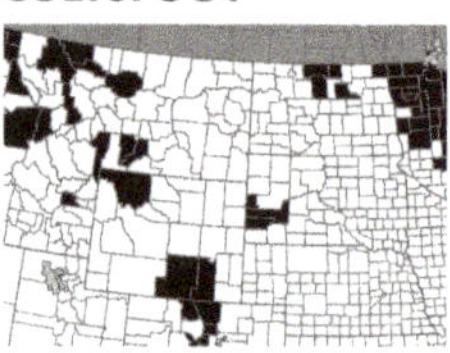

DESCRIPTION Similar to *Petasites frigidus* in habit; flowering stems 3–6 dm tall, thinly white-tomentose, sometimes glandular-puberulent as well. **Basal leaf blades** sagittate and toothed, 5–20 cm long, 3–30 cm wide, glabrous to thinly tomentose above, densely white-tomentose beneath; **cauline leaves** averaging longer and narrower than in *P. frigidus,* the lower ones often with abortive blades; petioles of basal leaves 1–3 dm long. **Heads** as in *P. frigidus.* May–June.

SYNONYM Now often treated as *Petasites frigidus* var. *sagittatus.*

HABITAT Fresh wet meadows, boggy and swampy woodlands.

WETLAND STATUS
GP FAC | MIDW FACW | WMTN FACW

Petasites sagittatus
ARROW-LEAF SWEET COLTSFOOT

Silphium perfoliatum L.

CUP PLANT

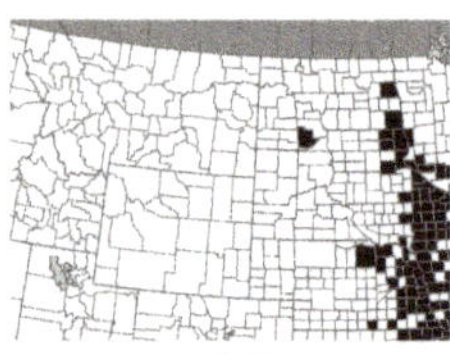

DESCRIPTION Stout, erect, rhizomatous perennial 8–20 dm tall; **stem** square, glabrous. **Leaves** simple, opposite, at least the upper ones connate-perfoliate, forming a cup around the stem, the lower leaves often short-petioled and connate by wings on the petioles, the blades ovate to deltate, 7–30 cm long, 3–15 cm wide, scabrous above, scabrous to hispidulous beneath, acute, coarsely toothed. **Heads** several to many in an open inflorescence, the disk 1.5–2.5 cm across; involucre 12–25 mm high, the bracts broadly ovate to elliptic, subequal, acute to obtuse, ciliolate on the margins; receptacle flat, chaffy, the outer bracts on the disk spatulate, transitional to narrower, more linear bracts inward; **ray florets** 20–30, pistillate, the ligules yellow, 1.5–2.5 cm long, the ovaries imbricate in 2–3 series, their styles branched, papillose; **disk florets** functionally male, their corollas pale to yellowish, their styles undivided, puberulent. **Achenes** flat, obovate, narrowly winged, 8–10 mm long, 5–6 mm wide; pappus none. July–Aug.

HABITAT River bottoms and swampy places.

WETLAND STATUS
GP FAC | MIDW FACW

Silphium perfoliatum
CUP PLANT

Solidago GOLDENROD

Erect perennials, rhizomatous or clumped from a thick caudex. **Leaves** simple, alternate, entire or toothed, sessile or the lower ones petiolate. **Heads** radiate, rather small and usually numerous, in a paniculiform, corymbiform or capitate inflorescence; involucres campanulate to turbinate, the bracts in few to several series, imbricate, more or less herbaceous, green-tipped with a chartaceous base; receptacle small, flat or convex, naked; **ray florets** pistillate, with a yellow ligule; **disk florets** perfect, their corollas yellow; style branches flattened, puberulent. **Achenes** subterete or angled, several-nerved; pappus of numerous white capillary bristles.

1 Involucres 2–4.5 mm high; cauline leaves flat, not clasping ... ***S. gigantea***

1 Involucres 5–6 mm high; cauline leaves conduplicate and clasping the stem ... ***S. riddellii***

Solidago gigantea Ait.
LATE GOLDENROD

DESCRIPTION Stems 5–12 dm tall, from stout, creeping rhizomes, often forming patches, mostly glabrous, puberulent on branches of the inflorescence. **Leaves** lanceolate to elliptic, 6–15 cm long, 1–4 cm wide, prominently 3-nerved, glabrous or sparsely pubescent on the 3 major veins beneath, acuminate, serrate, tapered to a subsessile or short-petiolate base. **Inflorescence** paniculiform, usually with numerous small heads on ascending to spreading or recurved-secund branches; involucre 2–4.5 mm high, the bracts linear, acute to blunt; **ray florets** usually 10–17, the ligules mostly 2–3 mm long. **Achenes** ca. 1.5 mm long, short-hairy. Late July–Oct.

HABITAT Wet meadows, ditches, streambanks, moist woods and thickets, also in drier places.

WETLAND STATUS
GP FAC | MIDW FACW | WMTN FACW

ADDITIONAL SPECIES **Canada goldenrod** (*Solidago canadensis* L.), is a common upland species that sometimes occurs in wet meadows and other moist places. It is similar to *S. gigantea,* but smaller in stature and densely short-pubescent on lower leaf surfaces and on the stem below the inflorescence.

Solidago gigantea
LATE GOLDENROD

Solidago riddellii Frank
RIDDELL'S GOLDENROD

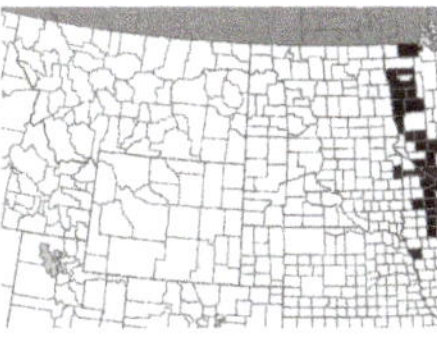

DESCRIPTION Stems 2–7(10) dm tall, clumped from a thick caudex, sometimes with rhizomes, glabrous to sparsely pubescent above. **Leaves** with blades lanceolate to linear-oblong, conduplicate, firm and

Solidago canadensis
CANADA GOLDENROD

glabrous, margins entire or with obscure, remote teeth, the **basal leaves** much larger than the cauline, often deteriorating early, their blades 10–20 cm long, 5–30 mm wide, on long, winged petioles usually exceeding the blade in length; **cauline leaves** smaller and numerous, becoming sessile and clasping upward, falcate-folded. **Inflorescence** corymbiform, often with corymbiform lateral branches; heads usually numerous and crowded, not secund on the branches; involucre 5–6 mm high, the bracts obtuse or rounded; **rays** 7–9, the ligules ca. 2 mm long. **Achenes** ca. 2 mm long, glabrous or nearly so. July–Oct.

SYNONYM *Oligoneuron riddellii* (Frank) Rydb.

HABITAT Fresh wet meadows and fens; reported from e SD, more common eastward into Minn and the Great Lakes region.

WETLAND STATUS
GP OBL | MIDW OBL

Sonchus arvensis L.
FIELD SOW THISTLE

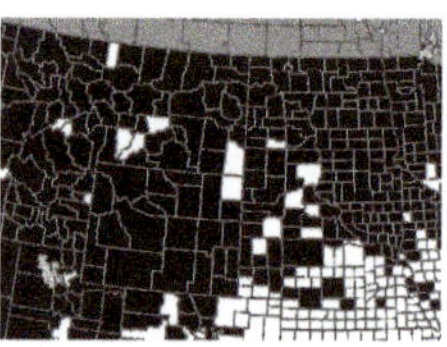

DESCRIPTION Stout, erect, milky-juiced perennial 5–15 dm tall, spreading by deep creeping roots; **stems** glabrous or with spreading gland-tipped hairs above. **Leaves** cauline, alternate, often larger and more crowded toward the base, lanceolate to oblong or oblanceolate, pinnately lobed to pinnatifid or unlobed, 10–30 cm long, 2–10 cm wide, acute to broadly rounded at the tip, prickly-margined, auriculate-clasping. **Inflorescence** containing several to many heads, open, corymbiform, the branches and involucres sometimes with spreading gland-tipped hairs; **heads** ligulate, yellow, many-flowered, 2–3.5 cm across; involucres campanulate, 10–22 mm high, sometimes thinly tomentose at the base, the bracts linear, imbricate, the outer ones much shorter than the inner; receptacle naked. **Achenes** reddish-brown, flattened, 2–3.5 mm long; pappus of numerous white capillary bristles, strongly accrescent. Late June–Aug.

SYNONYM *Sonchus uliginosus* Bieb.

HABITAT Wet meadows, shores, streambanks and other wet places; also fields, woods and roadsides; a cosmopolitan noxious weed originating in Europe.

WETLAND STATUS
GP FAC | MIDW FACU | WMTN FACU

Solidago riddellii
RIDDELL'S GOLDENROD

Sonchus arvensis
FIELD SOW THISTLE

Symphyotrichum ASTER

Annual or most often perennial herbs with simple, alternate leaves. **Heads** radiate (discoid in *S. ciliatum*), few to many, often showy, borne at the ends of leafy stems or branchlets, often aggregated in paniclelike inflorescences; **ray florets** pistillate, the ligules white to pink to various shades of blue or purple, mostly more than 0.5 mm wide; when rays are absent, the outermost florets on the disk pistillate with a slender, tubular corolla; involucral bracts in 2 or more series, usually imbricate, herbaceous and green-tipped, chartaceous at the base; receptacle naked, flat or slightly convex; **disk florets** perfect, the corollas yellow, white or red to purple; style branches flattened with minutely hairy appendages. **Achenes** mostly several-nerved, rarely 2-nerved; pappus of numerous capillary bristles.

NOTE The keys and description that follow include our species formerly described within the genus *Aster;* modern treatments separate our species into *Symphyotrichum* (7 species), *Almutaster* (1 species) and *Doellingeria* (1 species); these are all treated by this key.

1 Rays lacking; plants annual, taprooted **S. ciliatum**

1 Rays evident; plants perennial 2

2 Involucres, peduncles and upper stem glandular-pubescent 3

2 Involucres, peduncles and upper stem glabrous or pubescent but not glandular 4

3 Leaves lanceolate, clasping the stem **S. novae-angliae**

3 Leaves linear to linear-oblanceolate, not clasping **Almutaster pauciflorus**

4 Pappus a double row of capillary bristles, the outer row much shorter than the inner; heads clustered in a terminal, somewhat flat-topped inflorescence; rays white **Doellingeria umbellata**

4 Pappus a single row of equal capillary bristles; heads usually in paniclelike inflorescences; rays white to pink or blue 5

5 Leaves auriculate-clasping **S. puniceum**

5 Leaves sessile but not auriculate-clasping, only weakly clasping if at all 6

6 Leaves linear, 2–6 mm wide; rhizome slender, 0.5–1.5(2) mm thick; heads rather few (sometimes solitary) and uncrowded **S. boreale**

6 Leaves linear to linear-lanceolate, 3–25 mm wide; rhizome stouter, mostly 2–6 mm thick; heads usually many, often crowded 7

7 Veinlets of the leaf forming a conspicuous reticulum on the underside, outlining nearly equal-sided areolae **S. praealtum**

7 Veinlets of the leaf forming a rather inconspicuous reticulum on the underside, the areolae, if evident, clearly longer than broad **S. lanceolatum**

Symphyotrichum boreale (Torr. & Gray) Á. & D. Löve

RUSH ASTER

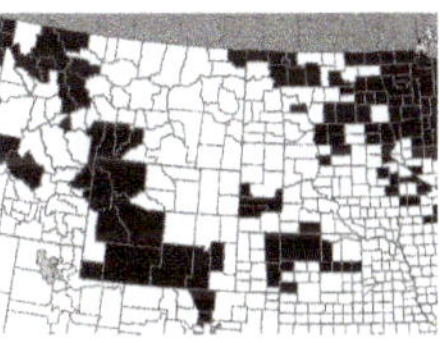

DESCRIPTION Slender erect perennial 3–8 dm tall, from slender rhizomes 0.5–1.5(2) mm thick, mostly glabrous except for lines of appressed pubescence decurrent from leaf bases; **stem** simple below, usually branched above in the inflorescence. **Leaves** sessile, linear, 4–11 cm long, 2–6 mm wide,

Symphyotrichum boreale
RUSH ASTER

acuminate, scabrous on the margin, sometimes slightly clasping at the base. **Heads** (1) few to several, seldom many, uncrowded, 1.5–2 cm across; involucre 5–7 mm high, the bracts imbricate, often with purplish tips and margins; **rays** usually 20–50, white to light blue or lavender, 7–15 mm long. Late July–Sept.

SYNONYMS *Aster junciformis* Rydb., *Aster junceus* Ait.

HABITAT Springs, fens, seepage areas, wet meadows and boggy places, typically where the soil remains saturated all year.

WETLAND STATUS
GP OBL | MIDW OBL | WMTN OBL

Symphyotrichum ciliatum (Ledeb.) Nesom

RAYLESS ASTER

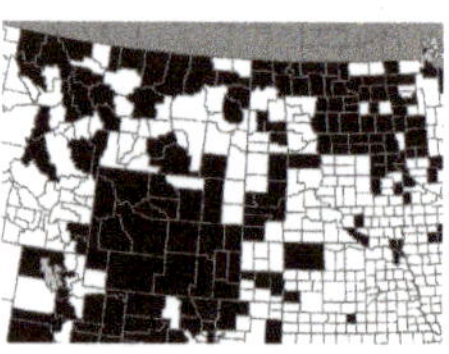

DESCRIPTION Taprooted annual 1.5–6 dm tall, simple and erect to branched and spreading, mostly glabrous. **Leaves** sessile, linear, 2–8 cm long, 2–5 mm wide, remotely ciliolate on the margins, acuminate. **Heads** usually numerous, hemispheric, 1–2 cm across, in an open-paniculate to spikelike inflorescence which commonly comprises the bulk of the plant; involucre 5–11 mm high, the bracts linear, equal or slightly imbricate, or the outer ones sometimes slightly longer than the inner; rays none, the outer pistillate florets with short, tubular corollas, more numerous than the perfect disk florets. **Achenes** flattened, 1.5–2 mm long, appressed-puberulent; pappus conspicuous, longer than the corollas, soft and copious. July–Sept.

SYNONYMS *Aster brachyactis* Blake, *Brachyactis ciliata* (Ledeb.) Ledeb.

HABITAT Shores, streambanks, wet meadows and flats, often where saline.

WETLAND STATUS
GP FACW | MIDW FAC | WMTN FACW

Symphyotrichum lanceolatum (Willd.) Nesom

MARSH ASTER

DESCRIPTION Erect or leaning perennial 3–12 dm tall, colonial from stout rhizomes mostly 2–6 mm thick, usually branched above; **stems** glabrous below, pubescent above in lines decurrent from the leaf bases. **Leaves** linear to linear-lanceolate, 3–15 cm long, 3–25 mm wide, acuminate, entire to shallowly toothed, smooth to scabrous on the margin, often slightly clasping the stem. **Heads** usually numerous, often crowded in a leafy panicle-like inflorescence, mostly 1.5–2.5 cm across; involucre 5–7.5 mm high, the bracts mostly not strongly imbricate in flowering heads, the outer bracts usually at least 2/3 as long as the inner ones in mature heads, sometimes to 1 mm or more wide, ciliolate on the margins; **rays** 20–50,

Symphyotrichum ciliatum
RAYLESS ASTER

Symphyotrichum lanceolatum
MARSH ASTER

white to pink or light blue, mostly 6–14 mm long; lobes of the disk corollas comprising 19–45% of the limb. Aug–Sept.

SYNONYMS *Aster hesperius* A. Gray, *Aster coerulescens* DC., *Aster simplex* Willd.

HABITAT Wet meadows, ditches, shores, streambanks, springs, seepage areas and other wet places; common and often ± weedy.

WETLAND STATUS
GP FACW | MIDW FAC | WMTN OBL

Symphyotrichum novae-angliae (L.) Nesom
NEW ENGLAND ASTER

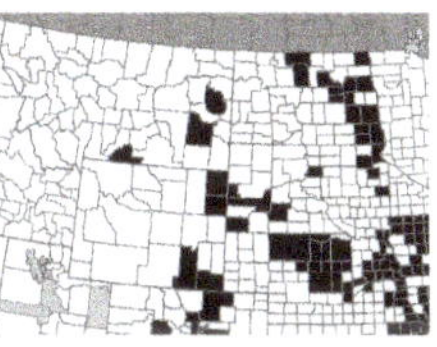

DESCRIPTION Stout erect perennial 4–10 dm tall, from a short rhizome or caudex; **stems** usually clustered, spreading hirsute, also finely glandular-pubescent upward. **Leaves** sessile, lanceolate, 3–7 cm long, 0.8–2.5 cm wide, scabrous or short appressed-hairy on the upper side, more softly pubescent beneath, acute, entire, strongly auriculate-clasping. **Heads** usually several to many, 1.5–3 cm across; involucre 7–12 mm high, the bracts slender, attenuate, glandular-pubescent, sometimes purplish; **rays** 45–100, blue-violet to reddish-purple, 1–2 cm long. Aug–Sept.

SYNONYM *Aster novae-angliae* L.

HABITAT Wet meadows, boggy areas, low prairie.

WETLAND STATUS
GP FACW | MIDW FACW | WMTN FACW

Symphyotrichum novae-angliae
NEW ENGLAND ASTER

Symphyotrichum praealtum (Poir.) Nesom
WILLOWLEAF ASTER

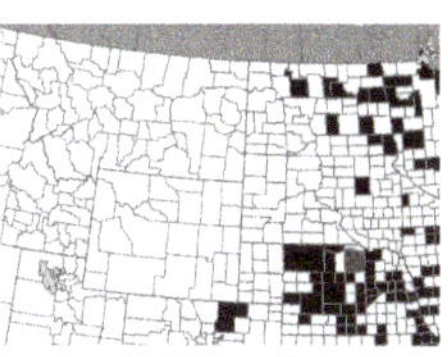

DESCRIPTION Quite similar to *S. lanceolatum,* differing mainly as follows: **Stem** usually rather uniformly pubescent, less often glabrate or with pubescence in lines decurrent from the leaf bases. **Leaves** thick-textured and firm, the veinlets on the lower surface dark and forming a conspicuous reticulum with nearly equal-sided areolae. Involucral bracts imbricate in several series, the outer much shorter than the inner, often reddish; **rays** blue, purple or sometimes whitish. Sept–Oct.

SYNONYM *Aster praealtus* Poir.

HABITAT Moist to wet meadows, floodplains, streambanks and thickets.

WETLAND STATUS
GP FACW | MIDW FACW | WMTN OBL

Symphyotrichum praealtum
WILLOWLEAF ASTER

Symphyotrichum puniceum (L.) Á. & D. Löve

SWAMP ASTER

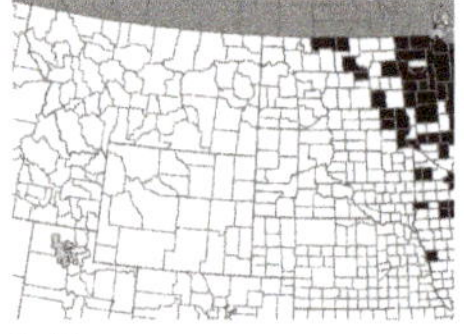

DESCRIPTION Erect perennial 4–12 dm tall, from short to long-creeping rhizomes; **stem** stout, branched above, pubescent in lines in the inflorescence, glabrous or sparingly hispid below the inflorescence. **Leaves** sessile, lanceolate, 5–20 cm long, 1–3.5 cm wide, reduced and crowded in the inflorescence, scabrous above, acuminate, entire to shallowly toothed, auriculate-clasping. **Inflorescence** leafy, crowded to rather open; heads usually many, 1.5–2.5 cm across; involucre 5–10 mm high, the bracts slender, scarcely or not at all imbricate, the outer ones occasionally enlarged and foliaceous; **rays** usually 30–60, blue to lavender or seldom white, 7–18 mm long. Aug–Sept.

SYNONYMS *Aster lucidulus* (A. Gray) Wieg., *Aster puniceus* L.

HABITAT Streambanks, pond margins and swampy places.

WETLAND STATUS
GP OBL | MIDW OBL

Tephroseris palustris (L.) Reichenb.

SWAMP RAGWORT

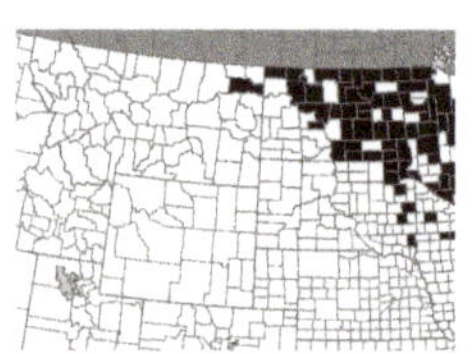

DESCRIPTION Fibrous-rooted annual or biennial 1.5–10 dm tall; **stems** hollow, especially toward the base, sparsely to densely villous. **Leaves** basically all similar in shape, oblong-linear to oblong-lanceolate or the lower ones often spatulate, 4–20 cm long, 0.5–6 cm wide, or the basal leaves occasionally larger, glabrous or villous in patches, blunt to rounded at the apex, entire to coarsely toothed or sinuate-toothed, sometimes crisped, the basal and lower leaves usually petiolate, often deciduous, the middle and upper cauline leaves sessile, with winged to auriculateclasping bases. **Heads** several to many, in one or more congested clusters, 1–1.5 cm across; involucre 4–8 mm high, the bracts pale and scarious toward the tip, darker at the base; **rays** pale yellow, 4–9 mm long. **Achenes** 1.5–2.5 mm long, glabrous; pappus bristles very fine and numerous, strongly accrescent. Late May–early Aug.

SYNONYM *Senecio congestus* (R.Br.) DC.

Symphyotrichum puniceum
SWAMP ASTER

Tephroseris palustris
SWAMP RAGWORT

HABITAT Shores and mud flats.

WETLAND STATUS
GP FACW | MIDW FACW

ADDITIONAL SPECIES The related **alkali-marsh ragwort,** *Senecio hydrophilus* Nutt., rarely occurs in the w part of our region, with records from Pennington Co., SD and se Wyo. The plant is hollow-stemmed like *T. congestus,* but is a glaucous perennial with glabrous stems and leaves and with the cauline leaves greatly reduced upward on the stem. *S. hydrophilus* favors marshy, alkaline sites, and can tolerate standing water.

Vernonia fasciculata Michx.
IRONWEED

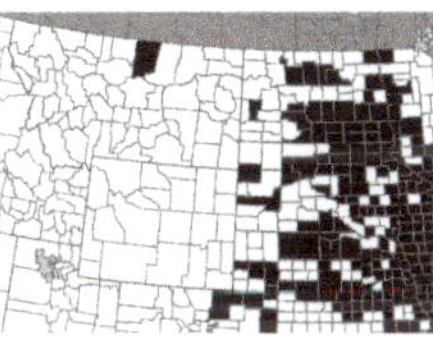

DESCRIPTION Stout, erect perennial 4–12 dm tall, from a thick-rootstock; **stems** single or clumped, often reddish or purplish, glabrous below to puberulent above, especially in branches of the inflorescence. **Leaves** simple, alternate, short-petioled, lanceolate to ovate-lanceolate, 4–15 cm long, 0.8–4.5 cm wide, smooth to scabrous above, dark-pitted beneath, acute to acuminate, denticulate to serrate, cuneate at the base. **Inflorescence** corymbiform, the heads usually numerous and crowded in flat-topped clusters; heads discoid; involucres campanulate, 6–9 mm high, the bracts imbricate, green with purple tips, obtuse to subacute, finely villous on the margins; receptacle flat, naked; disk florets all perfect, the corollas purple; style branches slender and tapered, puberulent. **Achenes** 3–4 mm long, strongly ribbed; pappus of numerous purplish to coppery capillary bristles. Mid July–Oct.

HABITAT Wet meadows, streambanks, ditches, flood plains and prairie swales.

WETLAND STATUS
GP FAC | MIDW FACW | WMTN FACW

Xanthium strumarium L.
COCKLEBUR

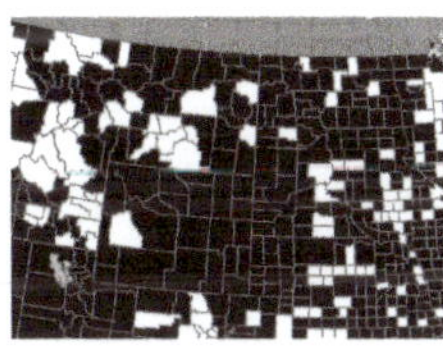

DESCRIPTION Weedy taprooted annual 2–8 dm tall, simple to much branched; **stem** rough due to coarse, appressed hairs, often dark brown spotted. **Leaves** simple, alternate, the blades broadly ovate to suborbicular or reniform, often weakly 3- to 5-lobed, 3–15 cm long, narrower to wider than long, scabrous, minutely punctate, bluntly toothed, obtuse to truncate or cordate at the base; petioles 3–10 cm long. **Heads** small, unisexual and dimorphic, in axillary clusters, the male florets in small heads above the larger female heads; male heads nearly spherical, many-flowered, ca. 5 mm in diameter; involucres of 1–3 series of separate bracts; receptacle cylindric, chaffy; florets

Senecio hydrophilus
ALKALI-MARSH RAGWORT

Vernonia fasciculata
IRONWEED

minute, their corollas brownish, filaments monadelphous, the nonfunctional pistil consisting mainly of an undivided style; **female heads** with a spiny involucre completely enclosing 2 florets, forming a conspicuous 2-chambered bur with hooked prickles; corolla none; **burs** ultimately yellow-brown to brown, ellipsoid, 1.5–3 cm long, with 2 prominent, hard, often incurved beaks at the tip, the body and spines of the bur typically covered with spreading hairs and stipitate glands. **Achenes** thick, one in each chamber of the bur. Late July–Sept.

HABITAT Shores, streambanks, wet meadows, sand bars, often where disturbed, also fields, roadsides and waste places; a cosmopolitan weed probably originating in America.

WETLAND STATUS
GP FAC | MIDW FAC | WMTN FAC

Xanthium strumarium
COCKLEBUR

Balsaminaceae
touch-me-not family

Impatiens TOUCH-ME-NOT

Erect to spreading, glabrous annuals with shallow weak roots and hollow, succulent stems. **Leaves** simple, alternate, exstipulate, the blades shallowly and often remotely serrate. **Flowers** perfect, irregular, yellow to orange-yellow, often reddish-brown spotted, pouchlike and spurred, hanging on the pedicels in few-flowered axillary racemes; sepals 3, petaloid, membranous, the upper 2 small, broadly obovate, cuspidate, the lower 1 large, saccate and spurred, broadly funnel-shaped, the spur usually recurved; petals 3, the upper petal often broader than long, the 2 lateral petals 2-lobed, each apparently derived by the fusion of a pair of petals; stamens 5, the anthers united around the stigma; ovary superior, 5-celled, each cell containing several ovules. **Fruit** a 5-valved, fusiform capsule, explosively dehiscent at maturity, scattering the seeds for some distance when jarred or touched.

1 Flowers orange-yellow, usually reddish-brown spotted; spur recurved parallel to the sac and 1/3 to 1/2 its length, the sac conic, longer than wide *I. capensis*

1 Flowers pale yellow, only faintly spotted if at all; spur recurved at a right angle to the sac and 1/5 to 1/4 its length, the sac broadly obtuse, about as wide or wider than long *I. pallida*

Impatiens capensis Meerb.
SPOTTED TOUCH-ME-NOT

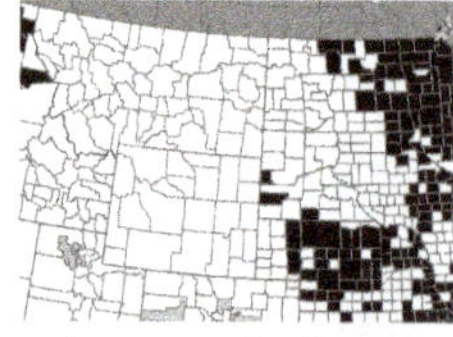

DESCRIPTION Plants 4–10 dm tall, usually branched and often spreading above. **Leaf blades** ovate to elliptic, 3–9 cm long, 1.5–4 cm wide, acute to obtuse and minutely apiculate at the tip, shallowly and remotely serrate, cuneate to rounded at the base; petioles longest on lower leaves, shorter upward, mostly 0.5–6 cm long. **Flowers** orange-yellow, unspotted or usually with reddish-brown spots, 1.5–2.5 cm long, the spur recurved parallel to the sac, 1/3 to 1/2 its length, the sac conic, longer than wide. **Capsules** ca. 2 cm long, explosively dehiscent. July–Sept.

SYNONYM *Impatiens biflora* Walt.

HABITAT Swamps, springs, streambanks, shores and boggy places, often where wooded.

WETLAND STATUS
GP FACW | MIDW FACW | WMTN FACW

Impatiens pallida Nutt. (*not illustrated*)

PALE TOUCH-ME-NOT

DESCRIPTION Very similar to *I. capensis*, but averaging larger, the **leaves** with blades to 12 cm long, 8 cm wide, more closely serrate than in *I. capensis*. **Flowers** pale yellow, unspotted or faintly reddish-brown spotted, 2–3(4) cm long, the spur recurved at a right angle to the sac, 1/5 to 1/4 its length, the sac broadly obtuse, about as wide or wider than long. **Capsules** as in *I. capensis*. July–Sept.

HABITAT Wooded floodplains and streambanks.

WETLAND STATUS
GP FACW | MIDW FACW

Impatiens capensis
SPOTTED TOUCH-ME-NOT

Betulaceae *birch family*

Monoecious trees or shrubs. **Leaves** alternate, stipulate, serrate, pinnately veined. **Flowers** much reduced and imperfect; the male and female flowers produced in separate catkins on the same individual. Flowers 3 per scale in the catkin; perianth none; stamens 2–10; carpels 2, each with a stigma and style, ovary 2-celled, 4-ovuled. **Fruit** a winged nutlet, 1-seeded by abortion.

1 Catkins in terminal clusters of 3-several; pistillate scales unlobed ***Alnus***

1 Catkins borne singly; pistillate scales 3-lobed ***Betula***

Alnus incana (L.) Moench

SPECKLED ALDER

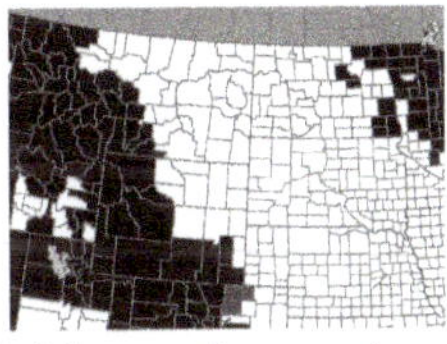

DESCRIPTION Shrub to 5 m tall; **twigs** reddish-brown, pruinose, mostly with conspicuous, light-colored lenticels. **Leaves** dark green and glabrous above, paler and pubescent on the veins beneath, ovate to elliptic, 5–14 cm long, 3–7 cm wide, serrate and shallowly lobed; petioles 1–2.5 cm long;

Alnus incana
SPECKLED ALDER

stipules caducous, irregularly lanceolate, 6–15 mm long, 1.5–5 mm wide, variably pubescent. **Catkins** in terminal clusters of 3-several; **male catkins** short-peduncled, elongate, 4–9 cm long, 7–9 mm thick; **female catkins** sessile, cylindric, in flower 5–6 mm long, 2–2.5 mm thick, the scales unlobed, in fruit becoming conelike, 13–18 mm long, 8–13 mm thick. **Nutlets** flat, slightly winged, obovate, 3 mm long, about as wide. Flowering May, fruiting Aug–Sept.

SYNONYM *Alnus rugosa* (Du Roi) Spreng

HABITAT Swampy areas and streambanks.

WETLAND STATUS
GP FACW | MIDW FACW

Betula glandulosa Michx.

RESIN BIRCH

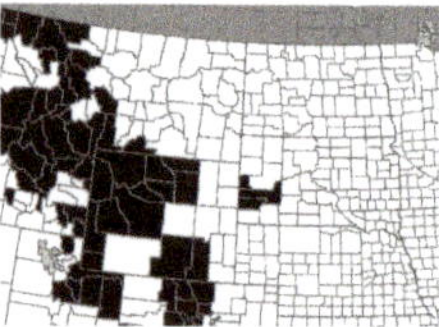

DESCRIPTION Erect, colonial shrub to 2 m tall; **bark** dull gray; **twigs** grayish-puberulent and dotted with resin glands, becoming reddish-brown and pruinose with age. **Leaves** dark green above, paler below, suborbicular to obovate, 20–35(70) mm long, 10–25(50) mm wide, glabrous with age, coarsely serrate, the teeth blunt or sharp; petioles 3–6 mm long; stipules caducous, ovate, 2–2.5 mm long, ciliate. **Catkins** produced singly from the buds; **male catkins** sessile, cylindric, 15–18 mm long, 1–2.5 mm thick; **female catkins** on peduncles 3–10 mm long, the peduncle often bearing a single petioled, toothed bract, the catkin cylindric, 7–18 mm long, ca. 5 mm thick; scales 3-lobed. **Nutlets** flat, winged, suborbicular to obovate, 1.5 mm long, 1.2–2 mm wide. Flowering late May–mid June, fruiting late July–Aug.

HABITAT Swamps, cold springs, bogs, seepage areas and streambanks.

WETLAND STATUS
GP OBL | WMTN OBL

NOTE North Dakota plants are treated as **bog birch** (*Betula pumila* L.), the common shrubby birch in the Great Lakes region. Plants have resinous young leaves and twigs, and catkins with scales averaging larger (3–4.5 mm long) than in *B. glandulosa*.

ADDITIONAL SPECIES In the Black Hills, **paper birch** (*Betula papyrifera* Marsh.), and **mountain or water birch** (*Betula occidentalis* Hook.), are often closely associated with streams. *B. occidentalis* in particular is common on low banks of Black Hills streams. Both of these birches are easily distinguished from bog birch. The peeling white bark and arborescent habit of paper birch are distinctive. Mountain birch is a large shrub or small tree with ovate leaves and lustrous, bronzy, non-peeling bark. Although both paper and mountain birch occur in other parts of our range, only in the Black Hills are they so consistently found next to water.

Betula glandulosa
RESIN BIRCH

Betula pumila
BOG BIRCH

Boraginaceae *borage family*

Annual or perennial herbs (those included here), often widely branching, with mostly alternate, simple, entire leaves and 1-sided scorpioid spikes or racemes of flowers. **Flowers** perfect, regular, 5-merous; calyx deeply lobed; corolla white, bluish or blue, conspicuous or not, lobed to near the middle or less, the lobes spreading, sometimes with 5 appendages (fornices) opposite the lobes, these more or less closing the throat of the corolla tube; stamens 5, epipetalous, included; pistil 2-carpellary, stigma sessile or on a short, terminal or gynobasic style, ovary superior, 4-celled, often 4-lobed. **Fruit** 4-lobed, splitting into 4 nutlets at maturity.

1 Leaves oblanceolate to obovate, glabrous; flowers showy, white or bluish, 5–10 mm wide ***Heliotropium***

1 Leaves linear, strigose; flowers inconspicuous, white, 1–2 mm wide ***Plagiobothrys***

Heliotropium curassavicum L.

SEASIDE HELIOTROPE

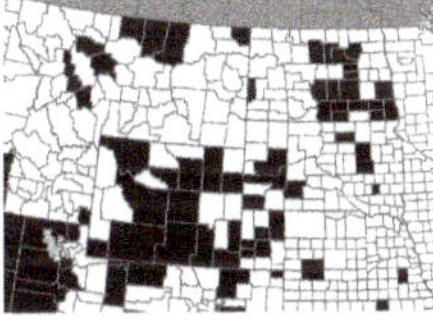

DESCRIPTION Succulent, glabrous, taprooted annual or short-lived perennial, 1–4 dm tall, decumbent to prostrate with spreading-ascending branches. **Leaves** alternate or an occasional pair subopposite, the blades oblanceolate to obovate, 2–6 cm long, 0.6–2(3) cm wide, obtuse to rounded at the tip, tapered to a subsessile to short-petioled base. **Flowers** closely spaced and secund on the 1–several branches of the scorpioid spikes or racemes, sessile or on pedicels to ca. 1 mm long; branches of the inflorescences usually less than 5 cm long but up to 10 cm long. Flowers showy, white or bluish with a yellow throat, 5–10 mm across, fornices lacking; calyx 2–4 mm long; corolla salverform, the spreading lobes about equaling or shorter than the tube; stigma as broad as the ovary, sessile or on a short terminal style. **Fruit** splitting into 4 nutlets at maturity, these 1.5–3 mm long, sometimes with corky ridges on the back. July–Sept.

HABITAT Saline or alkaline shores, flats and streambanks; introduced from tropical America.

WETLAND STATUS

GP OBL | MIDW OBL | WMTN OBL

ADDITIONAL SPECIES In the Black Hills, **true forget-me-not** (*Myosotis scorpioides* L.) is frequent around springs and cold streams. The attractive blue flowers with yellow centers are of about the same dimensions as those of *Heliotropium curassavicum,* but the plant is nonsucculent, with sparsely strigose foliage, decumbent or creeping stems that root at lower nodes and more open scorpioid cymes. This plant is an escaped ornamental from Eurasia and has become widely established in other parts of North America.

Heliotropium curassavicum
SEASIDE HELIOTROPE

Myosotis scorpioides
TRUE FORGET-ME-NOT

Plagiobothrys scouleri (Hook. & Arn.) I. M. Johnst.
POPCORN-FLOWER

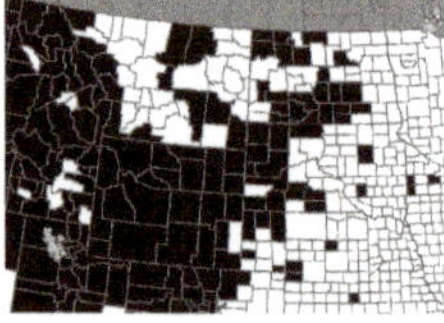

DESCRIPTION Low, strigose annual 0.5–3 dm tall, with several to many prostrate to ascending or suberect stems, or single-stemmed and erect. **Leaves** alternate or only the lowest 1-few pairs opposite, sessile, linear, 0.5–3 cm long, 1–3 mm wide, strigose to short-hispid, acute-tipped. **Flowers** very small, remotely spaced in irregularly bracteate false spikes or racemes which often run nearly the entire length of the stems; calyx deeply lobed, 2–4 mm long; corolla inconspicuous, white, slightly surpassing the calyx, 1–2 mm across, fornices minute but well-developed. **Nutlets** usually 4 (or 1–3 by abortion), ovate-lanceolate, 1.5–2 mm long, rugose on the back and also commonly tuberculate, sometimes with minute bristles. June–Aug.

HABITAT Shores, streambanks, ditches and dried flats of temporary ponds and marshes, usually in heavy clay soils.

SYNONYM *Plagiobothrys hispidulus* (Greene) I.M. Johnston

WETLAND STATUS
GP FACW | MIDW FACW | WMTN FACW

Plagiobothrys scouleri
POPCORN-FLOWER

Brassicaceae *mustard family*

Annual, biennial or perennial herbs with alternate, exstipulate, simple to compound **leaves**, the blades entire to dentate, sinuate, laciniate or pinnately lobed or divided. **Flowers** in terminal and lateral, ebracteate racemes, these indeterminate in development so that the lower portion is often in fruit while flowers continue to be produced at the tip; pedicels elongating in fruit. Flowers minute to conspicuous, perfect, regular; sepals 4, usually deciduous; **petals** 4, yellow, white or pink, clawed at the base; stamens 6, the outer 2 stamens shorter than the inner 4; pistil 1,2-carpellary, style 1, short, elongating in fruit, ovary superior, 2-celled, elongating or enlarging in fruit, maturing into an elongate to cylindrical or globose, sometimes curved, 2 (rarely 4)-valved capsule (termed a silique in those included here), the 2 halves separated by a membranous replum which persists on the pedicel after dehiscence, the seeds in 1 or 2 rows in each cell of the fruit.

---

1 Flowers white or pink — 2
1 Flowers yellow or yellowish-green — 4

2 Plants stoutly taprooted; siliques distinctly stipitate, borne on a stipe that extends about 1 mm or more beyond the tip of the pedicel — ***Thelypodium***
2 Plants fibrous-rooted, sometimes bulbous at the base as well; siliques sessile on the pedicels — 3

3 Seeds in 1 row in each cell of the silique; plants of wet habitats but not usually growing in water — ***Cardamine***
3 Seeds in 2 rows in each cell of the silique; plants usually growing in water — ***Nasturtium***

4 Siliques linear or curved-linear, 20–30x or more longer than wide; seeds in 1 row in each cell of the silique — ***Barbarea***
4 Siliques linear-elongate or cylindrical to globose, 1–8(10)x longer than wide; seeds in 2 rows in each cell of the silique — ***Rorippa***

Barbarea vulgaris R. Br.
WINTER CRESS

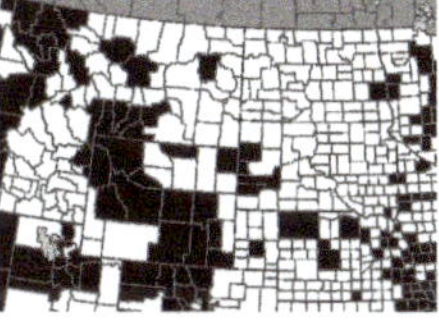

DESCRIPTION
Glabrous to sparsely hirsute, taprooted biennial or perennial 2–6 dm tall; simple below, branched above, often purplish at the base. **Basal leaves** all or mostly lyrate-pinnatifid and long-petioled, with a large, round to oblong terminal lobe and (0)1–4 pairs of small lateral lobes, 3–15 cm long including the petiole, 1–3 cm wide; **cauline leaves** similar to the basal ones but short-petioled to sessile and auriculate-clasping, reduced upward, the uppermost sinuate-lobed or angular-toothed, not pinnatifid. **Inflorescence** of few to many, terminal and lateral ebracteate racemes; pedicels 3–6 mm long. **Flowers** rather showy; sepals yellowish-green, oblong, 2–3 mm long; petals yellow, spatulate, 5–8 mm long. **Siliques** 2-valved, linear to curved-linear, 10–30 mm long, 0.5–1 mm thick, slightly 4-angled, the style persistent as a beak 1.5–3.5 mm long; seeds in 1 row in each cell of the fruit. Late May–June.

HABITAT Wet meadows, streambanks and alluvial bars; common in the Black Hills; introduced from Europe and naturalized throughout most of N America.

WETLAND STATUS
GP FACU | MIDW FAC | WMTN FAC

Barbarea vulgaris
WINTER CRESS

Cardamine BITTER CRESS
Annual, biennial or perennial plants, glabrous throughout or puberulent toward the base. **Leaves** petiolate to sessile, simple to pinnately compound, the basal leaves often differing in shape from the upper cauline leaves. **Flowers** small or medium in size, white; sepals green to yellowish, obtuse, caducou petals white, obovate to spatulate. **Siliques** 2-valved, linear, mostly straight, slightly flattened, 10–30x longer than wide, the seeds in a single row in each cell.

1 Leaves simple, the blades entire to sinuate-dentate; petals 6–12 mm long ... *C. bulbosa*

1 Leaves pinnately compound; petals 1.5–3 mm long ... *C. pensylvanica*

Cardamine bulbosa (Schreb.) B.S.P.
SPRING CRESS

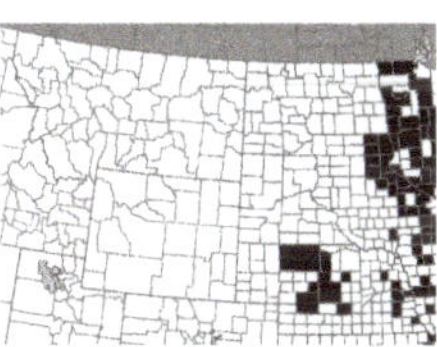

DESCRIPTION Erect perennial from a short thick tuber, 1.5–6 dm tall, the foliage watercress-flavored; **stems** single or few together, simple or sparingly branched above into lateral racemes, glabrous or puberulent near the base. **Leaves** simple, the basal leaves long-petioled, with blades oval to rotund, slightly sinuate to remotely crenate, rounded to cordate at the base; cauline leaves short-petioled to sessile upward, the blades oblong to elliptic, 2–7 cm long, 0.5–2.5 cm wide, irregularly sinuate to

Cardamine bulbosa
BITTER CRESS

sinuate-dentate, cuneate at the base. Sepals 2–4 mm long; petals 6–12 mm long. **Siliques** 10–25 mm long, 1–1.5 mm wide, on pedicels 1–3 cm long. Late May–June.

HABITAT Springs, fens, wet meadows and stream margins, where water is fresh.

WETLAND STATUS
GP OBL | MIDW OBL

Cardamine pensylvanica Muhl. ex Willd.
BITTER CRESS

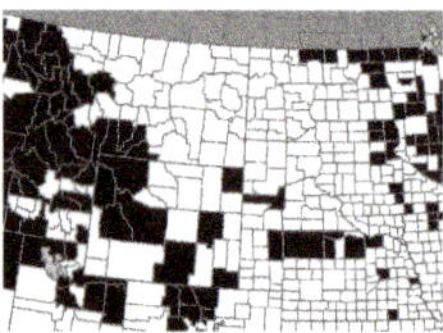

DESCRIPTION Annual or biennial 1–6 dm tall, forming a basal rosette of leaves in the first year when biennial; **stems** simple and erect to branched and spreading, usually pubescent toward the base. **Leaves** pinnately compound, divided into usually 2–5 pairs of lateral leaflets and a single terminal lobe, the lower leaves typically divided into broader segments than the upper ones; blades 2–7 cm long, 1–4 cm wide, the leaflets oblong to obovate or oblanceolate, entire to remotely dentate or lobed, the terminal leaflet the largest, usually 5–35 mm long, 4–15 mm wide; petioles shorter than the blades, decreasing in length upward. Sepals 1–1.5 mm long; petals 1.5–3 mm long. **Siliques** 10–30 mm long, 0.5–1 mm wide, on pedicels 2–15 mm long. June–July.

HABITAT Streambanks, springs and swamps.

WETLAND STATUS
GP FACW | MIDW FACW | WMTN FACW

Cardamine pensylvanica
BITTER CRESS

Nasturtium officinale Ait. f.
WATERCRESS

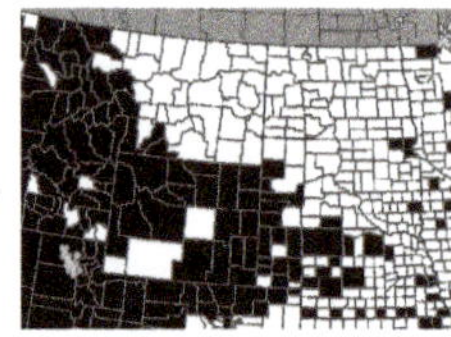

DESCRIPTION
Glabrous, fibrous-rooted, perennial aquatic; **stems** rather lax and trailing in water or on mud, freely rooting from lower nodes, erect or ascending toward the tips, 1–5 dm long. **Leaves** 4–13 cm long including the short to long petiole, 2–5 cm wide, pinnate or sometimes the earliest ones simple with only the terminal leaflet present; terminal leaflet ovate-cordate to subrotund or obovate, often much larger than the elliptic to obovate lateral leaflets, entire or weakly crenate. **Racemes** 1–several per stem, flat-topped and elongating in fruit. Sepals greenish-white, oblong, the outer pair saccate at the base, 1.5–2.5 mm long; petals white, sometimes tinged with purple, obovate, 3.5–5 mm long. **Siliques** linear, often curved, 10–25 mm long, ca. 2 mm thick, subterete, tipped with a short, truncate style beak 1 mm or less long; seeds in 2 rows in each cell of the fruit, conspicuously areolate. June–Oct.

SYNONYM *Rorippa nasturtium-aquaticum* (L.) Hayek

HABITAT Springs and streams where water is fresh; especially common in streams of the Black Hills; introduced from Eurasia and established throughout most of the USA and s Can.

WETLAND STATUS
GP OBL | MIDW OBL | WMTN OBL

NOTE This plant is the watercress of commerce and is often used as a salad green.

Nasturtium officinale
WATERCRESS

Rorippa YELLOW CRESS

Annual, biennial or perennial herbs, glabrous or variably pubescent. **Leaves** often forming a basal rosette in young plants, short-petiolate to sessile, simple, often deeply sinuate, pinnatifid or pinnately lobed to divided, otherwise entire, irregularly serrate, repand or laciniate. **Flowers** minute or small, yellow or yellowish-green; sepals green to yellowish, deciduous by fruiting time; petals yellow, often fading to whitish when dried, obovate to oblanceolate or spatulate, considerably shorter than to longer than the sepals. **Siliques** 2 (4)-valved, linear-elongate or more often cylindrical to globose, the seeds crowded in 2 irregular rows.

1 Petals shorter than or equal to the sepals 2

1 Petals longer than the sepals 4

2 Oldest siliques on the lower portion of the terminal raceme; petals usually longer than 1.2 mm; plants stout and erect, rarely decumbent or prostrate *R. palustris*

2 Oldest siliques on lower axillary racemes, or siliques nearly equal in age at corresponding points on the terminal and lateral racemes; petals usually shorter than 1.2 mm; plants prostrate to decumbent and spreading, seldom erect 3

3 Plants mostly taller than 3 dm, prostrate, decumbent or erect; pedicels ascending or strongly recurved, sometimes so that the raceme appears 1-sided; siliques acute to obtuse or sometimes truncate at the apex, the valves glabrous; leaves mostly entire to irregularly serrate or repand, only the lower deeply lobed *R. curvipes*

3 Plants mostly shorter than 3 dm, prostrate to decumbent; pedicels ascending to divergent; siliques truncate to obtuse or acute at the apex (where the style adjoins), but if acute, then the valves minutely papillate; leaves mostly lobed to the middle or nearly so *R. tenerrima*

4 Stems decumbent to prostrate; leaves and stems with few to many hemispherical trichomes; leaves pinnatifid *R. sinuata*

4 Stems erect; leaves and stems glabrous, puberulent or sparsely hirsute; leaves unlobed except for the auriculate base, or pinnately divided to the midrib 5

5 Cauline leaves pinnately divided to the midrib; siliques linear-elongate, 5–10 mm long *R. sylvestris*

5 Cauline leaves unlobed except for the auriculate-clasping base, otherwise irregularly serrate; siliques globose, 1–2 mm long *R. austriaca*

Rorippa austriaca (Crantz) Bess.
AUSTRIAN FIELD CRESS

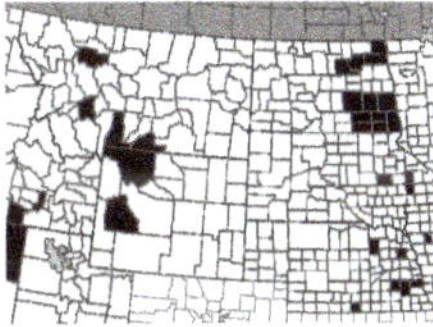

DESCRIPTION Stout, erect perennial 3–9 dm tall, from a fleshy taproot, often forming patches by creeping rootstocks, simple or branched from the base, branched toward the tip, glabrous to minutely strigose or puberulent on the leaves and stems. **Leaves** short-petioled near the base, becoming smaller, sessile and auriculate-clasping upward, the blades elliptic to oblanceolate, 3–10 cm long, 0.8–4 cm wide, obtuse to rounded at the tip, irregularly serrate to subentire, unlobed except for the auriculate-clasping base. **Racemes** terminal and in upper axils, often branched. Sepals yellowish-green, 1.5–2 mm long; petals yellow, 2–3 mm long, exceeding the sepals. **Siliques** rarely producing seed, globose, 1–2 mm long, the persistent

Rorippa austriaca
AUSTRIAN FIELD CRESS

style about equaling or exceeding the fruit body; pedicels spreading or a few recurved, 6–10(15) mm long. June–Aug.

HABITAT Low wet areas in and near fields; introduced from Europe, but showing little tendency to spread.

WETLAND STATUS
GP FACW | MIDW FACW | WMTN FAC

Rorippa curvipes Greene
BLUNT-LEAF YELLOW CRESS

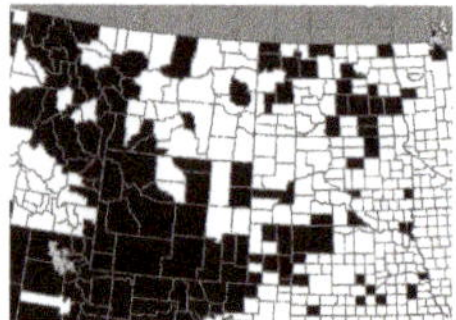

DESCRIPTION Annual or possibly a short-lived perennial, usually over 3 dm tall; stems prostrate to decumbent or occasionally erect, 1–5 dm long, single or branched from the base, sparingly to moderately hirsute in the lower portion. **Leaves** short-petioled to sessile, auriculate and clasping to nonauriculate and nonclasping, the blades oblong, obovate, spatulate or oblanceolate, mostly 3–10 cm long, 0.5–2(3) cm wide, glabrous on both surfaces or sparingly hirsute above, obtuse to acute at the tip, the margin entire to irregularly toothed or repand, or the lower leaves sometimes pinnately divided to the midrib, the lobe margins entire to slightly toothed, the terminal lobe acute to obtuse. **Racemes** terminal and axillary, all about the same age or the siliques somewhat older on the lower portion of the terminal raceme. Sepals greenish, 0.8–1.7 mm long, caducous; petals yellow, fading to whitish when dried, 0.5–1.2 mm long, mostly shorter than the sepals. **Siliques** short-cylindrical, 1.4–5 mm long, 1.5–2.5x longer than wide, straight or curved upward and inward toward the raceme axis, the valves glabrous; pedicels ascending or strongly recurved, sometimes so that the raceme appears 1-sided, mostly 2–6 mm long. July–Aug.

Rorippa curvipes
BLUNT-LEAF YELLOW CRESS

SYNONYMS *Rorippa obtusa* (Nutt.) Britt., in part; *Rorippa truncata* (Jeps.) Stuckey

HABITAT Shores, streambanks, mud flats.

WETLAND STATUS
GP OBL | MIDW FACW | WMTN FACW

Rorippa palustris (L.) Bess.
BOG YELLOW CRESS

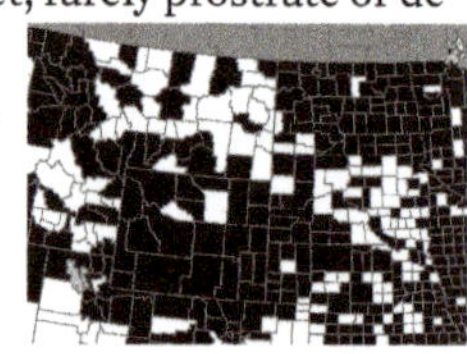

DESCRIPTION Annual, biennial or short-lived perennial; stems erect, rarely prostrate or decumbent, usually single from the base, simple or branched upward, 3–12 dm long, glabrous or sparingly to densely hirsute below, sparingly hirsute to glabrous above. **Leaves** short-petioled to sessile, auriculate and clasping to nonauriculate and nonclasping, the blades oblong to oblanceolate, 4–20(30) cm long, 1–5(8) cm wide, glabrous or sparingly to densely hirsute, narrowly to broadly acute at the apex, the margin irregularly serrate, incised, deeply cleft, repand or variously pinnate-divided. **Racemes** terminal

Rorippa palustris
BOG YELLOW CRESS

and axillary, the terminal one developing earliest, the oldest siliques on the lower portion of the terminal raceme. Sepals greenish, 1.2–2.5 mm long, caducous; petals yellow, drying whitish, 0.8–2.5 mm long, shorter than or equal to the sepals. **Siliques** globose to oblong or cylindrical, straight or slightly curved upward, 2–2.8 mm long, 1.2–3.4 mm wide, mostly 1–4x longer than wide, not at all to slightly tapering to the apex, truncate, obtuse or acute at the apex, the valves glabrous, the style 0.2–1.2 mm long; pedicels ascending, divergent or slightly to strongly recurved, 3–10 mm long. June–Sept.

SYNONYM *Rorippa islandica* (Oeder) Bourbas.

HABITAT Marshes, wet meadows, shores, streambanks, ditches and other wet places.

WETLAND STATUS

GP OBL | MIDW OBL | WMTN OBL

Rorippa sinuata (Nutt.) A. S. Hitchc.

SPREADING YELLOW CRESS

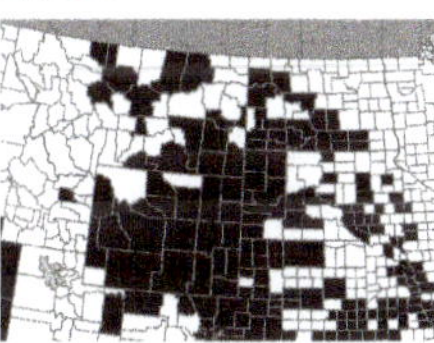

DESCRIPTION Decumbent to prostrate, branched perennial from rhizomes, the stems 1–5 dm long, sparsely to densely pubescent with hemispherical vesicular trichomes. **Leaves** sessile, auriculate and clasping to non-auriculate and non-clasping, the blades oblong to oblanceolate, 1.5–6 cm long, 0.4–2 cm wide, glabrate on the upper surface, sparsely to densely covered with vesicular trichomes on the midrib beneath, acute to mucronate, the margin shallow to deeply sinuate, pinnatifid to subpinnatifid, the lobes entire to slightly toothed. **Racemes** terminal and axillary, all about the same age or the axillary ones with the oldest siliques. Sepals yellowish-green, 2.7–4.5 mm long, caducous; petals bright yellow, 3–6 mm long, to 3 mm longer than the sepals. **Siliques** short to elongate-cylindrical, straight to strongly curved upward, 5–12 mm long, 1–2 mm thick, tapered to the style, the valves glabrous or roughened with hemispherical trichomes; pedicels strongly recurved to divergent or ascending, 4–10 mm long. June–Aug.

HABITAT Streambanks, ditches, wet meadows and other low places.

WETLAND STATUS

GP FACW | MIDW FACW | WMTN FAC

Rorippa sylvestris (L.) Bess.

CREEPING YELLOW CRESS

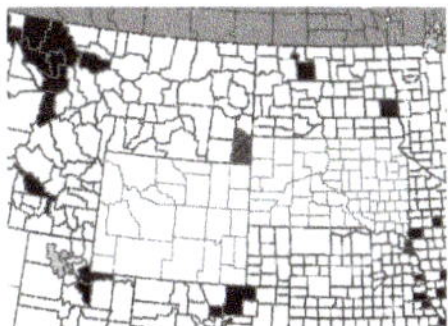

DESCRIPTION Rhizomatous and sometimes stoloniferous perennial 2–5 dm tall, glabrous or sparsely hirsute, the stems erect and branched. **Leaves** short to long-petioled at the base, becoming sessile upward, nonauriculate and nonclasping to auriculate and partly clasping, the blades of cauline leaves pinnately divided to the midrib, 3–12 cm long, 2–5 cm wide, glabrous, the lobes oblong or oblanceolate, entire to irregularly toothed or cleft, the blades of lower and basal leaves often larger with broader lobes. **Racemes** terminal and axillary, all about the same age or the oldest siliques on the lower portion of terminal racemes. Sepals yellowish-green, 2–2.5 mm long; petals yellow, 2.5–4 mm long, up to 2 mm longer than the sepals. **Siliques** linear-elongate, 4–10 mm

Rorippa sinuata
SPREADING YELLOW CRESS

Rorippa sinuata
SPREADING YELLOW CRESS

long, 0.5–1 mm wide, mostly 5–10x longer than wide, usually ascending on divergent to deflexed pedicels 5–12 mm long. June–Aug.

HABITAT Introduced to N America from Europe; potentially invasive in wetlands, e.g., streambanks, shores, ditches.

WETLAND STATUS
GP OBL | MIDW OBL | WMTN OBL

Rorippa tenerrima Greene

MODOC YELLOW CRESS

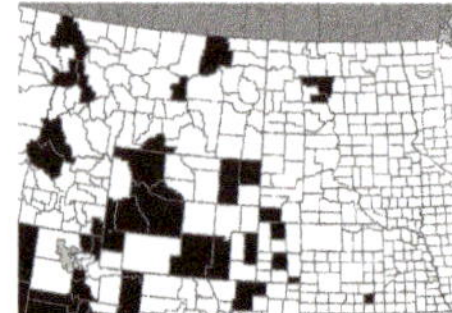

DESCRIPTION Glabrous, decumbent to prostrate annual, much-branched from the base, the stems 1–2 dm long. **Leaves** short-petioled, nonauriculate and nonclasping, the blades oblong or oblanceolate to spatulate, 2–5(8) cm long, 0.8–1.5 cm wide, lyrate-divided nearly to the midrib, rarely undivided, the lobes entire, the terminal lobe wider than the lateral ones, obtuse at the tip. **Racemes** terminal and axillary or all lateral, all about the same age or the lowest racemes with the oldest siliques. Sepals green, 0.7–1.2 mm long, caducous; petals yellow, fading to whitish when dried, 0.6–0.8 mm long, shorter than the sepals. **Siliques** elongate-cylindrical, slightly curved, tapering to the style, not constricted at the middle, 3–8 mm long, 0.8–2 mm wide, roughened with minute papillae; pedicels ascending, 1.5–4 mm long, usually rough with papillae. July–Sept.

SYNONYM *Rorippa obtusa* (Nutt.) Britt., in part.

HABITAT River bottoms and streambanks.

WETLAND STATUS
GP FAC | WMTN FAC

Rorippa tenerrima
MODOC YELLOW CRESS

Thelypodium integrifolium (Nutt.) Endl.

ENTIRE-LEAF THELYPODY

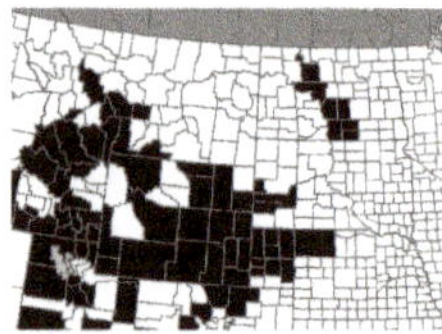

DESCRIPTION Erect, stoutly taprooted biennial 6–10 dm or more tall, glabrous, somewhat glaucous, simple below, usually freely branched above. **Leaves** simple, the basal ones usually withering early, 10–25 cm long, with blades ovate-lanceolate to narrowly lanceolate or oblanceolate, entire to sinuate-dentate; petioles about equaling the blades; cauline leaves sessile, linear-lanceolate and acute below, becoming linear-attenuate and reduced above, 3–10 cm long, 2–15 mm wide. **Racemes** flat-topped, elongating in fruit; pedicels slender, 4–12 mm long, ascending to spreading. Sepals whitish to purple-tinged, linear to linear-lanceolate, 3.5–5 mm long; petals white to pink or purple-tinged, spatulate, long-clawed, 6–9 mm long. **Siliques** linear, ascending, sometimes upcurved, 20–35 mm long, borne on a stipe that extends about 1 mm or more beyond the tip of the pedicel; seeds in 1 row in each cell of the fruit. June–Aug.

HABITAT Wet meadows, streambanks and flats, often where alkaline or saline.

WETLAND STATUS
GP FACW | WMTN FACW

Thelypodium integrifolium
ENTIRE-LEAF THELYPODY

Campanulaceae *bellflower family*

Perennial or biennial herbs with simple, alternate, exstipulate leaves. **Flowers** in terminal bracteate racemes or solitary and pedicelled in upper leaf axils, perfect, 5-merous, regular (*Campanula*) or irregular (*Lobelia*); calyx of 5 more or less equal sepals; corolla white to blue-violet, funnelform or bilabiate, in the latter case, split open along the upper side with the anthers protruding through the cleft; stamens 5, separate or united into a tube around the style; ovary inferior or partly so, 2- to 5-celled, many-ovuled; hypanthium usually prominent. **Fruit** a many-seeded capsule, opening laterally or at the top.

1 Flowers regular; stamens separate ***Campanula***

1 Flowers irregular; stamens united to form a tube around the style ***Lobelia***

Campanula aparinoides Pursh
MARSH BELLFLOWER

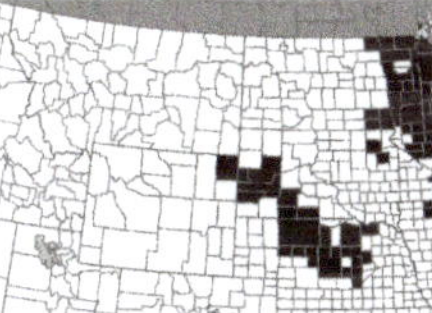

DESCRIPTION Perennial from slender rhizomes, the **stems** slender, weak, usually reclining on other plants, 2–6 dm long, 3-angled, usually roughened on the angles. **Leaves** sessile, linear or narrowly lanceolate, 1–8 cm long, 2–8 mm wide, reduced upward, often roughened on the margins and midvein beneath, acuminate at the tip, with very low remote teeth on the margins, tapered to the base. **Flowers** solitary on long, slender pedicels from the upper leaf axils, regular; sepals triangular to lanceolate, 1.5–5 mm long; corolla funnelform, pale blue to white, 4–12 mm long. **Capsule** opening near the base to release the seeds. July–Aug.

Campanula aparinoides
MARSH BELLFLOWER

HABITAT Fresh wet meadows, fens and boggy places.

WETLAND STATUS
GP OBL | MIDW OBL | WMTN OBL

Lobelia LOBELIA

Perennial or biennial plants with irregular flowers in terminal bracteate racemes. **Corolla** sometimes showy, white, pale blue or dark blue, often with white or yellow markings, bilabiate, the 2 lobes of the upper lip erect or projecting forward, the 3 lobes of the lower lip spreading, the corolla split to the base along the upper side between the lobes of the upper lip, the anthers projecting through the cleft; stamens fused to form an upward-arching tube around the style, the anthers usually colored, the lower 2 bearded at the tip; ovary 2-celled. **Capsule** dehiscing at the top.

1 Flowers large and showy, the corolla 16–30 mm long, dark blue with white stripes on the lower lip ***L. siphilitica***

1 Flowers small, not especially showy, the corolla 4–11 mm long, white to lavender or blue, often with a white or yellow center 2

2 Cauline leaves linear or narrowly lanceolate, 1–5 mm wide; racemes rather loose and open, the pedicels mostly 4–10 mm long ***L. kalmii***

2 Cauline leaves oblanceolate to lanceolate, 5–18 mm wide; racemes rather dense and spikelike, the pedicels 1–4 mm long ***L. spicata***

Lobelia kalmii L.
KALM'S LOBELIA

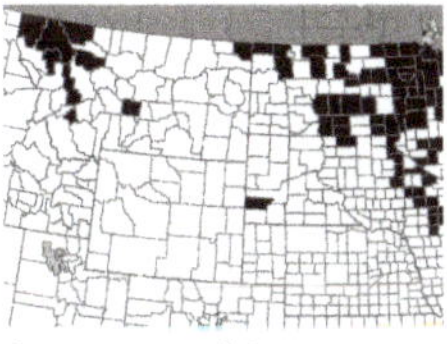

DESCRIPTION Small, erect, essentially glabrous biennial 1–4 dm tall, simple or sparingly branched above, often with a basal rosette of small obovate to oblanceolate leaves. **Cauline leaves** linear to narrowly lanceolate, 1–5 cm long, 1–5 mm wide, blunt to acute-tipped, remotely and obscurely toothed. **Racemes** rather loose and open, the flowers widely spaced on minutely bracteolate pedicels mostly 4- 10 mm long. Calyx lobes linear to linear-lanceolate, 1.5–5 mm long; corolla blue with a white or white-and-yellow center, 7–11 mm long. Late July–Sept.

HABITAT Fresh springs, fens, seepage areas and wet meadows.

WETLAND STATUS
GP OBL | MIDW OBL | WMTN OBL

Lobelia siphilitica L.
BLUE CARDINAL-FLOWER

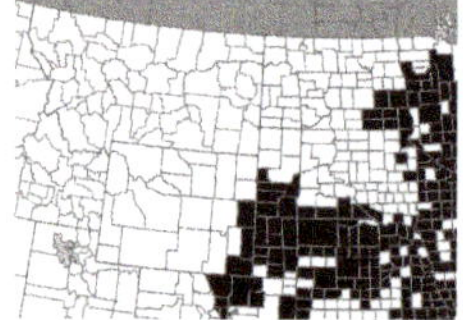

DESCRIPTION Erect perennial (1)2–8 dm tall, often flowering the first year; stems sparsely ciliate on decurrent leaf bases. **Leaves** sessile, oblong or elliptic to lanceolate or oblanceolate, 2–12 cm long, 1–3 cm wide, reduced upward and becoming bractlike in the inflorescence, acute or blunt-tipped, irregularly crenate-serrate, narrowed to the base. **Racemes** usually many-flowered (few-flowered on depauperate specimens), mostly 1–3 dm long; pedicels ascending, 4–10 mm long. Sepals triangular to lanceolate, 5–19 mm long, usually with narrow appendages toward the base; corolla showy, dark blue with the lower lip longitudinally blue-and-white striped, 16–30 mm long. Aug–Sept.

HABITAT Streambanks, shores, wet meadows and swampy places.

WETLAND STATUS
GP OBL | MIDW OBL | WMTN OBL

Lobelia spicata Lam.
PALE-SPIKE LOBELIA

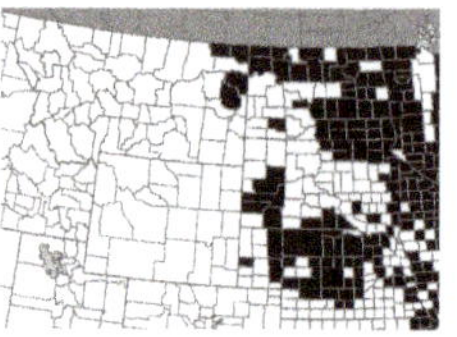

DESCRIPTION Erect perennial (1)2–6 dm tall simple or sometimes branched, pubescent especially toward the base. **Leaves** obovate to spatulate and often short-petiolate below, becoming oblanceolate upward on the stem, 1.5–5(7) cm long, 5–18 mm wide, puberulent mainly above, rounded at the apex, obscurely crenate to irregularly serrate, narrowed or long-tapered to the base. **Racemes** many-flowered, rather dense and spikelike; pedicels mostly 1–4 mm long, minutely bracteolate at the base. Calyx lobes linear to linear-lanceolate, 2–6 mm long; corolla white to pale blue, marcescent, 4–8 mm long. July–Aug.

HABITAT Fresh wet meadows, low prairie, springs, seepage areas and boggy places.

WETLAND STATUS
GP FAC | MIDW FAC | WMTN FAC

Lobelia kalmii
KALM'S LOBELIA

Lobelia siphilitica
BLUE CARDINAL-FLOWER

Lobelia spicata
PALE-SPIKE LOBELIA

Caryophyllaceae *pink family*

Annual or perennial herbs, often with a sticky, glandular pubescence. **Leaves** simple, opposite, sessile, entire, usually exstipulate. **Flowers** perfect, regular, solitary in forks of the stem or in terminal cymes; sepals 5; petals 5, often reduced or absent, often bilobed or deeply cleft, white or pinkish; stamens usually numbering the same or twice the number of the sepals, sometimes fewer; pistil 3- or 5-carpellary, styles 3 or 5, ovary superior, 1-celled. **Fruit** a many-seeded capsule, dehiscent by valves or terminally by teeth.

1 Stipules present, scarious; leaves rather succulent ***Spergularia***

1 Stipules none; leaves not succulent 2

2 Capsules laterally dehiscent by valves; styles 3 ***Stellaria***

2 Capsules terminally dehiscent by teeth; styles 5 ***Cerastium***

Cerastium MOUSE-EAR CHICKWEED

Low, erect to widely spreading annuals (in those treated here) with short, sticky pubescence. **Leaves** lanceolate, narrowly elliptic or oblanceolate, acute to obtuse, exstipulate. **Flowers** in compact to open, terminal cymes; sepals acute to blunt, scarious-margined; petals present or often absent, white, shorter than to exceeding the sepals, usually bilobed; stamens 5 or 10; carpels 5, styles 5. **Fruit** a many-seeded, straight or curved, membranous, cylindric capsule, terminally dehiscent by 10 teeth.

1 Pedicels 0.5–1.25x the length of the calyx in flower, to 3x the calyx length and straight or only slightly curved in age ***C. brachypodum***

1 Pedicels 1–3x the length of the calyx in flower, to 5x the calyx length and sharply curved below the calyx in age ***C. nutans***

Cerastium brachypodum (Engelm. ex Gray) B.L. Robins.

MOUSE-EAR CHICKWEED

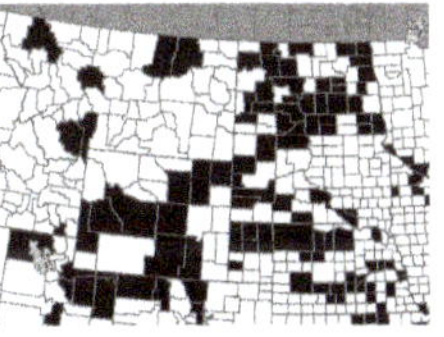

DESCRIPTION Short glandular-pubescent plant 0.5–3.5 dm tall; **stems** simple or branched from the base. **Leaves** 5–30 mm long, 1.5–5 mm wide. **Flowers** in rather compact cymes; sepals 2.5–4 mm long; petals, if present, shorter than to exceeding the sepals, often absent. **Capsules** 6–10 mm long; pedicels 0.5–1.25x the length of the calyx in flower, to 3x the calyx length and straight or only slightly curved in age. **Seeds** reddish-brown, angular-obovoid, papillate, 0.4–0.5 mm long. Late May–July.

HABITAT Wet alkali flats and drier places, often where sandy.

WETLAND STATUS
GP FACU | MIDW FACU | WMTN FAC

NOTE *C. brachypodum* is sometimes treated as a variety of the following species.

Cerastium nutans Raf.

NODDING MOUSE-EAR CHICKWEED

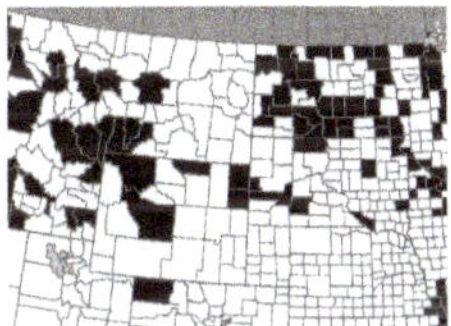

DESCRIPTION Quite similar to *Cerastium brachypodum* but attaining larger stature and usually more openly branched above, 1–5 dm tall. **Leaves** 8–60 mm long, 4–12 mm wide. **Flowers** in rather open and of-

Cerastium brachypodum
MOUSE-EAR CHICKWEED

ten widely branched cymes; sepals 2.5–5 mm long; petals as in *C. brachypodum.* **Capsules** 7–12 mm long; pedicels 1–3x the length of the calyx in flower, to 5x the calyx length and sharply curved below the calyx in age. **Seeds** as in the preceding, but 0.4–0.7 mm long. Late May–Aug.

HABITAT Shores, streambanks, springs, boggy places and wet woods.

WETLAND STATUS
GP FAC | MIDW FACU | WMTN FACU

Spergularia marina (L.) Griseb.
SALTMARSH SAND-SPURRY

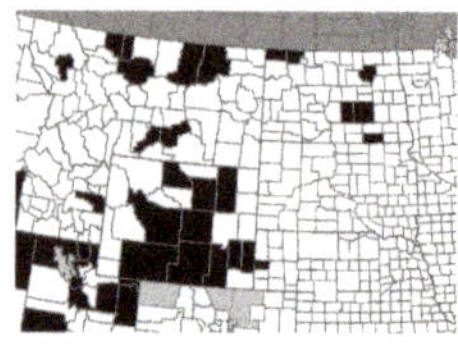

DESCRIPTION Low decumbent to erect annual, simple to diffusely branched, 5–20 cm tall, glabrous or glandular-pubescent. **Leaves** rather succulent, linear, blunt to mucronate at the tip, 5–40 mm long, 0.5–1.5 mm wide; stipules present, broadly deltate to round or reniform, scarious. **Flowers** usually numerous, in a widely spreading, bracteate cyme, the inflorescence comprising most or nearly all of the plant; bracts similar to the leaves though smaller; flowers subsessile or on pedicels 1–10 mm long with age; sepals obtuse to rounded, 2–5 mm long; petals white or more often pinkish, 1/2 to nearly as long as the sepals; stamens 2–5; carpels 3, styles 3. **Capsules** 3-valved, ovoid, equaling or surpassing the calyx, mostly 3–6 mm long. **Seeds** light to dark brown, obovoid to nearly round, compressed, 0.5–0.9 mm long, smooth or slightly rough, sometimes winged. July–Sept.

HABITAT Wet alkali flats and shores; occasional in n and e ND.

WETLAND STATUS
GP FACU | MIDW FACU

Stellaria STITCHWORT

Low, spreading or erect perennials (in those treated here), mostly glabrous; **stems** slender, 4-angled. **Flowers** solitary in forks of the stem or in rather sparsely flowered terminal cymes; pedicels filiform. Sepals green with scarious margins; petals white, bilobed to deeply cleft; stamens 10 or sometimes fewer; carpels 3, styles 3. **Fruit** an ovoid or oblong capsule, laterally dehiscent by 6 valves.

1 Flowers solitary in forks of the stem, not subtended by scarious bracts ***S. crassifolia***
1 Flowers in terminal cymes, subtended by small scarious bracts **2**

2 Inflorescence widely branched, the pedicels spreading or reflexed with age; leaves spreading to ascending, widest at or above the middle ***S. longifolia***
2 Inflorescence rather narrow, the pedicels erect to ascending; leaves strongly ascending, widest near the base ***S. longipes***

Cerastium nutans
NODDING MOUSE-EAR CHICKWEED

Spergularia marina
SALTMARSH SANDY-SPURRY

Stellaria crassifolia Ehrh.

FLESHY STITCHWORT

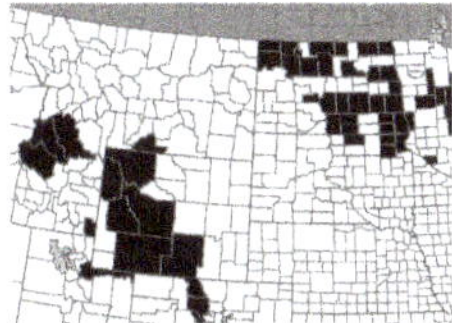

DESCRIPTION **Stems** decumbent and matted to erect, often supported by surrounding vegetation, freely branched, 8–30 cm long. **Leaves** elliptic to lanceolate or oblanceolate, acute or blunt, narrowed at the base, 5–25 mm long, 1–3 mm wide. **Flowers** solitary in forks of the stem, on pedicels 1–5 cm long; sepals 2.5–4 mm long; petals surpassing the sepals. **Capsules** ovoid, equaling or exceeding the calyx, 3.5–5 mm long. **Seeds** reddish brown, oblong-orbicular, 0.7–1 mm long. June–July.

HABITAT Springs, boggy areas and stream margins.

WETLAND STATUS

GP OBL | MIDW FACW | WMTN FACW

Stellaria longifolia Muhl. ex Willd.

LONG-LEAVED STITCHWORT

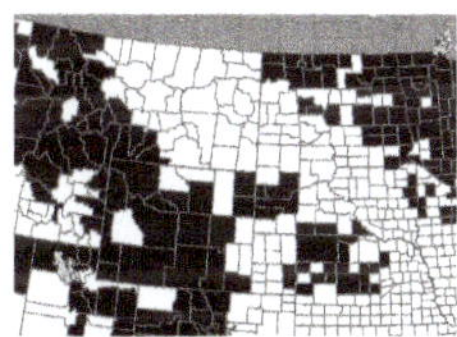

DESCRIPTION **Stems** weak, decumbent to ascending, often sprawling among surrounding vegetation, prominently 4-angled, usually freely branched, 1–4.5 dm long. **Leaves** spreading to ascending, linear to narrowly lanceolate or oblanceolate, widest at or above the middle, acute at both ends, 1–5 cm long, 0.8–5(8) mm wide. Inflorescence widely branched, the pedicels spreading or reflexed with age, 1–8 cm long, subtended at the base by a pair of small scarious bracts. **Flowers** few to many; sepals 3–5 mm long, acute; petals equaling or slightly

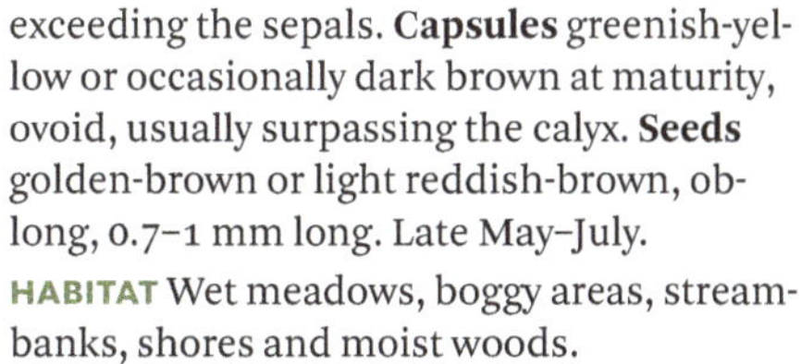

exceeding the sepals. **Capsules** greenish-yellow or occasionally dark brown at maturity, ovoid, usually surpassing the calyx. **Seeds** golden-brown or light reddish-brown, oblong, 0.7–1 mm long. Late May–July.

HABITAT Wet meadows, boggy areas, streambanks, shores and moist woods.

WETLAND STATUS

GP FACW | MIDW FACW | WMTN FACW

Stellaria longipes Goldie

LONG-STALK STARWORT

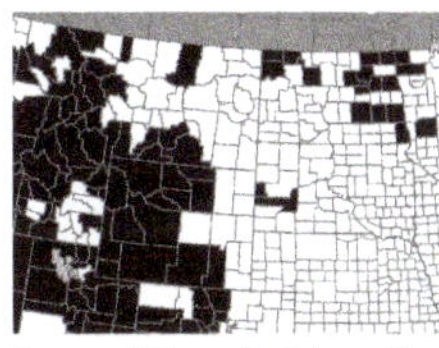

DESCRIPTION **Stems** erect or decumbent, often densely matted, 5–25 cm long, from slender rhizomes. **Leaves** strongly ascending, rather stiff and shiny, linear or linear-lanceolate, widest near the base, acute at the tip, 1–2.5(4) cm long, 0.8–3(5) mm wide. **Inflorescence** similar to that of *S. longifolia* but rather narrow, the pedicels erect to ascending, 1–3(6) cm long. **Flowers** 1–several per stem, seldom more; sepals 3–5 mm long, acute or obtuse; petals slightly shorter to slightly longer than the sepals. **Capsules** stramineous to shiny purplish at maturity, ovoid, surpassing the calyx. **Seeds** light to dark reddish-brown, oblong to oval, 0.7–1 mm long. Late May– July.

HABITAT Wet meadows, moist ditches and thickets.

WETLAND STATUS

GP OBL | MIDW OBL | WMTN FACW

Stellaria crassifolia
FLESHY STITCHWORT

Stellaria longifolia
LONG-LEAVED STITCHWORT

Stellaria longipes
LONG-STALK STARWORT

Ceratophyllaceae *hornwort family*

Ceratophyllum demersum L.

COONTAIL

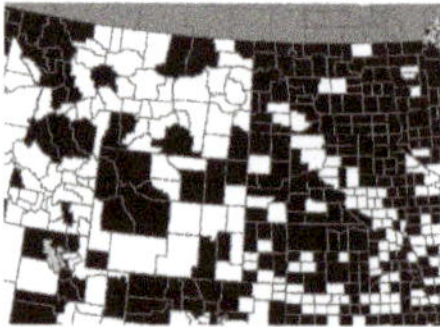

DESCRIPTION Rootless perennial aquatic, entirely submersed, free-floating or more often anchored by the older decaying portion of the stem. **Stems** delicate, freely branching, the living portion usually no more than a few dm long, often fragmenting as water warms during summer. **Leaves** sessile, whorled, 5–12 at each node, once or twice dichotomously branched into filiform segments, 5–30 mm long, sparsely to conspicuously spinulose, spreading to curving upward, more crowded toward the branch tips to give the coontail appearance. **Flowers** minute, imperfect, solitary and sessile in the leaf axils, both male and female flowers subtended by an involucre or perianth of 8–15 greenish segments; male flowers containing 10–16 stamens; female flowers each consisting of a single unicarpellary pistil. **Fruit** a dark olive, ellipsoid achene, usually with 1 terminal and 2 basal spinelike projections, the achene body 4–5 mm long. June–Sept.

HABITAT Quiet water of lakes, ponds, marshes and streams; common, often abundant.

WETLAND STATUS

GP OBL | MIDW OBL | WMTN OBL

Ceratophyllum demersum
COONTAIL

Cornaceae *dogwood family*

Cornus sericea L.

RED OSIER

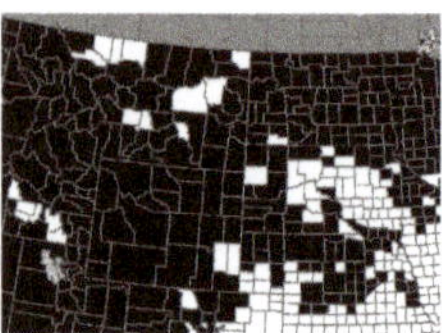

DESCRIPTION Many-stemmed shrub 1.5–3 m tall; main **stems** branched above, the young branches and first-year twigs dark red, older branches yellowish; new growth glabrous to strigulose with 2-branched hairs. **Leaves** opposite, simple, green above, whitish beneath, ovate to broadly lanceolate, mostly 4–15 cm long, 1–7 cm wide; petioles 5–25 mm long. **Flowers** crowded in dense, flat-topped, terminal cymes, perfect, regular, ca. 4 mm across; sepals 4, minute to 0.5 mm long; petals 4, white, 2–3 mm long; stamens 4; carpels 2, style 1, ovary inferior. **Fruit** a white drupe, 6–9 mm in diameter; stone brown with yellow stripes. Late May–Aug.

SYNONYMS *Cornus alba* var. *alba, Cornus stolonifera* Michx.

HABITAT Shores, streambanks, floodplains, moist wooded slopes, springs, fens and other wet or moist habitats.

WETLAND STATUS

GP FACW | MIDW FACW | WMTN FACW

ADDITIONAL SPECIES Other dogwoods are encountered on the eastern edge of our range but are less often associated with wetland habitats than *C. sericea:* **pale dogwood** (*Cornus amomum* P. Mill.), **rough-leaved dogwood** (*C. drummondii* C. A. Mey.), and **gray dogwood** (*C. racemosa* Lam.). All three are found in moist to dry woodlands, often on floodplains, seldom where wetter.

Cornus sericea
RED OSIER

Droseraceae *sundew family*

Drosera rotundifolia L.

ROUND-LEAVED SUNDEW

DESCRIPTION Small perennial (or annual), insectivorous bog plant. **Leaves** in a basal rosette, circinate in bud, broadly obovate to rotund or depressed-obovate, 2–10 mm long, about as wide or wider, glandular-viscid, covered with long reddish, gland-tipped hairs, tapering to a stout petiole 2–5(9) cm long. **Flowers** 2-many, in a simple or forked, mostly 1-sided, racemelike cyme, borne on a naked scape usually 10–25 cm tall; sepals (4)5, connate below, narrowly oblong, 4–6 mm long, erose at the tip; petals (4)5, white, marcescent, oblong, slightly to considerably exceeding the calyx; stamens (4)5, slightly shorter than the petals; carpels 3(–5), deeply cleft, ovary slightly inferior. **Fruit** a many-seeded, 3(–5)-valved capsule, surpassing the calyx. **Seeds** light brown, finely striate and shiny, fusiform, 1–1.5 mm long. July–Aug.

HABITAT Swamps and bogs.

WETLAND STATUS

GP OBL | MIDW OBL | WMTN OBL

Drosera rotundifolia
ROUND-LEAVED SUNDEW

Elatinaceae *waterwort family*

Small, usually freely branched annuals of shores, flats or shallow water. **Leaves** simple, opposite, entire or glandular-serrate, with small membranous stipules. **Flowers** axillary, 1-few per axil, small and nonshowy, perfect, regular, hypogynous; sepals and petals (2)3 or 5 (in those included here); stamens numbering the same or 2x the number of petals; styles 3 or 5, ovary 3- or 5-celled. **Fruit** a capsule; seeds few to many, pitted.

1 Plants glandular-pubescent; flowers 5-merous ***Bergia***

1 Plants glabrous; flowers (2)3-merous ***Elatine***

Bergia texana (Hook.) Walp.

TEXAS BERGIA

DESCRIPTION Plant glandular-pubescent throughout, often reddish, simple to usually branched from the base, the **stems** decumbent and then ascending toward the tips, 1–4 dm long. **Leaves** opposite but crowded above to appear whorled, elliptic to oblong-lanceolate, 2–4 cm long, to 1.5 cm wide, acute, glandular-serrate, tapered to a short-petioled base; stipules subulate, serrate, ca. 1 mm long. **Flowers** 1–3 in the axils, short-pedicelled; sepals 5, 2–3.5 mm long, acuminate, scarious-margined with a thick,

Bergia texana
TEXAS BERGIA

green midvein; petals 5, white, oblong, shorter than the sepals; stamens 5 or 10; ovary 5-celled. **Capsule** globose, 2–3 mm in diameter, firm-textured; **seeds** brown, shiny, elliptic-oblong and slightly curved, 0.3–0.5 mm long, obscurely pitted. July–Oct.

HABITAT Muddy or sandy shores and flats.

WETLAND STATUS
GP OBL | MIDW OBL | WMTN OBL

Elatine rubella Rydb.

WATERWORT

DESCRIPTION Plant glabrous, small, matted, rather fleshy, densely branched from the base, the branches sprawling or floating, often rooting at the nodes, 2–15 cm long. **Leaves** opposite, sessile, obovate to oblanceolate, 3–12 mm long, 1–3 mm wide, mostly truncate or emarginate at the tip, entire; stipules minute. **Flowers** sessile and solitary in the leaf axils, 1.5–2 mm across, (2)3-merous; ovary 3-celled. **Capsule** globose, 1–1.5 mm in diameter, membranous; **seeds** many, golden-yellow, oblong, ca. 0.5 mm long, ridged and obscurely pitted in rows. July–Sept.

SYNONYM *Elatine triandra* Schkuhr

HABITAT Mud flats or in shallow water of lakes and ponds.

WETLAND STATUS
GP OBL | MIDW OBL | WMTN OBL

Elatine rubella
WATERWORT

Fabaceae *bean family*

Perennial herbs and shrubs (those included here) with pinnately compound, alternate, usually stipulate leaves; **leaflets** 3-many, the terminal one sometimes modified as a tendril (*Lathyrus*). **Inflorescences** of densely to loosely flowered, simple or branched racemes. **Flowers** perfect, irregular, hypogynous to somewhat perigynous; calyx regular to strongly irregular, 5-lobed; corolla of 5 separate lobes (only 1 lobe in *Amorpha*), these unequal, the upper median lobe termed the standard, exterior to and larger than the others, the 2 lateral petals termed the wings, exterior to the lowest ones which are called the keel petals, the latter partly coherent to enclose the stamens and style; stamens 10, monadelphous or diadelphous in a 9 + 1 arrangement; pistil 1, simple, with a single stigma and style, ovary 1-celled, maturing into a legume (pod) which is dehiscent by 2 sutures or indehiscent, sometimes constricted between the seeds (*Desmodium*) and ultimately breaking into 1-seeded joints or segments.

---

1 Leaves twice-pinnate; flowers (or fruits) in dense globose clusters **_Desmanthus_**

1 Leaves once-pinnate; flowers (or fruits) in simple or branched racemes 2

2 Shrub; corolla 1-lobed, only the standard present **_Amorpha_**

2 Herbs; corolla 5-lobed 3

3 Plants climbing; leaves with a tendril at the tip; fruits smooth **_Lathyrus_**

3 Plants erect; leaves without tendrils; fruits with hooked hairs or prickles 4

4 Leaflets 3; petals reddish-purple, drying dark blue; fruit with hooked hairs, constricted between the seeds **_Desmodium_**

4 Leaflets 7–21; petals cream or creamy-white; fruit with stout hooked bristles, not constricted between the seeds **_Glycyrrhiza_**

Amorpha fruticosa L.

FALSE INDIGO

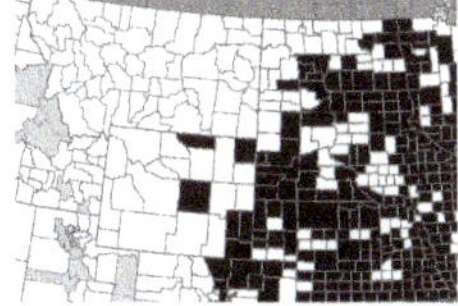

DESCRIPTION Branching shrub mostly 1–3 m tall; twigs tan to gray. **Leaves** once-pinnate, the blade oblong, 6–16 cm long; leaflets 9–27, elliptic to obovate, 1–4 cm long, 4–28 mm wide, glabrous above, puberulent and sometimes punctate beneath, on petiolules 1–3 mm long; petioles 2–5 cm long; stipules none. **Inflorescence** of 2-several, terminal, dense, spikelike racemes, these 2–15 cm long, the raceme axis and pedicels puberulent; bracts lanceolate, shorter than the calyx, pubescent, deciduous; pedicels 1–2 mm long. **Flowers** dark purple, irregular; calyx campanulate, unequally and shallowly 5-toothed, 1.5–3 mm long, often glandular, the lowest tooth somewhat longer and narrower than the others; corolla 1-lobed, only the standard present, folded to enclose the stamens, 3–5 mm long; stamens 10, the filaments united near the base. **Fruits** indehiscent, oblong, curved upward, 5–7 mm long, strongly glandular, 1- to 2-seeded. Flowering June, fruiting late July–Sept.

HABITAT Wet meadows, streambanks, shores, ditches and floodplains.

WETLAND STATUS

GP FACW | MIDW FACW | WMTN FACW

Amorpha fruticosa
FALSE INDIGO

Desmanthus illinoensis (Michx.) MacM.

PRAIRIE MIMOSA

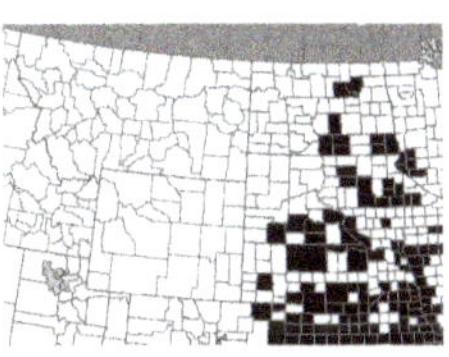

DESCRIPTION Erect perennial 3–10 dm tall; **stems** strongly ribbed, glabrous to hirsutulous. **Leaves** twice pinnate, 3–10 cm long; pinnae 6–12 pairs, 1.5–4 cm long; leaflets very small, oblong, 2–5 mm long, 0.5–1 mm wide,ciliolate, mucronate; stipules setaceous, 5–10 mm long. **Inflorescence** a small globose cluster of minute flowers; peduncles 2–6 cm long. **Flowers** whitish to greenish, regular; calyx campanulate, 5-toothed; corolla of 5 petals, separate or slightly united at the base; stamens 5, distinct, long-exserted. **Fruits** in a dense, subglobose head, flat, strongly curved, 1–2.5 cm long, 4–7 mm wide, containing few to several seeds. Late July–Sept.

HABITAT Confined to sandy or gravelly lakeshores in the n part of our region, becoming more common and occupying a variety of moist to dry habitats in the c and s parts.

WETLAND STATUS

GP FACU | MIDW FACU | WMTN FACU

Desmodium canadense (L.) DC.

CANADA TICKCLOVER

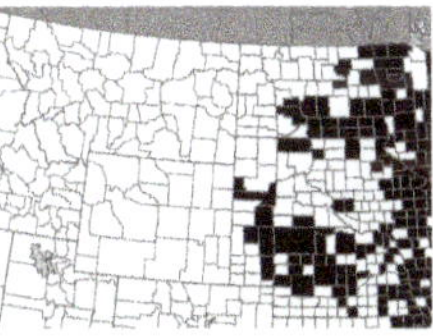

DESCRIPTION Erect perennial 6–15 dm tall; **stem** stout, simple or branched, strongly ribbed and hirsute above. **Leaves**

Desmanthus illinoensis
PRAIRIE MIMISA

pinnately 3-foliate, the leaflets dark green above, pale beneath, ovate-lanceolate to lanceolate or occasionally obovate, 3–9 cm long, 1–3.5 cm wide, puberulent above, more densely so beneath, acute to obtuse, sometimes apiculate; petioles 2–20 mm long; stipules linear-subulate, 4–9 mm long, ciliate. **Inflorescence** of 1 terminal raceme or a panicle of several racemes, the terminal one longest, 6–20 cm long, pubescent on the-axis; bracts brownish, lanceolate, appressed-hairy, deciduous; pedicels 4–8 mm long. **Flowers** reddish-purple, drying dark blue, irregular; calyx purple, deeply 5-toothed, somewhat irregular, 4–6 mm long, the lowest sepal longest; corolla 5–9 mm long; stamens diadelphous in a 9 + 1 arrangement. **Fruits** constricted between the 2-several seeds, 1.5–5 cm long, beset with hooked hairs, ultimately breaking apart into 1-seeded segments which are 3–5 mm wide. July–Sept.

HABITAT Streambanks, pond margins, ditches, floodplains and swampy places.

WETLAND STATUS
GP FAC | MIDW FACU | WMTN FACU

Glycyrrhiza lepidota Pursh
WILD LICORICE

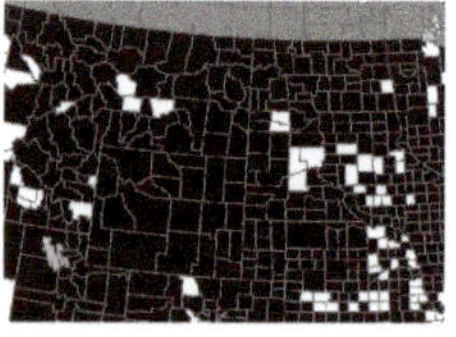

DESCRIPTION Perennial herb 3–10 dm tall, from long creeping rhizomes, often forming patches, puberulent or glabrous, glandular-punctate with yellowish or brownish translucent glands; **stem** simple below, woody at the base, usually with short lateral branches above. **Leaves** pinnate, mostly 8–18 cm long including the rather short petiole; leaflets 7–21, lanceolate to oblong-lanceolate or seldom elliptic, 1.5–5 cm long: 5–16 mm wide, often smaller on later developed leaves, apiculate, glandular-punctate on both surfaces; petiolules mostly 1–2 mm long; stipules brownish, lanceolate, 3–7 mm long, deciduous. **Racemes** axillary, spikelike, many-flowered, on peduncles 1–7 cm long; bracts deciduous, breaking off to leave the cupulate base; pedicels 1 mm or less long. Calyx tubular-campanulate in the lower half, 5–6 mm long, glandular-stipitate on the outside, the upper 2 lobes united for 1/2 or more of their length; petals cream or creamy white, the standard 10–14 mm long, wings and keel shorter; stamens 10, diadelphous in a 9 + 1 arrangement. **Fruits** brown, indehiscent, ellipsoid, 1–2 cm long, densely covered with hooked bristles, the style usually persistent as a terminal beak ca. 3 mm long. Flowering July–Aug, fruiting late July–Sept, the fruits commonly persisting into late fall.

HABITAT Shores, streambanks, wet meadows, floodplains, moist prairies, ditches and drier habitats.

WETLAND STATUS
GP FACU | MIDW FACU | WMTN FAC

Desmodium canadense
CANADA TICKFLOWER

Glycyrrhiza lepidota
WILD LICORICE

Lathyrus palustris L.
MARSH VETCHLING

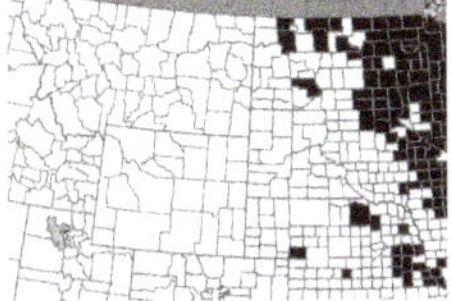

DESCRIPTION Rhizomatous, climbing perennial, clinging to surrounding vegetation by tendrils; **stems** strongly 2-winged, 3–10 dm long. **Leaves** pinnately compound, 5–10 cm long, the terminal leaflet modified as a tendril, otherwise leaflets in 2–4 pairs, linear to lanceolate, occasionally elliptic, 2–7 cm long, 3–25 mm wide, apiculate; stipules prominent, semisagittate, 1–3 cm long. **Flowers** in lateral racemes, 2–6 per raceme, reddish-purple, rarely whitish, drying blue to blue-violet; calyx irregular, 7–10 mm long, the lowest lobe considerably longer than the others; corolla 12–20 mm long; stamens diadelphous in a 9 + 1 arrangement; style bearded along the inner side. **Fruits** resembling a pea pod, flat, dehiscent, many-seeded, 3.5–5 cm long. June–Aug.

HABITAT Wet meadows, streambanks and boggy or swampy places.

WETLAND STATUS
GP FACW | MIDW FACW | WMTN OBL

Lathyrus palustris
MARSH VETCHLING

Gentianaceae *gentian family*

Late summer and fall-flowering annual and perennial herbs. **Leaves** opposite, entire, sessile, exstipulate. **Flowers** 1-many, clustered and sessile or terminal and solitary on long peduncles, blue to bluish-purple, often greenish or whitish on the outside, rather large and showy, perfect, regular, 4- or 5-merous; calyx tubular, obconic or funnelform, the lobes often unequal; corolla tubular, obconic, funnelform or campanulate, sometimes plicate between the lobes, convolute in bud, the plicate membrane erose to toothed or lacerate at the summit; stamens 4 or 5, epipetalous, alternate with the corolla lobes; stigma 2-lobed, style short, ovary superior, 1-celled, ellipsoid to cylindric. **Fruit** a 2-valved, many-seeded capsule enclosed by the marcescent corolla.

---

1 Flowers (or flower) 4-merous; plants annual — ***Gentianopsis***

1 Flowers 5-merous; plants annual (*Eustoma*) or perennial (*Gentiana*) — 2

2 Corolla with plicate folds between the lobes — ***Gentiana***

2 Corolla without folds or pleats between the lobes — ***Eustoma***

Eustoma grandiflorum (Raf.) Shinners
CATCHFLY PRAIRIE-GENTIAN

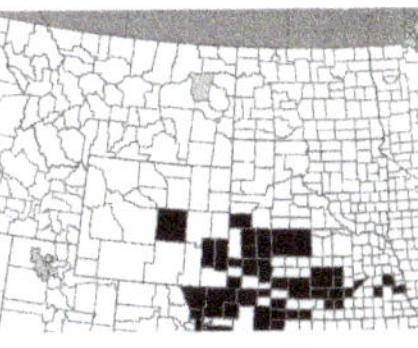

DESCRIPTION Plants annual (or short-lived perennials), erect, 2–6 dm tall, **stems** 1 or several. **Leaves** opposite, oblong to lance-ovate, glaucous, 3-nerved, 1.5–7.5 cm long, 0.3–5 cm wide. Inflorescence of 2–6 flowers, on pedicels to 6 cm long. **Flowers** 5-merous; calyx deeply cleft, the lobes keeled and narrowly lance-shaped, 2–3 mm wide; corolla campanulate, deeply cleft, blue-purple to sometimes pink or whitish; stamens 5 or 6; stigma 2-lobed. **Capsule** ellipsoid, to 2 cm long. July–Sept.

SYNONYM *Eustoma exaltatum* subsp. *russellianum* (Hook) Kartesz

HABITAT Moist to wet meadows, alkaline prairies.

WETLAND STATUS
GP FACW | MIDW FACW | WMTN OBL

Gentiana GENTIAN

Perennials with thick, fibrous roots. **Flowers** large, lavender to bluish-purple, usually greenish or yellowish-green and blue on the outside, clustered in the upper part of the plant, appearing axillary, each subtended by a pair of small bractlike leaves, 5-merous; calyx obconic to funnelform, the lobes unequal; corolla tubular to obconic, shallowly lobed, plicate between the lobes, the plicate membrane erose to toothed or lacerate at the summit, less than to slightly exceeding the lobes; stamens 5, the anthers sometimes fused around the ovary. **Seeds** flattened and winged with a pale margin.

1 Corolla lobes distinct, 2.5–6 mm long, the plicate membrane between the lobes toothed at the summit; stems puberulent on the decurrent leaf bases **G. affinis**

1 Corolla lobes low and indistinct, continuous with the erose plicate membrane; stems glabrous **G. andrewsii**

Gentiana affinis Griseb.

PLEATED GENTIAN

DESCRIPTION **Stems** 1–several from the base, 1–4 dm tall, simple or with short, erect branches above in the inflorescence, puberulent on the decurrent leaf bases especially in the upper part. **Leaves** lanceolate to elliptic-lanceolate, or the lower ones sometimes ovate to elliptic, 1–4 cm long, 0.3–1.5 cm wide, thick-textured and roughened, bluntly acute to obtuse at the tip, rounded and not clasping at the base. **Flowers** (1) few to many, 2–3.5 cm long; calyx tube funnelform, 4–8 mm long, the lobes unequal, erect to ascending, linear-lanceolate, less than 1/2 to nearly as long as the tube; corolla tubular to obconic, the lobes distinct, acute to rounded, 2.5–6 mm long, the plicate membrane between the lobes toothed at the summit; anthers free, not fused around the ovary. **Seeds** roughly oval to elliptic, 1–1.5 mm long. Aug–Sept.

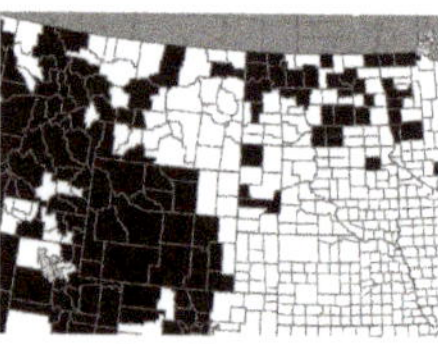

HABITAT Wet meadows, shores, springs, seepage areas and low prairie.

WETLAND STATUS

GP FACU | MIDW FACU | WMTN FACU

ADDITIONAL SPECIES *Gentiana puberulenta* Pringle, which is similar to *G. affinis,* typically grows in drier situations, but a few collections have come from low prairie. *G. puberulenta* is distinguished from *G. affinis* by its larger flowers (3.5–5 cm long), longer calyx tube (ca. 1 cm long), sub-equal calyx lobes and longer corolla lobes (4–8 mm long).

Gentiana andrewsii Griseb.

CLOSED GENTIAN

DESCRIPTION **Stems** erect, single or few together, 1.5–6 dm tall, simple, glabrous. **Leaves** lanceolate to broadly elliptic-lanceolate, 4–9 cm long, 0.6–3 cm wide, smooth, acute at the tip, rounded to acute at the base. **Flowers** (1) few to many, clustered in upper leaf axils, 2.5–4.5 cm long;

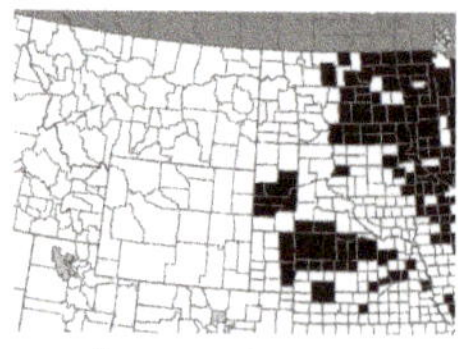

Eustoma grandiflorum
CATCH-FLY PRAIRIE GENTIAN

Gentiana affinis
PLEATED GENTIAN

calyx tube obconic, 10–14 mm long, the lobes unequal, variable in shape, mostly lanceolate to ovate-lanceolate, less than 1/2 to as long as the tube; corolla tubular, remaining closed, the lobes low and indistinct, rounded, continuous with and slightly shorter than to about equaling the erose-margined, plicate membrane; anthers fused around the ovary at anthesis. **Seeds** as in *G. affinis* except larger, 1.5–2 mm long. Aug–Sept.

HABITAT Wet meadows, low prairie and ditches.

WETLAND STATUS
GP FAC | MIDW FACW | WMTN FAC

Gentianopsis FRINGED GENTIAN

Slender annuals with a small taproot. **Flowers** 1–several (many), often large, purple to bluish-purple, sometimes tinged with white on the outside, long-peduncled, terminal on the main axis and branches, 4-merous; calyx obconic to funnelform, with 2 opposite lobes longer and more slender than the other 2 opposite lobes; corolla tubular to campanulate, deeply lobed, lacking a plicate membrane between the lobes, the lobes erose or fringed at the tips and sometimes down the sides; stamens 4, the anthers free. **Seeds** somewhat flattened, densely covered with irregular vesicles.

Gentiana andrewsii
CLOSED GENTIAN

1 Leaves ovate to ovate-lanceolate, 2–5x longer than wide; corolla lobes fringed across the tip with linear fringe-segments which are 2–6 mm long ***G. crinita***

1 Leaves linear to linear-lanceolate, 10–15x longer than wide; corolla lobes erose-toothed across the tips, often fringed on the sides ***G. virgata***

Gentianopsis crinita (Froel.) Ma
GREATER FRINGED-GENTIAN

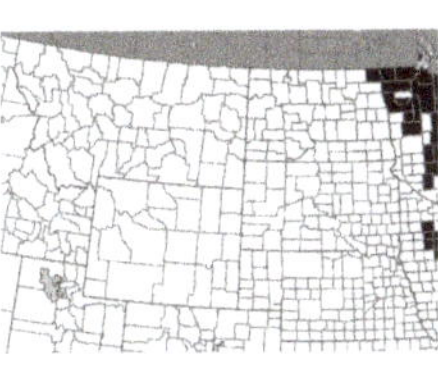

DESCRIPTION Plants erect, 1–5 dm tall, usually branched above. **Basal leaves,** when persistent, smaller than the cauline leaves, spatulate to oblanceolate; **cauline leaves** ovate to ovate-lanceolate, 2–5 cm long, 0.5–2.5 cm wide, 2–5x longer than wide, acute at the tip, rounded to subcordate at the base, usually clasping. **Flowers** 3.5–6 cm long; calyx tube funnelform, 12–20 mm long, the broader pair of lobes oblong to ovate, hyaline-margined, the narrower pair usually longer and lance-attenuate; corolla funnelform to campanulate, the lobes fringed across the tip (and often down the sides) with linear fringe-segments 2–6 mm long. Aug–Sept.

SYNONYM *Gentiana crinita* Froel.

HABITAT Wet meadows, streambanks, wet woods and boggy places; rare and barely entering our range in ne ND.

WETLAND STATUS
GP OBL | MIDW OBL

Gentianopsis crinita
GREATER FRINGED GENTIAN

Gentianopsis virgata (Raf.) Holub
LESSER FRINGED-GENTIAN

DESCRIPTION Quite similar to *G. crinita*, differing mainly as follows: **Leaves** linear to linear-lanceolate, (1.5)2–6 cm long, (1)2–7 mm wide, 10–15x longer than wide, bluntly acute, tapered slightly to a nonclasping base. **Flowers** (1.5)2.5–5 cm long; calyx tube obconic, 6–15 mm long; corolla tubular to obconic, the lobes erose-toothed across the tips, often fringed on the sides. Aug–Sept.

SYNONYMS *Gentiana procera* Froel., *Gentianopsis procera* (Holm.) Ma

HABITAT Calcareous springs and fens.

WETLAND STATUS
GP OBL | MIDW OBL | WMTN OBL

Gentianopsis virgata
LESSER FRINGED-GENTIAN

Grossulariaceae
gooseberry family

Ribes CURRANT, GOOSEBERRY

Rather low shrubs with erect to spreading or reclining stems, the **stems** smooth or armed with nodal spines and sometimes with internodal prickles as well. **Leaves** alternate, often fascicled on short, lateral branches, the blades palmately veined, palmately 3- to 5-lobed and toothed, rotund to angular in outline; stipules absent or adnate to the petioles. **Flowers** 1-few in short clusters or few to many in bracteate racemes, produced on short, axillary branches, greenish to white or yellow, perfect, regular, epigynous. Sepals 5, erect, spreading or reflexed, arising from the summit of a tubular to saucer-shaped hypanthium; petals 5, erect to spreading, shorter than the sepals, inserted at or near the top of the hypanthium; stamens 5, inserted an the hypanthium, alternate with the petals; pistil 2-carpellary, styles 2, separate to below the middle or united nearly to the stigmas, ovary mostly inferior, 1-celled with 2 parietal placentae. **Fruit** a many-seeded berry, usually with the dried floral remains persisting at the tip.

1 Stems armed with nodal spines ***R. missouriense***
1 Stems lacking spines **2**

2 Leaves dotted beneath with shiny, yellow to brown, resinous glands ***R. americanum***
2 Leaves without resinous glands ***R. triste***

Ribes americanum P. Mill.
WILD BLACK CURRANT

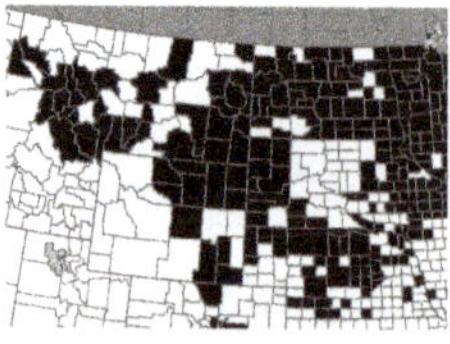

DESCRIPTION Shrub 1–1.5 m tall, the branches erect to spreading; **stems** unarmed, the new growth closely pubescent; **twigs** gray to grayish-brown and glabrous, blackish with age, often with low corky wings decurrent from the nodes. **Leaves** 3-lobed and usually with 2 shallow basal lobes, 2–6(9) cm long, 2.5–8(11) cm wide, dotted with shiny, yellow to brown resinous glands especially beneath, glabrous or puberulent above, hirsute beneath, coarsely serrate or crenate-serrate, broadly rounded to cordate at the base; petioles shorter than to longer than the blade, ciliate and pubescent. **Inflorescences** racemose, 6- to 15-flowered, drooping, 3–8 cm long, the axis short-pubescent; bracts linear-lanceolate, exceeding the pedicels; pedicels 2–5 mm long. **Flowers** greenish-white to cream-colored, 8–12 mm long; sepals erect to spreading, oblong, 4–5 mm long, rounded; petals creamy white, 2–3 mm long; stamens about equaling the petals; styles united nearly to the tip; hypanthium campanulate in flower, tubular in fruit, 3–4.5 mm long. **Fruit** a black berry, 6–1 0 mm in diameter. Flowering May–June, fruiting July–Aug.

HABITAT Marsh and lake borders, wet meadows, streambanks, floodplains and moist woods.

WETLAND STATUS

GP FACW | MIDW FACW | WMTN FAC

Ribes missouriense Nutt.

MISSOURI GOOSEBERRY

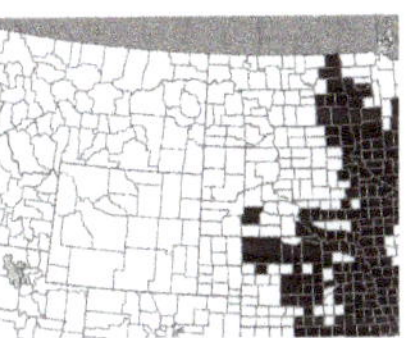

DESCRIPTION Shrub 1–1.5 m tall, the **branches** arching, armed with 1–4 stout spines at each node, these 5–15 mm long, rarely with some internodal prickles as well; new growth closely pubescent; **twigs** gray to yellowish-gray , exfoliating with age. **Leaves** 3-lobed and usually with 2 small basal lobes, 1–5 cm long, 1.5–6 cm wide, short-pubescent to nearly glabrous above, hirsute beneath, crenate-serrate, obtuse to subcordate at the base; petioles shorter than to about equaling the blade, densely pubescent and often with a few glandular hairs. **Inflorescences** of short clusters containing 2–5 flowers, on lateral branches 1–3 cm long; bracts ovate, ciliate, shorter than the pedicels; pedicels 3–10 mm long. **Flowers** greenish to white; sepals erect to eventually reflexed, 4–6 mm long, linear-oblong; petals erect, white to cream-colored, 2–3.5 mm long; stamens strongly exserted at anthesis, ca. 2x the length of the sepals; styles fused from about 1/2 to their entire length; hypanthium short-cylindric, 1–3 mm long. **Fruit** brown or purple at maturity, 6–11 mm in diameter. Flowering May–June, fruiting late June–Aug.

HABITAT Streambanks, floodplains and moist to dry woods.

WETLAND STATUS

GP UPL | MIDW UPL | WMTN UPL

ADDITIONAL SPECIES *Ribes lacustre* (Pers.) Poir. occurs in wet places in the Black Hills. Similar to *R. missouriense* in its spiny habit but differs as follows: flowers in racemes rather than short clusters; hypanthium saucer-shaped and spreading; ovary and fruit beset with glandular bristles and hairs.

Ribes americanum
WILD BLACK CURRANT

Ribes missouriense
MISSOURI GOOSEBERRY

Ribes triste Pall.
SWAMP CURRANT

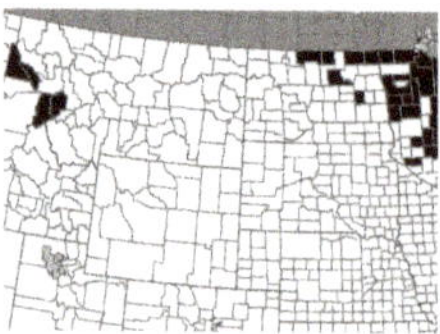

DESCRIPTION Low, straggling shrub 0.4–1 m tall, the spreading **branches** usually rooting adventitiously along the lower portion; **stems** unarmed, the new growth pubescent and with some short, stipitate glands, soon glabrous and gray to brownish-gray, turning dark brown and exfoliating with age. **Leaves** broadly 3(5)-lobed, 2–11 cm long, 2.5–11 cm wide, glabrous or nearly so above, short-pubescent at least on the veins beneath, coarsely dentate-serrate, broadly cordate at the base, the lobes broadly triangular, projecting forward; petioles shorter than to about equaling the blade, short-pubescent and often glandular, ciliate toward the base. **Inflorescence** racemose, 5- to 12-flowered, drooping, 2–9 cm long, the axis short-pubescent, often glandular; bracts broadly cordate, shorter than the pedicels; pedicels 1–4 mm long. **Flowers** green or purple-tinged; sepals spreading, broadly rounded, 1–2 mm long and about as wide; petals erect, red to purple, ca. 1 mm long; stamens about equaling the petals; styles united 1/3 to 1/2 of their length; hypanthium saucer-shaped, 0.5–1 mm long. **Fruit** a red berry 8–10 mm in diameter. Flowering May, fruiting June–July.

HABITAT Bogs and swampy woods.

WETLAND STATUS
GP OBL | MIDW OBL

Ribes triste
SWAMP CURRANT

Haloragaceae
water milfoil family

Myriophyllum WATER MILFOIL

Rooted aquatic or amphibious perennials with flexuous submerged **stems** and mostly whorled **leaves**, these pinnately divided into filiform segments. **Flowers** small, sessile in the axils of the upper reduced leaves or bracts, also usually subtended laterally by a pair of minute bracteoles, emersed at anthesis, in terminal, interrupted spikes. Flowers imperfect or perfect, the male flowers above the female in the spike, with perfect flowers, if any, in the middle portion of the spike; calyx inconspicuous, 4-parted in the male flowers, appressed to the ovary and minutely toothed at the summit in female or perfect flowers, the teeth usually deciduous in fruit; corolla of 4 membranous petals, reduced or absent in female flowers; stamens 8 or 4; pistil 4-carpellary, stigmas 4, sessile or nearly so, papillose-tufted, ovary inferior, 4-celled. **Fruit** nutlike, 4-lobed, eventually splitting into 4 mericarps.

---

1 Upper floral bracts entire; lower bracts entire, serrate or pinnatisect, not more than 2x longer than their subtended flowers ***M. sibiricum***

1 Upper floral bracts entire to serrate or pinnatisect; lower bracts serrate or pinnatisect, usually more than 2x longer than their subtended flowers **2**

2 Upper bracts shorter than or equaling their subtended flowers, only the lower bracts exceeding their flowers; stamens 8 ***M. verticillatum***

2 All bracts conspicuously exceeding their subtended flowers; stamens 4 **3**

3 Bracts laminate, mostly oblong-lanceolate to ovate-lanceolate, pectinate below to serrate above in the spike, the lowermost distinctly different from the foliage leaves ***M. heterophyllum***

3 Bracts not laminate, mostly linear, finely pinnatisect below to pinnatifid or shallowly toothed to subentire above in the spike, the lowermost similar to the foliage leaves ***M. pinnatum***

Myriophyllum heterophyllum Michx.

TWO-LEAF WATER-MILFOIL

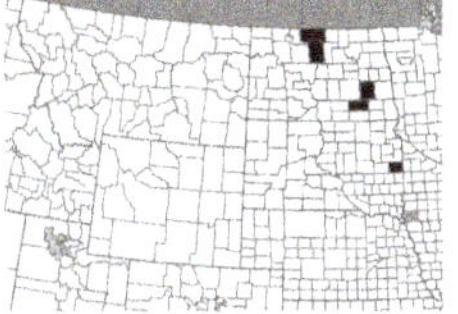

DESCRIPTION Submersed aquatic with simple or branching, rather robust **stems** to 8 mm thick, usually red-tinged, 3–10 dm or more long; winter buds produced at the plant base or from rhizomes, not clavate, green. **Foliage leaves** in whorls of 4–6 or some scattered, 1.5–5 cm long, with 5–14 divisions on each side of the midrib. **Flowering spikes** green to reddish, 5–30(40) cm long; bracts in whorls of 4 or some occasionally offset, smaller than the foliage leaves and distinct from them, laminate, oblong-lanceolate to ovate-lanceolate, 3–12 mm long, serrate or the lower ones usually pectinate but with a laminate central portion, with 4–8 ascending teeth or segments, eventually reflexed; bracteoles triangular to ovate, spinulose, to 1.2 mm long. **Flowers** perfect and imperfect; petals of male and perfect flowers pale, 1–3 mm long; stamens 4. **Fruits** olive, subglobose, 1.5–2 mm long, the mericarps rounded or with 2 undulate keels on the dorsal side, otherwise smooth to papillate, conspicuously beaked with the recurved stigma. June–Aug.

HABITAT Stream pools and ponds.

WETLAND STATUS
GP OBL | MIDW OBL

Myriophyllum heterophyllum
TWO-LEAF WATER-MILFOIL

Myriophyllum pinnatum (Walt.) B.S.P.

CUT-LEAF WATER-MILFOIL

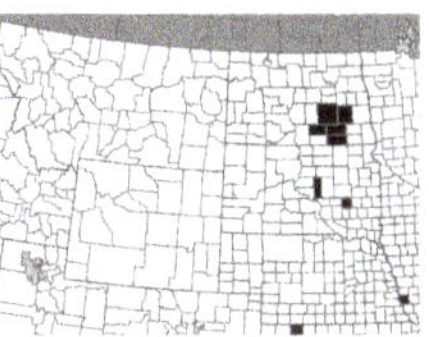

DESCRIPTION Aquatic or amphibious plant with **stems** freely branching and rooting inmud when stranded or greatly elongating in water. **Foliage leaves** mostly in whorls of 3–5, partly scattered on the stem, mostly 1–3 cm long, with 3–6 remote capillary segments along each side of the midrib; **emersed leaves**, including those subtending flowers, 0.5–2 cm long, much surpassing the flowers and fruits, pinnatifid to shallowly toothed or seldom subentire, usually with 2–4 ascending segments or teeth on each side, the lowermost floral bracts grading into the foliage leaves; bracteoles bluntly triangular, ca. 1 mm long. **Flowers** perfect or imperfect; petals purplish, rounded and short-clawed, 1.5–2 mm long; stamens 4. **Fruits** pale, cubic-ovoid, 1.3–1.8 mm long, the mericarps with flat sides and 2 tuberculate dorsal ridges, beaked. July–Aug.

HABITAT Shallow water or mud of marshes and shores.

WETLAND STATUS
GP OBL | MIDW OBL

Myriophyllum sibiricum Komarov

COMMON WATER-MILFOIL

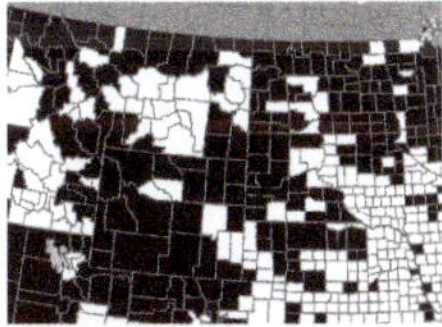

DESCRIPTION Submersed aquatic with **stems** simple to freely branched, whitened when dried, elongate and

Myriophyllum pinnatum
CUT-LEAF WATER-MILFOIL

flexuous, 2–10(15) dm long; winter buds with reduced blackish leaves and shortened internodes often produced from lower nodes in late summer and fall, present through spring. **Leaves** in whorls of 3–4, pinnately dissected, with 5–10(12) filiform segments on each side of the midrib, (0.7)1–3 cm long; lower and middle nodes mostly 1 cm or more apart. **Flowering spikes** red or reddish-purple, clearly distinct from the submersed portion of the stem, 3–10(14) cm long, the floral bracts much smaller than the leaves, oblong to obovate, all or at least the upper bracts entire, mostly shorter than the flowers and fruits, only the lower ones sometimes denticulate and slightly longer; bracteoles ovate, entire. **Flowers** imperfect, the upper male and the lower female; petals present in male flowers, lacking in female flowers, oblong-obovate, concave, 1.5–3 mm long; stamens 8, the yellowish-green anthers conspicuous at anthesis. **Fruits** olive, subglobose, 2–4 mm long, the mericarps rounded on the back, smooth or rugulose. June–Sept.

SYNONYM *Myriophyllum exalbescens* Fern.

HABITAT Shallow to deep water of lakes, ponds, marshes, ditches and sluggish streams; common and often abundant, the numerous reddish spikes often conspicuous on the water surface.

WETLAND STATUS
GP OBL | MIDW OBL | WMTN OBL

ADDITIONAL SPECIES **Eurasian water-milfoil** (*Myriophyllum spicatum* L.) is introduced and spreading in the USA, including the Great Plains region. It is very similar to *M. sibiricum* but differs in having leaves more finely divided, with 12–24 filiform segments on each side of the midrib, and larger floral bracts, among other traits. The more aggressive nature of **Eurasian milfoil** makes it an invasive threat in recreational waters.

Myriophyllum sibiricum
COMMON WATER-MILFOIL

Myriophyllum spicatum
EURASIAN WATER-MILFOIL

Myriophyllum verticillatum L.
WHORLED WATER-MILFOIL

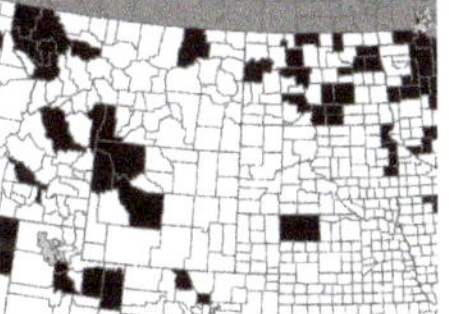

DESCRIPTION Quite similar to *M. sibiricum,* often more robust with **stems** 5–25 dm long. **Leaves** in whorls of 4–5, with 9–13 filiform segments along each side of the midrib, 1–4.5 cm long; lower and middle nodes mostly less than 1 cm apart; winter buds (turions) present fall to early spring, clavate. **Flowering spikes** 4–12 cm long, the floral bracts much smaller than the leaves, pectinate, mostly exceeding the flowers; bracteoles minute or absent, palmately 7-lobed. **Flowers** perfect or the lower female and the upper male; petals reduced in female flowers, otherwise spoon-shaped, obtuse, to ca. 2.5 mm long; stamens 8. **Fruits** brownish, subglobose, 2–3 mm long, the mericarps rounded on the back, smooth or somewhat roughened. June–Sept.

HABITAT In much the same habitats as *M. sibiricum,* but restricted to fresh water.

WETLAND STATUS
GP OBL | MIDW OBL | WMTN OBL

Myriophyllum verticillatum
WHORLED WATER-MILFOIL
(with turion)

Hypericaceae
st. john's-wort family

Perennial and annual glabrous herbs (in those included here), usually simple below, branched above in the inflorescence. **Leaves** simple, opposite, sessile and often clasping at the base, glandular-punctate with brown or black glands, especially beneath. **Flowers** few to many in terminal cymes or clusters, or some also in lateral clusters from the upper axils, perfect, regular; sepals 5; petals 5, yellow or pink to greenish or purplish; stamens 9–35, free or united below by their filaments into 3 or more fascicles; pistil 3-carpellary, styles 3, ovary superior, maturing as a 3-valved, many-seeded capsule.

1 Petals yellow; stamens 15–35, free ... *Hypericum*

1 Petals pink to greenish or purplish; stamens 9(12), fascicled in 3 groups ... *Triadenum*

Hypericum majus (A. Gray) Britt.
GREATER ST. JOHN'S-WORT

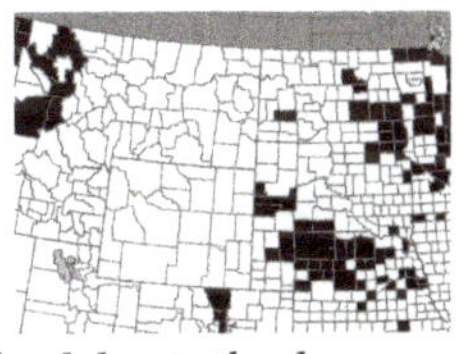

DESCRIPTION Annual or often perennial from short leafy stolons at the plant base. **Stems** upright, simple or branching above, 1–5 dm tall, ridged due to the decurrent leaf bases. **Leaves** lanceolate to oblong, 1–4 cm long, 2–11 mm wide, glandular-punc-

Hypericum majus
GREATER ST. JOHN'S-WORT

tate, with minute brownish-translucent glands, 5- to 7-nerved from the base, with 3 major veins continuous to the obtuse or rounded tip, broadly rounded to somewhat cordate and clasping at the base. **Flowers** few to numerous in terminal cymes. Sepals lanceolate, mostly 3–5(7) mm long, acute; petals yellow, about equaling the sepals, somewhat marcescent and shriveling to ca. 1/2 the length of the sepals; stamens 15–35, free; styles 1 mm or less long, each with a capitate stigma, ovary 1-celled. **Capsule** often purplish, ovoid-conic, 5–7 mm long; seeds yellowish, narrowly ellipsoid, ca. 0.5 mm long. Mid July–Sept.

HABITAT Sandy wet meadows, shores and ditches.

WETLAND STATUS
GP FACW | MIDW FACW | WMTN FACW

Triadenum fraseri (Spach) Gleason
MARSH ST. JOHN'S-WORT

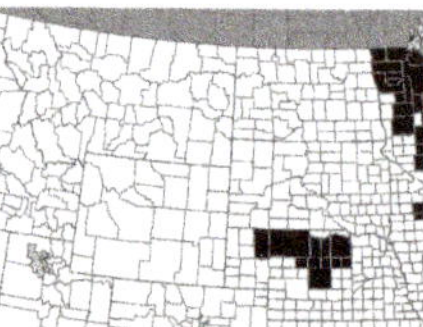

DESCRIPTION Erect perennial from creeping rhizomes, 3–6 dm tall, the **stem** simple or branched above. **Leaves** elliptic-ovate to oblong, 3–6 cm long, 1–3 cm wide, pinnately veined, glaucous and brown or black dotted beneath, rounded or emarginate at the tip, broadly rounded to cordate-clasping at the base. **Flowers** in terminal and axillary clusters of usually few to several, short-pedicelled. Sepals oblong-ovate, 2.5–5 mm long, obtuse or rounded; petals pink to greenish or purplish, 5–10 mm long; stamens 9(12), united below into 3 fascicles, the fascicles alternating with orange glands; styles 0.5–1.5(2) mm long, ovary 3-celled. **Capsule** purplish, ovoid to oblong, 7–12 mm long, blunt-tipped; seeds brown, oblong-ellipsoid, ca. 1 mm long. July–Aug.

SYNONYM *Hypericum fraseri* (Spach) Steud.

HABITAT Marshes and shores; Neb (Sand Hills).

WETLAND STATUS
GP OBL | MIDW OBL

Triadenum fraseri
MARSH ST. JOHN'S-WORT

Lamiaceae *mint family*

Perennial herbs, some aromatic, with square **stems** and simple, opposite, serrate or incised-toothed **leaves. Flowers** axillary or in terminal heads or spikes, often appearing verticillate, perfect, nearly regular to irregular; calyx 5-toothed or sometimes bilabiate, the lobes equal to unequal; corolla white, pink, blue or purple, sometimes spotted, often bilabiate, 5-lobed or 4-lobed by fusion, the lobes very unequal to subequal; stamens 4 or 2, epipetalous; pistil 2-carpellary, style slender, gynobasic, 2-cleft at the tip, ovary superior, 4-celled and 4-lobed, splitting into 4, 1-seeded nutlets at maturity.

1 Corolla regular or nearly so, the lobes subequal 2

1 Corolla distinctly irregular, the lobes unequal 3

2 Stamens 2; plants not strongly aromatic *Lycopus*

2 Stamens 4; plants strongly mint-scented *Mentha*

3 Upper lip of the corolla lacking, the lower lip prominent, the other 4 corolla lobes positioned on its lateral margins *Teucrium*

3 Both upper and lower lips of the corolla well-developed 4

4 Calyx with a rounded protuberance on the upper side, 2-lipped, not toothed *Scutellaria*

4 Calyx without a dorsal protuberance, toothed, 2-lipped or not 5

5 Flowers short-pedicelled, borne singly in the axils of short bracts, the inflorescence comprised of 1–several, terminal and lateral racemes *Physostegia*

5 Flowers sessile and whorled, in a continuous or interrupted terminal spike 6

6 Calyx strongly 2-lipped, unequally 5-toothed; spike dense, continuous *Prunella*

6 Calyx not strongly 2-lipped, with 5 equal or subequal teeth; spike interrupted *Stachys*

Lycopus BUGLEWEED

Erect to decumbent perennials, nonaromatic or only faintly aromatic, often stoloniferous and/or tuberiferous. **Leaves** sessile or short-petioled, coarsely serrate or incised-toothed, reduced upward, usually punctate. **Flowers** small, not showy, clustered in the axils of middle and upper leaves, often appearing verticillate, white to pinkish, the calyx and corolla often punctate on the outside; calyx regular, 5-toothed; corolla nearly regular, 4-lobed, the tube short, hairy internally at the throat, the upper lobe formed by the fusion of 2 lobes, tending to be broader than the other 3 lobes, often emarginate; functional stamens 2, exserted, the upper pair absent or reduced to staminodes. **Nutlets** broadened upward to a truncate or rounded crest, sometimes undulate or slightly tuberculate on the crest, nearly flat on the outer surface, with a corky ridge on each margin and often across the top, the inner surface convex and punctate with yellowish-viscid glands.

1 Calyx lobes broad, triangular to ovate, soft, obtuse to bluntly acute, 0.5–1 mm long, shorter than to about equaling the nutlets *L. uniflorus*

1 Calyx lobes slender, firm, sharply pointed, 1–3 mm long, surpassing the nutlets 2

2 Nutlets rounded apically on the outer margin, mostly 1–1.5 mm long; leaves short-petioled to subsessile, the blades irregularly incised-toothed to subpinnatifid *L. americanus*

2 Nutlets truncate apically on the outer margin, mostly 1.5–2 mm long; leaves sessile, the blades rather regularly and coarsely serrate *L. asper*

Lycopus americanus Muhl. ex Bart.

AMERICAN BUGLEWEED

DESCRIPTION Simple or more often branched perennial 2–8 dm tall, from nontuberiferous rhizomes, glabrous or strigulose in the upper part. **Leaves** short-petioled to subsessile, the blades coarsely and irregularly incised-toothed to subpinnatifid with the lowest teeth largest, lanceolate to linear-lanceolate in outline, 3–8 cm long, 1–4 cm wide, acute at the tip, cuneate at the base. **Calyx lobes** slender, firm, sharply pointed, 1–2 mm long, distinctly surpassing the nutlets at maturity; **corolla** white, sometimes pink to purple-dotted, weakly surpassing the calyx. **Nutlets** rounded apically on the outer margin, mostly 1–1.5 mm long, the corky ridge continuous over the rounded apex. July–Sept.

HABITAT Marshes, wet meadows, shores, streambanks, ditches, springs and other wet places.

WETLAND STATUS
GP OBL | MIDW OBL | WMTN OBL

Lycopus asper Greene
ROUGH BUGLEWEED

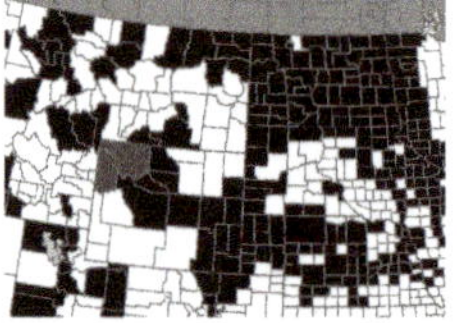

DESCRIPTION Rhizomatous and usually stoloniferous perennial 2–8 dm tall, arising from thick tubers on the rhizomes, simple or branched, short-pubescent at least above. **Leaves** sessile, elliptic-lanceolate to oblong-lanceolate or linear-lanceolate, 3–9(11) cm long, 0.5–3(4) cm wide, acute-tipped, coarsely and rather regularly serrate with ascending teeth, tapered to a broad or rarely narrow base. **Calyx lobes** as in the preceding, 1–3 mm long; **corolla** as in *L. americanus.* **Nutlets** truncate apically on the outer margin, sometimes somewhat toothed apically, mostly 1.5–2 mm long, the corky ridge often rather indistinct, mainly lateral. July–Sept.

HABITAT In the same habitats as *Lycopus americanus* and often found growing with it.

WETLAND STATUS
GP OBL | MIDW OBL | WMTN OBL

Lycopus uniflorus Michx.
NORTHERN WATER-HOREHOUND

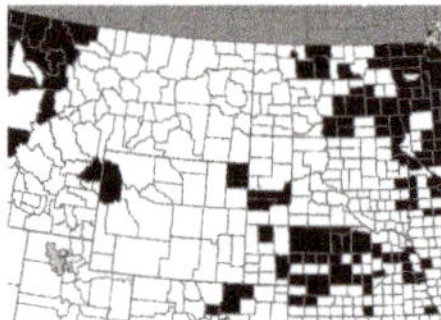

DESCRIPTION Similar to *Lycopus asper* in habit, 1–5 dm tall, puberulent at least above. **Leaf blades** ovate-lanceolate to oblong-lanceolate, 4–10 cm long, 1.5–4 cm wide, acute to shoracuminate, irregularly and coarsely dentate, cuneate to the short-petioled or subsessile base. **Calyx lobes** broad, triangular to ovate, soft, obtuse to bluntly acute, 0.5–1 mm long, shorter than to about equaling the nutlets; **corolla** white or pinkish, 2.5–3.5 mm long, clearly surpassing the calyx. **Nutlets** mostly 1.2–1.5 mm long, the outer apical margin subtruncate and shallowly toothed. July–Sept.

HABITAT Swamps, streambanks and springs, where water is fresh.

WETLAND STATUS
GP OBL | MIDW OBL | WMTN OBL

Mentha arvensis L.
COMMON MINT

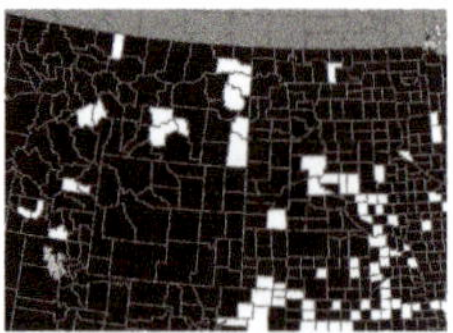

DESCRIPTION Erect to decumbent rhizomatous perennial 2–10 dm tall, frequently stoloniferous, strongly mint-scented; **stems** simple or more often branched and spreading, strigose at least on the angles,

Lycopus americanus
AMERICAN BUGLEWEED

Lycopus asper
ROUGH BUGLEWEED

villous at the nodes. **Leaves** short-petioled, the blades ovate to lanceolate or elliptic-lanceolate, 2–6(9) cm long, 0.6–2.5(4) cm wide, glabrous to strigulose on the nerves, strongly punctate, acute to acuminate at the tip, serrate, cuneate to rounded at the base. **Flowers** small, light pink to lavender, crowded in bracteate, verticillate clusters in middle and upper leaf axils, the bracts linear-subulate to linear-lanceolate; pedicels 1–3 mm long. Calyx regular, 5-toothed, 2–3 mm long, strigulose and glandular; corolla nearly regular to slightly bilabiate, 4–6 mm long, glandular on the outside, 4- or 5-lobed, the upper 2 lobes partly to completely fused; stamens 4, strongly exserted; style exserted. **Nutlets** obovoid to ellipsoid, 0.7–1 mm long, smooth, enclosed by the persistent calyx. July–Sept.

HABITAT Wet meadows, marshes, ditches, streambanks, springs and other wet places; common.

WETLAND STATUS
GP FACW | MIDW FACW | WMTN FACW

Lycopus uniflorus
NORTHERN WATER-HOREHOUND

Mentha arvensis
COMMON MINT

Physostegia OBEDIENT-PLANT

Stout, erect, perennial herbs; **stems** arising singly from slender rhizomes or from a slender caudex, simple throughout or branched at the base of the inflorescence. **Leaves** sessile or the lowermost sometimes short-petioled, elliptic to lanceolate or oblanceolate, mostly serrate. **Flowers** showy, pinkish-lavender to purple-lavender, short-pedicelled and borne singly in the ails of short bracts, arranged in 1–several terminal and lateral, spikelike racemes. Calyx regular or nearly so, tubular-campanulate, 5-toothed, the teeth much shorter than the tube; corolla bilabiate, much longer than the calyx, the upper lip hoodlike, entire to emarginate, the lower lip about equaling or shorter than the upper, shallowly 3-lobed; stamens 4, included or slightly exserted under the upper corolla lip; stigma 2-lobed, style about equaling the stamens. **Nutlets** ovoid, unequally trigonous, smooth.

1 Upper leaves not clasping the stem, broadest at the middle or above; corolla glabrous or sparsely puberulent but lacking stipitate glands; common species **_P. virginiana_**

1 Upper leaves clasping the stem, some of them usually broadest near the base of the blade; corolla usually with some stipitate glands (often lacking in *P. ledinghamii*); rare species in our area 2

2 Corolla 14–23 mm long, often lacking stipitate glands; nutlets 2.8–4 mm long **_P. ledinghamii_**

2 Corolla 9–16 mm long, usually with stipitate glands; nutlets 2.1–3.3 mm long **_P. parviflora_**

Physostegia ledinghamii (Boivin) Cantino
LEDINGHAMS'S FALSE DRAGONHEAD

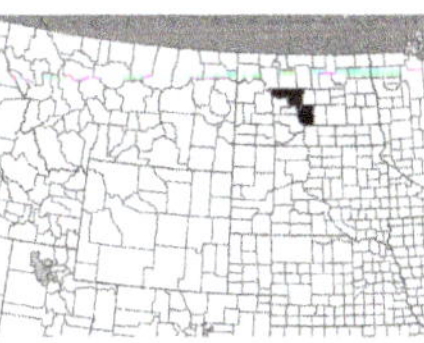

DESCRIPTION Similar to and in most respects intermediate between the following two species, distinguished only by a combination of characters: **Leaves** serrate, the upper ones clasping the stem. **Corolla** 14–23 mm long, puberulent and usually lacking stipitate glands. **Nutlets** 2.8–4 mm long. July–Aug. (*not illustrated*)

HABITAT River and streambanks, shores, floodplains and ditches; rare, with records from Burleigh and McLean Cos., ND.

WETLAND STATUS
GP OBL

NOTE Because it is essentially reproductively isolated from *P. virginiana* and *P. parviflora*, recognizing *P. ledinghamii* as a distinct species is well justified despite morphological overlaps with our two species below.

Physostegia parviflora Nutt. ex Gray
WESTERN FALSE DRAGONHEAD

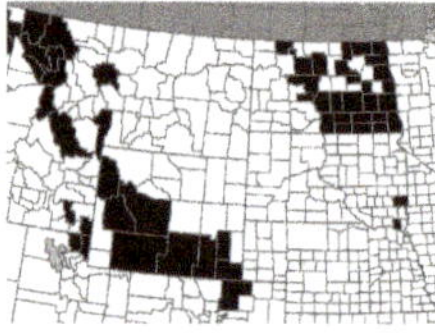

DESCRIPTION **Stems** 2–8 dm tall, from a slender caudex or short rhizome. **Leaves** lanceolate to oblong-lanceolate, rarely ovate, 4–12 cm long, 0.7–2 cm wide, acute to obtuse at the tip, remotely serrate or dentate to subentire, rounded to subcordate and clasping at the base. **Racemes** 5–10 cm long; bracts ovate to lanceolate or elliptic-lanceolate, to about 1/2 as long as the calyx. **Calyx** 3–5(6) mm long in flower, to 7 mm long in fruit, sparsely stipitate glandular, the teeth triangular, 1–1.5 mm long; **corolla** lavender to purple-lavender, 9–16 mm long, puberulent and usually with stipitate glands. **Nutlets** 2.1–3.3 mm long. July–Aug.

HABITAT Wet meadows, shores and streambanks.

Physostegia parviflora
WESTERN FALSE DRAGONHEAD

WETLAND STATUS
GP FACW | MIDW FACW | WMTN FACW

Physostegia virginiana (L.) Benth.
OBEDIENT-PLANT

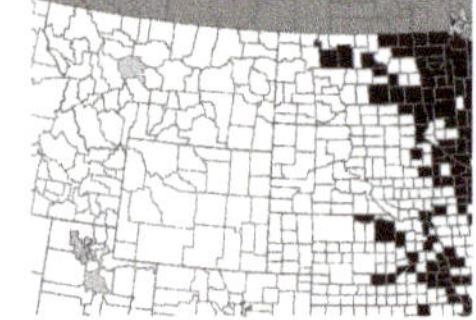

DESCRIPTION **Stems** 4–15 dm tall, from buds along slender rhizomes. **Leaves** elliptic-lanceolate to oblong or oblanceolate, 4–15(18) cm long, 1–4(5) cm wide, acute to acuminate at the tip, serrate on the margins, tapered to slightly widened to a sessile base, not clasping. **Racemes** 5–20 cm long; bracts ovate to ovate-lanceolate, about 1/2 as long as the calyx. **Calyx** 4–8 mm long in flower, to 10 mm long in fruit, often with some stipitate glands, the teeth triangular, 1–2 mm long; **corolla** deep lavender to white with purple spots, (13)17–25 mm long, puberulent to glabrous, lacking stipitate glands. **Nutlets** 2.5–3.2 mm long. July–Sept.

HABITAT Wet meadows, shores, fens, floodplains and wet woodlands.

Physostegia virginiana
OBEDIENT-PLANT

WETLAND STATUS
GP FACW | MIDW FACW | WMTN FACW

NOTE This plant is sometimes cultivated in gardens for its attractive flowers.

Prunella vulgaris L.
SELF-HEAL

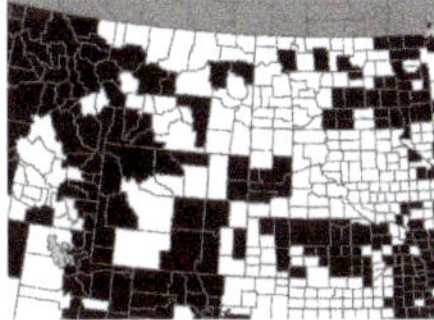

DESCRIPTION Rather low perennial 0.5–4 dm tall from a small caudex or short rhizome; **stems** erect to decumbent, simple or branched, glabrous or sparingly villous below to more densely villous above, especially on the angles. **Leaves** principally cauline, smaller basal leaves commonly present; blades of cauline leaves lanceolate to ovate-lanceolate, mostly 3–5 cm long, 0.8–2.2 cm wide, bluntly acute to obtuse, entire to obscurely serrate; smaller basal leaf blades ovate, subcordate to rounded at the base; petioles mostly 1–3 cm long, villous. **Flowers** rather small, crowded in a short-cylindric bracteate spike 1.5–6 cm long, ca. 1.5 cm thick; bracts reniform with a cuspidate to caudate tip, mostly shorter than the subtended calyces, ciliate. **Calyx** green or purple, bilabiate, 7–10 mm long, pubescent mainly on lobe margins, the upper lip broad, shallowly 3-lobed, the lower lip slender, cleft into 2 narrow teeth, slightly shorter than the upper lip; **corolla** blue-violet, seldom pink or white, bilabiate, the tube about equaling the calyx, the upper lip hoodlike, arched over the stamens, extending 4–8 mm beyond the calyx, the lower lip shorter, with 2 small lateral lobes and a larger, erose middle lobe; stamens 4, barely exserted, the filaments bifid at the tip; style exserted. **Nutlets** flattened-obovoid, 1.5–2 mm long. July–Sept.

Prunella vulgaris
SELF-HEAL

HABITAT Moist or wet places around springs, streams, seepage areas or fens; also in various drier habitats; locally common in the Black Hills.

WETLAND STATUS
GP FAC | MIDW FAC | WMTN FACU

Scutellaria SKULLCAP

Erect to spreading, rhizomatous and sometimes stoloniferous perennials. **Leaves** subsessile or petioled, the blades ovate to lanceolate or elliptic, nearly entire to serrate, thin-textured. **Flowers** blue or blue with white markings, rarely pink or white, borne singly (paired at the nodes) in the axils of middle and upper leaves or borne in axillary racemes, short-pedicelled; **calyx** 2-lipped, not toothed, with a rounded protuberance or pouch on the upper side, the 2 calyx lips separating at maturity to release the enclosed nutlets; **corolla** 2-lipped, pubescent on the outer surface, the upper lip concave and hoodlike, the lower lip nearly flat, 3-lobed, the tube straight or curved upward; stamens 4, ascending into the upper corolla lip. **Nutlets** ovoid, golden-brown, verrucose, raised on a gynophore.

---

1 Flowers paired at the nodes, borne in the axils of normal leaves; corolla 15–24 mm long ***S. galericulata***

1 Flowers in elongate, bracteate racemes which are axillary or at the tips of axillary branches; corolla 5–8 mm long ***S. lateriflora***

Scutellaria galericulata L.
MARSH SKULLCAP

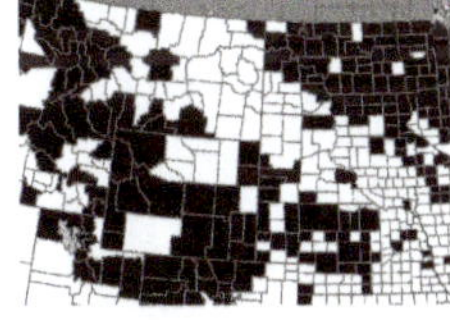

DESCRIPTION Plants 1–6(9) dm tall from slender rhizomes; **stems** simple or branched, erect or spreading, puberulent at least on the angles in the upper part. **Leaves** scarcely petioled, the blades lanceolate to ovate-lanceolate or elliptic, 2–6.5 cm long, 0.6–2.5 cm wide, glabrous or nearly so above, densely puberulent below, bluntly acute, nearly entire to irregularly crenate-serrate, rounded to cordate at the base. **Flowers** paired at the nodes, borne singly in the axils of normal leaves; pedicels 1–3 mm long; calyx 3–6 mm long; corolla blue marked with white, 15–24 mm long, the tube gently curved. June–Aug.

HABITAT Shores, streambanks, marshes, wet meadows, ditches, springs and other wet places.

WETLAND STATUS
GP OBL | MIDW OBL | WMTN OBL

Scutellaria lateriflora L.
BLUE SKULLCAP

DESCRIPTION Plants 1.5–6 dm tall, usually branched and spreading, the **stems** glabrous or puberulent on the angles above. **Leaf blades** ovate to ovate-lanceolate, 3–8 cm long, 1.5–5 cm wide, glabrous or nearly so, acute to acuminate at the tip, crenate-serrate to serrate on the margins, cuneate to subcordate at the base; petioles 0.5–3 cm long. **Flowers** in elongate, bracteate racemes which are axillary or at the tips of axillary branches; pedicels mostly less than 1 mm long; **calyx** 1.5–2.5 mm long in flower, to 4 mm long in fruit; **corolla** blue, rarely pink or white, 5–8 mm long, the tube straight. July–Aug.

HABITAT Shores, streambanks, springs, meadows, swampy places and moist woods.

Scutellaria galericulata
MARSH SKULLCAP

WETLAND STATUS
GP FACW | MIDW OBL | WMTN FACW

Stachys HEDGE-NETTLE

Stout, erect, rhizomatous perennials. **Leaves** sessile or short-petioled, the blades ovate-lanceolate to elliptic or oblong-lanceolate, crenate-serrate, reduced to bracts in the inflorescence. **Flowers** somewhat showy, pink to lavender, in terminal, interrupted spikes, appearing whorled in rather evenly spaced clusters; **calyx** nearly regular, with 5 equal or subequal, lance-triangular to lance-subulate teeth, the teeth ca. 1/3 to 1/2 the length of the calyx; **corolla** whitish to pink with purplish spots or mottling, distinctly 2-lipped, the upper lip concave, entire to emarginate, the lower lip spreading or deflexed, 3-lobed; stamens 4, ascending under the upper lip. **Nutlets** dark brown, ovoid, ca. 2 mm long, loosely enclosed by the persistent calyx.

1 Stems pubescent on the sides as well as on the angles ***S. pilosa***
1 Stems glabrous or pubescent with retrorse bristles only on the angles ***S. tenuifolia***

Scutellaria lateriflora
BLUE SKULLCAP

Stachys pilosa Nutt.
HEDGE-NETTLE

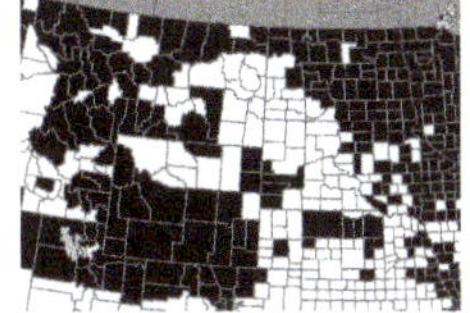

DESCRIPTION Plants (2)3–8 dm tall; **stems** simple or branched, hairy on the angles and shortpubescent on the sides, the short hairs glandular in the inflorescence. **Leaves** sessile or short-petioled, the blades ovate-lanceolate to lanceolate or elliptic, 4–13 cm long, 2–5 cm wide, softly pubescent on both surfaces, acute to acuminate, finely crenate-serrate, rounded to truncate or subcordate at the base. **Calyx** 5–8 mm long, covered with stout eglandular hairs and shorter glandular ones; **corolla** 9–13 mm long. June–July.

SYNONYM *Stachys palustris* var. *pilosa.*

HABITAT Marshes, wet meadows, ditches, shores, streambanks, and other wet or moist places.

WETLAND STATUS
GP FACW | MIDW FACW | WMTN FACW

Stachys pilosa
HEDGE-NETTLE

Stachys tenuifolia Willd.
SMOOTH HEDGE-NETTLE

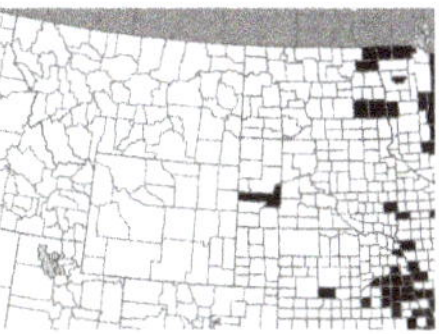

DESCRIPTION Similar to *S. pilosa,* 4–10 dm tall; **stems** glabrous or retrorsely bristled on the angles, the sides glabrous, bearded at the nodes. **Leaves** sessile or on petioles to 2 cm long, the blades ovate-lanceolate to oblong-lanceolate, 6–14 cm long, 2–6 cm wide, glabrous or with a few scattered hairs, acute to acuminate, serrate on the margins, rounded to cordate at the base. **Calyx** 5–7 mm long, usually with eglandular hairs only; **corolla** as in *S. pilosa*. June–Sept.

HABITAT Alluvial woods and streambanks.

WETLAND STATUS
GP FACW | MIDW OBL | WMTN FACW

Teucrium canadense L.
AMERICAN GERMANDER

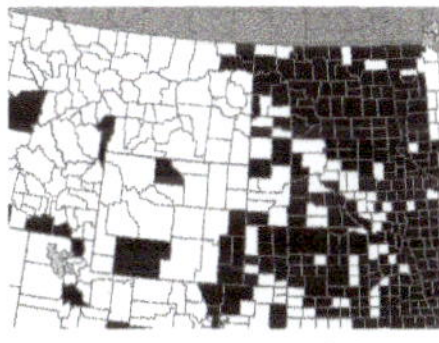

DESCRIPTION Rhizomatous, sometimes stoloniferous perennial 3–10 dm tall; **stems** simple or branched, spreading pilose to decurved pubescent. **Leaves** with petioles mostly 5–15 mm long, the blades

Stachys tenuifolia
SMOOTH HEDGE-NETTLE

ovate-lanceolate to lanceolate or oblong-lanceolate, 4–12(16) cm long, 1.5–4(6) cm wide, finely pubescent especially beneath, acute-tipped, finely serrate, cuneate to rounded at the base. **Inflorescence** of 1-many continuous, terminal, bracteate racemes; bracts narrowly lanceolate, mostly surpassing the calyces; pedicels 2–4 mm long. **Flowers** purple to lavender or pink, rarely white; **calyx** nearly regular, purple or green, 4.5–7 mm long, covered with long silky hairs and very short glandular ones, the lobes triangular, acute to acuminate, 1/3 to 1/2 the length of the calyx; **corolla** irregular, 10–13 mm long, very finely glandular-pubescent, upper lip absent, lower lip prominent, the other 4 corolla lobes positioned on its lateral margins; stamens 4, arched over the corolla. **Nutlets** golden, ovoid, 1.5–2.4 mm long, glandular. July–Sept.

HABITAT Marshes, wet meadows, ditches, shores, streambanks, and other wet or moist places.

WETLAND STATUS
GP FACW | MIDW FACW | WMTN FAC

Teucrium canadense
AMERICAN GERMANDER

Lentibulariaceae
bladderwort family

Utricularia BLADDERWORT

Free-floating aquatics with elongate, lax **stems** clothed in alternate, submersed **leaves** which are finely dissected into many linear segments, some or all of these bearing **bladders** which trap tiny aquatic invertebrates; stem apices commonly producing free-floating winter buds in late summer and autumn. **Flowers** 2-many in scapose, bracteate racemes held above the water surface, perfect, irregular; calyx 2-parted nearly to the base into an upper and lower segment, the upper one somewhat broader; corolla yellow, bilabiate, the upper lip erect, subentire or slightly 2-lobed, the lower lip entire or slightly 3-lobed, prominently arched toward the base to form a conspicuous or inconspicuous palate, the tube prolonged backward into a spur; stamens 2, inserted near the base of the tube; stigma unevenly 2-lobed, style short or obsolete, ovary superior, maturing into a many-seeded capsule.

1 Ultimate leaf segments filiform, the segments progressively narrower in successive branchings; lower lip of the corolla little if any longer than the upper *U. macrorhiza*

1 Ultimate leaf segments flat, nearly to fully as wide as the primary ones, with a distinct midvein; lower lip of the corolla about 2x as long as the upper 2

2 Bladders borne on specialized branches distinct from the dissected leaves; leaf segments spinulose-toothed *U. intermedia*

2 Bladders borne on ordinary leaves; leaf segments entire, sometimes toothed at the tip *U. minor*

Utricularia subulata
ZIG-ZAG BLADDERWORT

Utricularia intermedia Hayne

FLAT-LEAF BLADDERWORT

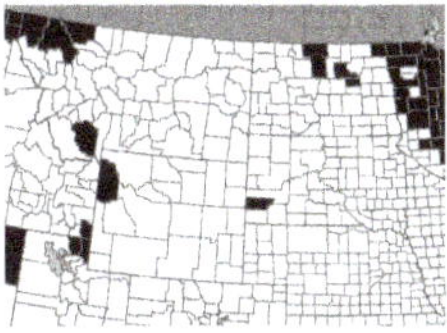

DESCRIPTION Stems very slender, usually creeping along the bottom in shallow water. **Leaves** numerous, commonly trichotomous at the base and then 1–3x dichotomous, 0.5–2 cm long, the segments slender, flat, not much reduced with each branching, the ultimate segments blunt-tipped; **bladders** borne on specialized branches distinct from the leaves, 2–4 mm wide; winter buds ovoid or ellipsoid, 5–7 mm long. **Flowers** usually 2–4 in a lax raceme; peduncle 6–20 cm long; corolla with a lower lip mostly 8–12 mm long, the palate well-developed, nearly 2x as long as the upper lip; spur nearly as long as the lower lip; fruiting pedicels suberect. July–Aug.

HABITAT Shallow water of springs, bogs and swamps.

WETLAND STATUS
GP OBL | MIDW OBL | WMTN OBL

Utricularia macrorhiza Le Conte

COMMON BLADDERWORT

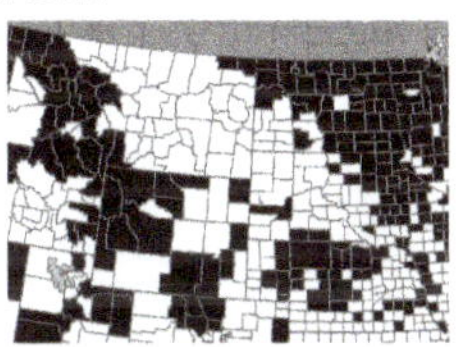

DESCRIPTION Stems free-floating, often extensive. **Leaves** numerous, mostly dichotomous at the base and then repeatedly and unequally dichotomous, 1–5 cm long, the segments more or less terete, progressively reduced with branching, the ultimate segments filiform, attenuate; **bladders** borne on segments of ordinary leaves, 1–4 mm wide; winter buds ovoid or ellipsoid, 1–2 cm long. **Flowers** usually 6–20 in a lax raceme borne on a stout peduncle 6–25 cm long; corolla with a lower lip mostly 10–20 mm long, sometimes much smaller on later flowers, the palate well-developed, the upper lip about equaling the lower one; spur ca. 2/3 as long as the lower lip; fruiting pedicels recurved. June–Aug.

SYNONYM *Utricularia vulgaris* L.

HABITAT Shallow, standing water of lakes, ponds, marshes and ditches, often among rushes or cattails.

WETLAND STATUS
GP OBL | MIDW OBL | WMTN OBL

ADDITIONAL SPECIES Zigzag bladderwort (*Utricularia subulata* L.), a species mostly of the Atlantic Coast, is known from Holt and Rock counties, Neb; its yellow flowers are single atop a slender, leafless scape.

Utricularia minor L.

LESSER BLADDERWORT

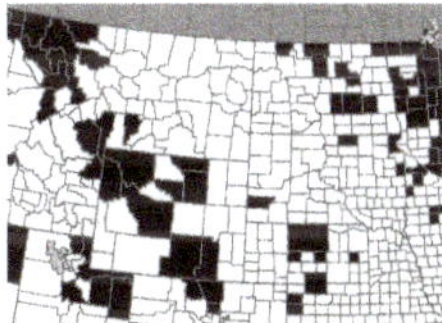

DESCRIPTION Similar in habit to *U. intermedia*. **Leaves** numerous, commonly trichotomous at the base and then dichotomous or irregularly 1–3x divided, mostly 0.3–1 cm long, the segments slender, flat,

Utricularia intermedia
FLAT-LEAF BLADDERWORT

Utricularia macrorhiza
COMMON BLADDERWORT

entire, sometimes toothed at the tip, not much reduced with each branching, the ultimate segments strongly acuminate; **bladders** borne on ordinary leaves, 1–2 mm wide; winter buds obovoid to globose, 2–5 mm long. **Flowers** usually 2–9 in a lax raceme; peduncle 4–15 cm long; corolla with a lower lip 4–8 mm long, 2x as long as the upper lip, the palate scarcely developed; spur small, to 1/2 as long as the lower lip; fruiting pedicels recurved. July–Aug.

HABITAT Shallow water of fens and fresh marshes.

WETLAND STATUS
GP OBL | MIDW OBL | WMTN OBL

Utricularia minor
LESSER BLADDERWORT
(flower, bladders)

Linderniaceae
false pimpernel family

Lindernia dubia (L.) Pennell
FALSE PIMPERNEL

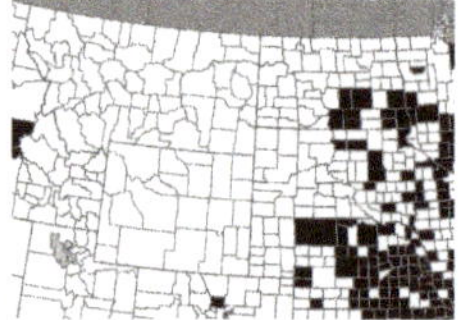

DESCRIPTION Small, usually widely branched annual 0.5–2 dm tall, similar to *Gratiola neglecta* in overall appearance and sometimes confused with it. **Leaves** simple, opposite, sessile, ovate to elliptic or obovate, 5–30 mm long, 2–10 mm wide, entire to denticulate, narrowed to rounded or cordate at the base. **Flowers** borne singly in the axils of the leaves, on slender pedicels 0.5–2.5 cm long; calyx regular, comprised of 5 distinct, linear to linear-lanceolate sepals 3–5 mm long; corolla light blue-violet, tubular-campanulate, 4–10 mm long, bilabiate, the upper lip erect, shallowly 2-lobed, the lower lip somewhat deflexed, shallowly 3-lobed, much wider than the upper lip; stamens 2, with a lower pair of filament-like staminodes also present, these inserted at the middle of the corolla tube. **Capsule** ovoid to ellipsoid, 3.5–6 mm long, membranous. Aug–Sept.

SYNONYM *Lindernia anagallidea* (Michx.) Pennell.

HABITAT Mud flats and shores of temporary ponds and marshes.

WETLAND STATUS
GP OBL | MIDW OBL | WMTN OBL

Lindernia dubia
FALSE PIMPERNEL

Lythraceae
loosestrife family

Annual or perennial herbs, sometimes woody at the base. **Leaves** simple, entire, opposite or both opposite and alternate or rarely some whorled, sessile or nearly so, exstipulate. **Flowers** single or few clustered in leaf axils or in terminal, bracteate, spikelike inflorescences, subsessile or short-pedicelled, subtended by a pair of small bracteoles on the pedicels, these sometimes deciduous with age. Flowers perfect, regular or somewhat irregular, often di- or trimorphic with respect to stamen and style lengths; calyx tubular, cylindrical or campanulate to globose, strongly nerved, the calyx lobes 4 or 6, alternating with longer appendages in the sinuses between the lobes; petals 4 or 6, separate, pink or purple, crumpled, deciduous; stamens numbering as many as or 2x (rarely to 3x) as many as the petals; stigma capitate, style simple, slender, ovary superior, free of the calyx tube or cup. **Fruit** a many-seeded capsule.

1 Flowers 6-merous, in terminal, bracteate, spikelike inflorescences; calyx tube cylindrical; plants perennial ***Lythrum***

1 Flowers 4-merous, axillary; calyx tube campanulate to globose; plants annual **2**

2 Leaves of middle and upper stem attenuate at the base; flowers solitary in the axils ***Rotala***

2 Leaves of middle and upper stem auriculate or cordate and clasping at the base; flowers (1)3-many in the axils ***Ammannia***

Ammannia TOOTHCUP

Usually small, simple to widely branched, glabrous annuals. **Leaves** opposite or an occasional pair subopposite, linear to oblong or oblanceolate, mostly auriculate or cordate and clasping at the base. **Flowers** small, 4-merous, (1)3-many in axillary cymes, these sessile or pedunculate. Calyx tube urceolate to campanulate in flower, subglobose in fruit, often strongly 8-ribbed in flower, the 4 lobes low and broadly triangular, alternating with 4 short, thickened appendages or these sometimes absent; petals normally 4, early deciduous, lavender to rose pink or rose-purple, drying purple; stamens 4(–12), exserted; style persistent, exserted in fruit. **Capsules** globose, smooth, membranous, irregularly dehiscent; seeds numerous, ovoid, ca. 1 mm long.

1 Plants slender, with axillary cymes borne on filiform peduncles 3–9 mm long; flowers many, with mostly at least 7 per axil; capsules mostly 2.5 mm or less in diameter ***A. auriculata***

1 Plants robust, with axillary cymes sessile or borne on stout peduncles to 4(–9) mm long; flowers mostly 3–5 per axil; capsules mostly 3.5 mm or more in diameter **2**

2 Inflorescences sessile or nearly so, mostly with 1–3 flowers per axil; petals pale lavender and anthers yellow in fresh flowers; capsules 4–6 mm in diameter ***A. robusta***

2 Inflorescences on short to prominent peduncles, rarely sessile, mostly with 3–5 flowers per axil; petals deep rose-purple or rose with a deep purple midvein and anthers deep yellow in fresh flowers; capsules mostly 3.5–5 mm in diameter ***A. coccinea***

Ammannia auriculata Willd. (*not illus.*)

EARED REDSTEM

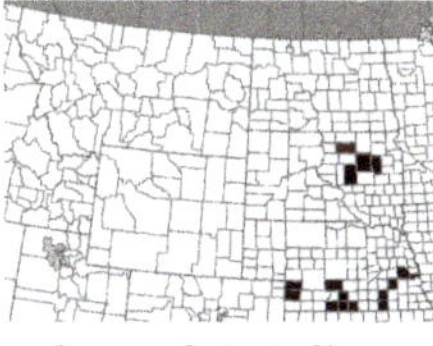

DESCRIPTION Slender annual 0.5–5(8) dm tall, simple or branched above the base, the branches ascending. **Leaves** linear-lanceolate to linear-oblong, 1–4(6.5) cm long, 1.5–6(10) mm wide, auriculate-clasping at the base. **Inflorescences** of simple to compound, rather open cymes, peduncled from the axils, with (1)3–12(15) flowers per axil, most with at least 7 flowers; peduncles filiform, 3–9 mm long; pedicels 1–3(6) mm long. Calyx tube 1–3 mm long in flower, the lobes alternating with short appendages or the appendages absent; petals 4, deep rose-purple, ca. 1.5 mm long; stamens 4(–8), anthers deep yellow when fresh; styles 1–3 mm long in fruit. **Capsules** 1.5–3 mm in diameter, equaling or exceeding the calyx. Jul–Oct.

HABITAT Muddy shores, flats and low spots in fields where water stands temporarily.

WETLAND STATUS
GP OBL | MIDW OBL | WMTN OBL

Ammannia coccinea Rottb.
VALLEY REDSTEM

DESCRIPTION Erect, simple or freely branched annual 0.5–5(10) dm tall, the branches ascending, or basal branches, when present, spreading. **Leaves** linear-lanceolate to linear-oblong or rarely elliptic to spatulate, 2–8 cm long, 2.5–15 mm wide, auriculate or cordate and clasping at the base or cuneate on lowermost leaves. **Inflorescences** of rather compact, axillary cymes, nearly sessile to pedunculate, with (1)3–5(–14) flowers per axil; peduncles stout, to 9 mm long; pedicels to 2 mm long. **Calyx tube** (2.5)3–5 mm long, the lobes alternating with thickened appendages of about the same length; **petals** 4(5), deep rose-purple, sometimes with a deep purple midvein at the base, ca. 2 mm long; stamens 4(–7), anthers deep yellow when fresh; styles 1.5–3 mm long in fruit. **Capsules** 3.5–5 mm in diameter at maturity, equaling or exceeding the calyx. July–Oct.

HABITAT Exposed mud of ponds, marshes and streambanks

WETLAND STATUS
GP OBL | MIDW OBL | WMTN OBL

NOTE See note following *A. robusta*.

Ammannia coccinea
VALLEY REDSTEM

Ammannia robusta Heer & Regel
GRAND REDSTEM

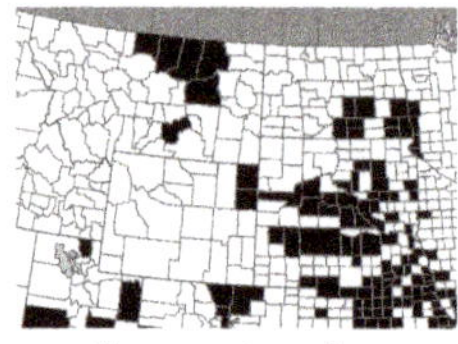

DESCRIPTION Simple to widely branching annual 0.5–5(10) dm tall, the lower branches, when developed, usually decumbent, often as long as the erect main stem. **Leaves** linear-lanceolate to linear-oblanceolate, rarely elliptic to spatulate, 1.5–8 cm long, 2–15 mm wide, mostly auriculate-clasping at the base, sometimes cuneate on lowermost leaves. **Inflorescences** compact, sessile or essentially so, with 1–3(5) flowers per axil. **Calyx tube** ca. 3.5 mm long in flower, the lobes alternating with thickened appendages of about the same length; **petals** 4(–8), pale lavender, sometimes with a deep rose midvein at the base, ca. 2.5 mm long; stamens 4(–12), anthers pale yellow to yellow. **Capsules** 4–6 mm in diameter at maturity, enclosed by or equaling the calyx. July–Oct.

HABITAT Same habitats as *Ammannia coccinea.*

WETLAND STATUS
GP OBL | MIDW OBL | WMTN OBL

NOTE The great majority of plants in our region formerly identified as *Ammannia coccinea* are actually *A. robusta*.

Ammannia robusta
GRAND REDSTEM

Lythrum LOOSESTRIFE

Erect, perennial herbs, sometimes rather woody at the base; **stems** usually with ascending branches above, prominently 4-angled in the upper part. **Leaves** usually mostly opposite below and becoming alternate upward, rarely whorled, sessile; linear-lanceolate to ovate-lanceolate, reduced to bracts in the inflorescence. **Flowers** in showy, terminal, bracteate, spikelike inflorescences, 1–several in the axils of the bracts, regular or somewhat irregular, dimorphic or trimorphic, the stamens and styles of 2 or 3 different lengths. **Calyx tube** cylindrical, green-striate due to the 8–12 strong nerves, the lobes alternating with slender appendages; **petals** 6, purple; stamens 6 or 12; ovary 2-celled. **Capsule** ovoid, firm, septicidal, enclosed by the calyx tube.

---

1 Flowers solitary in the axils of the bracts; calyx tube glabrous; stamens usually 6 ***L. alatum***

1 Flowers mostly 2-several in the axils of the bracts; calyx tube pubescent; stamens usually 12, with 6 exserted and 6 included by the calyx tube ***L. salicaria***

Lythrum alatum Pursh
WINGED LOOSESTRIFE

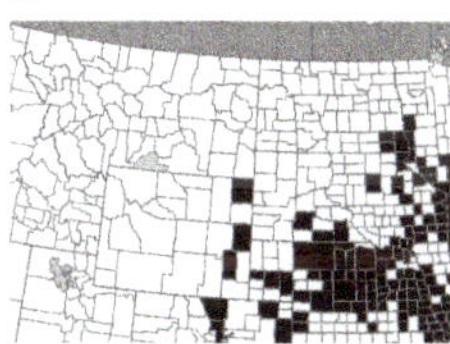

DESCRIPTION Glabrous, rhizomatous perennial 2–8 dm tall; **stems** simple or usually branched above. **Leaves** usually opposite below and alternate above, ovate-lanceolate to lanceolate, 1–4 cm long, 3–10 mm wide, acute at the tip, rounded at the base. **Flowers** solitary in the axils of the bracts, dimorphic, either the stamens or the style exserted; calyx tube 4–6 mm long, glabrous, with appendages ca. 2x as long as the calyx lobes; petals 3–7 mm long; stamens usually 6. July–Aug.

SYNONYM *Lythrum dacotanum* Nieuw.

HABITAT Wet meadows and ditches, especially in sandy areas.

WETLAND STATUS
GP OBL | MIDW OBL | WMTN OBL

Lythrum salicaria L.
PURPLE LOOSESTRIFE

DESCRIPTION Rarely glabrous to usually pubescent perennial 6–12 dm tall; **stems** often much-branched above with the ascending branches terminating in the showy, spikelike inflorescences. **Leaves** opposite to subopposite, rarely whorled, becoming alternate and bractlike in the inflorescence, linear-lanceolate to lanceolate, 3–1 0 cm long, 0.5–2 cm wide, acute at the tip, obtuse to cordate at the base. **Flowers** mostly 2-several in the ails of the bracts, trimorphic, the stamens and styles of 3 different lengths; **calyx tube** 4–6 mm long, sparsely to densely pubescent, with appendages 2–3x longer than the calyx lobes; **petals** 7–10 mm long; stamens usually 12. July–Aug.

HABITAT Introduced from Europe as an ornamental, escaping to wet ditches, marshes, streambanks and floodplains.

WETLAND STATUS
GP OBL | MIDW OBL | WMTN OBL

NOTE The introduction of purple loosestrife to North America has proven very detrimental to wetlands of the USA and Canada. It is an exceedingly aggressive (albeit attractive) weed of marshes that successfully outcompetes native wetland species. The problem is most acute from central Minn eastward to the Atlantic Coast. Unfortunately the plant is of little or no value to wildlife and appears to have no natural enemies here.

Lythrum alatum
WINGED LOOSESTRIFE

Rotala ramosior (L.) Koehne

TOOTHCUP

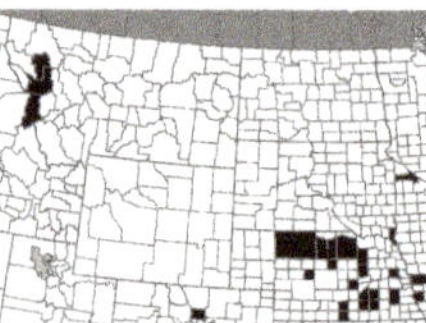

DESCRIPTION Small, glabrous, simple to freely branched annual to 4 dm tall, the branches spreading to ascending. **Leaves** opposite, linear to oblong or oblanceolate, 1–5 cm long, 2–12 mm wide, attenuate to a subpetiolate base, not clasping. **Flowers** solitary, sessile in the axils, 4(–6)-merous. **Calyx tube** campanulate to subglobose, 2.5–5 mm long, not strongly nerved, the broad lobes alternating with thickened appendages of about equal length; **petals** 4(–6), white to pink, barely exceeding the calyx; stamens 4(–6), included or barely exserted; style none or short, not exserted, stigma capitate. **Capsule** globose, 2- to 4-locular, included by the calyx, minutely cross-striate on the outer wall; seeds ovoid, 1 mm or less long. July–Oct.

HABITAT Muddy or sandy shores, low spots in fields and other temporarily flooded places; uncommon.

WETLAND STATUS

GP OBL | MIDW OBL | WMTN OBL

Rotala ramosior
TOOTHCUP

Lythrum salicaria
PURPLE LOOSESTRIFE

Menyanthaceae *buckbean family*

Menyanthes trifoliata L.

BUCKBEAN

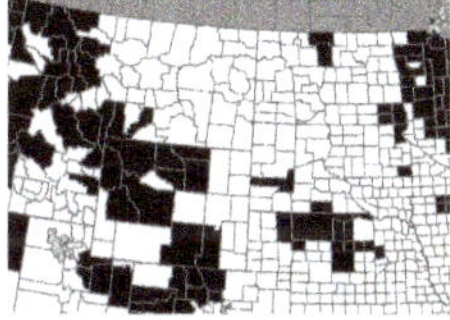

DESCRIPTION Glabrous, rhizomatous perennial with flowering scapes 1.5–3.5 dm tall; rhizome thick, covered with old leaf bases. **Leaves** all basal, with the sheathing petiole bases arranged alternately on the rhizomes, the blades palmately 3-foliate, the leaflets elliptic to oblong or oblanceolate to obovate, 3–10 cm long, 1–5 cm wide, entire or sometimes coarsely undulate-dentate. Inflorescence a scapose, bracteate raceme, surpassing the leaves; bracts mostly 3–5 mm long; pedicels 4–20 mm long. **Flowers** perfect, regular, 5(4–6)-merous, often dimorphic, some flowers with exserted stamens and included style, others with exserted style and included stamens; calyx deeply divided, the oblong-ovate lobes 1.5–3 mm long; corolla whitish or pinkish, salverform, lobed to near or below the middle, 8–12 mm long, the lobes eventually recurved, conspicuously fringed on the inner surface; stamens usually 5, epipetalous; stigma 2-lobed, style elongate, ovary ca. 1/3 inferior, 1-celled. **Fruit** a globose, corky-walled capsule 6–10 mm in diameter, rupturing irregularly, containing many shiny, yellowish-brown seeds. June–Aug.

HABITAT Fens and bogs.

WETLAND STATUS
GP OBL | MIDW OBL | WMTN OBL

Menyanthes trifoliata
BUCKBEAN

Nymphaeaceae *water lily family*

Aquatic perennials with large floating and often some emersed leaves; **stem** a thick, fleshy, submerged rhizome anchored in the substrate, the older portions decaying behind the growing apex. **Leaves** arranged in a close spiral on the rhizome, subpeltate, the blades large and leathery, oblong to oval or rotund in outline but with a sinus behind the petiole attachment to the blade; smaller, thin textured submersed leaves sometimes present, especially early in the growing season; petioles elongate, stout and tough. **Flowers** solitary on long peduncles, borne at or above the water surface, white or yellow, 4–20 cm across, perfect, regular, hypogynous to nearly epigynous; **sepals** 4–6, quite petaloid, green or greenish on the outside, white or yellow on the inside, when yellow usually reddish toward the base; **petals** numerous, either white, large and showy, or yellow, small and inconspicuous, spirally arranged, gradually passing into the stamens; stamens numerous, with flattened and often broadened filaments; carpels several to many, fused into a compound ovary, stigmas radiating from the center of the disk-like summit of the ovary. **Fruit** fleshy and leathery, many-seeded, eventually breaking open under water.

1 Flowers yellow; leaf blades oblong to oval or subsagittate — *Nuphar*

1 Flowers white; leaf blades rotund or nearly so — *Nymphaea*

Nuphar variegatum Engelm.

YELLOW WATER LILY, SPATTERDOCK

DESCRIPTION Stout plants from thick, yellowish, branched rhizomes. **Leaf blades** floating or some often emersed, oblong to oval or subsagittate, (1)1.5–4 dm long, the basal lobes divergent to slightly overlapping; earliest leaves submersed, thin and reddish; petioles flattened and winged on the upper side. **Flowers** yellow, 4–6 cm across; sepals usually 6(5), spirally arranged and overlapping, greenish to yellow, usually reddish toward the base on the inside, oblong; petals oblong to spatulate, much smaller than the sepals, usually shorter than the stamens; anthers 4–7 mm long, filaments flat; stigmatic rays usually 7–16. **Fruit** somewhat ovoid, mostly 2–4 cm long; **seeds** ovoid, to 5 mm long. June–Aug.

SYNONYMS *Nuphar advenum* Ait., *Nuphar luteum* (L.) Sibth. & Small

HABITAT Shallow to deep water of quiet streams, lakes and ponds.

WETLAND STATUS
GP OBL | MIDW OBL | WMTN OBL

Nymphaea WATER LILY

Large-flowered plants from stout rhizomes, these sometimes with lateral tubers. **Leaf blades** nearly always floating, seldom some emersed, rotund or nearly so, with a narrow V-shaped sinus behind the petiole attachment to the blade; petioles not flattened or winged. **Flowers** white and showy, 7–25 cm across, usually opening in morning and closing in afternoon, remaining open on cool days; sepals 4, greenish; petals numerous and overlapping, white, gradually passing into the stamens; stamens numerous, the outer ones with broadened, petaloid filaments, anthers yellow, the spiral of petals and stamens encroaching up the sides of the ovary so that the flower is nearly epigynous; ovary concave at its summit with a rounded protuberance projecting from the center, stigmas usually 10–25, radiating from the center, overarched by finger-like projections around the margin of the stigmatic disk. **Fruit** subglobose, covered with the persistent petal and stamen bases, maturing under water; **seeds** numerous, each enveloped by a saclike aril.

1 Petals elliptic, broadest near the middle, tapered to a subacute tip; flowers usually fragrant; rhizome lacking tubers **_N. odorata_**

1 Petals oblanceolate to spatulate, broadest above the middle, obtuse to rounded at the tip; flowers scarcely if at all fragrant; rhizome with lateral tubers **_N. odorata_ subsp. _tuberosa_**

Nymphaea odorata Ait.
FRAGRANT WHITE WATER LILY

DESCRIPTION Rhizome elongate, without lateral tubers. **Leaf blades** to 25 cm across, green above, usually purple or red-tinged beneath; petioles not striped. **Flowers** pleasantly fragrant while open, 7–12 cm across; sepals often purplish on the back; petals elliptic, tapered to a subacute tip. **Seeds** ellipsoid, ca. 2 mm long. June–Sept.

HABITAT Quiet waters of ponds, lakes, and streams.

WETLAND STATUS
GP OBL | MIDW OBL | WMTN OBL

Nymphaea odorata subsp. *tuberosa* (Paine) Wiersma & Hellquist
WHITE WATER LILY

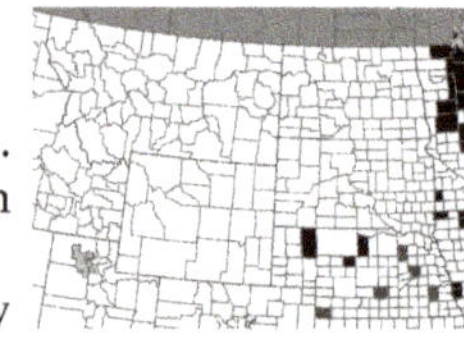

DESCRIPTION Rhizome producing knotty, lateral tubers. **Leaf blades** 10–30 cm across, green above and beneath or rarely dull purple beneath; petioles usually striped. **Flowers** odorless or scarcely fragrant, 8–25 cm across; sepals green on the back; petals oblanceolate to spatulate, obtuse to rounded at the tip. **Seeds** 2.8–4.4 mm long. June–Sept.

SYNONYM *Nymphaea tuberosa* Paine

HABITAT Same habitats as *Nymphaea odorata.*

WETLAND STATUS
GP OBL | MIDW OBL

Nuphar variegatum
YELLOW WATER LILY, SPATTERDOCK

Nymphaea odorata
FRAGRANT WHITE WATER LILY

Onagraceae
evening primrose family

Perennial and annual herbs (those included here), often flowering in the first year when perennial. **Leaves** simple, alternate, opposite or frequently both on the same plant, entire or toothed, sessile or short-petioled, exstipulate. **Flowers** solitary in the axils of leaves or leaflike bracts, sessile or pedicelled, often arranged in distinct terminal inflorescences (bracteate spikes or racemes), perfect, regular, 4-merous, epigynous, the sepals, petals and stamens arising from the summit of the ovary or from the rim of a short floral tube prolonged beyond the ovary; sepals free; petals free, white, yellow or pinkish to rose-purple; stamens 8 or 4; stigma capitate or often 4-lobed, style 1, ovary 4-celled, often 4-angled, sometimes long and slender to appear like a pedicel or an extension thereof. **Fruit** a dehiscent, 4-valved capsule or sometimes weakly dehiscent; **seeds** many, with or without a tuft of hairs (coma).

1 Petals minute or absent, greenish when present; seeds without a coma *Ludwigia*

1 Petals white, pinkish or rose-purple; seeds with or without a coma *Epilobium*

Epilobium WILLOWHERB

Fibrous-rooted perennial herbs (those included here), often producing leafy rosettes or subterranean turions from the base late in the growing season, these sessile or on lateral rhizomes or stolons. **Leaves** simple, opposite, alternate or usually opposite below and becoming alternate above, sessile or short-petioled, exstipulate. **Flowers** small, white to pink, solitary in the axils of upper reduced leaves, arranged in simple or branched, terminal racemes. Floral tube short or absent; sepals 4; petals 4, white to pink, notched at the tip; stamens 8, in 2 series of 4, the inner 4 stamens much shorter than the outer 4; stigma sessile, undivided, ovary linear-elongate, appearing like a continuation of the pedicel, 4-celled, maturing into a linear, 4-valved capsule, splitting from the tip to release numerous brown, ellipsoid **seeds** which bear a terminal tuft of fine hairs (the coma).

1 Plants annual; seeds without a coma *E. campestre*

1 Plants perennial, often flowering the first year; seeds with a coma 2

2 Leaves linear to linear-lanceolate, 1–5(7) mm wide, the margins entire, revolute *E. leptophyllum*

2 Leaves lanceolate to ovate-lanceolate, mostly 0.8–3 cm wide, the margins shallowly serrate, not revolute 3

3 Leaves acute or acuminate to a rather blunt tip, each margin with mostly 15–30 teeth or fewer; flower buds obtuse to rounded or slightly pointed at the tip; coma white or nearly so, attached to a short, broad, flattened beak at the tip of the seed *E. ciliatum*

3 Leaves long-acuminate to a slender, pointed tip, each margin of the main cauline leaves mostly with 30–75 teeth; flower buds tipped with the 4 projecting or divergent sepal tips; coma cinnamon-colored at maturity, sessile on the beakless rounded tip of the seed *E. coloratum*

Epilobium campestre (Jepson) Hoch & W.L. Wagner
SMOOTH WILLOWHERB

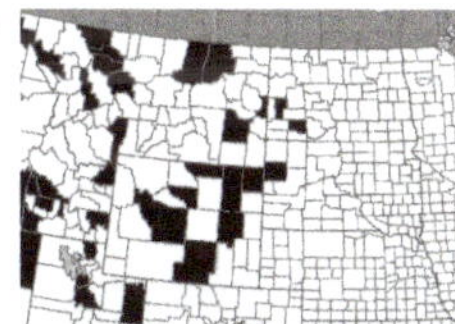

DESCRIPTION Low annual 1–3 dm tall, glabrous below, stigulose or villous with white hairs above, usually

Epilobium campestre
SMOOTH WILLOWHERB

branched from the base, the **branches** suberect to decumbent, often rooting at the lower nodes. **Lowest leaves** opposite, glabrous, narrowly connate at the base, otherwise leaves alternate, sessile, lanceolate to elliptic-lanceolate, 8–20(30) mm long, 3- 6(9) mm wide, sparsely denticulate, loosely villous or strigulose to glabrate, often ciliate. Flowering terminal portion of stems erect, the **flowers** solitary and sessile in the axils of leaflike bracts, the flowers and fruits often hidden by the bracts. **Floral tube** prolonged 0.3–1 mm beyond the ovary, short funnelform; sepals 4, erect, 0.7–1.8 mm long; **petals** 4, pinkish to rose-purple, 1–3 mm long, deeply bilobed; stamens 8, with 4 long ones inserted at the mouth of the floral tube opposite the sepals and 4 shorter ones inserted below the mouth of the tube opposite the petals; stigma 4-lobed, ovary 4-celled. **Capsule** elongate, 6–8 mm long, slightly curved, pointed, dehiscent in the upper portion; **seeds** in 1 row of 6–14 per locule, brownish, 1–1.3 mm long. June–July.

SYNONYM

Boisduvalia glabella (Nutt.) Walp.

HABITAT Muddy banks and flats of streams and temporary ponds.

WETLAND STATUS

GP FACW | WMTN OBL

Epilobium ciliatum Raf.

AMERICAN WILLOWHERB

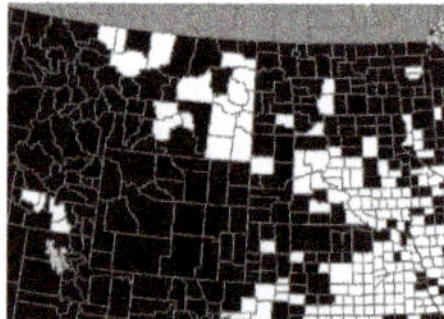

DESCRIPTION Simple and erect to branched and spreading perennial, flowering the first year, producing overwintering above-ground leafy rosettes or subterranean fleshy turions at the base in autumn, 1.5–8(12) dm tall; **stems** glabrous below, sparingly to densely pubescent above, with short eglandular hairs intermixed with long glandular hairs, especially pubescent on the decurrent leaf bases and in the inflorescence. **Leaves** opposite below or for much of the length of the stem, usually alternate above, at least in the inflorescence, sessile or with short winged petioles to 6 mm long, the blades lanceolate to ovate-lanceolate, 2.5–9(12) cm long, 0.8–3(4.5) cm wide, acute or acuminate to a rather blunt tip, irregularly and shallowly serrate, mostly with 15–30 teeth on each margin, rounded at the base. **Floral tube** 0.5–2.5 mm long; sepals ovate, acute, 1.5–5 mm long, not projecting or divergent in bud, the tips forming an obtuse to rounded or slightly pointed bud apex; **petals** whitish to pink, 1.5–10(12) mm long, strongly notched; pedicels mostly 3–10 mm long. **Capsules** linear, 3–7(10) cm long, pubescent; **seeds** mostly 0.8–1.6 mm long, longitudinally striate with hyaline crests or ridges, usually not papillate, the coma white or nearly so, attached to a short, broad, flattened beak at the tip of the seed. July–Sept.

Epilobium ciliatum
AMERICAN WILLOWHERB

SYNONYMS *Epilobium glandulosum* Lehm. var. *adenocaulon* (Hausskn.) Fern., *Epilobium adenocaulon* Hausskn.

HABITAT Shores, streambanks, marshes, wet meadows, springs, seepage areas, ditches and other wet places.

WETLAND STATUS

GP FACW | MIDW FACW | WMTN FACW

ADDITIONAL SPECIES Also reported for the Black Hills and Wyoming are three species similar to though apparently much less common than *E. ciliatum* in the Great Plains: ***Epilobium halleanum*** Hausskn., ***E. hornemannii*** Reichenb. and ***E. saximontanum*** Hausskn. They may be distinguished from *E. ciliatum* and each other by the following key:

1 Plants clumped or cespitose, forming short, leafy, above-ground shoots at the base; stem decumbent to ascending at the base; leaves petiolate ***E. hornemannii***

1 Plants not clumped or only loosely clumped, forming sessile, leafy rosettes or fleshy turions; stem erect at the base; leaves sessile to occasionally petiolate 2

2 Plants 0.3–19 dm tall, forming leafy rosettes or large, subterranean turions; seeds 0.8–1.6(1.9) mm long, longitudinally striate with hyaline crests or ridges but lacking distinct papillae ***E. ciliatum***

2 Plants 0.2–6 dm tall, lacking rosettes, forming only compact subterranean turions; seeds 1. 1–1.6 mm long, distinctly papillate, the papillae often in longitudinal rows 3

3 Leaves sessile, clasping, mostly narrowly ovate, denticulate; turions fleshy, elongate; capsules subsessile, appressed; seed collar conspicuous below the coma ***E. saximontanum***

3 Leaves petiolate or subsessile, not clasping, lanceolate or narrower, subentire or denticulate; turions compact, round; capsules on pedicels 0.8–3.8 cm long; seed collar inconspicuous ***E. halleanum***

Epilobium coloratum Biehler

PURPLE-LEAVED WILLOWHERB

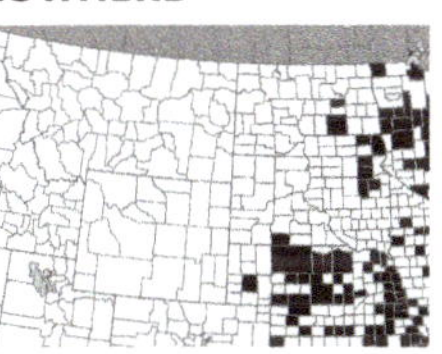

DESCRIPTION Similar to *Epilobium ciliatum,* averaging larger, mostly 5–10 dm tall, simple below and bushy-branching above in the inflorescence, producing basal, leafy rosettes in autumn; **stems** glabrous below, puberulent above with incurved hairs mainly on the decurrent leaf bases. **Leaves** mostly opposite, becoming alternate and reduced in the inflorescence, the main cauline leaves distinctly short-petioled, the blades elongate-lanceolate, mostly 4–15 cm long, 0.5–3 cm wide, long-acuminate to a slender, pointed tip, serrulate, with mostly 30–75, low, rather sharp teeth on each margin, rounded at the base. **Floral tube** 0.3–1 mm long; sepals ovate to lanceolate, acute to cuspidate, 1.5–3 mm long, projecting or divergent in bud: petals pink, 3–5 mm long, strongly notched; pedicels to 10 mm long. **Capsules** linear, 3–4.5(6) cm long; **seeds** 1.2–1.7 mm long, papillate on the surface, the coma cinnamon-colored at maturity, sessile on the beakless rounded tip of the seed. July–Sept.

Epilobium coloratum
PURPLE-LEAVED WILLOWHERB

HABITAT Shores, springs, swamps and fens.

WETLAND STATUS
GP OBL | MIDW OBL

Epilobium leptophyllum Raf.

NARROW-LEAVED WILLOWHERB

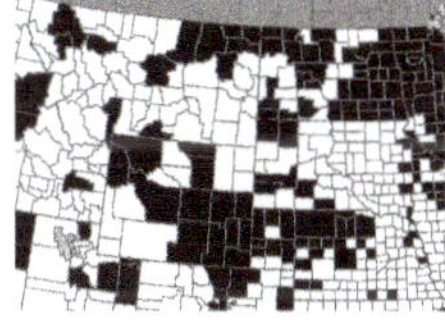

DESCRIPTION Erect perennial 2–8 dm tall, usually simple from the base, branching above, densely puberulent upward with incurved hairs, producing filiform stolons at the base in late summer and autumn, the stolons usually reddish, with remote pairs of minute scales and terminating in ovoid, fleshy turions. **Leaves** usually opposite below and alternate above, sessile, linear to linear-lanceolate, gradually reduced upward, 1–6(7.5) cm long, 1–5(7) mm wide, usually densely puberulent with incurved hairs, acute to blunt-tipped, revolute, cuneate at the base. **Floral tube** 0.8–2 mm long; sepals ovate-lanceolate, acute, 1.5–4.5 mm long;

Epilobium leptophyllum
NARROW-LEAVED WILLOWHERB

petals white to pinkish, 3–7 mm long, notched; pedicels mostly 0.5–2 cm long. **Capsules** linear, 2.5–8 cm long; **seeds** 1–2.2 mm long, the coma white to tawny. July–Sept.

HABITAT Shores, streambanks, springs, fens, seepage areas and ditches, often in shallow water.

WETLAND STATUS
GP OBL | MIDW OBL | WMTN FACW

ADDITIONAL SPECIES *Epilobium palustre* L., which is very similar to *E. leptophyllum*, is known from the Black Hills region of SD, nw Neb, and e Mont. The two species are distinguished as follows:

1 Leaves linear to lanceolate or oblong, subglabrous to sparsely strigulose above; inflorescence often nodding in bud, eglandular — ***E. palustre***

1 Leaves linear, rarely wider, densely strigulose above; inflorescence erect or nearly so, often with a mixture of strigulose and glandular hairs — ***E. leptophyllum***

Ludwigia SEEDBOX

Perennial, fibrous-rooted herbs with floating, creeping or ascending to erect **stems**, sometimes stoloniferous. **Leaves** simple, opposite or alternate, entire, subsessile or wing-petioled; stipules minute, deciduous. **Flowers** solitary in the axils, sessile or nearly so; floral tube not prolonged beyond the ovary; sepals 4, green, persistent; petals 4 or none, minute and greenish when present; stamens 4; stigma unlobed, capitate, style short; ovary obconic to cylindric, often 4-angled, usually with 2 bracteoles at or toward the base. **Capsules** 4-celled, many-seeded, dehiscent longitudinally or by a terminal pore, sometimes irregularly dehiscent; **seeds** lacking a coma.

1 Leaves opposite — ***L. palustris***

1 Leaves alternate — ***L. polycarpa***

Ludwigia palustris (L.) Ell.
MARSH SEEDBOX

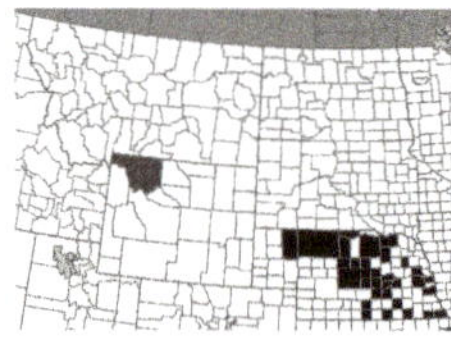

DESCRIPTION **Stems** creeping and rooting at the nodes or floating, simple to freely branched, 1–5 dm long, glabrous or with few scattered hairs on younger portions. **Leaves** opposite, the blades elliptic to ellipticovate, 3–25(40) mm long, 4–20 mm wide, shiny green or reddish, acute to acuminate, tapered at the base to a winged petiole 3–25 mm long. **Sepals** deltate, 1–2 mm long; **petals** none; bracteoles at base of the ovary to 1 mm long or absent. **Capsule** elongate-globose, somewhat 4-angled, with a longitudinal green band on each angle, 2- 5 mm long, 2–3 mm thick; **seeds** whitish or yellowish-brown, curved-oblong, 0.5- 0.9 mm long. June–Sept.

HABITAT Shallow water or mud of ponds, lakes, streams, ditches and around springs.

WETLAND STATUS
GP OBL | MIDW OBL

Ludwigia palustris
MARSH SEEDBOX

Ludwigia polycarpa Short & Peter
MANY-SEED SEEDBOX

DESCRIPTION **Stems** erect or ascending, simple to freely branched, usually 4-angled, 1–9 dm tall, glabrous, producing leafy stolons from the base in autumn. **Leaves** alternate, the main cauline leaves with blades lanceolate to oblanceolate, 3–12 cm long, 5–15 mm wide, tapered to an acute tip and a sessile or wing-petioled base, the petiole to 8 mm long. **Sepals** lanceolate-deltate, 2.5–4 mm long; **petals** greenish and minute or absent; bracteoles arising from above the base of the ovary, linear-lanceolate, 2–5 mm long. **Capsule** short-cylindric to somewhat obconic, roundly 4-sided or shallowly grooved below each sepal, 4–7 mm long, 3–5 mm thick; **seeds** light yellow, slightly curved, 0.6–0.9 mm long, puncticulate. June–Sept.

HABITAT Borders of marshes, lakes, ponds, streams and in wet depressions.

WETLAND STATUS
GP OBL | MIDW OBL

Ludwigia polycarpa
MANY-SEED SEEDBOX

Orobanchaceae
broom-rape family

Annual, biennial, or perennial herbs; some genera fleshy, without green color, parasitic on the roots of other plants. **Leaves** opposite, alternate, or the leaves reduced to scales. **Flowers** mostly perfect, single or few from leaf axils, or numerous in clusters at ends of stems or leaf axils, usually with a distinct upper and lower lip; calyx 2–5-lobed or toothed, persistent in fruit; petals 4–5 (petals sometimes absent); stamens usually 4, inserted on the corolla tube. **Fruit** a several- to many-seeded 2-valved capsule. Includes former members of Scrophulariaceae.

1 Most leaves deeply pinnately divided; corolla cream-colored or yellow ***Pedicularis***

1 Stem leaves toothed or entire, not deeply pinnately lobed; corolla reddish to pinkish-purple ***Agalinis***

Agalinis **GERARDIA**

Shallowly fibrous-rooted annuals, often hemiparasitic; foliage dark green and tending to blacken upon drying; **stems** slender, erect, branched, 4-angled. **Leaves** opposite or subopposite, sessile, often with fascicles of smaller leaves in the axils, linear to linear-lanceolate, sparingly to conspicuously scabrous on the upper surface. **Flowers** solitary from the ads of upper leaves, on slender pedicels; calyx nearly regular, 5-lobed, campanulate in flower to hemispheric in fruit, somewhat accrescent, lobes broadly triangular at the base with an acuminate tip; corolla 5-lobed and weakly bilabiate, reddish to pinkish-purple, obliquely campanulate, the lobes shorter than the tubular portion, finely fringed with cilia; stamens 4, the lower pair longer; style simple. **Capsule** globose or nearly so, surpassing the calyx; **seeds** numerous, dark brown to blackish, triangular to trapezoidal, reticulate.

1 Pedicels 2–5 mm long, shorter than to as long as the calyx; corolla 20–32 mm long ***A. purpurea***

1 Pedicels 8–20 mm long, longer than the calyx; corolla 8–15 mm long ***A. tenuifolia***

Agalinis purpurea (L.) Penn.
PURPLE FALSE FOXGLOVE

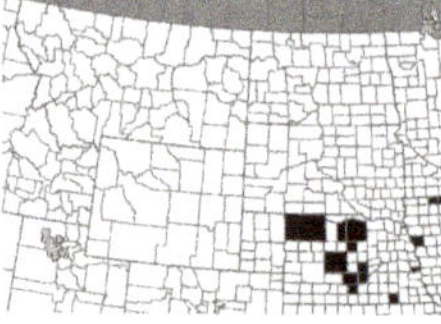

DESCRIPTION Plant 1.5–8(12) dm tall, glabrous to sparsely scaberulous, usually much branched and spreading above. **Leaves** spreading to upcurved or curled, linear, 1–4(5.5) cm long, 0.8–2(3) mm wide, acute; fascicles of smaller leaves often present. Pedicels spreading, 2–5 mm long, shorter than to equaling the calyx. **Calyx** 4–5.5(6) mm long, the lobes 1–2.2 mm long; **corolla** (17)20–32 mm long, the lobes spreading, 5–9 mm long. **Capsule** subglobose, 4–6 mm in diameter; seeds 0.7–1.2 mm long. Aug–Sept.

SYNONYM *Gerardia purpurea* L.

HABITAT Wet meadows, shores and ditches, usually where sandy.

WETLAND STATUS
GP FACW | MIDW FACW

Agalinis tenuifolia (Vahl) Raf.
COMMON FALSE FOXGLOVE

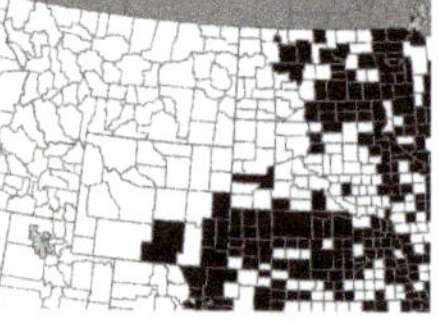

DESCRIPTION Slender, erect plant (0.5)1–5 dm tall, usually branched. **Leaves** spreading to arched-ascending, linear, 1–5 cm long, 1–3 mm wide, acute, scaberulous on the upper surface; fascicles of smaller leaves present or not. Pedicels ascending to spreading, filiform, 8–20 mm long, longer than the calyx. **Calyx** 3.5–5.5 mm long, the lobes 1–2 mm long; **corolla** 8–15 mm long, the lobes 3–5 mm long. **Capsule** globose, 4–6 mm in diameter; **seeds** 0.7–0.9 mm long. Aug–Sept.

SYNONYM *Gerardia tenuifolia* Vahl.

HABITAT Wet meadows, low prairie, shores, streambanks and ditches, usually where sandy.

WETLAND STATUS
GP FAC | MIDW FACW | WMTN FAC

Pedicularis LOUSEWORT

Erect perennials with opposite or alternate pinnatifid leaves and terminal, bracteate spikes or spikelike racemes of yellow flowers. **Calyx** tubular or campanulate, oblique, obscurely lobed or with 2 prominent lateral lobes; **corolla** pale yellow, strongly irregular, bilabiate, the upper lip strongly concave or arched, entire, sometimes with a pair of lateral teeth near the tip, lower lip about equal to or shorter than the upper, with 2 longitudinal folds; stamens 4, didynamous, included by the upper lip. **Capsule** ovate to oblong, laterally flattened, loculicidally dehiscent, splitting only or mainly along the upper side.

1 Calyx entire or nearly so; upper lip of the corolla with 2 lateral teeth near the tip ***P. canadensis***

1 Calyx with 2 crenate, lateral lobes; upper lip of the corolla entire ***P. lanceolata***

Agalinis purpurea
PURPLE FALSE FOXGLOVE

Agalinis tenuifolia
COMMON FALSE FOXGLOVE

Pedicularis canadensis L.

WOOD BETONY

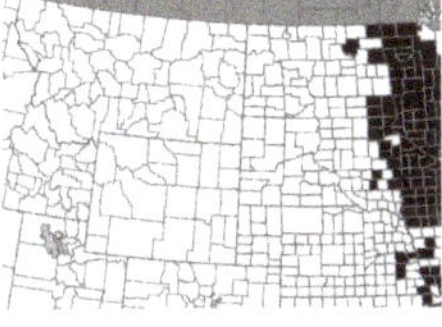

DESCRIPTION Plants loosely tufted with clustered stems 1–3 dm tall. **Leaves** basal and cauline, alternate, the basal and lower leaves longer-petioled and more prominent than the upper ones, the blades oblanceolate in outline, 3–6 cm long, 0.8–2.5 cm wide, rounded at the tip, pinnatifid with the lobes crenate-margined, divided more than halfway to the midrib, tapered to the petiole. **Flowers** sessile or subsessile, in dense spikes 2–15 cm long, the spike axis densely pubescent; bracts obovate to oblanceolate, deciduous; calyx very oblique, entire or nearly so, 6–9 mm long; corolla 17–25 mm long, the upper lip with 2 lateral teeth near the tip. **Capsule** oblong, ca. 2x the length of the calyx. Late May–June.

HABITAT Wet meadows and low prairie, often where sandy.

WETLAND STATUS
GP FACU | MIDW FACU | WMTN FAC

Pedicularis canadensis
WOOD BETONY

Pedicularis lanceolata Michx.

SWAMP LOUSEWORT

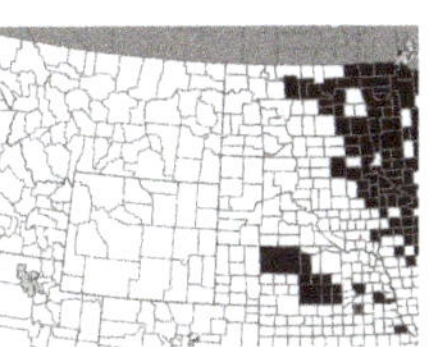

DESCRIPTION **Stems** simple or few-branched, 3–6 dm tall, glabrous or ciliate on the margins of the leaf bases and onto the stem. **Leaves** mostly opposite, sessile or subsessile, oblanceolate to oblong or linear-lanceolate, 4–9 cm long, 1–2 cm wide, bluntly acute at the tip, pinnatifid and finely crenate, tapered to rounded at the base. **Flowers** sessile to subsessile in terminal and sometimes upper axillary spikes 2–10 cm long; bracts 3- to 5-lobed and crenate, the terminal lobe largest; calyx oblique, 9–12 mm long, including the 2 lateral crenate lobes which are 2–3 mm long; corolla 18–23 mm long, the upper lip entire. **Capsule** obliquely ovoid, mostly included by the calyx. July–Sept.

HABITAT Wet meadows, fens and springs, where water is fresh.

WETLAND STATUS
GP OBL | MIDW OBL

Pedicularis lanceolata
SWAMP LOUSEWORT

Parnassiaceae
grass-of-parnassus family

Parnassia GRASS-OF-PARNASSUS

Single or clumped, glabrous perennials. **Leaves** all basal except 1 sessile leaf usually present near orbelow the middle on each scape, the blades entire; petioles winged at the base by the elongate brownish stipules. **Flowers** rather showy, white, solitary on the scapes; calyx usually adnate to the ovary in the connate lower portion, the sepals acute to rounded; petals white, strongly nerved; functional stamens 5, inserted on the hypanthium opposite the sepals and alternating with 5 staminodes which are opposite the petals, staminodes dilated from the base and divided into many filamentlike segments tipped with glandular knobs, shorter than to slightly exceeding the functional stamens; pistil 4-carpellary, stigmas 4, sessile or nearly so, ovary superior to slightly inferior; hypanthium very short. **Capsule** 4-valved, containing numerous oblong, angular seeds.

1 Staminodes 3-parted; sepals with a narrow hyaline margin *P. glauca*

1 Staminodes (5)7- to many-parted; sepals herbaceous throughout 2

2 Leaf blades ovate to subrotund, broadly rounded to usually cordate at the base, not decurrent on the petiole; bract-leaf often clasping; petals 7- to 11-nerved *P. palustris*

2 Leaf blades elliptic to elliptic-ovate, tapered to and decurrent on the petiole; bract-leaf not clasping; petals usually 5(7)-nerved *P. parviflora*

Parnassia glauca Raf.
FEN GRASS-OF-PARNASSUS

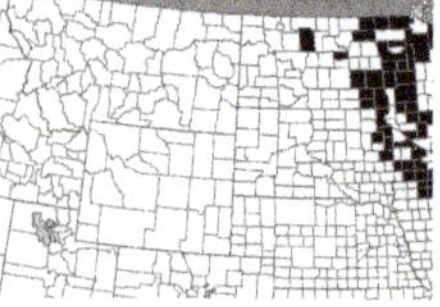

DESCRIPTION Plants with scapes 1–4 dm tall. **Leaf blades** rotund-ovate to ovate, 2–7 cm long, 1–5 cm wide, somewhat coriaceous, rounded to subacute at the tip, rounded to truncate or subcordate at the base, decurrent on the upper part of the petiole; bract-leaf usually present, clasping or not. **Sepals** oblong to oval, 2.5–5.5 mm long, herbaceous with a narrow hyaline margin, 3- to 7-nerved, rounded; **petals** oblong to oval, 9–18 mm long, 7- to 9-nerved; staminodes 3-parted from below the middle, shorter than to slightly exceeding the stamens. July–Sept.

HABITAT Calcareous fens and wet meadows.

WETLAND STATUS
GP OBL | MIDW OBL

Parnassia palustris L.
MARSH GRASS-OF-PARNASSUS

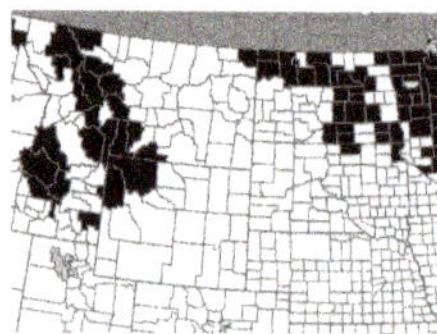

DESCRIPTION Plants with scapes 1.5–4 dm tall. **Leaf blades** ovate to subrotund, 1–3 cm long, broadly acute to obtuse or rounded at the tip, rounded to usually cordate at the base, not decurrent on the petiole; bract-leaf usually cordate-clasping on the scape. **Sepals** lanceolate to oblong-lanceolate, 4–11 mm long, herbaceous throughout, 5- to 9-nerved, acute; **petals** ovate to elliptic-ovate, 8–15 mm long, usually 1.5–2x longer than the sepals, 7- to 11-nerved; staminodes (7)9- to many-parted toward the dilated tip, 5–9 mm long. July–Sept.

HABITAT Calcareous fens, shores, streambanks and wet meadows.

WETLAND STATUS
GP OBL | MIDW OBL | WMTN OBL

Parnassia glauca
FEN GRASS-OF-PARNASSUS

Parnassia parviflora DC.
SMALL-FLOWER GRASS-OF-PARNASSUS

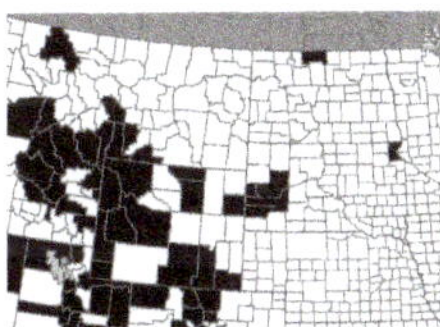

DESCRIPTION
Plants with scapes 0.5–3 dm tall. **Leaf blades** elliptic to elliptic-ovate, 0.6–3 cm long, obtuse to rounded at the tip, basally tapered to and decurrent on the petiole; bract-leaf not clasping. **Sepals** narrowly lanceolate to oblong, 3–7 mm long, herbaceous throughout, usually 5-nerved; **petals** obovate to elliptic, 4–10 mm long, mostly 1–1.5x longer than the sepals, usually 5(7)-nerved; staminodes 5–7(9)- parted at the slightly dilated tip, shorter than the stamens. July–Sept.

HABITAT Calcareous fens, wet meadows and streambanks.

WETLAND STATUS
GP OBL | MIDW OBL | WMTN OBL

Parnassia parviflora
SMALL-FLOWER GRASS-OF-PARNASSUS

Parnassia palustris
MARSH GRASS-OF-PARNASSUS

Penthoraceae
ditch-stonecrop family

Penthorum sedoides L.
DITCH STONECROP

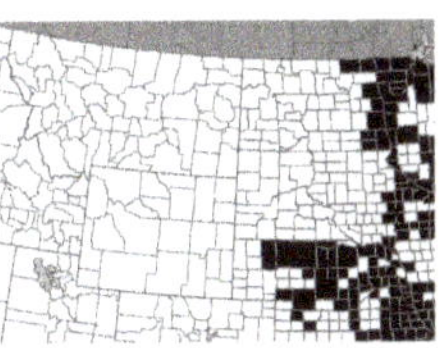

DESCRIPTION Erect, rhizomatous, perennial herb 1–6 dm tall, frequently red-tinged, glabrous below, glandular-pubescent in the inflorescence. **Leaves** simple, alternate, sessile or on petioles to 1 cm long; blades elliptic-lanceolate to oblanceolate, 2–10 cm long, 0.5–3 cm wide, acute to acuminate, rather finely serrate, cuneate at the base. **Flowers** in terminal, branched scorpioid racemes; pedicels 0.5–3 mm long. Flowers star-shaped, perfect, regular, perigynous, 3–4 mm across in flower, 5–7 mm across in fruit; **sepals** 5, green, narrowly triangular, 0.8–2 mm long, fused below to form a short, broad hypanthium; **corolla** none; stamens 10; pistils 5, simple, fused toward the base and laterally to form a ring, sharing a large, central axile placenta, each pistil maturing into an obliquely circumscissile **capsule**; **seeds** numerous, tawny to reddish-brown, ellipsoid, ca. 0.5 mm long, minutely papillate. July–Sept.

HABITAT Streambanks, shores and ditches where water is fresh.

WETLAND STATUS
GP OBL | MIDW OBL | WMTN OBL

Penthorum sedoides
DITCH STONECROP

Phrymaceae *lopseed family*

Mimulus MONKEY-FLOWER

Rhizomatous and sometimes stoloniferous perennials (or annual, in part, in *M. guttatus*) of various growth habits. **Leaves** opposite, sessile or petiolate, subentire to shallowly serrate. **Flowers** solitary on pedicels from the leaf axils or in terminal, leafy-bracteate racemes, often rather large and showy; calyx tubular, regular to irregular, the lobes shorter than the tube; corolla slightly to strongly bilabiate, yellow or blue-violet, lobes of the upper lip erect to reflexed, lobes of the lower lip spreading or deflexed, the lower lip often arched or ridged in the throat so that the orifice is mostly or completely closed; stamens 4, in 2 pairs of differing lengths; stigmas 2, distinct. **Capsule** cylindric, loculicidal.

1 Flowers blue-violet to lavender — 2
1 Flowers yellow — 3

2 Leaves sessile; pedicels longer than the calyx — ***M. ringens***
2 Leaves petiolate; pedicels shorter than the calyx — ***M. alatus***

3 Stems weak, spreading or creeping; corolla 9–15 mm long — ***M. glabratus***
3 Stems stout, erect; corolla 25–45 mm long — ***M. guttatus***

Mimulus alatus Ait. (*not illustrated*)
SHARPWING MONKEY-FLOWER

DESCRIPTION Erect to ascending, glabrous perennial 3- 7 dm tall, stoloniferous at the base; **stem** simple or branched above, 4-angled and winged. **Leaves** petiolate, reduced upward, the blades broadly lanceolate to ovate, 5–12 cm long, 2.5–4 cm wide, acute to acuminate, serrate, tapered to a narrowly winged petiole 1–2 cm long. **Flowers** blue-violet to lavender, on pedicels 2–8 mm long, to 14 mm long in fruit, shorter than the calyx; calyx regular, 11–18 mm long, the lobes narrow and subulate, 0.8–2.6 mm long; corolla strongly bilabiate, 20–28 mm long, nearly closed at the throat, the upper lip erect to reflexed, the lower lip longer and spreading. **Capsule** oblong-ovoid, 8–11 mm long. July–Sept.

HABITAT Streambanks and floodplains, often where shaded.

WETLAND STATUS
GP OBL | MIDW OBL

Mimulus glabratus H.B.K.
ROUNDLEAF MONKEY-FLOWER

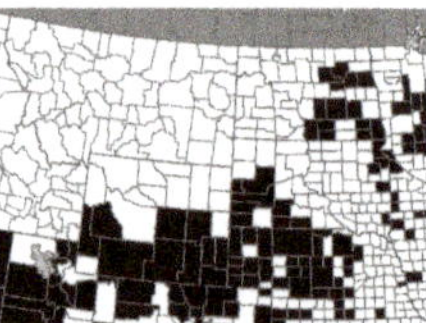

DESCRIPTION Low, branching and spreading perennial, the **stems** rather succulent, mostly prostrate and rooting at the nodes, ascending at the tips, 0.5–5 dm long, often forming extensive mats. **Leaves** sessile or the lower ones with winged petioles, the blades subrotund to rotund-ovate or reniform, 0.8–3 cm across, weakly to distinctly dentate, pubescent when young, glabrous with age. **Flowers** yellow, on pedicels 10–40 cm long; calyx accrescent, irregular, somewhat bilabiate or oblique, 5–9 mm long, the upper lobe much enlarged, the lateral and lower lobes very low or lacking; corolla obscurely bilabiate, 9–15 mm long, the throat open and bearded. **Capsule** ovoid, 5–6 mm long. July–Aug.

SYNONYM *Mimulus geyeri* Torr.

HABITAT Cold springs, seepage areas and banks of spring-fed streams.

WETLAND STATUS
GP OBL | MIDW OBL | WMTN OBL

Mimulus glabratus
ROUNDLEAF MONKEY-FLOWER

Mimulus guttatus DC.

COMMON YELLOW MONKEY-FLOWER

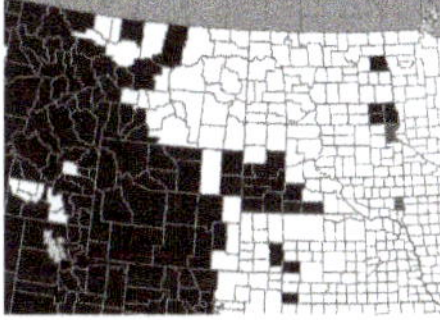

DESCRIPTION Annual or perennial 0.5–6 dm tall, with stolons or rhizomes, the main **stem** simple or branched, glabrous or glandular-puberulent above. **Leaves** petioled below to sessile and clasping above, reduced to bracts in the inflorescence, the blades variable in shape, subrotund or broadly ovate to obovate, sometimes broadly elliptic, 2–8 cm long, 1–4 cm wide, glabrous or pubescent, irregularly dentate. **Flowers** yellow, showy, in leafy-bracteate, loosely flowered, terminal racemes; pedicels 1–2.5 cm long; calyx accrescent, irregular, 10–17 mm long, the upper lobe largest, the lower lobes projected upward in fruit; corolla often spotted with reddish-brown, strongly bilabiate, 25–45 mm long, bearded at the throat. **Capsule** flattened, ovate, about equaling the calyx tube. July–Sept.

HABITAT Margins of springs and spring-fed streams, mostly in the Black Hills; becoming common westward in USA, sparingly introduced as an ornamental and escaped in e USA.

WETLAND STATUS

GP OBL | MIDW OBL | WMTN OBL

Mimulus guttatus
COMMON YELLOW MONKEY-FLOWER

Mimulus ringens L.

ALLEGHENY MONKEY-FLOWER

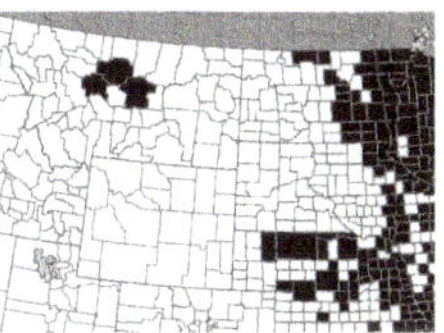

DESCRIPTION Erect or occasionally decumbent perennial 3–8 dm tall, from stout rhizomes, sometimes stoloniferous at the base, glabrous throughout; **stem** simple or branched, 4-angled, sometimes narrowly winged on the angles. **Leaves** sessile, reduced and sometimes bractlike above in the inflorescence, oblong to lanceolate or oblanceolate, 4–12 cm long, 1–3.5 cm wide, acute to acuminate at the tip, shallowly serrate, auriculate-clasping at the base. **Flowers** blue-violet to lavender, on slender pedicels 1.5–4.5 cm long, longer than the calyx; calyx regular, angular, 10–18 mm long, the lobes slender, subulate, 3–5 mm long; corolla strongly bilabiate, 20–30 mm long, nearly closed at the throat, the upper lip erect and reflexed, the lower lip longer, deflexed. **Capsule** broadly oblong, about equaling the calyx tube. June–Aug.

HABITAT Streambanks, oxbow marshes and wooded floodplains.

WETLAND STATUS

GP OBL | MIDW OBL | WMTN OBL

Mimulus ringens
ALLEGHENY MONKEY-FLOWER

Plantaginaceae *plantain family*

Annual or perennial herbs. **Leaves** simple, entire, all from base of plant. **Flowers** perfect in a narrow spike (*Plantago*), each flower subtended by bracts, or single-sexed, the staminate and pistillate flowers on same plant (*Littorella*); flower parts mostly in fours. **Fruit** a capsule opening at tip.

NOTE Plantaginaceae is the current name for the family that encompasses not only the plantains with their reduced flowers, but also the related larger-flowered genera formerly placed in the Scrophulariaceae, as well as highly reduced aquatics, such as *Hippuris* (former Hippuridaceae) and *Callitriche* (former Callitrichaceae).

1 Plants aquatic or exposed on wet shores 2
1 Plants of mostly moist to drier habitats, never truly aquatic 5

2 Leaves in a basal rosette ***Limosella***
2 Leaves whorled or opposite 3

3 Leaves in whorls of 6–12 (usually 9) ***Hippuris***
3 Leaves opposite 4

4 Plants small; flowers solitary in the leaf axils, corolla absent ***Callitriche***
4 Plants larger; flowers 1–4 in the leaf axils, corolla present ***Bacopa***

5 Flowers tiny, lacking a corolla, or corolla regular and scarious ***Plantago***
5 Flowers usually conspicuous, with both calyx and corolla, the corolla petal-like, usually bilaterally symmetrical 6

6 Corolla often nearly regular, the tube shorter than the lobes (usually a flat limb); flowers in axillary racemes or solitary in axils of bracts or leaves; sepals 4 ***Veronica***
6 Corolla 2-lipped, the tube longer than the lobes; flowers mostly solitary in axils of opposite leaves; sepals 5 7

7 Leaves pinnately lobed ***Leucospora***
7 Leaves not pinnately lobed ***Gratiola***

Bacopa rotundifolia (Michx.) Wettst.

WATER HYSSOP

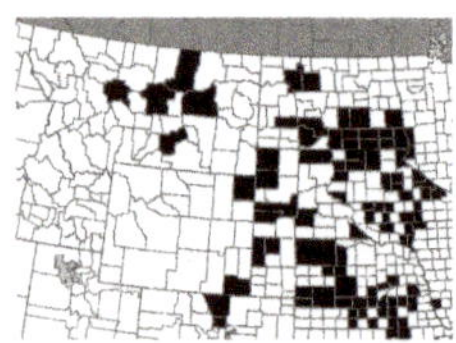

DESCRIPTION Small, amphibious perennial with **stems** 0.5–4 dm long, floating in shallow water or sprawling on mud and rooting at the nodes, emersed or ascending at the tips, pubescent on emersed portions. **Leaves** opposite, sessile, rotund-ovate to rotund-obovate, 1–3.5 cm long, 0.7–2.5 cm wide, conspicuously palmate-veined with 7–13 nerves, rounded at the apex, clasping at the base. **Flowers** axillary, 1–several per axil, on stout, pubescent, ultimately recurved pedicels 4–15 mm long; calyx of 5 unequal, distinct and imbricate sepals 4–5 mm long; corolla white with a yellow throat, nearly regular, narrowly campanulate, 6–10 mm long, the 5 lobes shorter than the tube; stamens 4. **Capsule** globose to subglobose, about equaling the calyx. July–Sept.

SYNONYM *Hydranthelium rotundifolium* (Michx.) Pennell.

HABITAT Mud flats and shallow water of ponds and marshes.

WETLAND STATUS
GP OBL | MIDW OBL | WMTN OBL

Callitriche WATER STARWORT

Rather small and delicate, perennial aquatics with weak, filiform, leafy stems and fibrous roots, seldom growing on mud. **Leaves** small, all submersed or the upper ones floating, simple, opposite; submersed leaves sessile, clasping or connate by a narrow ridge or wing, linear, 1-nerved, entire except for the shallowly bidentate apex; floating leaves,

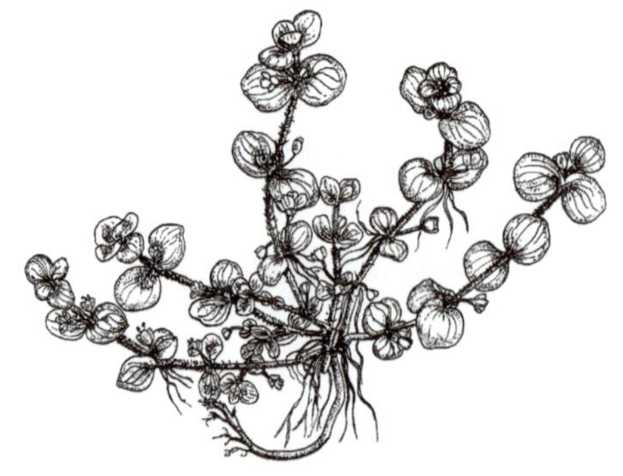

Bacopa rotundifolia
WATER HYSSOP

when present, clustered in terminal rosettes or partly in separate pairs, obovate to oblanceolate or spatulate, 3- to 5-nerved, rounded, tapered at the base to a flat petiole. **Flowers** minute, inconspicuous, sessile and solitary in middle and upper leaf axils, naked or subtended by a pair of hyaline bracts, usually imperfect (the plants polygamo-monoecious), each consisting of 1 stamen or 1 pistil or both; pistil 2-carpellary, styles 2, deciduous, ovary flattened, oval to orbicular, 4-celled, furrowed down the middle of each face and partitioned from each edge toward the middle, separating at maturity into 4 flattened, 1-seeded nutlets. Formerly placed in family Callitrichaceae.

1 Plants with linear submersed leaves only, these clasping, not connected at their bases; flowers not subtended by a pair of hyaline bracts ***C. hermaphroditica***

1 Plants often with some obovate to spatulate floating leaves; submersed leaves connected at their bases by a ridge or narrow wing; flowers subtended by a pair of hyaline bracts **2**

2 Fruit longer than broad, its segments narrowly wing-margined at least toward the summit, pitted in rows ***C. palustris***

2 Fruit about as broad as long, its segments rounded on the back, not wing-margined, irregularly pitted ***C. heterophylla***

Callitriche hermaphroditica
AUTUMN WATER-STARWORT

Callitriche hermaphroditica L.
AUTUMN WATER-STARWORT

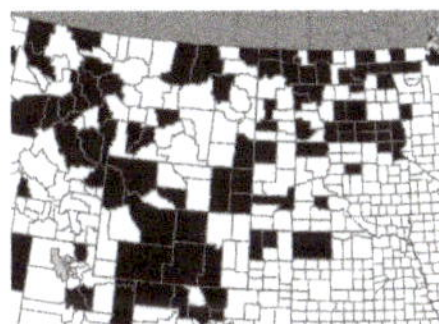

DESCRIPTION Submersed aquatic with **stems** 0.5–3 dm long. **Leaves** all submersed, linear, 3–12 mm long, 0.5–1.3 mm wide, 1-nerved, shallowly bidentate at the tip, clasping at the base, opposite leaf bases not connected. **Flowers** naked, not subtended by hyaline bracts. **Fruits** as wide or wider than high, 1–1.3 mm high, 1–1.5 mm across. June–Sept.

HABITAT Shallow to moderately deep water of lakes, ponds, marshes, ditches and sluggish streams.

WETLAND STATUS
GP OBL | MIDW OBL | WMTN OBL

Callitriche heterophylla Pursh
LARGE WATER-STARWORT

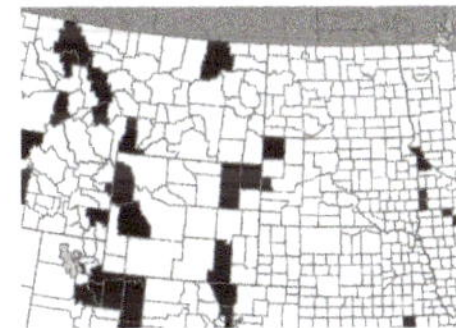

DESCRIPTION Submersed aquatic or seldom stranded on mud, with **stems** 0.5–2 dm long. **Leaves** usually of 2 basic types, with linear, submersed leaves below on the stems and obovate to spatulate floating leaves upward, the leaves seldom all of one type; submersed leaves 10–20 mm long, 0.5–1.2 mm wide, 1-nerved, bidentate at the tip, the pairs connected at their bases by a ridge or narrow wing; floating leaves in terminal rosettes or partly scattered along the stem, 6–15 mm long, 3–7 mm wide, 3- to 5-nerved, rounded at the tip, attenuate to the

Callitriche heterophylla
LARGE WATER-STARWORT

base, connected at their bases like the linear leaves, the transition from submersed to floating leaf types gradual. **Flowers** subtended by a pair of hyaline bracts, these deciduous. **Fruits** 0.6–1.2 mm long, about as wide, slightly broader toward the tip than the base, the segments convex on the face and rounded on the back, not wing-margined, irregularly pitted on the surface. June–Sept.

HABITAT Shallow water or mud of springs and stream pools.

WETLAND STATUS
GP OBL | MIDW OBL | WMTN OBL

Callitriche palustris L.
VERNAL WATER-STARWORT

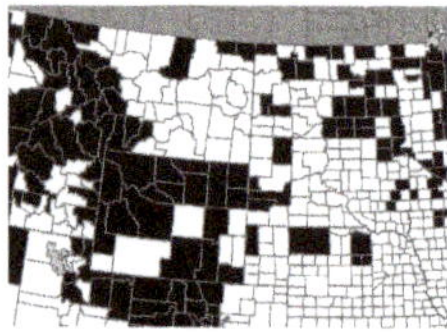

DESCRIPTION Submersed or occasionally amphibious aquatic, sometimes growing on mud, with **stems** 0.5–2 dm long. **Leaves** typically of 2 types; submersed leaves mostly linear, 5–20 mm long, 0.3–1 mm wide, shallowly bidentate at the tip, clasping at the base with a ridge or narrow, membranous wing connecting the leaf bases; floating leaves seldom absent, mostly in terminal rosettes, obovate to oblanceolate or spatulate, 3–10 mm long, 1.5–5 mm wide, 3- to 5-nerved, rounded at the tip, tapered to a flat petiole and connected by a ridge at their bases; leaves transitional between the submersed and floating leaves usually present. **Flowers** subtended by a pair of hyaline bracts, these soon deciduous. **Fruits** higher than wide, 0.8–1.7 mm long, 0.6–1.4 mm wide, narrowly wing-margined at least toward the summit, pitted in longitudinal rows. June–Sept.

Callitriche palustris
VERNAL WATER-STARWORT

SYNONYM *Callitriche verna* L.

HABITAT In the same habitats as *Callitriche hermaphroditica;* sometimes stranded and persisting on mud.

WETLAND STATUS
GP OBL | MIDW OBL | WMTN OBL

Gratiola neglecta Torr.
HEDGE HYSSOP

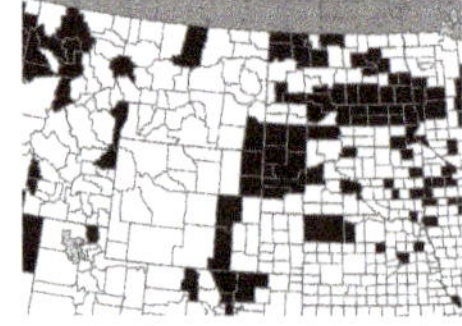

DESCRIPTION Small, erect to decumbent annual 0.3–2.5 dm tall, usually widely branched, glandular-pubescent at least in the upper part. **Leaves** opposite, sessile, variable in shape, linear to lanceolate or oblanceolate, 5–25 mm long, 1–12 mm wide, acute-tipped, entire to sinuate-toothed, clasping at the base. **Flowers** solitary in the leaf axils, subtended by a pair of narrow bractlets which equal or surpass the calyx, borne on slender, divergent, glandular-pubescent pedicels 9–20 mm long; calyx of 5 unequal, distinct sepals, these elongate, accrescent, 3–

Gratiola neglecta
HEDGE HYSSOP

6 mm long; corolla white to yellow, tubular, obscurely bilabiate, shallowly lobed, 6–10 mm long; stamens 2, a pair of very minute staminodes also sometimes present. **Capsule** broadly ovoid, 4-valved, with both septicidal and loculicidal dehiscence, 3–5 mm in diameter. July–Sept.

HABITAT Mud flats, shores of temporary ponds and marshes.

WETLAND STATUS
GP OBL | MIDW OBL | WMTN OBL

Hippuris vulgaris L.

MARE'S-TAIL

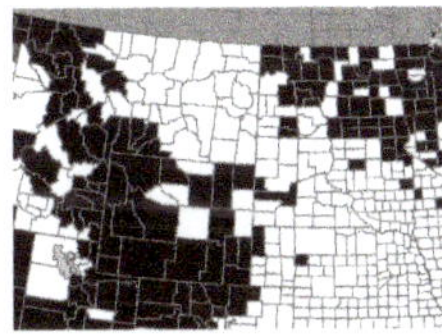

DESCRIPTION Aquatic or amphibious perennial from stout, spongy rhizomes, 1–5 dm tall above the mud or water surface; **stems** simple, erect or sometimes curved and recurved, densely clothed by the closely spaced whorls of leaves. **Leaves** in whorls of 6–12, sessile, the emersed leaves linear to linear-oblong, 0.8–2.5 cm long, 0.5–3 mm wide, blunt-tipped; submersed leaves, when present, thin and flaccid, mostly more elongate than the emersed leaves. **Flowers** perfect, minute, sessile and solitary in the upper axils, often not produced; calyx and corolla lacking; stamen 1, attached at the summit of a hypanthium; style 1, filiform, appearing alongside the stamen, ovary inferior, completely enclosed by the hypanthium, I-celled, 1-ovuled. **Fruit** nutlike, ellipsoid, 1.5–2 mm long, 0.8–1 mm thick. June–Sept.

HABITAT Shallow water or mud of marshes, lakes, streams and ditches.

WETLAND STATUS
GP OBL | MIDW OBL | WMTN OBL

Hippuris vulgaris
MARE'S-TAIL

Leucospora multifida (Michx.) Nutt.

NARROW-LEAF PALESEED

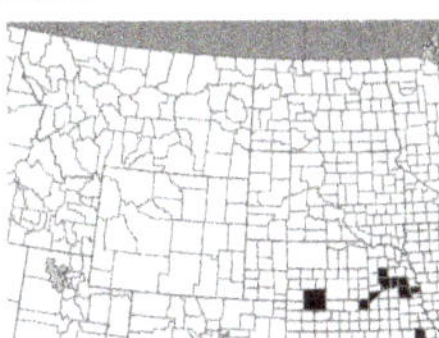

DESCRIPTION Small, taprooted annual mostly 1–2 dm tall, diffusely branched and spreading, glandular-pubescent throughout. **Leaves** opposite, pinnatifid to weakly bipinnatifid, with 1–3 pairs of narrow lateral lobes or teeth, 1–2.5 cm long. **Flowers** axillary, 1–2 per axil, on slender pedicels 3–7 mm long; calyx 5-parted nearly to the base, the lobes linear-subulate, 2.5–4 mm long in flower, to 5.5 mm long in fruit; corolla tubular, shallowly 5-lobed, pink or lavender with a greenish-yellow tube and throat, 2.5–5 mm long; stamens 4, didynamous, included; style exserted, stigma 2-lobed. **Capsule** oblong-ovoid, 4-valved, 3–5 mm long; **seeds** numerous, greenish-yellow, ovoid, 0.2–0.4 mm long, longitudinally ridged and grooved, faintly reticulate. June–Sept.

HABITAT Shores and streambanks, often where sandy.

WETLAND STATUS
GP FACW | MIDW FACW

Leucospora multifida
NARROW-LEAF PALESEED

Limosella aquatica L.

MUDWORT

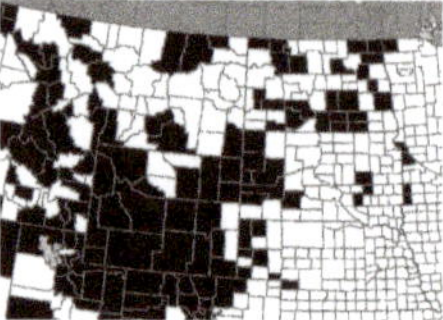

DESCRIPTION Very small amphibious annual 3–10 cm tall, with the leaves and flowers arising from the tufted base, often extensively stoloniferous. **Leaves** linear, usually dilated at the tip to form an emersed or floating blade, these elliptic to obovate or oblanceolate, 8–25 mm long, 2–8 mm wide, rounded to acute at the tip, tapered to the elongate petiole, some or rarely all the leaves linear throughout, without blades. **Flowers** small, not showy, regular, borne singly on peduncles arising from the plant base, the peduncles shorter than the leaves; calyx campanulate, 1.5–3 mm long, the lobes about equal, triangular, mostly 1/2 to equaling the length of the calyx tube; corolla minute, inconspicuous, white or pinkish, campanulate, membranous, slightly exceeding the calyx; stamens 4. **Capsule** ovoid, 2–2.5 mm long. June–Sept.

HABITAT Streambanks, shores and mud flats of temporary ponds and marshes.

WETLAND STATUS

GP OBL | MIDW OBL | WMTN OBL

Plantago PLANTAIN

Low, acaulescent, perennial herbs (those included here) with simple, petiolate leaves in basal rosettes and numerous small flowers in long, scapose, bracteate spikes. **Flowers** greenish to strarnineous, perfect, 4-merous, regular, or the calyx slightly irregular; sepals scarious-margined; corolla salverform, membranous, persistent, the tube enclosing the top of the capsule, the lobes spreading to reflexed; stamens 4, epipetalous, exserted; style strongly exserted, stigmatic for most of its length, ovary superior. **Fruit** a circumscissile capsule containing 2-many seeds.

1 Bracts subtending the flowers rounded on the back; corolla lobes 1 mm or more long; seeds 2–4 per capsule, 2–3 mm long ***P. eriopoda***

1 Bracts subtending the flowers keeled on the back; corolla lobes less than 1 mm long; seeds 6–30 per capsule, 0.8–1.5 mm long ***P. major***

Plantago eriopoda Torr.

ALKALI PLANTAIN

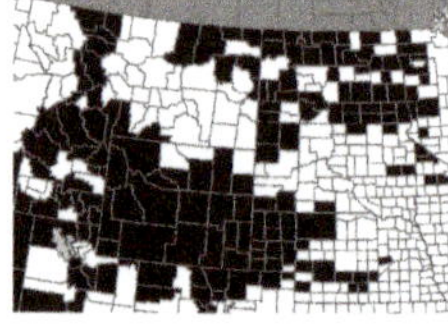

DESCRIPTION Plants with 1–several stout tap roots and scapes 1–4 dm tall. **Leaf blades** narrowly elliptic or elongate, tapered at both ends, 8–18 cm long, 0.5–3.5 cm wide, coriaceous; petioles mostly 1/2 to as long as the blade, tufted with coppery hairs at the base. **Spikes** 5–20 cm long, the axis pubescent; bracts and sepals scarious-margined, rounded on the back, 2–2.5 mm long, the sepals elliptic, concave; corolla lobes 1 mm or more long. **Capsule** ellipsoid, 3–4 mm long; seeds 2–4 per capsule, elliptic, flattened, 2–3 mm long. June–July.

HABITAT Alkaline and saline meadows, flats, ditches and streambanks.

WETLAND STATUS

GP FAC | MIDW FAC | WMTN FACW

Limosella aquatica
MUDWORT

Plantago eriopoda
ALKALI PLANTAIN

Plantago major L.
COMMON PLANTAIN

DESCRIPTION Fibrous-rooted plants with scapes 1–4 dm tall. **Leaf blades** narrowly to broadly ovate, 4–22 cm long, 3–12 cm wide, pubescent when young, remaining pubescent on the principal veins or becoming glabrous at maturity, obtuse to rounded at the tip, entire or irregularly and shallowly toothed, sometimes slightly crisped, abruptly tapered to the petiole; petioles shorter to longer than the blades. **Spikes** mostly 5–30 cm long, the scape sparsely pubescent; bracts and sepals keeled on the back, the bracts shorter than the sepals; sepals obovate, 1.5–2 mm long, widely scarious-margined; corolla lobes less than 1 mm long. **Capsule** ovoid, 2.5–3.5 mm long; seeds 6–30 per capsule, oblong-angular, 0.8–1.5 mm long. July–Sept.

HABITAT Shores, streambanks, flats and ditches, also in disturbed places, lawns and gardens; common; introduced from Europe and established as a weed throughout much of N America.

WETLAND STATUS
GP FAC | MIDW FAC | WMTN FAC

Veronica SPEEDWELL

Annuals or perennials with entire to serrate, opposite leaves, or the leaves becoming alternate in the inflorescence. **Flowers** usually numerous in axillary or terminal racemes; calyx accrescent, deeply 4-parted, the sepals often variable in shape; corolla white, pink, blue or violet, 4-lobed, slightly bilabiate, rotate, the tube much shorter than the limb; stamens 2; styles exsert, persistent in fruit, stigma capitate. **Capsule** somewhat to strongly flattened, loculicidal, often lobed or notched at the tip.

Plantago major
COMMON PLANTAIN

---

1 Racemes terminal; pedicels 2 mm or less long ... ***V. peregrina***

1 Racemes axillary; pedicels longer than 2 mm ... 2

2 Leaves short-petioled ... ***V. americana***

2 Leaves sessile, often clasping ... 3

3 Capsules strongly flattened, conspicuously notched at the tip; seeds 5–9 per locule, 1.2–1.8 mm long; leaves (3)4–20x longer than wide ... ***V. scutellata***

3 Capsules turgid, slightly notched at the tip; seeds numerous, 0.5 mm or less long; leaves 1.5–5x longer than wide ... ***V. anagallis-aquatica***

Veronica americana (Raf.) Schwein. ex Benth.
BROOKLIME

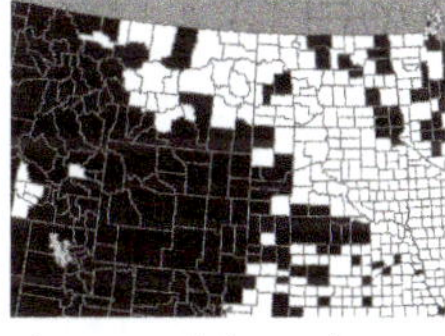

DESCRIPTION Glabrous, rhizomatous perennial from shallow rhizomes; stems erect to procumbent, 1–6 dm long. **Leaves** opposite, short-petiolate, the blades ovate-lanceolate to lanceolate, or the lower ones elliptic, 1.5–8 cm long, 0.6–3 cm wide, bluntly acute at the tip or the lower ones rounded, subentire to serrate, rounded to subcordate at the base. **Racemes** axillary, mostly 10- to 25- flowered; pedicels divaricate, 4–10 mm long; corolla blue, 5–10 mm across. **Capsule** suborbicular, rounded to

Veronica americana
BROOKLIME

slightly notched at the apex, 2.5–4 mm long, about as wide or slightly wider, the style 2.5–3.5 mm long; seeds numerous, ca. 0.5 mm long. June–Aug.

HABITAT Springs and fresh streams.

WETLAND STATUS
GP OBL | MIDW OBL | WMTN OBL

Veronica anagallis-aquatica L.
WATER SPEEDWELL

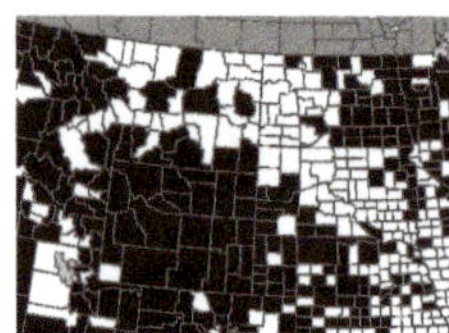

DESCRIPTION Erect to spreading, fibrous-rooted biennial or short-lived perennial 1–5 dm tall, glabrous or nearly so. **Leaves** sessile, opposite, elliptic to elliptic-obovate or elliptic-oblong, 2–10 cm long, 0.6–5 cm wide, 1.5–3x longer than wide, bluntly acute to obtuse at the tip, entire to serrate, mostly clasping. **Racemes** axillary, many-flowered; pedicels ascending or upcurved, 3–8 mm long; corolla light blue, violet or white, ca. 5 mm across. **Capsule** orbicular, turgid, rounded or scarcely notched at the tip, 2.5–4 mm long, about as wide, the style 1.5–2.5 mm long; seeds numerous, 0.5 mm or less long. June–Sept.

SYNONYM Includes *Veronica catenata* Pennell

HABITAT Stream margins and low areas on floodplains, often emergent in shallow water; introduced from Eurasia and naturalized throughout most of N America.

WETLAND STATUS
GP OBL | MIDW OBL | WMTN OBL

Veronica anagallis-aquatica
WATER SPEEDWELL

Veronica peregrina L.
PURSLANE SPEEDWELL

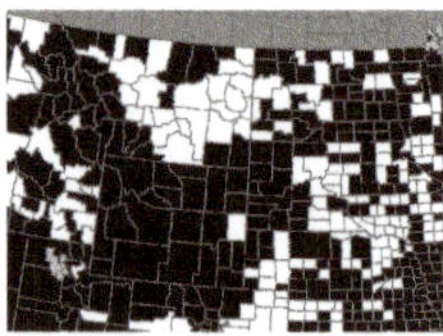

DESCRIPTION Small, simple to diffusely branched annual 0.5–3 dm tall, erect or curved at the base, conspicuously glandular-pubescent, especially above. **Leaves** sessile, opposite near the base, becoming alternate in the inflorescence, oblong to linear-oblong or oblanceolate, 5–25 mm long, 1–5(9) mm wide, rounded at the tip, entire to irregularly toothed. **Flowers** subsessile in terminal, lax racemes which comprise most of the plant, the leaflike bracts gradually reduced upward; pedicels to 1–2 mm long in fruit; calyx 2–5 mm long; corolla inconspicuous, white or whitish, ca. 2 mm across. **Capsule** obcordate, notched at the tip, 2.5–4 mm long, 3.5–5 mm wide. June–Aug.

HABITAT Mud flats, shores, ditches, temporary ponds and swales.

WETLAND STATUS
GP FACW | MIDW FACW | WMTN OBL

Veronica peregrina
PURSLANE SPEEDWELL

Veronica scutellata L.
MARSH SPEEDWELL

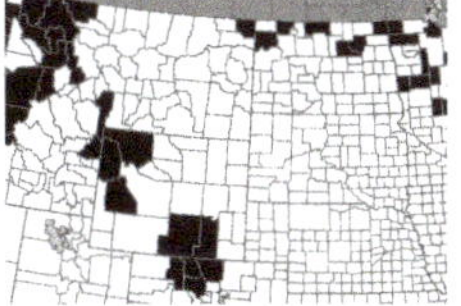

DESCRIPTION Erect to reclining, rhizomatous perennial, glabrous throughout (in our region), the **stems** rather weak, 1–4 dm long. **Leaves** sessile, opposite, linear to lanceolate, 3–8 cm long, 2–18 mm wide, (3)4–20x longer than wide, acute to attenuate, entire or obscurely and remotely denticulate, tapered to the sessile base. **Racemes** axillary, mostly 5- to 20-flowered; pedicels divaricate to recurved, 5–17 mm long; calyx 1.5–2 mm long; corolla bluish, 6–10 mm across. **Capsule** flattened, conspicuously notched, appearing 2-lobed, 2.5–4 mm long, 4–5 mm wide. June–Aug.

HABITAT Marshes, springs, streambanks and swales, where water is fresh.

WETLAND STATUS
GP OBL | MIDW OBL | WMTN OBL

Veronica scutellata
MARSH SPEEDWELL

Polygonaceae *smartweed family*

Annual and perennial herbs with small flowers often aggregated in conspicuous inflorescences. **Leaves** simple, alternate, the blades sometimes undulate-crisped along the margin, otherwise entire; stipules united to form a membranous or papery sheath (the ocrea) around the stem at each node. **Flowers** borne in spikelike racemes or in small axillary clusters (*Persicaria*), or in dense, terminal panicles of racemes in which the flowers are densely whorled on the branches (*Rumex*), in the latter case, the inflorescence often comprising much of the upper part of the plant. Flowers perfect (in those included here), regular; perianth consisting of a calyx only; in *Persicaria* the sepals petaloid or at least so on the margins, whitish to pink or sometimes yellowish, in a cycle of (4)5(6); in *Rumex* the sepals herbaceous, green to brown, in 2 cycles of 3, the 3 inner sepals accrescent, becoming broadly winged and appressed to each other, persisting to enclose the achene; stamens 4–8; pistil 1, 2- or 3-carpellary, styles 2–3, ovary superior, 1-celled. **Fruit** a trigonous or lenticular achene.

---

1 Flowers in spikelike racemes or in groups of 1-few in the axils of ordinary leaves, whitish to pink or greenish, the calyx petaloid at least on the margins ***Persicaria***

1 Flowers in dense terminal panicles of racemes in which flowers are in dense whorls, green to brown, the calyx herbaceous ***Rumex***

Persicaria SMARTWEED, KNOTWEED, TEAR-THUMB

Erect to sprawling, often weedy annuals and perennials, the **stems** often swollen at the nodes. **Leaves** sessile to short-petioled, the blades generally lanceolate to oblanceolate or elliptic (sagittate in *P. sagittata*); stipules fused to form a tubular or 2-lobed ocrea which sheaths the stem above each node, the ocreae (pl.) membranous or chartaceous, often lacerate or completely broken up at older nodes, sometimes fringed with bristles. **Flowers** small, greenish, whitish or pinkish, usually borne in spikelike racemes, or (in *Polygonum ramosissimum*) the flowers in axillary positions, with 1–few flowers in the axils of ordinary leaves; in those species with flowers in racemes, the racemes terminal or both terminal and axillary, loosely to densely flowered, the flowers fascicled in the axils of small bracts (ocreolae), the pedicels included by or barely surpassing the ocreolae. Sepals usually 5(4–6), petaloid throughout or at least around the margins, greenish-white to deep pink (green with yellowish or pinkish margins in *Polygonum ramosissimum*), somewhat accrescent, united near or well below the middle; stamens 8 or fewer; styles 2–3. **Achenes** brown to black, lenticular or trigonous, sometimes both in the same inflorescence, apiculate, surrounded by the persistent sepals.

NOTE *Polygonum ramosissimum* is retained in the following key and descriptions; our remaining species, formerly within genus *Polygonum,* are now placed in *Persicaria.*

---

1 Leaves sagittate; stems armed with reflexed prickles **P. sagittata**

1 Leaves generally lanceolate to oblanceolate or elliptic; stems unarmed 2

2 Flowers strictly axillary, with 1-few in the upper ails; leaves jointed at the base **Polygonum ramosissimum**

2 Flowers all or mostly in spikelike racemes; leaves not jointed at the base 3

3 Racemes strictly terminal, solitary or paired (rarely 3 or 4) at the tips of stems; plants perennial; flowers rose pink **P. amphibia**

3 Racemes terminal and axillary, often numerous on the stems; plants annual (except *P. hydropiperoides*); flowers white, greenish-white or pink 4

4 Ocreae entire or lacerate 5

4 Ocreae fringed with bristles 7

5 Outer sepals strongly 3-nerved, each nerve ending in an anchor-shaped fork; racemes nodding to erect **P. lapathifolia**

5 Outer sepals with faint, irregularly forked nerves; racemes erect 6

6 Stamens and styles unequal, either the stamens or styles exserted conspicuously beyond the perianth **P. bicornis**

6 Stamens and styles equal or nearly so, both included in or equaling the perianth **P. pensylvanica**

7 Calyx conspicuously glandular-punctate on the outside 8

7 Calyx not glandular-punctate, or inconspicuously glandular on the inner sepals only 9

8 Ocreae gibbous, concealing cleistogamous flowers; achenes brown to dark brown, dull **P. hydropiper**

8 Ocreae cylindrical; cleistogamous flowers none; achenes black, smooth and shiny **P. punctata**

9 Plants perennial; racemes slender, often interrupted, commonly more than 3 cm long; achenes all trigonous **P. hydropiperoides**

9 Plants annual; racemes rather thick and dense, mostly continuous, rarely to 3 cm long; achenes mostly lenticular, some often trigonous **P. persicaria**

Persicaria amphibia (L.) S.F. Gray p.p.
WATER SMARTWEED

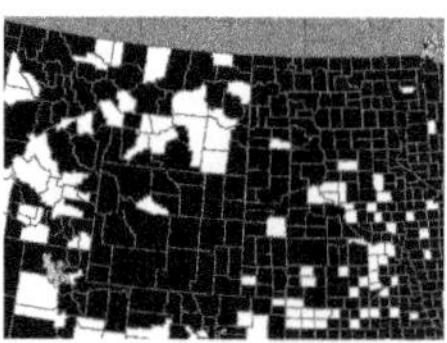

DESCRIPTION Aquatic, amphibious or terrestrial perennial, quite variable depending on the water regime, glabrous when aquatic, commonly with spreading to appressed pubescence when terrestrial, more often flowering when in water. **Stems** trailing through water or over mud, usually branched, with the tips erect and emersed, to 1 m long, freely rooting at the nodes, or in drier situations, the stems erect from an underground portion, usually simple, to 8 dm tall. **Leaves** of submersed plants commonly floating, with leathery, glabrous blades on petioles 1–8 cm long; leaves of emersed plants often pubescent, sessile or with petioles to 5 mm long; leaf blades elliptic, elliptic-oblong or elliptic-lanceolate, 4–12(18) cm long, 1–4 cm wide, acute to obtuse or rounded at the tip, subcordate to cuneate at the base; ocreae membranous, 0.5–2 cm long, truncate at the summit or often with a spreading herbaceous margin on emersed plants, glabrous to strigose-hispid. **Racemes** terminal, 1–2, rarely 3 or 4, globose to short-cylindric, 1–3.5(5) cm long, 1–1.5(2) cm thick; peduncles 1.5–6 cm long, glabrous or pubescent, seldom glandular. **Flowers** rose pink; calyx 5-lobed to below the middle, 4–6 mm long; stamens 5, included or exserted; style , branches 2. **Achenes** brown to nearly black, lenticular, 1.9–2.5 mm long. July–Sept.

SYNONYMS *Polygonum amphibium* L., *Polygonum coccineum* Muhl. ex Willd., *Polygonum natans* Eat.

HABITAT Meadows, marshes, springs, fens, streams, ponds and lakes, usually where water is fresh; occasional.

WETLAND STATUS
GP OBL | MIDW OBL | WMTN OBL

NOTE Some authors distinguish *P. coccinea* from *P. amphibia;* in our region, *P. coccinea* is a facultative wetland plant, often spreading into upland habitats adjacent to wetlands, while *P. amphibia* is an obligate wetland plant.

Persicaria bicornis (Raf.) Nieuwl.
PINK SMARTWEED

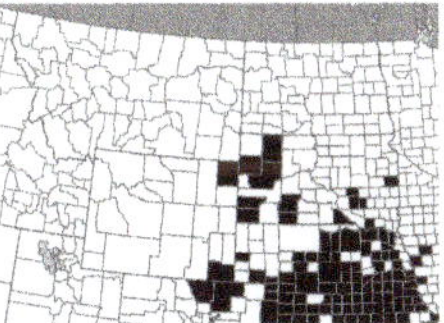

DESCRIPTION Very similar to *Persicaria pensylvanica;* differing in having heterostylic flowers which are consistently pink; **racemes** appearing fringed due to the extruded stamens and styles, some flowers with the stamens exserted from the perianth and the styles included, others with the stamens included and the styles exserted; calyx 3–4 mm long, conspicuously exceeding the achene. **Achenes** lenticular, concave on both sides and usually with a slight elevation in the center of the concavity. July–Sept.

SYNONYMS
Polygonum bicorne Raf., *Polygonum longistylum* Small.

HABITAT Shores, streambanks, ditches and other wet, disturbed habitats.

WETLAND STATUS
GP FACW | MIDW FACW | WMTN FACW

Persicaria hydropiper (L.) Delarbre
WATER PEPPER

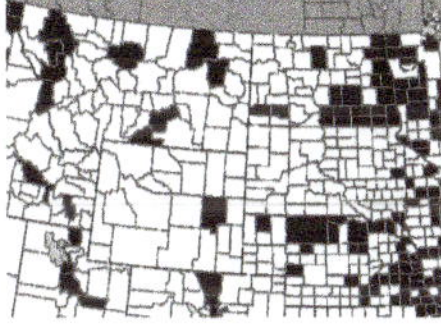

DESCRIPTION Erect to decumbent, simple to freely branched annual 2–8 m tall; **stems** often reddish, sometimes rooting at the lower nodes, foliage and flowers peppery to the taste. **Leaf blades** lanceolate, 3–8 cm long, 0.5–2 cm wide, glabrous except for the strigulose margins, narrowed to a blunt or rounded tip, cuneate at the base, subsessile or short-petioled; ocreae membranous, 5–15 mm long, fringed with bristles, those on the upper half of the stem gibbous, concealing

Persicaria amphibia
WATER SMARTWEED

Persicaria bicornis
PINK SMARTWEED

cleistogamous flowers which are like those in the racemes. Racemes often many, loosely flowered, elongate, interrupted in the lower portion, 2–9 cm long. **Flowers** greenish-white to greenish-pink; calyx 2.5–4 mm long, strongly glandular-punctate, usually 4-parted to slightly below the middle; stamens usually 4 or 6, included by the calyx; style branches 2 or 3. **Achenes** brown to dark brown, dull, lenticular or trigonous, 2–3 mm long. Late July–Sept.

SYNONYM *Polygonum hydropiper* L.

HABITAT Shores and streambanks; introduced from Europe.

WETLAND STATUS
GP OBL | MIDW OBL | WMTN OBL

Persicaria hydropiperoides (Michx.) Small
MILD WATER PEPPER

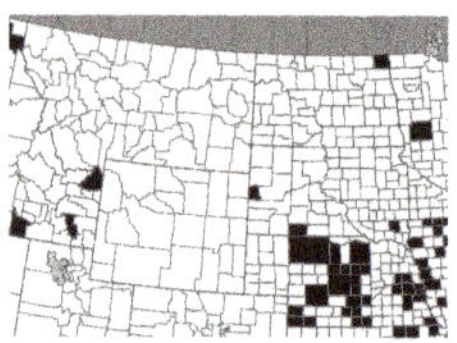

DESCRIPTION Erect to decumbent, usually branched perennial, nearly glabrous to strigose throughout; **stems** commonly rooting at the lower nodes when decumbent, to 1 m long. **Leaf blades** linear-lanceolate to lanceolate, 4–12 cm long, 0.3–2.5 cm wide, nearly glabrous to strigose, especially pubescent on the midvein and margins, gradually tapering to a narrow, blunt to acute tip, cuneate at the base, short-petioled to subsessile; ocreae membranous, 5–15 mm long, usually strigose, fringed with bristles. **Racemes** 2-several, slender, often interrupted in the lower portion, 1–6 cm long. **Flowers** greenish to white or pinkish; calyx 2–3 mm long, 5-parted to just below the middle, eglandular or only the inner sepals slightly glandular; stamens 8, included; style branches 3. **Achenes** black, shiny, trigonous with concave sides, 2–3 mm long. Aug–Sept.

SYNONYM *Polygonum hydropiperoides* Michx.

HABITAT Wet soil or mud of meadows, marshes, ditches, streambanks and shores.

WETLAND STATUS
GP OBL | MIDW OBL | WMTN OBL

Persicaria lapathifolia (L.) S.fac. Gray
PALE SMARTWEED

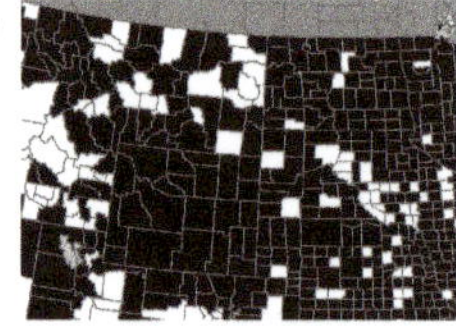

DESCRIPTION Erect to decumbent, simple to widely branching annual 2–10 dm tall. **Leaf blades** lanceolate or elliptic-lanceolate, 4–1 7(20) cm long, 0.5–3.5 (5) cm wide, glabrous above, glabrous to minutely strigose and often glandular-punctate beneath, occasionally finely white-tomentose beneath, acute or acuminate to the narrow, blunt tip, cuneate at the base, on short petioles to 2 cm long; ocreae 5–20 mm long, entire to lacerate, glabrous, not fringed with bristles.

Persicaria hydropiper
WATER PEPPER

Persicaria hydropiperoides
MILD WATER PEPPER

Persicaria lapathifolia
PALE SMARTWEED

Racemes usually numerous, densely flowered, cylindric, erect or often nodding, 0.8–5 cm long, 0.5–1 cm thick. **Flowers** greenish-white or white to deep pink; calyx 2.5–3.5 mm long, 4- or 5-lobed to below the middle, the outer 2 sepals strongly 3-nerved, each nerve ending in an anchor-shaped fork; stamens usually 6, included; style branches 2. **Achenes** brown to dark brown, lenticular, concave on the sides, ca. 2 mm long, including the style beak. July–Sept.

SYNONYM *Polygonum lapathifolium* L.

HABITAT Marshes, wet meadows, ditches, shores, streambanks and other places where water stands temporarily; common

NOTE The spikelike racemes are usually erect and shorter on smaller plants. The flowers can be strongly pink.

WETLAND STATUS

GP OBL | MIDW FACW | WMTN FACW

Persicaria maculosa S.F. Gray

LADY'S THUMB

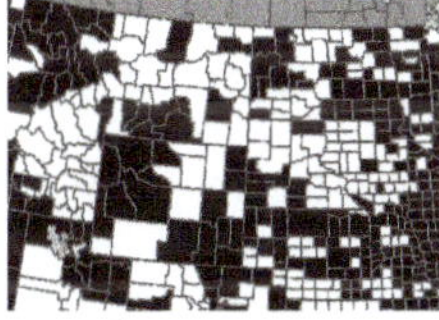

DESCRIPTION Erect to spreading, simple to branched annual 2–8 dm tall, the **stems** often reddish. **Leaf blades** lanceolate to elliptic-lanceolate, 3–15 cm long, 0.5–3 cm wide, glabrous to sparsely strigose, usually puncticulate, especially beneath, acute to long-acuminate, cuneate at the base, subsessile or on petioles to 1 cm long; ocreae 5–15 mm long, fringed with bristles, glabrous to strigose. **Racemes** usually few to many, globose to cylindric, 0.5–2.5 cm long, 0.5–1 cm thick. **Flowers** light to deep pink; calyx 2.5–3 mm long, 5-lobed to near the middle; stamens 6, included; style branches 2 or 3. **Achenes** black, shiny, lenticular or some often trigonous, 2–2.5 mm long. July–Sept.

SYNONYM *Polygonum persicaria* L.

HABITAT Springs, shores, streambanks and ditches, where water is fresh; introduced from Europe and now throughout N America.

WETLAND STATUS

GP FACW | MIDW FACW | WMTN FACW

Persicaria maculosa
LADY'S THUMB

Persicaria pensylvanica (L.) M. Gómez

PENNSYLVANIA SMARTWEED

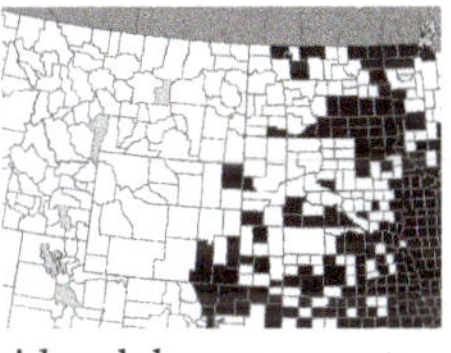

DESCRIPTION Erect, simple to widely branching annual 3–12 dm tall. **Leaf blades** ovate-lanceolate to lanceolate, 3–15 cm long, 0.7–4 cm wide, glabrous except for the minutely strigose margin, sometimes puncticulate on the lower surface, acute or acuminate to the narrow, blunt tip, cuneate at the base; petioles short to 2.5 cm long; ocreae 0.5–1.5 cm long, entire to lacerate, glabrous, not fringed with bristles. **Racemes** usually few to many, dense, cylindric, erect, 1–3 cm long, 0.7–1 cm thick; peduncles glan-

Persicaria pensylvanica
PENNSYLVANIA SMARTWEED

dular-pubescent. **Flowers** pink to nearly white; calyx 3–4.5 mm long, 5-lobed to well below the middle, the outer sepals with faint, irregularly forked nerves; stamens 8 or fewer, included; style branches 2. **Achenes** dark brown to black, shiny, lenticular, concave on the sides, 2.5–3 mm long, including the beak. July–Aug.

SYNONYM *Polygonum pensylvanicum* L.

HABITAT Streambanks, shores and ditches; occasional to common.

WETLAND STATUS
GP FACW | MIDW FACW | WMTN FACW

Persicaria punctata (Ell.) Small
WATER SMARTWEED

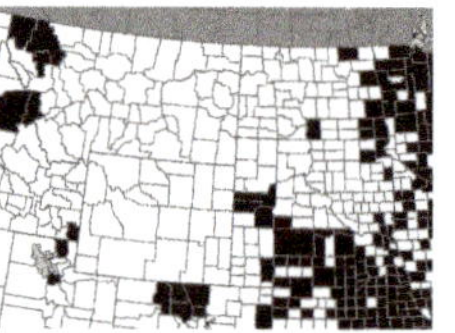

DESCRIPTION Simple to branched, erect to spreading annual 3–10 dm tall. **Leaf blades** lanceolate, 4–10 cm long, 1–2 cm wide, glabrous except for the minutely strigose margin, usually puncticulate, especially beneath, acute or acuminate to the blunt tip, cuneate to the short-petioled base; ocreae 5–15 mm long, fringed with bristles, glabrous to strigose. **Racemes** usually many, slender, elongate, loosely flowered, interrupted in the lower portion, to 10 cm long. **Flowers** greenish-white; calyx 3–4 mm long, strongly glandular-punctate, 5-parted to about the middle; stamens 6–8, included; style branches 2–3. **Achenes** dark brown to black, shiny, lenticular or trigonous, 2–3 mm long. Aug–Sept.

SYNONYM *Polygonum punctatum* Ell.

HABITAT Shores and streambanks.

WETLAND STATUS
GP OBL | MIDW OBL | WMTN OBL

Persicaria punctata
WATER SMARTWEED

Persicaria sagittata (L.) Gross
TEAR-THUMB, ARROW-VINE

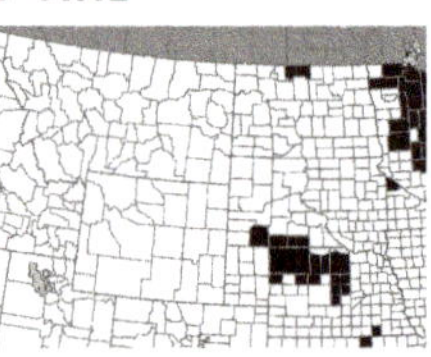

DESCRIPTION Slender annual; **stems** weak, usually reclining on surrounding vegetation, to 2 m long, strongly 4-angled, armed with reflexed prickles on the angles, the prickles usually continuous onto the petioles, leaf midribs and peduncles. **Leaves** mostly long-petioled below, becoming short-petioled upward on the stem, the blades lanceolate to elliptic and sagittate, 3–10 cm long, 0.8–2.5 cm wide, acute, the basal lobes straight or incurved; ocreae 5–10 mm long, sparsely ciliate. **Racemes** terminal and axillary on slender peduncles, globose, 5–10 mm long. **Flowers** whitish or pinkish; calyx 2.5–3 mm long, 5-parted to below the middle; stamens 8, included; style branches 3. **Achenes** brown to black, shiny, trigonous, 2–3 mm long. July–Sept.

SYNONYM *Polygonum sagittatum* L.

HABITAT Swamps, springs and wet meadows.

WETLAND STATUS
GP OBL | MIDW OBL | WMTN OBL

Persicaria sagittata
TEAR-THUMB, ARROW-VINE

Polygonum ramosissimum Michx.

BUSHY KNOTWEED

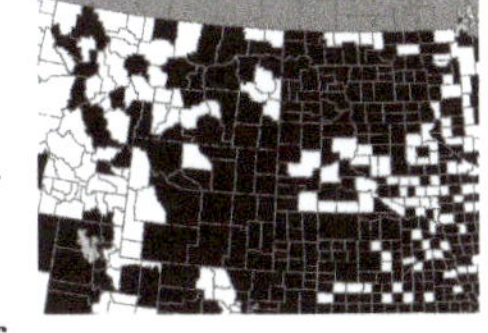

DESCRIPTION Erect, sparingly to usually freely branched, taprooted annual 2–9 dm tall, the **stems** strongly ribbed. **Leaf blades** elliptic to mostly narrowly elliptic or nearly linear, 8–40 mm long, 2–15 mm wide, acute to obtuse, often revolute, cuneate and jointed at the subsessile or short-petioled base; ocreae membranous with 2 acute lobes, soon lacerate and breaking into brownish fibers. **Flowers** in axillary positions near the tips of branches, 1-few per axil; calyx 5(6)-parted, 3–4.5 mm long, the lobes united for about 1/3 of their length or less, green with yellowish, whitish or slightly pinkish margins, the outer 3 cucullate and enclosing the inner lobes; stamens usually 5, included; style branches 3. **Achenes** dark brown, dull to shiny, sharply trigonous, mostly 3–3.5 mm long and enclosed by the calyx, those produced later in the season often larger and distended beyond the calyx; pedicels mostly enclosed by the ocreae, 1–3 mm long. July–Sept.

SYNONYM *Polygonum prolificum* (Small) Robins.

Polygonum ramosissimum
BUSHY KNOTWEED

HABITAT

Shores and exposed flats, especially where alkaline.

WETLAND STATUS

GP FACW | MIDW FACU | WMTN FAC

NOTE Other *Polygonum* spp. of the knotweed group are sometimes found in the same habitats as *P. ramosissimum,* but are more characteristic of disturbed upland habitats. Most typical of this group is ***Polygonum aviculare*** L. which differs from *P. ramosissimum* in its spreading to erect habit; calyx lobes pink-margined, more or less flat and not cucullate at the tips; achene unequally trigonous, with 2 convex sides and 1 narrower concave side. Another is ***Polygonum erectum*** L., a spreading plant with broadly elliptic leaves; calyx united near the middle and constricted above to appear bottle-shaped.

Rumex DOCK

Erect to spreading, often weedy perennials (annual in *R. fueginus*), usually from stout, often few-branched taproots, often tall in stature, some with a basal clump of large leaves, the **stems** often tinged with purple. **Leaves** generally oblong to lanceolate, cuneate to rounded, truncate or cordate at the base, flat to undulate-crisped along the margins, usually petioled, the lowest leaves often long-petioled; ocreae usually membranous, brittle when dry. **Inflorescence** consisting of 1 or more, often large, dense, terminal, leafy-bracteate panicles of racemose branches in which the flowers are crowded in whorls, the entire inflorescence often comprising much of the upper part of the plant. **Flowers** small and numerous, green, eventually turning brown; calyx herbaceous, the sepals in 2 cycles of 3, the outer 3 sepals remaining small after anthesis, the inner 3 sepals (the valves) expanding, becoming winged, the margins veiny, entire to sinuate, dentate or spinulose-toothed, the valves loosely coherent around the achene with the margins appressed to each other, giving the appearance of a 3-winged fruit, the midvein of the valve often swollen to produce a grain-like tubercle on the back; stamens 6; styles 3. **Achenes** brown, sharply trigonous, apiculate, enclosed by the persistent calyx.

1 Margins of mature valves coarsely to spinulose-toothed 2
1 Margins of mature valves entire to sinuate or shallowly erose, not toothed 3

2 Plants perennial from a stout taproot; margins of the valves coarsely toothed ***R. stenophyllus***
2 Plants annual, fibrous-rooted or with a slender taproot; margins of the valves deeply dissected into spinulose teeth ***R. fueginus***

3 Valves, or at least one of them, with a grainlike tubercle on the back 4
3 Valves lacking grainlike tubercles 9

4 Pedicels without a visibly swollen joint; tubercles 3, the base of the tubercle distinctly above the base of the valve ***R. britannica***
4 Pedicels with a visibly swollen joint below the middle or near the base; tubercles 1–3, the base of the tubercle even with the base of the valve 5

5 Plant branched and spreading, leafy throughout, with clusters of leaves or short branches in the axils of the main cauline leaves, the leaves not greatly reduced upward on the stem; blades flat to slightly undulate 6
5 Plant usually simple or seldom few-stemmed, the stem(s) erect, unbranched below the inflorescence, with large basal leaves and smaller leaves upward on the stem; blades undulate to undulate-crisped; valves ovate-deltate to rotund-ovate or reniform, rounded to cordate at the base 7

6 Principal leaves ovate-lanceolate to elliptic-lanceolate, mostly not more than 4x longer than broad; tubercle usually well-developed on only 1 of the 3 valves ***R. altissimus***
6 Principal leaves oblong-lanceolate to linear-lanceolate, mostly more than 4x longer than broad; tubercles well developed on 1–3 of the valves ***R. triangulivalvis***

7 Tubercles 1–3, well-developed, the largest at least 1/2 as long as the valve ***R. crispus***
7 Tubercle 1, poorly developed, the largest 1/3 or less as long as the valve 8

8 Valves 3.5–5 mm long ***R. pseudonatronatus***
8 Valves (5)6–9 mm long ***R. patientia***

9 Pedicels lacking a visibly swollen joint; principal leaf blades truncate to cordate at the base ***R. occidentalis***
9 Pedicels with a visibly swollen joint near the base; principal leaf blades acute to rounded at the base 10

10 Plant usually simple or seldom few-stemmed, the stem(s) erect, unbranched below the inflorescence, with large basal leaves and smaller leaves upward on the stem, the blades undulate-crisped ***R. pseudonatronatus***
10 Plant branched and spreading, leafy throughout, with clusters of leaves or short branches in the axils, the leaves not greatly reduced upward on the stem, flat to slightly undulate 11

11 Principal leaves ovate-lanceolate to elliptic-lanceolate, mostly not more than 4x longer than wide ***R. altissimus***
11 Principal leaves oblong-lanceolate to linear-lanceolate, mostly more than 4x longer than wide ***R. triangulivalvis***

Rumex altissimus Wood

PALE DOCK

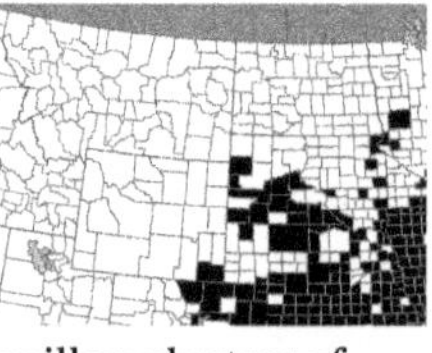

DESCRIPTION Quite similar to *R. triangulvalvis,* 3–10 dm tall, usually branched from the base and with short branches or axillary clusters of leaves above. **Leaf blades** ovate-lanceolate to elliptic-lanceolate, 6–18 cm long, 1.5–5.5 cm wide, mostly not more than 4x longer than wide, flat to slightly undulate-crisped along the margin, acute to acuminate at the tip, cuneate to obtuse or rounded at the base. **Panicles** 1–3 dm long, the branches erect to ascending, rather few; pedicels shorter than to about equaling the length of the valves, swollen-jointed near the base. Valves triangular to rotund-ovate, 4–5(6) mm long, about as wide, bluntly acute or obtuse at the tip, entire or erose-margined, truncate to subcordate at the base; tubercle well-developed usually on only 1 of the 3 valves, although sometimes present on none or all 3 of the valves, mostly 1/2–2/3 the length of the valve; **achenes** brown, 2.3–3 mm long. June–Aug.

HABITAT Wet meadows, marshes, shores, streambanks, ditches and other wet places.

WETLAND STATUS

GP FAC | MIDW FACW | WMTN FAC

Rumex britannica L. (*not illustrated*)

GREAT WATER DOCK

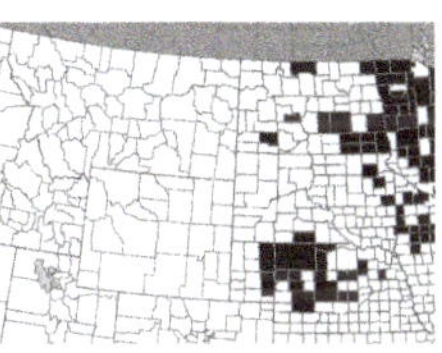

DESCRIPTION Very similar to *R. occidentalis,* differing as follows: basal **leaf blades** oblong-lanceolate, flat, acute to rounded at the base. Pedicels obscurely jointed below the middle, the joint not swollen; valves orbicular to rotund-ovate, truncate at the base, 5–8 mm long and about as wide, entire to denticulate; tubercles 3, ca. 1/2 the length of the valve, the base of the tubercle distinctly above the base of the valve.

HABITAT Springs, fens and fresh marshes; uncommon and scattered.

WETLAND STATUS

GP OBL | MIDW OBL | WMTN OBL

Rumex crispus L.

SOUR DOCK, CURLED DOCK

DESCRIPTION Stout, erect, usually single-stemmed perennial mostly 5–10 dm tall, from a thick taproot. **Leaves** basal and cauline, the basal leaves large and long-petioled, often drying early, the stem leaves decreasing in size and shorter-petioled upward: blades elliptic to oblong-lanceolate, those of the basal leaves 10–30 cm long, 1–5 cm wide, conspicuously undulate-crisped, cuneate to rounded at the base. **Panicle** large, the branches erect to ascending, intermixed with some linear leaves; pedicels mostly 1.5–2x the length of the valves, swollen-jointed below the middle. Valves cordate-deltate to ovate-deltate, mostly 3–5 mm long and about as wide, entire to scarcely erose; tubercles usually 3 (or absent from 1 or 2 of the valves), often unequal, the largest 1/2 or more the length of the valve; **achenes** brown, 2–2.5 mm long. July–Sept.

Rumex altissimus
PALE DOCK

Rumex crispus
SOUR DOCK, CURLED DOCK

HABITAT Wet meadows, shores, ditches and other low areas; common, weedy; introduced from Eurasia, now naturalized throughout the USA and s Canada and most of the world.

WETLAND STATUS
GP FAC | MIDW FAC | WMTN FAC

NOTE Other species are often confused with *Rumex crispus*. Correct identification requires close inspection of the winglike valves surrounding the achene.

Rumex fueginus Phil.
GOLDEN DOCK

DESCRIPTION Stout, erect annual, freely branched upward and often from the base, 0.5–7 dm tall. **Leaves** mostly cauline, reduced in size upward, the blades oblong-lanceolate to oblong-linear, 5–20 cm long, 0.5–4 cm wide, the margins flat to sometimes undulate, cuneate to truncate or sometimes cordate at the base; petiole to as long as the blades on the lower leaves, shorter upward. **Panicle** rather open, the branches spreading-ascending, leafy, puberulent or papillose, the verticils of flowers conspicuously separated, especially in the lower part; pedicels jointed near the base, slightly to greatly exceeding the valves in length. Valves ovate, acuminate-tipped, 2–3 mm long, the margins deeply incised into (1)2–4 spinulose teeth which are 1–3(6) mm long; tubercles 3, linear-lanceolate, running about 1/2 the length to the acuminate tip of the valve; **achenes** light brown, 1–1.5 mm long. July–Sept.

SYNONYM *Rumex maritimus* L.

HABITAT Marshes, shores, streambanks and ditches, where water is fresh to brackish.

WETLAND STATUS
GP FACW | MIDW FACW | WMTN FACW

Rumex fueginus
GOLDEN DOCK

Rumex occidentalis S. Wats.
WESTERN DOCK

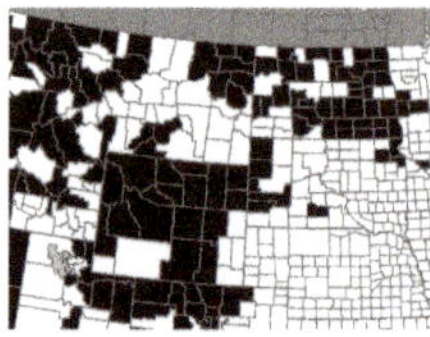

DESCRIPTION Stout, taprooted perennial 5–15(20) dm tall, usually single-stemmed and unbranched below the panicle, with large, long-petioled basal **leaves** and smaller, shorter-petioled cauline leaves. Basal leaf blades oblong-ovate to oblong-lanceolate, 10–40 cm long, ca. 1/4 to 1/3 as wide, flat to somewhat crisp-margined, truncate to cordate at the base. **Panicle** large, rather narrow, sparsely leafy-bracteate in the lower portion, the branches ascending; pedicels mostly 1–2x the length of the valves, obscurely jointed near the middle. Valves ro-

Rumex occidentalis
WESTERN DOCK

tund-ovate to triangular-ovate, cordate at the base, 3–10 mm long and about as wide, nearly entire to denticulate toward the base, lacking tubercles; **achenes** 3.5–4.5 mm long. July–Aug.

HABITAT Wet meadows, ditches and other moist or wet places.

WETLAND STATUS

GP OBL | MIDW OBL | WMTN FACW

Rumex patientia L.

PATIENCE DOCK

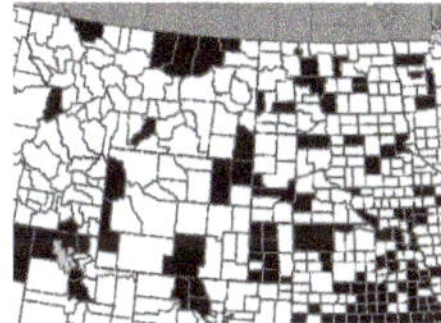

DESCRIPTION Similar to *Rumex crispus* but averaging larger and paler green in color, 8–20 dm tall; stems 1-few, erect. **Basal leaves** long-petioled, the blades ovate-lanceolate to oblong-lanceolate, (2)3–4x longer than wide, to 15 cm wide, flat to undulate, broadly cuneate to truncate at the base; cauline leaves reduced upward. **Panicle** large and dense, with a few reduced leafy bracts, the branches ascending; pedicels equaling or longer than the valves, swollen-jointed near or below the middle. Valves rotund-ovate, strongly cordate at the base and ultimately forcing the small outer sepals into a reflexed position, (5)6–9 mm long (including lobes of the cordate base), at least as wide, entire to denticulate; tubercle on one of the valves only, to about 1/3 the length of the valve and only about 116 as wide; **achenes** brown, 3–3.5 mm long. June–July.

Rumex patientia
PATIENCE DOCK

HABITAT Wet meadows, ditches, shores and waste places; introduced from Eurasia and locally established in many parts of N. America.

WETLAND STATUS

GP UPL | MIDW UPL | WMTN UPL

Rumex pseudonatronatus Borbas

FIELD DOCK

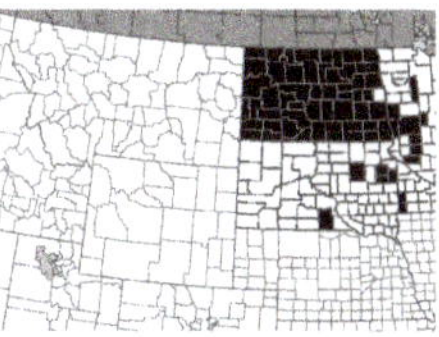

DESCRIPTION Quite similar to *R. crispus.* **Leaf blades** oblong to oblong-lanceolate, widest near the middle, undulate to undulate-crisped, cuneate at the base. **Panicle** as in *R. crispus;* pedicels mostly 1–2x the length of the valves, swollen-jointed near the base. Valves rotund-ovate to reniform, mostly 3.5–5 mm long, about as wide, entire, lacking tubercles or with a weakly developed tubercle on one of the valves; **achenes** 2–3 mm long. June–Sept.

HABITAT

Wet meadows, ditches, shores and other low areas, often where disturbed; introduced from Europe and naturalized over much of the n USA and s Canada.

WETLAND STATUS

GP UPL | MIDW UPL

Rumex pseudonatronatus
FIELD DOCK

Rumex stenophyllus Ledeb.

NARROW-LEAF DOCK

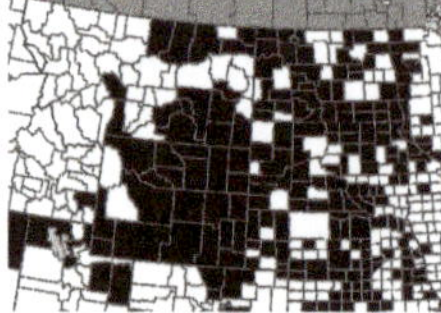

DESCRIPTION Very similar to *R. crispus* but the **leaves** not as undulate-crisped, otherwise differing as follows: Valves triangular to rotund-ovate in outline, 3.5–5.5 mm long, 4–5.5 mm wide, coarsely toothed on the margins, entire at the triangular tip; tubercles 3, ellipsoid, mostly 1/3 to 1/2 the length of the valve; **achenes** brown, 2–2.5(3) mm long. July–Sept.

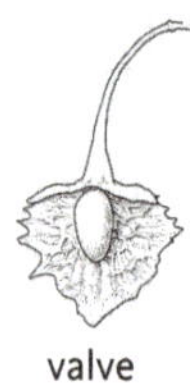

valve

HABITAT Wet meadows, shores, ditches, streambanks and disturbed places, often where alkaline or saline; introduced from Europe to e N America.

WETLAND STATUS

GP FACW | MIDW FACW | WMTN FACW

Rumex triangulivalvis (Danser) Rech. f.

MEXICAN DOCK, WILLOW-LEAVED DOCK

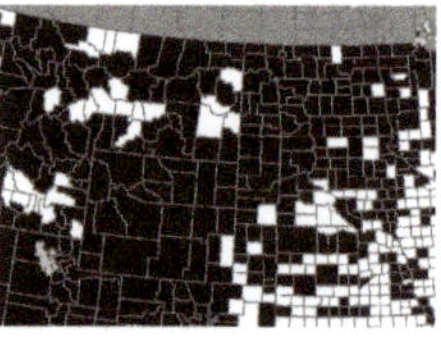

DESCRIPTION Perennial from a stout taproot, 2–8 dm tall, usually branched and spreading from the base, also with short branches or axillary clusters of leaves above, rather leafy throughout except in the panicles. **Leaves** mostly cauline, not greatly reduced upward, the blades linear-lanceolate to oblong-lanceolate, 5–16 cm long, 0.8–2.5(4) cm wide, mostly more than 4x longer than broad, flat to slightly undulate, acute to acuminate at the tip, cuneate to slightly rounded at the base; petioles short, mostly 1–3 cm long, the leaves becoming subsessile upward. **Panicles** 1–3 dm long, leafy-bracteate only in the lower part, the branches ascending to divergent; pedicels shorter than to 1.5x the length of the valves, swollen-jointed near the base. Valves triangular, 3–5(6) mm long, 3–4 mm wide, entire to shallowly sinuate-dentate; tubercles absent or present on 1, 2 or all 3 of the valves, 1/2 to 2/3 as long as the valve; **achenes** brown, ca. 2 mm long. June–Aug.

SYNONYMS *Rumex mexicanus* Meisn., *Rumex salicifolius* Weinm.

HABITAT Wet meadows, marshes, ditches, shores, streambanks and other low areas, often where water is brackish.

WETLAND STATUS

GP FACW | MIDW FACW | WMTN FAC

Rumex stenophyllus
NARROW-LEAF DOCK

Rumex triangulivalvis
MEXICAN DOCK, WILLOW-LEAVED DOCK

Primulaceae *primrose family*

Perennial herbs; **leaves** simple, exstipulate, entire, opposite or mostly so (sometimes alternate above in *Lysimachia maritima;* occasionally appearing whorled in *Lysimachia quadriflora* due to leaf fascicles in the axils), or the leaves all basal. **Flowers** perfect, regular, axillary and single or in dense axillary racemes or in a terminal umbel; calyx deeply (3–)5(–9)-parted, often nearly to the base; corolla (absent in *Glaux*) deeply (3)5(–9)-lobed, rotate or salverform; stamens typically 5, epipetalous and opposite the corolla lobes (free and alternate with the sepals in *Glaux*); staminodes sometimes present, alternating with the stamens; stigma capitate at the tip of the single, slender style, ovary superior, 1-celled, with free-central placentation. **Fruit** a few- to many-seeded, 5-valved capsule.

---

1 Leaves all basal, strongly whitened beneath ***Primula***

1 Leaves cauline, green on both surfaces ***Lysimachia***

Lysimachia LOOSESTRIFE

Typically erect, rhizomatous perennials. **Leaves** opposite (occasionally appearing whorled in *L. quadriflora* due to axillary leaf fascicles), sessile or petiolate, the blades generally ovate, ovate-lanceolate, lanceolate or linear, the petioles or leafbases frequently fringed with cilia. **Flowers** single from the leaf axils, or numerous in pedunculate, axillary racemes; calyx green, deeply (3-)5(–9)-parted nearly to the base; corolla yellow to white or pinkish, deeply (3-)5(–9)-parted, rotate, the tube very short; stamens adnate to the corolla near the base, sometimes alternating with small, membranous staminodes. **Capsules** globose to broadly ovoid, tardily dehiscent from the apex, containing few to many angular seeds which cohere to the placenta.

---

1 Flowers white to pinkish, solitary and sessile in the leaf axils; corolla absent, the perianth comprised of a petaloid calyx ***L. maritima***

1 Flowers yellow, solitary and pedicellate from the axils, or in axillary racemes; calyx and corolla both present 2

2 Flowers clustered in dense axillary racemes; foliage punctate with dark glands ***L. thyrsiflora***

2 Flowers single from the axils, borne on slender pedicels; foliage not punctate 3

3 Cauline leaves sessile, linear, 1.5–6 mm wide ***L. quadriflora***

3 Cauline leaves distinctly petioled or subsessile, the blades ovate to lanceolate, 6–60 mm wide 4

4 Leaf blades ovate to ovate-lanceolate, 20–60 mm wide; petioles distinct, fringed with cilia for the entire length ***L. ciliata***

4 Leaf blades ovate-lanceolate to lanceolate, 6–20 mm wide; petioles of middle and upper cauline leaves rather poorly distinguished from the blade, especially shortened upward, winged, ciliate at least in the lower half ***L. hybrida***

Lysimachia ciliata L.
FRINGED LOOSESTRIFE

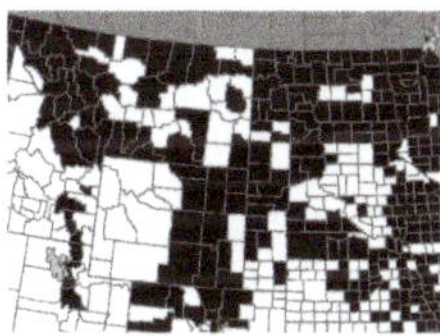

DESCRIPTION Plants erect, simple or sparingly branched above, 3–10 dm tall, from long-spreading rhizomes. **Leaves** distinctly petioled, the blades dark to bright green above, slightly paler beneath, ovate to ovate-lanceolate, 4–13 cm long, 2–6 cm wide, acute to acuminate, short-ciliate on the mar-

Lysimachia ciliata
FRINGED LOOSESTRIFE

gin, subcordate to rounded or somewhat attenuate at the base; petioles 0.5–4 cm long, ciliate for the entire length. **Flowers** single from the upper axils, on pedicels 2–6(8) cm long; calyx lobes lanceolate, 4–8 mm long; corolla lobes rotund to obovate, 5–12 mm long, finely erose, apiculate; anthers 2–3.5 mm long; staminodes present, inconspicuous, narrowly triangular, 1–2 mm long, membranous. **Capsules** 4–7 mm in diameter, containing many angular, dark brown to black seeds 1–2 mm long. July–Aug.

HABITAT Shores, streambanks, wet meadows, ditches, floodplains, moist woods and thickets.

WETLAND STATUS

GP FACW | MIDW FACW | WMTN FACW

Lysimachia hybrida Michx.

LOWLAND YELLOW-LOOSESTRIFE

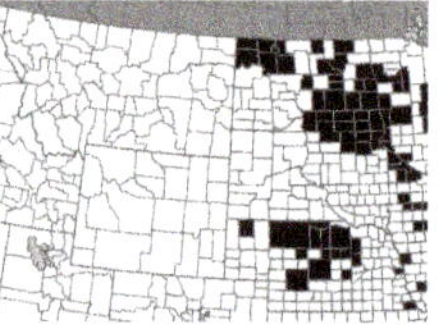

DESCRIPTION Erect or sometimes reclining, rhizomatous perennial 1.5–7 dm tall, simple or occasionally branched from the base, usually branched above. **Leaves** opposite and petioled below, becoming subverticillate and subsessile above, the basal and lower leaves usually not persistent; blades green above, the same or only slightly paler beneath, ovate-lanceolate to lanceolate, 2–7 cm long, 0.6–2 cm wide, acute, tapered to the subsessile or petiolate base; petioles (0.2)0.6–3(4) cm long, longest on the lower leaves, ciliate at least on the lower half, often over the entire length but more sparingly toward the blade. **Flowers** solitary from the axils, usually appearing clustered above due to close spacing of the nodes; pedicels 0.8–4 cm long; calyx lobes lanceolate, 3–6 mm long; corolla lobes rotund to obovate, 4–9 mm long, weakly erose, apiculate; anthers 1.5–2 mm long; staminodes as in *L. ciliata,* 1.2–1.7 mm long. **Capsules** 3.5–5 mm in diameter; seeds several to many, black, angular, 1–1.5 mm long. July–Aug.

SYNONYM *Lysimachia verticillata* Greene

HABITAT Wet meadows, marshes, ditches and shores, often in shallow water.

WETLAND STATUS

GP OBL | MIDW OBL | WMTN OBL

Lysimachia hybrida
LOWLAND YELLOW-LOOSESTRIFE

Lysimachia maritima (L.) Galasso, Banfi & Soldano

SEA MILKWORT

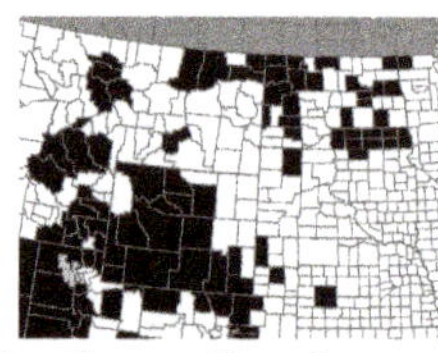

DESCRIPTION Low, glabrous, usually glaucous perennial from shallow rhizomes, 3–25 cm tall; **stems** leafy, simple and erect to branched and spreading. **Leaves** opposite or mostly so, sometimes becoming alternate above, sessile, rather succulent, el-

Lysimachia maritima
SEA MILKWORT

liptic to oblong or oblanceolate, 3–20 mm long, 1- 5 mm wide, obtuse to subacute. **Flowers** small, solitary and sessile in the leaf axils, white to pinkish; calyx campanulate, 3–4 mm long, lobed to about the middle, the lobes petaloid, rounded; corolla absent; stamens free of the calyx, alternate with the calyx lobes, about equal to or slightly exceeding the calyx. **Capsules** ovoid to globose, 2.5–3 mm long; seeds several, black, roughly elliptic and flattened, 1–1.5 mm long, coherent to the placenta. Mid-June–Sept.

SYNONYM *Glaux maritima* L.

HABITAT Wet meadows, seepage areas, stream margins and flats, where alkaline or saline.

WETLAND STATUS
GP OBL | MIDW OBL | WMTN OBL

Lysimachia quadriflora Sims
FOUR-FLOWER YELLOW-LOOSESTRIFE

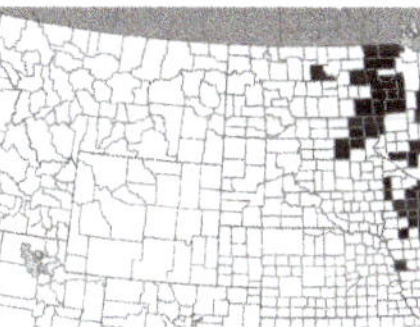

DESCRIPTION Slender erect perennial 2–8 dm tall, from slender rhizomes which commonly form lateral offshoots of basal rosettes. **Leaves** opposite, sometimes appearing whorled due to leaves fascicled in the axils; lower (and rosette) leaves, when persistent, petioled, with elliptic to obovate blades 2–3 cm long, 0.5–1 cm wide; cauline leaves sessile, linear, 2–7(9) cm long, 1.5–6 mm wide, acute, revolute along the margin, cuneate and sometimes ciliate at the base. **Flowers** solitary in upper leaf axils on pedicels 1–4 cm long; calyx lobes lanceolate, 3.5–6 mm long; corolla lobes oval to obovate, 7–12 mm long, erose or entire, apiculate; anthers ca. 2 mm long; staminodes as in *L. ciliata*. **Capsules** 3–5 mm in diameter, containing many black angular seeds ca. 1.2 mm long. July–Aug.

SYNONYM *Lysimachia longifolia* Pursh.

HABITAT Wet meadows and pond margins, usually where sandy.

WETLAND STATUS
GP FACW | MIDW OBL

Lysimachia quadriflora
FOUR-FLOWER YELLOW-LOOSESTRIFE

Lysimachia thyrsiflora L.
WATER LOOSESTRIFE

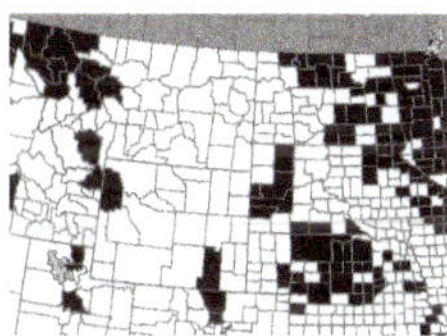

DESCRIPTION Stout erect perennial from rather thick rhizomes, 3–7 dm tall, usually simple or occasionally branched from lower nodes, strongly punctate throughout with dark glands; **stems** glabrous or brownish-villous in patches. **Leaves** opposite, sessile, the lower ones reduced, scalelike and scarious, the main cauline and upper leaves lanceolate, elliptic-lanceolate or oblanceolate, 4–13 cm long, 0.6–3.5 cm wide, glabrous above, glabrous or sparsely villous beneath, acute to acuminate, often blunt, cuneate at the base. **Flowers** small and numerous, in dense axillary racemes, the racemes bracteate, globose to ovoid, 1–3 cm

Lysimachia thyrsiflora
WATER LOOSESTRIFE

long, 1–2 cm thick, on peduncles 2.5–5.5 cm long; bracts linear-subulate, 2–5 mm long; pedicels 0.5–4 mm long. Calyx strongly punctate, deeply (3)5–7(9)-parted, the lobes lance-subulate, 1.5–3 mm long; corolla lobes usually equal in number to the calyx lobes, often streaked or punctate, linear, 3–5 mm long; stamens usually 5–7, much surpassing the corolla, anthers 0.5–0.8 mm long, on slender filaments; staminodes none. **Capsules** 2–4 mm in diameter, strongly punctate; seeds few, cocoa-colored, 1.2–1.5 mm long. June–Aug.

HABITAT Fens, bogs, springs, marshes, wet meadows and shores, where water is fresh, usually growing in shallow water.

WETLAND STATUS
GP OBL | MIDW OBL | WMTN OBL

Lysimachia thyrsiflora
WATER LOOSESTRIFE

Primula incana M.E. Jones

SILVERY PRIMROSE

DESCRIPTION

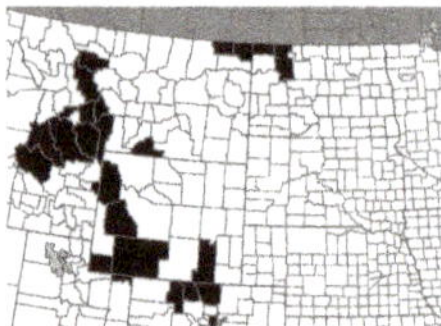

Scapose plant 6–40 cm tall, the foliage variously farinose; **leaves** in a basal rosette, strongly white-farinose beneath and usually less so above, elliptic to oblanceolate or spatulate, 2–6 cm long, 0.5–2 cm wide, shallowly denticulate to subentire, sessile or with winged petioles mostly 3–6 cm long. **Flowers** 3–12 in an umbel; bracts several, linear-lanceolate, 5–10 mm long, gibbous at the base; pedicels about equaling the bracts. Calyx farinose to some degree, 6–8 mm long, 5-lobed for ca. 1/3 of its length, the lobes obtuse to subacute; corolla lilac, salverform, the tube 8–11 mm long, the lobes deeply 2-lobed, 2–3 mm long; stamens attached in the upper 1/3 of the corolla tube. **Capsule** ellipsoid, about equaling to slightly exceeding the calyx; seeds strongly angular, 0.5–0.7 mm long, reticulate.

HABITAT Wet meadows, springs and shores, often where alkaline.

WETLAND STATUS
GP FACW

Primula incana
SILVERY PRIMROSE

Ranunculaceae *buttercup family*

Aquatic to terrestrial herbs (those included here). **Leaves** simple to palmately lobed or compound, alternate or mostly to entirely basal, usually petioled, exstipulate or stipulate. **Flowers** perfect, regular, hypogynous; sepals usually 5, occasionally more, herbaceous and greenish or petaloid and colored, often deciduous; petals 5 or none, seldom more than 5, usually yellow or white; stamens usually numerous; pistils several to many, simple, ripening into beaked achenes or follicles.

---

1 Pistils containing several ovules, maturing into follicles; leaves shallowly toothed, not divided into filaments or lobes ***Caltha***

1 Pistils containing 1 ovule, maturing into achenes; leaves all or mostly divided into lobes or filaments (except in *Ranunculus cymbalaria*, *R. flammula* and *Myosurus*, which have undivided leaves) 2

2 Sepals spurred at the base; achenes borne in a spikelike cluster on an elongate receptacle ***Myosurus***

2 Sepals not spurred; achenes borne in a globose to short-cylindric cluster 3

3 Flowers white, 2–5 cm across; leaves all basal except for the leaflike involucres subtending the flowers, deeply lobed ***Anemone***

3 Flowers yellow or white, but if white, then the flowers less than 2 cm across and the leaves finely dissected into filamentous segments; leaves cauline or basal ***Ranunculus***

Anemone canadensis L.

MEADOW ANEMONE

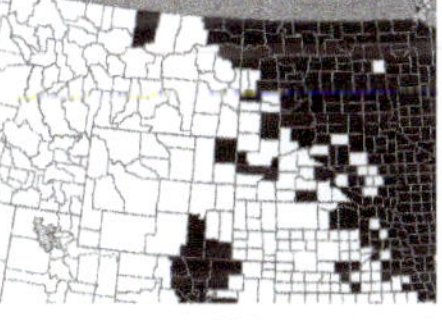

DESCRIPTION Erect perennial 1–6 dm tall, from slender rhizomes, often forming dense patches; **stems** simple below the inflorescence, setose. **Leaves** all basal and long-petioled except for the 2–3 sessile, leaflike involucres subtending the inflorescence; blades deeply 3- to 5-lobed and sharply toothed as well, rotund to reniform in outline, 4–15 cm across, sericeous, especially beneath; petioles 5–35 cm long. **Flowers** 1–3, white and showy, 2–5 cm across, lateral flowers, when present, subtended by smaller involucres than the central one; peduncles 3–12 cm long; sepals (4)5(6), petaloid, white, obovate, often unequal, 10–25 mm long; petals none; stamens numerous; pistils many, maturing into achenes, styles pubescent. **Achenes** in a globose head, obovate to suborbicular, 2.5–4.5 mm long, about as wide; style beak 2–4 mm long, pubescent. June–July, fruiting into Aug.

SYNONYMS *Anemonastrum canadense* (L.) Mosyakin, *Anemonidium canadense* (L.) Á. & D. Löve

HABITAT Wet meadows, low prairie, ditches, floodplains, moist woods and thickets.

WETLAND STATUS
GP FACW | MIDW FACW | WMTN FAC

Caltha palustris L.

MARSH MARIGOLD, COWSLIP

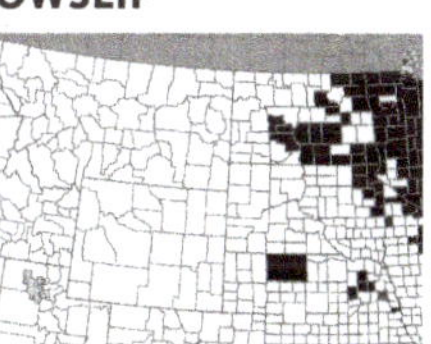

DESCRIPTION Loosely clumped, coarsely rooted perennial 2–6 dm tall; **stems** hollow, rather succulent, glabrous throughout. **Leaves** simple, the blades rotund-cordate, with the basal lobes separate or overlapping, 4–10(15) cm across, the margin shallowly dentate; membranous stipule-like appendages present at the nodes. **Flowers** few to several terminating each stem,

Anemone canadensis
MEADOW ANEMONE

showy, 2–4 cm across; sepals 4–9, bright yellow, petaloid, 12–20 mm long; petals none; stamens many; pistils 4–12, with very short styles. **Follicles** somewhat recurved and splitting lengthwise along the inside wall, 1–1.5 cm long. Flowering May–early June, fruiting June–early July.

HABITAT Swamps, springs, boggy areas and along fresh streams.

WETLAND STATUS
GP OBL | MIDW OBL | WMTN OBL

Myosurus minimus L.
MOUSETAIL

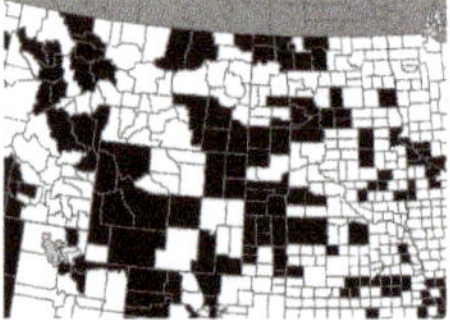

DESCRIPTION Small, glabrous, acaulescent annual 4–15 cm tall, with fibrous roots and a basal tuft of narrow, linear **leaves** mostly less than 1 mm wide. **Flowers** small, inconspicuous, usually few to many, borne singly above the leaves on slender peduncles, becoming more conspicuous in fruit; sepals greenish, usually 5, erect, spurred downward at the base, deciduous, 1.5–3.5 (5.5) mm long, the blade lightly 3-nerved on the back, usually longer than the spur; petals usually 5, sometimes none, inconspicuous, whitish or pinkish, about equaling the sepals; stamens 5–10; pistils many, borne on an elongate receptacle, ripening as a slender spike of usually more than 100, closely adherent, angular **achenes**; the achene bodies 1–2 mm long, with a sharp dorsal keel that extends beyond the body as an inconspicuous beak to 0.5 mm long, the beaks appressed in the spike; mature spike of achenes 1–6 cm long. Late April-June.

HABITAT Wet to moist places, sometimes in shallow water, usually where water stands only temporarily; scattered and probably more common than records indicate, as the plant is early and easily overlooked.

WETLAND STATUS
GP FACW | MIDW FACW | WMTN OBL

ADDITIONAL SPECIES A similar species, **bristly mousetail** (*Myosurus apetalus* C. Gay), is known from ND and Wyo. This plant also occurs in temporarily wet places. It differs from *M. minimus* mainly in the following ways: sepals 1-nerved or rarely with 2 faint lateral nerves; achenes 20–50(90), rather loose in the spike, the achene bodies with a broad low keel and usually 2 marginal ridges on the back, the beaks conspicuous, 0.5–1.5 mm long, divergent from the spike; mature spike of achenes 0.5–2.5 cm long.

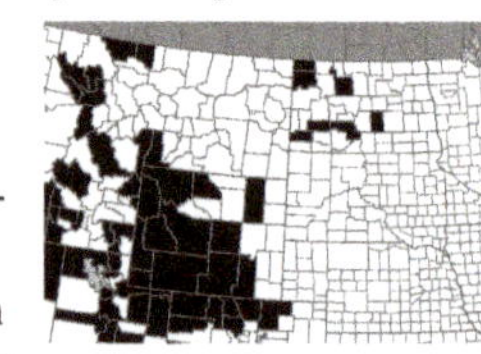

Ranunculus BUTTERCUP, CROWFOOT

Aquatic, semiaquatic and terrestrial perennials and annuals, some amphibious. **Stems** erect to procumbent, sometimes floating in water, often rooting at the nodes in some spp., branching or simple. **Leaves** simple or more often ternately compound to finely dissected, often variable on the same individual in this regard, alternate or mostly basal or entirely basal; nearly sessile to long-petioled,

Myosurus minimus
MOUSE-TAIL

Myosurus apetalus
BRISTLY MOUSE-TAIL

the petioles often dilated at the base (probably representing stipules), especially in submersed spp. **Flowers** from terminal buds, sometimes appearing axillary, often showy, borne above the water surface in aquatic spp.; sepals usually 5, greenish, deciduous; petals 5, seldom more, yellow or white, often fading to whitish with age, usually bearing a minute nectary pit covered by a scale toward the base; stamens usually 10-many; pistils usually numerous; receptacle conic to cylindric, glabrous or pubescent. **Achenes** usually many in a hemispheric, globose or cylindric head; achene body thick or flattened, the coat smooth, ridged, papillate or striate, glabrous or pubescent, tipped with a terminal or lateral style beak.

---

1 Flowers white; leaves finely divided into filiform segments; plants normally submersed 2

1 Flowers yellow; leaves simple to deeply lobed or finely divided into narrow, flat segments; plants submersed or emergent 3

2 Achenes (7)15–25; achene body averaging 1.5 mm long; achene beak prominent, 0.7–1.1 mm long (often shorter when dried) ***R. longirostris***

2 Achenes 30–45(80); achene body averaging 1.25 mm or less long; achene beak 0.2–0.5 mm long (often nearly beakless when dried) ***R. subrigidus***

3 Leaves all simple and entire or crenate to crenate-lobed 4

3 All, or at least the cauline leaves deeply lobed, divided or compound 5

4 Leaves ovate to round or reniform, crenate to crenate-lobed; achenes striate with longitudinal ribs ***R. cymbalaria***

4 Leaves elliptic to lanceolate or linear, entire to denticulate; achenes not striate ***R. flammula***

5 Basal and cauline leaves distinctly different in shape, the basal leaves mostly entire or crenate, not lobed or divided, the cauline leaves deeply divided ***R. cardiophyllus***

5 Basal and cauline leaves essentially similar in shape and form, all variously lobed, deeply divided or compound, none merely crenate 6

6 Achenes turgid, without a sharp border 7

6 Achenes flattened, with a sharp or winglike border 9

7 Petals 4–14 mm long; achenes 1.2–2.5 mm long, beaked; plants aquatic or amphibious perennials 8

7 Petals 3–5 mm long; achenes 0.8–1.2 mm long, nearly beakless; plants weedy annuals, usually emersed ***R. sceleratus***

8 Achene margin thickened and corky below the middle; petals mostly 6–14 mm long ***R. flabellaris***

8 Achene margin rounded but not thickened, petals mostly 4–8 mm long ***R. gmelinii***

9 Petals 7–16 mm long; anthers longer than 1 mm; stems often recurved and rooting at the nodes ***R. hispidus***

9 Petals 2–5 mm long; anthers less than 1 mm long; stems not rooting at the nodes 10

10 Petals distinctly shorter than the sepals; heads of achenes ovoid-cylindric to cylindric ***R. pensylvanicus***

10 Petals equaling or longer than the sepals; heads of achenes ovoid to globose ***R. macounii***

Ranunculus cardiophyllus Hook. (*not illus.*)
HEART-LEAF BUTTERCUP

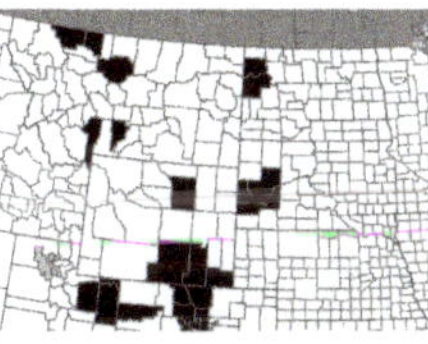

DESCRIPTION Erect, pilose to glabrate perennial with fibrous roots, 1.5–4 dm tall; **stem** single, simple or branched above, 1- to several-flowered. **Leaves** basal and cauline, the **basal leaves** with petioles 2–12 cm long, the previous year's petioles often persistent at the base of the stem; blades of the basal leaves simple, reniform to mostly ovate-cordate, 2–6 cm long, about as wide, crenate, one or two sometimes shallowly to deeply cleft; **cauline leaves** few, subsessile to sessile, deeply parted into several linear lobes. **Sepals** 5, yellowish, often tinged with purple, spreading, 6–10 mm long, pilose dorsally, petaloid on the margins; **petals** 5 or sometimes a few more, yellow, 8–15 mm long; stamens 35–80; receptacle oblong-ovoid, 4–14 mm long in fruit, hairy. **Achenes** 20–100 in an oblong-cylindric head 8–15 mm long, 7–9 mm thick; achene body obovate, turgid with an inconspicuous margin, 1.5–2.5 mm long, puberulent; beak straight or recurved at the tip, 0.5–1 mm long. June–July.

HABITAT Seepage areas, alpine meadows.

WETLAND STATUS

GP FACW WMTN FACW

NOTE This plant is vegetatively similar to **early wood buttercup**, *Ranunculus abortivus* L., which is common in moist woods throughout the region, but the flowers of *R. cardiophyllus* are much larger and showier.

Ranunculus cymbalaria Pursh

SEASIDE BUTTERCUP

DESCRIPTION Low, extensively stoloniferous perennial, often forming dense mats, 3–15(25) cm tall, glabrous or sparsely pubescent mainly on the scapes and petioles. **Leaves** all basal, the blades ovate to round or reniform, cordate or truncate at the base, 5–22(40) mm long, 4–20(35) mm wide, crenate to shallowly lobed, often with 3 prominent lobes at the tip. **Scapes** surpassing the leaves, simple or sparingly branched with 1–several small yellow flowers. **Sepals** 5, greenish-yellow, spreading, 3–5 mm long, deciduous; **petals** usually 5(–12), yellow, turning whitish with age, 3–8 mm long; stamens usually 10–30; receptacle 2–4 mm long in flower, 4–10 mm long in fruit, pubescent. **Achenes** usually 40–150 in a cylindric head 3-lo(15) mm long, 3–4(6) mm thick; achene body turgid, cuneate-oblong, longitudinally nerved, ca. 1.5 mm long; beak triangular, straight, 0.3 mm long. June–Oct.

Ranunculus cymbalaria
SEASIDE BUTTERCUP

SYNONYMS *Halerpestes cymbalaria* (Pursh) Greene

HABITAT Wet meadows, streambanks, shores, ditches and seepage areas, often where brackish; common.

WETLAND STATUS

GP OBL | MIDW OBL | WMTN OBL

NOTE Plants reproduce freely by stolons.

Ranunculus flabellaris Raf.

YELLOW WATER-CROWFOOT

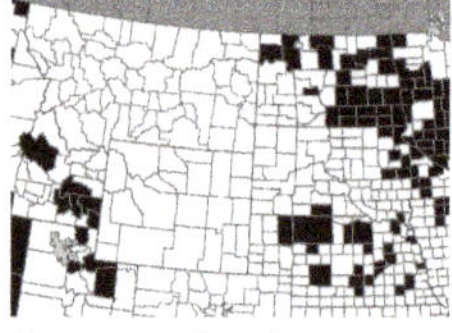

DESCRIPTION Amphibious perennial, usually submersed but occasionally stranded on mud, glabrous or rarely pubescent (when emersed. **Stems** floating or, when stranded, erect from a decumbent base, rooting at the lower nodes, branching, 3–7 dm long. **Leaves** all cauline; blades finely triternately dissected, the divisions narrow and flat, 1–2 mm or less wide, not as finely dissected on emersed plants, semicircular to reniform in outline, 1.5–10 cm long, 2–12 cm

Ranunculus flabellaris
YELLOW WATER-CROWFOOT

wide; petioles composed of the stipular leaf bases on upper leaves, 3–8 mm long, often extending beyond the stipular base and much longer on lower leaves. **Flowers** 1–several terminating each stem, bright yellow; **sepals** 5, greenish-yellow, spreading, 5–8 mm long, early deciduous; **petals** 5–8, yellow, 7–15 mm long; stamens 50–80; receptacle 2–3 mm long in flower, 5–7 mm long in fruit, pubescent. **Achenes** 50–75 in a globose to ovoid head 7–10 mm long, 5–8 mm thick; achene body obovate, 1.5–2 mm long, the margin thickened and corky below the middle; beak broad and flat, 1–1.5 mm long. Late May–July, occasionally in Aug and Sept.

HABITAT Fresh water or mud of ditches, slow streams, marshes and ponds.

WETLAND STATUS
GP OBL | MIDW OBL | WMTN OBL

Ranunculus flammula L.
LESSER SPEARWORT

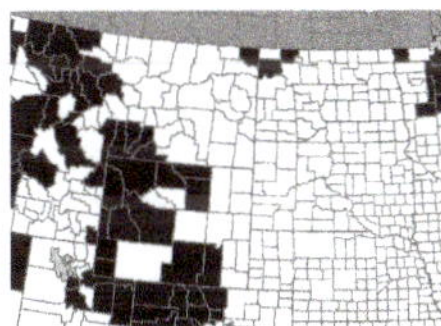

DESCRIPTION Low, stoloniferous perennial, often appressed-hairy. **Stems** decumbent to prostrate, rooting at the nodes, simple to sparingly branched, the upright tips 4–15 cm tall, 1- to several-flowered. **Leaves** clustered at rooting nodes, reduced and shorter-petioled on upper portions of the stem; blades simple, elliptic to lanceolate or linear, 1–5 cm long, 1.5–7(20) mm wide, entire or very slightly toothed, tapered to slender petioles mostly 5–15 mm long. **Sepals** 5, yellowish-green, ovate, 1.5–3(5) mm long, strigose; **petals** 5, yellow, obovate, 3–5 mm long; stamens 25–50; receptacle obovoid, ca. 1 mm long in fruit, glabrous. **Achenes** 10–25 in a subglobose head 2.5–3.5 mm long, 3–4.5 mm thick; achene body turgid, obovate, 1.3–1.7 mm long, inconspicuously margined, smooth, the beak 0.2–0.6 mm long. June–Aug.

HABITAT Marshes and muddy shores.

WETLAND STATUS
GP FACW | MIDW FACW

Ranunculus flammula
LESSER SPEARWORT

Ranunculus gmelinii DC.
LESSER YELLOW WATER-CROWFOOT

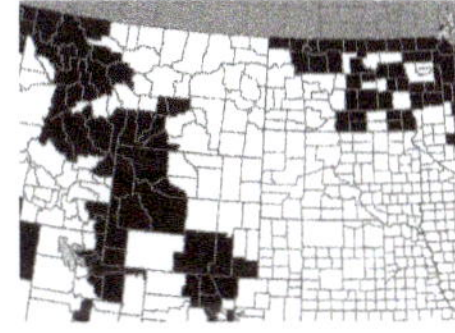

DESCRIPTION Similar in habit to *R. flabellaris* but usually emersed, glabrous to hirsute. **Stems** usually procumbent and rooting at the nodes, floating when submersed, sparsely branched, 1–5 dm long. **Leaves** all cauline or a few long-petioled basal leaves present, the upper leaves commonly floating on submersed plants; blades deeply 3-lobed or dissected, the divisions forked 2–3 times or sometimes dissected into flat segments when submersed, but not triternately dissected, pentagonal in outline, usually 0.8–2 cm long, 1.5–2.5 cm wide, or submersed leaves often larger, to 6–9 cm across; petioles mostly 1–4 cm long, the stipular bases 3–6 mm long. **Flowers** usually 1–several terminating each stem, seldom more, yellow; **sepals** 5, greenish-yellow, spreading, 2.5–6 mm long, deciduous with or before the petals; **petals** 5–8, rarely more, entire or sometimes lobed, 4–8 mm long; stamens 20–40; receptacle 1–2 mm long in flower, 3–6 mm long in fruit, short-pubescent. **Achenes** 50–70 in a globose

Ranunculus gmelinii
LESSER YELLOW WATER-CROWFOOT

to ovoid head 7–10 mm long, 5–8 mm thick; achene body obovate, 1–1.5 mm long, the margin rounded or inconspicuously keeled, not corky-thickened although the basal and ventral portions of the pericarp callous-thickened, the beak broad and thin, 0.5–0.8 mm long, slightly recurved. Late May–July, occasionally in Aug and Sept.

HABITAT Occurring in the same habitats as *R. flabellaris,* but more often where water is only temporary.

WETLAND STATUS
GP FACW | MIDW FACW

Ranunculus hispidus Michx.
MARSH BUTTERCUP

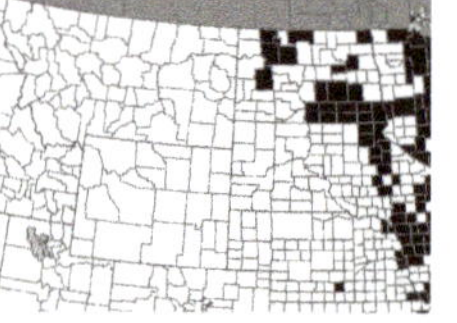

DESCRIPTION Nearly glabrous to strongly hirsute perennial 2–7 dm tall. **Stems** erect to ascending, some eventually reflexed and rooting at an upper node, acting as stolons, to 10 dm long, with spreading or somewhat deflexed hairs. **Leaves** basal and cauline, the basal leaves larger and longer-petioled than the cauline ones; blades simple and 3-lobed on earliest leaves, otherwise ternately compound, broadly ovate-cordate in outline, 3–14 cm long, 4–20 cm wide, the lobes or leaflets themselves 2- or 3-lobed or cleft and irregularly toothed, appressed-hairy mainly on the veins; petioles 3–30 cm long, with pubescence like that of the stems; stipular bases 10–40 mm long. **Flowers** 1–several per flowering stem, the pedicels appressed-hairy; **sepals** 5, yellowish-green, spreading, 5–11 mm long, appressed-hairy, deciduous before the petals; **petals** 5 (rarely to 10), yellow, fading white, 7–16 mm long; stamens mostly 40–70, the anthers more than 1 mm long; receptacle 2–3 mm long in flower, 4–8.5 mm long in fruit, hispidulous. **Achenes** usually 15–30 or more in an ovoid to globose head 6–12 mm long, 7–12 mm thick; achene body obovate, 2–4.5 mm long, glabrous, narrowly to broadly winged around the margin, the beak straight, 1.5–3 mm long. Late May–early July.

Ranunculus hispidus
MARSH BUTTERCUP

HABITAT Wet meadows and woods, springs, boggy areas, swamps, shores and streambanks, where water is fresh.

WETLAND STATUS
GP FACW | MIDW FAC

Ranunculus longirostris Godr.
WHITE WATER-CROWFOOT

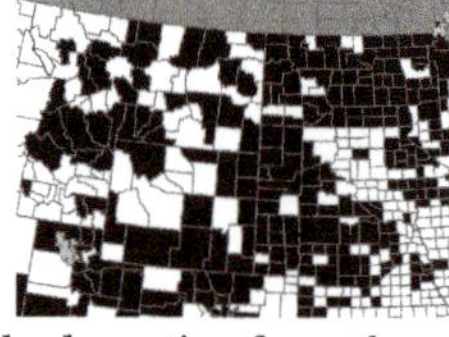

DESCRIPTION Submersed, mostly glabrous perennial. **Stems** floating, elongate, flexuous, mostly 3–8 dm long, simple or sparingly branched, rooting from the lower nodes. **Leaves** all cauline, the blades finely divided into filiform segments, once or twice trichotomous, then dichotomous, globular in outline, reniform when flattened, 1–2 cm long, 1.5–3 cm wide; petioles consisting

Ranunculus longirostris
WHITE WATER-CROWFOOT

of the inflated stipular leaf bases, 2–4 mm long, glabrous or pubescent. **Flowers** solitary from the axils in the upper portion of the stem, white; **sepals** 5(6), purplish-green, spreading, 2–3.5 mm long, deciduous shortly before the petals; **petals** 5, white, suffused with yellow at the base, 4–9 mm long; stamens 10–20. **Achenes** (7)15–25, in a hemispheric to globose head 3–5 mm long, 3–6 mm thick; achene body obovoid, transversely ridged, glabrous or slightly hispid, averaging 1.5 mm long, the beak prominent, slender and straight, 0.7–1.1 mm long (often shorter when dried); receptacle 1–2 mm long, pubescent; peduncles recurved in fruit, 1–5 cm long. June–July, occasionally in Aug–Sept.

SYNONYM *Ranunculus aquatilis* var. *diffusus*.

HABITAT Slow streams, ponds, marshes and water-filled ditches, usually in calcareous water.

WETLAND STATUS
GP OBL | MIDW OBL | WMTN OBL

NOTE *R. subrigidus* is very similar but the heads with more achenes, and the achene beaks shorter.

Ranunculus macounii Britt.

MACOUN'S BUTTERCUP

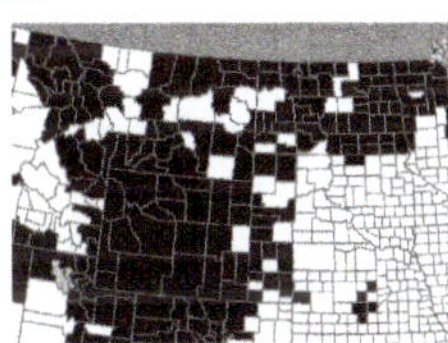

DESCRIPTION Sparsely to densely hirsute annual or short-lived perennial 2–7 dm tall. **Stems** hollow, erect or decumbent, dichotomously branched 1-few times, the branches rebranching and terminating in few to several flowers. **Leaves** basal and cauline, the basal leaves usually larger and longer-petioled than the cauline ones; blades deltoid in outline, simple and 3-lobed or more often compound and divided into 3 segments which themselves may be 3-lobed, 4–14 cm long, 6–16 cm wide, glabrous to hirsute on both surfaces, the ultimate segments coarsely and irregularly toothed; stipular bases 5–25 mm long, mostly 2–3 cm long on basal leaves. **Sepals** 5, yellowish, reflexed, 3–5(7) mm long, deciduous, glabrous or pilose; **petals** 5, yellow, equaling or longer than the sepals, 3–6(8) mm long; stamens 15–35, anthers less than 1 mm long; receptacle 1–2 mm long in flower, 4–6 mm long in fruit. **Achenes** 30–50 in an ovoid to globose head 7–12 mm long, 8–12 mm thick; achene body flattened, obovate, 2–3(3.5) mm long, smooth or shallowly pitted, glabrous, narrowly keeled on the margin; beak stout, slightly curved or straight, 1 mm long. June–July.

HABITAT Wet meadows, shores, streambanks, ditches and other wet places.

WETLAND STATUS
GP OBL | MIDW OBL | WMTN OBL

Ranunculus macounii
MACOUN'S BUTTERCUP

Ranunculus pensylvanicus L. f.

BRISTLY CROWFOOT

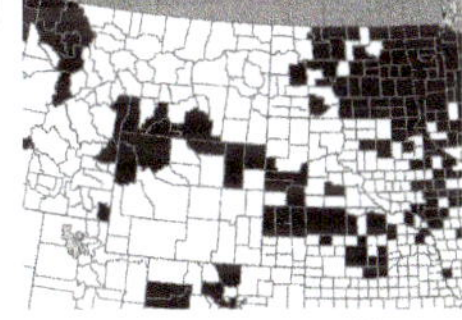

DESCRIPTION Hirsute annual or short-lived perennial 4–10 dm tall, resembling the preceding. **Stems** hollow, erect, branching and flowering like *R. macounii.*

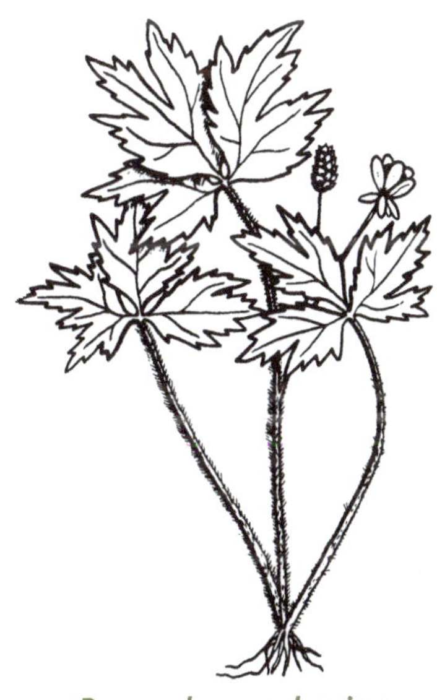

Ranunculus pensylvanicus
BRISTLY CROWFOOT

Basal **leaves** often withering early, larger and longer-petioled than the cauline leaves; blades of basal and cauline leaves pinnately compound, 4–12 cm long, 4–15 cm wide, appressed-hairy to rarely glabrous, the leaflets lobed and coarsely toothed, tapered to the slender bases, the terminal one 3-parted, the lateral ones 2- or 3-parted; stipular bases 1–4 cm long. **Sepals** 5, yellowish, reflexed, (3)4–5 mm long, deciduous, sparsely hirsute; **petals** 5, pale yellow, fading whitish, distinctly shorter than the sepals, (1.5)2–3 mm long; stamens 15–20, anthers less than 1 mm long; receptacle 2 mm long in flower, 5–13 mm long in fruit, short-pubescent. **Achenes** 60–80 in an ovoid-cylindric to cylindric head 10–15 mm long, 6–9 mm thick; achene body flattened, obovate, (1.5)2–2.5 mm long, smooth and glabrous, keeled on the margin; beak stout, deltoid, 0.6–0.9 mm long. June–Aug.

HABITAT Occurring in the same habitats as *R. macounii* but intolerant of brackish conditions.

WETLAND STATUS

GP FACW | MIDW OBL | WMTN FACW

Ranunculus sceleratus L.

CURSED CROWFOOT

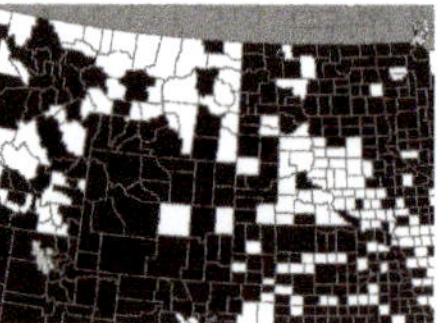

DESCRIPTION Weedy annual, glabrous or rarely hirsute, 1–5 dm tall or the stem to 10 dm long when submersed. **Stems** erect, inflated and hollow, sparsely to profusely branched, especially above. **Leaves** basal and cauline, the basal leaves often longer-petioled and less deeply dissected than the cauline ones, often floating when submersed; blades deeply 3-parted or divided, broadly truncate or cordate at the base, distally rounded, 1–6 cm long, 3–8 cm wide, the primary lobes or divisions lobed or divided, the ultimate lobes obtuse to rounded; petioles 2–15 cm long or much longer when submersed; stipular bases dilated, membranous, 5–12 mm long. **Flowers** numerous; **sepals** 5, yellowish-green, reflexed, 2–3 mm long, deciduous with the **petals**, glabrous; petals 5, light yellow, fading white, 3–5 mm long; stamens usually 10–25; receptacle 1.5–3 mm long in flower, 3–10 mm long in fruit. **Achenes** 40–300 in a cylindric-ovoid or rarely globose head 4–11 mm long, 3.5–7 mm thick; achene body obovoid, 0.8–1.2 mm long, glabrous, obscurely keeled on the margins, somewhat corky-thickened around the edges; beak minute and blunt. June–Sept.

Ranunculus sceleratus
CURSED CROWFOOT

HABITAT Shores, streambanks, mud flats, wet meadows, ditches, marshes and other wet places.

WETLAND STATUS

GP OBL | MIDW OBL | WMTN OBL

Ranunculus subrigidus W. Drew

SHORT-BEAK WHITE WATER-CROWFOOT

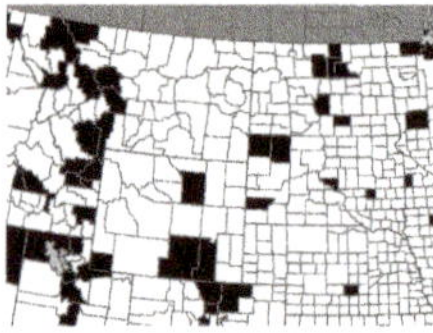

DESCRIPTION Very similar to *R. longirostris,* occasionally stranded on mud in late summer and assuming a semiterrestrial growth form. **Stems** usually 2–6 dm long, sometimes over 10 dm long in clear deep water, or only to 1 dm long when stranded, simple or sparingly branched, rooting at the lower nodes. **Leaves** all cauline or some basal on semiterrestrial forms, the blades finely divided as in *R. longirostris,* roughly globular in outline, usually 1–3 cm across, often smaller, the filamentous leaf segments flattened on stranded plants; petioles consisting of the stipular leaf bases or extending slightly beyond the dilated base, 2–5 mm long, glabrous or pubescent. **Flowers** as in *R. longirostris.* **Achenes** 30–45(80) in a globose head 3–5 mm long, 3–5 mm thick; achene body obovoid, transversely wrinkled, hispidulous on the back, 1–1.5 mm long, averaging 1.25 mm or less long, the beak 0.2–0.5 mm long (often nearly beakless when dried); receptacle 1–2 mm long, pubescent; peduncles strongly recurved in fruit, (1) 3–1 0 cm long. June–Aug.

SYNONYMS

Ranunculus aquatilis var. *diffusus, Ranunculus circinatus* var. *subrigidus*

HABITAT

Marshes, lakes, ponds, water-filled ditches and slow-moving streams.

WETLAND STATUS

GP OBL | MIDW OBL | WMTN OBL

NOTE The white water-crowfoots, *R. longirostris* and *R. subrigidus,* are especially conspicuous in June when the white blossoms cover the water surface of some Great Plains marshes.

Ranunculus subrigidus
SHORT-BEAK WHITE WATER CROWFOOT

Comarum palustre
BOG CINQUEFOIL

Rosaceae *rose family*

A large, diverse family commonly divided into several distinct subfamilies, many species cultivated for fruit and ornament. Those treated here herbs and shrubs with alternate, simple or compound **leaves,** the leaves usually stipulate (stipules absent in Spiraea). **Flowers** solitary or usually in cymose or paniculate inflorescences, perfect, regular, perigynous; sepals 5, sometimes alternating with 5 or more somewhat smaller bractlets, the **sepals** attached around the rim of a saucerlike, disklike or cupulate hypanthium; **petals** 5, usually yellow, white or pink, frequently small and inconspicuous; stamens 15-many, seldom as few as 10, inserted near the rim of the hypanthium; carpels 5-many, seldom fewer than 5, separate, the ovaries ripening as an aggregate of achenes or follicles, these often enclosed by the persistent hypanthium.

---

1 Shrub with simple leaves; flowers white; fruit of 5 or fewer follicles **Spiraea**

1 Herbs with compound leaves; flowers yellow or yellowish to pinkish; fruit an aggregate of many achenes 2

2 Styles conspicuous, elongating after flowering and persistent in fruit, jointed **Geum**

2 Styles inconspicuous, not elongating after flowering, usually deciduous 3

3 Petals very dark red; plants woody at the base **Comarum**

3 Petals yellow; plants herbaceous throughout **Potentilla**

Comarum palustre L.

BOG CINQUEFOIL

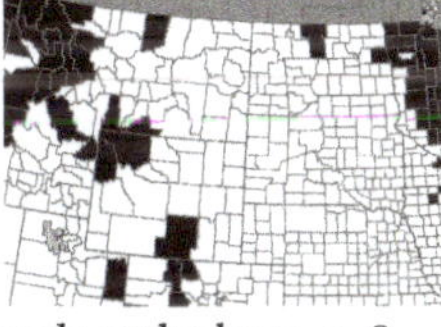

DESCRIPTION Perennial from a long, stout rhizome; **stems** ascending to sprawling or floating, often rooting at the nodes, sparingly branched, woody at the base, 3–8 dm long, glabrous below, pubescent above, the hairs often gland-tipped. **Leaves** cauline, long-petioled below to subsessile above, pinnate to subpalmate, with (3)5–7 leaflets, these oblong to elliptic, 3–10 cm long, 0.8–3.5 cm wide, glaucous beneath, sharply serrate; stipules winging the petioles of lower leaves, becoming shorter, broader and foliaceous upward. **Flowers** few to many in open cymes, single or paired from the axils; **sepals** dark red or purplish at least on the inside, ovate to lanceolate, acuminate, 6–20 mm long; **petals** 5(10), very dark red, elliptic to oblanceolate or spatulate, apiculate or cuspidate at the tip, 3–5 mm long; stamens ca. 25, dark red; styles laterally attached. **Achenes** reddish to golden brown, obliquely ovoid, plump, 1–1.2 mm long, smooth. June–Aug.

SYNONYM *Potentilla palustris* (L.) Scop.

HABITAT Bogs and swamps.

WETLAND STATUS
GP OBL | MIDW OBL | WMTN OBL

Geum rivale L.

WATER OR PURPLE AVENS

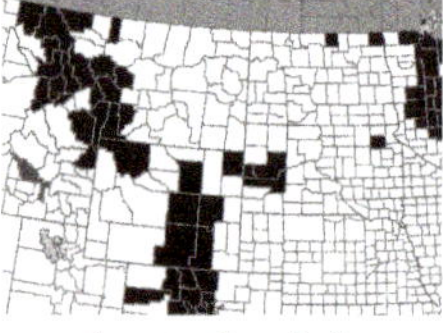

DESCRIPTION Erect perennial herb from a stout rhizome, 3–6(10) dm tall, sparingly hirsute throughout, also puberulent above with some short, glandular hairs. Principal **leaves** basal, pinnately compound, 1–4.5 dm long including the stipular-winged petiole; leaflets (5)7–15, the terminal 1–3 much larger than the others, shallowly lobed and coarsely dentate, terminal leaflet broadly cuneate-obovate to subrotund in outline, 2.5–10 cm long, 3–12 cm wide; cauline leaves 2- 5, much reduced upward, pinnate below to 3-lobed above, stipules foliaceous. **Flowers** 3–9 in a cymose inflorescence, nodding to erect, the pedicels densely glandular-puberulent and hirsute; **sepals** 5, sometimes initially green but always purple in flower, ascending in flower to spreading or reflexed in fruit, triangular, 6–10 mm long, acute to acuminate, alternating with 5 shorter, linear-oblong bractlets; **petals** 5, yellowish to pinkish with purple veins, erect, broadly retuse, tapered to a clawed base, about equaling to a bit shorter than the sepals; stamens numerous; carpels many, separate, styles long and slender with a hooked joint above the middle, the portion above the joint deciduous, the slender lower portion persistent and curved or deflexed in fruit, 6–10 mm long, hirsute and glandular-puberulent; receptacle short-cylindric, on a short stipe; hypanthium purple, saucer-shaped, 3–4.5 mm long, hirsute and glandular-puberulent. **Fruit** an aggregate of long-beaked achenes, the fruiting head globose; achene bodies fusiform, 3–4 mm long, hirsute. May–July.

HABITAT Swampy and boggy places, fresh wet meadows.

WETLAND STATUS
GP FACW | MIDW OBL | WMTN FACW

ADDITIONAL SPECIES **Yellow avens** (*Geum aleppicum* Jacq.) is a more common species that occurs throughout our region and is

Comarum palustre
BOG CINQUEFOIL

Geum rivale
WATER OR PURPLE AVENS

sometimes encountered on streambanks, alluvial deposits, and in wet meadows (though it is most characteristic of moist woodlands). It differs from *Geum rivale* in its typically larger stature, green calyx, and distinctly yellow petals.

Potentilla CINQUEFOIL

Annual and perennial herbs, erect, decumbent, sprawling or stoloniferous. **Leaves** pinnately or palmately compound, alternate or mostly basal, petiolate to subsessile, stipulate; leaflets 3-many, oblong to elliptic or obovate, serrate. **Flowers** perfect, regular, perigynous; calyx of 5 persistent sepals, these alternating with several to many, entire to toothed bractlets which are usually narrower and shorter than the sepals, the sepals and bractlets fused at the base to form a saucer-shaped hypanthium; corolla of usually 5 yellow petals which are attached near the summit of the hypanthium; stamens many, usually 10–25; pistils numerous, simple, styles simple, laterally or nearly basally attached to the ovary, deciduous. **Fruit** an aggregate of many small, smooth to ridged achenes, surrounded by the persistent hypanthium.

1 Plants extensively stoloniferous; leaves white-tomentose beneath ... *P. anserina*
1 Plants lacking stolons; leaves green, glabrous to hirsute beneath ... 2

2 Leaves pinnately compound with 7–11 leaflets ... *P. supina*
2 Leaves palmately compound with 3–7 leaflets ... 3

3 Achene surface ridged at maturity; petals 2.5–4 mm long, nearly equaling the sepals; leaflets 3 (rarely 5) ... *P. norvegica*
3 Achene surface smooth; petals 1–1.5(2) mm long, much smaller than the sepals; leaflets 3–7 ... *P. rivalis*

Potentilla anserina L.
SILVERWEED

DESCRIPTION Low, extensively stoloniferous perennial from a stout rootstock; **leaves** basal except for a few clustered on the stolons, pinnately compound with numerous leaflets, small leaflets often alternating with larger ones; blade oblanceolate in outline, 0.5–3(5) dm long, including the petiole which may be 1/2 the total leaf length, 2–8 cm wide, green and glabrous to grayish-green and sericeous above, densely white-tomentose beneath; **leaflets** elliptic to oblong or obovate, 1.5–5 cm long, 0.5–1.8 cm wide, greatly reduced downward, deeply serrate, the teeth sharply ascending; stipules prominent as brownish membranous wings on the basal portion of the petiole. **Flowers** yellow, rather showy, solitary from the leafy nodes of the stolons, on peduncles 4–15(25) cm long; **sepals** ovate, acuminate, white-sericeous on the outside; **petals** elliptic to obovate or nearly rotund, 5–10 mm long; stamens 20–25; pistils numerous, styles attached laterally on the ovary. Mature **achenes** golden brown, obliquely ovoid, usually corky with ridges or furrows. June–Aug.

SYNONYM *Argentina anserina* (L.) Rydb.

HABITAT Wet meadows, ditches, shores, streambanks and mud flats.

WETLAND STATUS
GP FACW | MIDW FACW | WMTN OBL

Potentilla norvegica L.
NORWEGIAN CINQUEFOIL

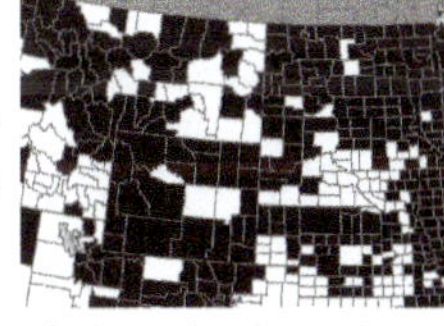

DESCRIPTION Hirsute, taprooted annual or biennial (1)2–7 dm tall; **stems** erect to decumbent, simple or branching from the base, branched above in the inflorescence. **Leaves** mostly cauline, long-petioled below to subsessile above, palmately

Potentilla anserina
SILVERWEED

compound with 3 (rarely 5) leaflets, the **leaflets** elliptic to oblanceolate or obovate, 1.5–7 cm long, 0.8–4 cm wide, coarsely crenate-serrate; stipules foliaceous, ovate, mostly 1–2.5 cm long, entire or usually toothed. **Flowers** usually numerous and crowded in terminal cymes, not showy; **sepals** ovate-lanceolate, 4–6 mm long, the bractlets about as long or longer; **petals** yellow, obovate, 2.5–4 mm long, usually 3/4 to about as long as the sepals; stamens ca. 20; styles terminal. **Achenes** tan to brown, ovate, 0.6–1 mm long, the surface longitudinally ridged at maturity. June–Aug.

HABITAT Wet meadows, streambanks, ditches, shores and a variety of drier habitats.

WETLAND STATUS
GP FAC | MIDW FAC | WMTN FAC

Potentilla rivalis Nutt.
BROOK CINQUEFOIL

DESCRIPTION Erect to spreading, hirsute, taprooted annual or biennial 1.5–9 dm tall, simple or branched from the base, branched above. **Leaves** mostly cauline, long-petioled below to subsessile above, palmately compound, with 3–7 leaflets, or the lower leaves closely pinnate, the **leaflets** obovate to elliptic or oblanceolate, 1.5–5 cm long, 0.5–2.5 cm wide, coarsely serrate; stipules ovate, usually toothed, mostly 0.5–1.5 cm long. **Flowers** numerous in leafy, branched cymes, not showy; **sepals** ovate-triangular, 2.5–6 mm long, the bractlets sometimes longer; **petals** yellow, obovate to oblanceolate, 1–1.5(2) mm long, ca. 1/2 or less as long as the sepals; stamens 10–15; styles terminal. **Achenes** yellowish, ovoid-reniform, 0.6–0.8 mm long, smooth. June–Aug.

SYNONYM *Potentilla millegrana* Engelm.

HABITAT Wet meadows, shores, ditches, streambanks and flats.

WETLAND STATUS
GP FACW | MIDW FACW | WMTN FACW

ADDITIONAL SPECIES **Biennial cinquefoil** (*Potentilla biennis* Greene) is a similar species of moist woodlands and streambanks known from c and w SD and Wyo (and more common in w USA). It differs from *P. rivalis* as follows: averaging smaller, 3–6 dm tall, the stems and leaves with fine glandular hairs as well as longer eglandular hairs; leaves all 3-foliate; calyx mealy-glandular.

Potentilla supina L.
BUSHY CINQUEFOIL

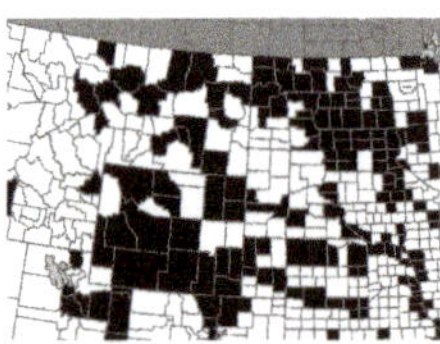

DESCRIPTION Erect to decumbent, taprooted annual or short-lived perennial 1.5–7 dm tall, glabrous below to hirsute above; **stems** simple or branched from the base, diffusely branched above. **Leaves** mostly cauline, long-petioled below, shorter petioled above, pinnately compound, with 7–11 leaflets, the smaller, sessile bractlike leaves of the inflorescence mostly ternate; **leaflets** elliptic to obovate, 0.8–4 cm long, 0.5–2 cm wide, serrate; stipules prominent, ovate, entire to serrate, mostly 0.5–1.5 cm long. **Flowers** usually numerous in open to dense cymes, not showy; **sepals** ovate-triangular,

Potentilla norvegica
NORWEGIAN CINQUEFOIL

Potentilla rivalis
BROOK CINQUEFOIL

acute to abruptly acuminate, 2.5–4 mm long, the bractlets sometimes longer; **petals** yellow, obovate, 2.5–3.5 mm long, about equaling the sepals; stamens (10–15)20; styles terminal. **Achenes** brownish, obliquely obovate, ca. 1 mm long, apically ridged, often corky-thickened in the lower half. June–Sept.

SYNONYM *Potentilla paradoxa* Nutt.

HABITAT Shores, ditches, floodplains and flats, often where sandy or gravelly.

WETLAND STATUS
GP FACW | MIDW FACW | WMTN FACW

Spiraea alba Du Roi
WHITE MEADOWSWEET

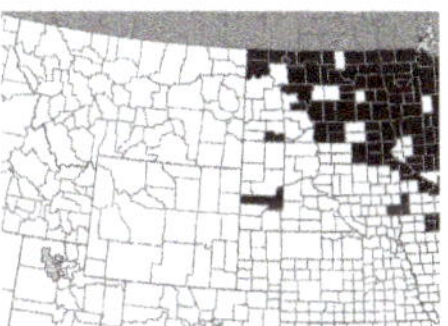

DESCRIPTION Small, erect shrub 0.4–1 m tall, often in colonies; **stems** puberulent when young, eventually brown to reddish-brown and glabrous. **Leaves** dark green, alternate, often rather crowded, narrowly elliptic to oblanceolate, 2.5–8 cm long, 0.8–1.5(3) cm wide, glabrous to puberulent, acute to obtuse, serrate with ascending teeth, cuneate to somewhat rounded at the base; petioles short, to 1 cm long; stipules none. **Flowers** numerous in a terminal, oblong to pyramidal panicle 0.5–2.5 dm long, the branches and tiny bracts puberulent. **Sepals** 5, broadly triangular, 1–1.5 mm long, puberulent; **petals** 5, white to slightly pinkish, subrotund, 1.5–3.5 mm long; stamens 25–50, the filaments persisting and forming a fringe around the inside of the hypanthium; carpels 5, seldom fewer, separate, styles 0.5–1.3 mm long; hypanthium cupulate, 1–1.8 mm long. **Fruit** a group of 5 (seldom fewer), 2- to several-seeded follicles, these 2.5–3.5 mm long and extending well above the hypanthium at maturity. Flowering late June–Aug, fruiting Aug–Sept.

HABITAT Wet meadows, streambanks, marshes and swamps, often in sandy soils.

WETLAND STATUS
GP FACW | MIDW FACW | WMTN FACW

Potentilla supina
BUSHY CINQUEFOIL

Spiraea alba
WHITE MEADOWSWEET

Rubiaceae *madder family*

Galium BEDSTRAW

Ascending to reclining, often matted perennials from slender rhizomes (those included here), with slender, 4-angled **stems** and small, simple, entire **leaves** in whorls of 4–6. **Flowers** minute, perfect, regular, 1-few in axillary or terminal cymes, the peduncles and pedicels short and slender. **Calyx** none; **corolla** white, inconspicuous, rotate, 3- or 4-lobed; stamens equaling the number of corolla lobes, inserted on the tube; styles 2, short, ovary inferior, 2-celled and 2-lobed, the 2 carpels maturing as dry, globose fruit segments which separate at maturity, sometimes one of the carpels abortive.

1 Corolla lobes 3, obtuse; leaves scaberulous on the margin but not hispidulous *G. trifidum*

1 Corolla lobes 4, acute; leaves hispidulous on the margin 2

2 Mature fruit segments 1–1.5 mm in diameter; fruiting pedicels 1–2.5(4) mm long; leaves eventually reflexed, 1–3 mm wide *G. labradoricum*

2 Mature fruit segments 2.5–3.5 mm in diameter; fruiting pedicels 5–10 mm long; leaves ascending to spreading, (2)3–5(6) mm wide *G. obtusum*

Galium labradoricum (Wieg.) Wieg.
LABRADOR BEDSTRAW

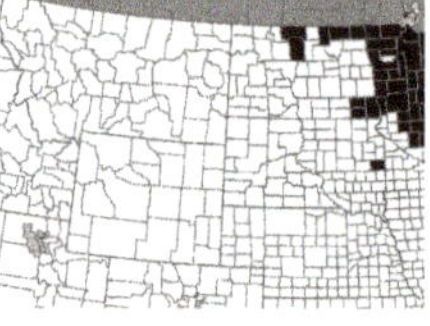

DESCRIPTION **Stems** 1–3 dm long, simple or branched above, pubescent only at the nodes, smooth or scaberulous on the angles. **Leaves** in whorls of 4, soon recurved or deflexed, linear-oblanceolate, 8–15 mm long, 1–3 mm wide, hispidulous on the margins, smooth on the midvein or mostly so, blunt-tipped, tapered to the base. **Inflorescences** few, mostly terminal, 3-flowered, soon overtopped by ascending lateral branches. Corolla lobes 4, 1–1.5 mm long, acute. **Fruit segments** black, 1–1.5 mm in diameter; fruiting pedicels 1–2.5(4) mm long. June–July.

HABITAT Bogs, fens and swamps.

WETLAND STATUS
GP OBL | MIDW OBL

Galium obtusum Bigel.
BLUNTLEAF BEDSTRAW

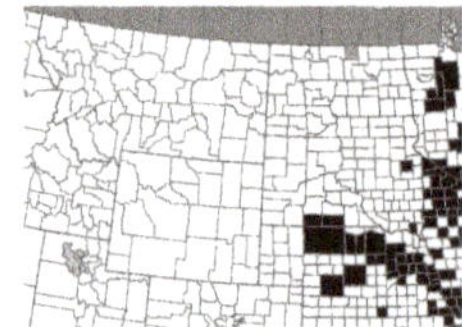

DESCRIPTION **Stems** 2–6 dm long, branched throughout or mainly from the base, pubescent at the nodes, otherwise glabrous. **Leaves** in whorls of 4(–6), ascending to spreading, linear to lanceolate, oblanceolate or elliptic-oblong, 10–25 (30) mm long, 3–5 mm wide, hispidulous on the margins and slightly so on the midrib beneath, often slightly revolute along the margins, obtuse-tipped. **Inflorescences** terminal on main stems and branches, not overtopped by lateral branches. Corolla lobes 4,1–1.3 mm long, acute. **Fruit segments** black, 2.5–3.5 mm in diameter at maturity; fruiting pedicels 5–10 mm long. May–July.

HABITAT Wet meadows, streambanks, ditches, wet thickets and floodplains.

WETLAND STATUS
GP FACW | MIDW FACW

Galium trifidum L.
THREE-PETAL BEDSTRAW

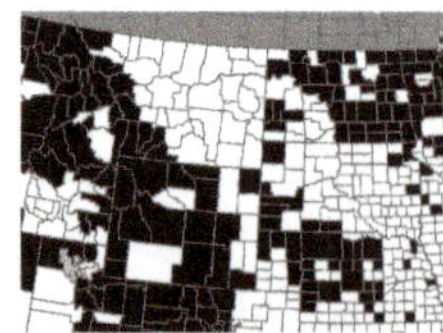

DESCRIPTION **Stems** 2–6 dm long, freely branching, retrorsely scaberulous on the angles. **Leaves** in whorls of 4–6, mostly

Galium labradoricum
LABRADOR BEDSTRAW

spreading, linear to narrowly elliptic or oblanceolate, 5–20 mm long, 1–3.5 mm wide, often scaberulous on the margin and on the midrib of the underside, blunt-tipped, narrowed to the base. **Inflorescences** usually many, axillary and terminal, 1- to 3-flowered. Corolla lobes 3, ca. 0.5 mm long, obtuse. **Fruit segments** black, 1–2 mm in diameter. June–Aug.

HABITAT Springs, seepage areas, shores, streambanks and swampy or boggy places, where water is fresh.

WETLAND STATUS
GP OBL | MIDW FACW | WMTN FACW

Galium obtusum
BLUNTLEAF BEDSTRAW

Galium trifidum
SMALL BEDSTRAW

Salicaceae *willow family*

Dioecious spring-flowering trees and shrubs with simple, alternate, usually stipulate leaves, the stipules often deciduous. **Flowers** in erect or pendulous, bracteate catkins, the bracts small and scalelike, often deciduous, each flower provided with either 1 or 2 enlarged basal glands (*Salix*) or an oblique, cup-shaped disk (*Populus*) positioned just inside the bract; perianth none; **male flowers** each consisting of 2-many stamens; **female flowers** comprised of a single pistil; carpels 2 or 4; stigmas equaling the number of carpels, sessile or with a common style, ovary 1-celled, ovules many. **Fruit** a many-seeded capsule, dehiscent by 2–4 valves; seeds minute, tufted with long, white silky hairs.

1 Large tree; leaf blades deltoid-ovate; flowers subtended by an obliquely cup-shaped disk; catkin bracts fimbriate; stamens (30)40–80; capsules 3- or 4-valved **Populus**

1 Shrubs and trees; leaf blades ovate, elliptic, lanceolate, linear-lanceolate, obovate or oblanceolate; flowers subtended by 1 or 2 enlarged glands; catkin bracts entire; stamens 2 or 3–8(12); capsules 2-valved **Salix**

Populus deltoides Bartr. ex Marsh.
EASTERN COTTONWOOD

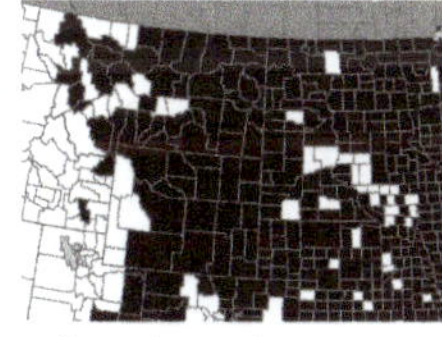

DESCRIPTION Large tree 20–30(40) m tall, with a massive trunk often 1 m or more in diameter, divided into large ascending branches near the base, forming a large rounded crown; **bark** gray, deeply furrowed; **twigs** olive-brown to yellowish, turning grayish with age; leaf buds covered by several bud scales, tan, ovoid, very resinous. **Leaves** light green, deltoid-ovate, mostly 4–10 cm long, 4–11 cm wide, caudate-acuminate at the tip, finely to coarsely crenate-serrate, obtuse to broadly truncate or cordate at the base; petioles flattened at the junction with the blade, 3–10 cm long; stipules minute, caducous. **Catkins** loosely flowered, pendulous; bracts fimbriate, caducous; flowers subtended by a cupshaped disk 1.5–4 mm wide; **male catkins** dark red, soon deciduous; male flowers of

(30)40–80 stamens; **female catkins** greenish, 7–13 cm long in flower, to 20.5 cm long in fruit; female flowers with stigmas expanded and spreading, platelike. **Capsules** 3- or 4-valved, elliptic-ovoid, 6–15 mm long. Flowering late Apr-May, fruiting June–July.

HABITAT Floodplains, stream courses, shores, wet meadows, ditches and ravines, also commonly planted in shelter belts.

WETLAND STATUS
GP FAC | MIDW FAC | WMTN FAC

ADDITIONAL SPECIES Among the species of *Populus*, cottonwood is the best known and the one most overwhelmingly associated with wetlands; however, **balsam poplar** (*P. balsamifera* L.) and its occasional hybrid with cottonwood, called **balm-of-gilead** (*P. x jackii* Sarg.) are sometimes found in lowland areas, especially in the eastern and northern parts of our region. Also in northern North Dakota, **aspen** (*P. tremuloides* L.) is often associated with wetland basins, although elsewhere it is typically upland in occurrence. A hybrid between *P. deltoides* and *P. angustifolia* James, called **lanceleaf cottonwood** (*P. x acuminata* Rydb.) is of uncommon occurrence along stream courses in the western portion.

Populus balsamifera
BALSAM POPLAR

Populus deltoides
EASTERN COTTONWOOD

Salix WILLOW

Shrubs and trees of typically wet or moist habitats. **Leaves** variable in shape, from ovate to lanceolate or linear-lanceolate, or obovate to oblanceolate, the margins serrate, crenate-serrate or entire; petioles glandular or glandular-viscid at the summit in some species; stipules persistent or caducous, occasionally lacking. **Catkins** sessile or on leafy branchlets, erect to pendulous, often precocious; bracts entire, usually pubescent, often apparently ciliate on the margins. **Flowers** each subtended by 1 or 2 enlarged basal glands; **male flowers** of commonly 2 or 3–8(12) stamens, the filaments sometimes connate; **female flowers** each comprised of a 2-carpellary pistil, stigmas 2- or 4-lobed, styles well-developed or none. **Capsules** 2-valved, sessile or stipitate.

1 Plants with female catkins 2
1 Plants vegetative or with male catkins, but with fully expanded leaves 19

2 Capsules glabrous (ovary pubescent in *S. interior*, but glabrous at maturity) 3
2 Capsules pubescent 13

3 Petioles bearing conspicuous, irregularly lobate glands at or near the attachment to the blade 4
3 Petioles lacking glands or sometimes with minute vestiges of glands, or the petioles only glandular-viscid when young (often persistently glandular-viscid in *S. fragilis*) 6

4 Leaves ovate-lanceolate, green on both surfaces, paler beneath but not white-glaucous; capsules 4–6.5 mm long at maturity 5
4 Leaves elliptic-lanceolate, white-glaucous beneath; capsules 7–10 mm long at maturity ***S. serissima***

5 Leaves acute to short-acuminate, dark green and glossy above, thick and rather leathery ***S. pentandra***
5 Leaves mostly long-acuminate, bright green and semi-glossy above, not especially thick ***S. lucida***

6 Leaves entire **S. pedicellaris**

6 Leaves conspicuously or inconspicuously toothed 7

7 Leaves linear-lanceolate, mostly 8–20x longer than wide; rhizomatous shrub often forming dense thickets **S. interior**

7 Leaves usually less than 10x longer than wide; trees and nonrhizomatous shrubs 8

8 Shrubs or small trees up to 5(7) m tall; catkins emerging before or with the leaves, sessile or on short branchlets with a few small leaves; bracts brown to nearly black, persistent after capsule maturity 9

8 Trees eventually 10–20 m tall; catkins emerging after the leaves, on leafy branchlets; bracts yellowish-green or pale yellow, deciduous before capsule maturity 11

9 Catkins sessile; leaves ovate to obovate **S. pseudomonticola**

9 Catkins sessile or often on short branchlets bearing a few small leaves; leaves lanceolate to somewhat oblanceolate 10

10 Twigs gray-brown to dark brown, closely gray-pubescent the first year and often into the second **S. eriocephala**

10 Twigs yellow or yellowish-gray to yellowish-brown, glabrous **S. lutea**

11 Capsules on stipes 1–2 mm long; native tree **S. amygdaloides**

11 Capsules on stipes 1 mm or less long; introduced trees frequently escaping 12

12 Twigs olive to brown; petioles glandular-viscid near the summit on vigorous shoots **S. fragilis**

12 Twigs golden-yellow to orange; petioles lacking glands or with minute traces of glands only **S. alba**

13 Catkins emerging and maturing ahead of the leaves 14

13 Catkins emerging and maturing with the leaves 16

14 Twigs of the previous year gray-pubescent or mostly so; leaves persistently gray-pubescent beneath (rarely glabrate); catkins 1–3(4) cm long; shrub usually of rather dry sites, often where sandy **S. humilis**

14 Twigs of the previous year glabrous (rarely pubescent in *S. discolor*); mature leaves glabrous beneath at maturity; catkins 2–6(9) cm long; shrubs of wet habitats **S. discolor**

15 Capsules nearly sessile or on stipes to 1 mm long **S. planifolia**

15 Capsules on stipes 1.5–4 mm long **S. discolor**

16 Leaves entire or merely crenate-serrate, the shallow teeth unevenly distributed around the margin 17

16 Leaves evenly serrate or mostly so 18

17 Leaves persistently white-tomentose beneath, linear-oblong to oblong or narrowly lanceolate; stipes ca. 1 mm long **S. candida**

17 Leaves grayish-pubescent to glabrate beneath, elliptic to narrowly ovate or narrowly obovate; stipes 2–5 mm long **S. bebbiana**

18 Capsules 8–10 mm long, gray-tomentose; leaves somewhat paler green beneath but not glaucous, mostly 2–3.5x longer than wide **S. maccalliana**

18 Capsules 5–7 mm long, closely pubescent mostly toward the base; leaves white-glaucous beneath, mostly 3–6x longer than wide **S. petiolaris**

19 Petioles bearing lobate glands at or near the attachment to the blade; leaves finely glandular-serrate; stamens 3–8(12) 20

19 Petioles lacking glands or sometimes with minute vestiges of glands, or the petioles only glandular-viscid, in which case the leaves are narrowly lanceolate to lanceolate; leaves serrate or entire, occasionally glandular-serrate; stamens 2, except 4–7 in *S. amygdaloides* 23

20 Leaves elliptic-lanceolate, white-glaucous beneath **S. serissima**

20 Leaves ovate-lanceolate, green on both surfaces, paler beneath but not white-glaucous 21

21 Leaves acute to short-acuminate, glossy above, thick and rather leathery **S. pentandra**

21 Leaves mostly long-acuminate, semi-glossy above, not especially thick **S. lucida**

22 Leaves linear-lanceolate, mostly 8–20x longer than wide, entire to remotely serrulate; rhizomatous shrub often forming dense thickets **S. interior**

22 Leaves broader in proportion to their length; non-rhizomatous shrubs and trees 23

23 Leaves acuminate, gradually or abruptly tapered to a long, slender tip 24

23 Leaves acute, obtuse, rounded or only short-acuminate at the tip 28

24 Leaves dark green and shiny above; twigs often brittle and easily snapping off at the base; introduced trees to 20 m tall 25

24 Leaves yellowish-green to dark green and dull above; twigs flexible, not easily snapping off at the base; native shrubs or trees to 12 m tall 26

25 Leaves coarsely serrate, with 4–6 glandular teeth per cm of leaf margin; petioles glandular-viscid at the summit; twigs olive to brown **S. fragilis**

25 Leaves more finely serrate, with 7–10 teeth per cm of leaf margin; petioles not glandular-viscid at the summit, or with only minute vestiges of glands; twigs golden-yellow to orange **S. alba**

26 Leaves ovate-lanceolate to lanceolate, mostly long-acuminate with tail- like tips; petioles commonly recurved; branchlets flexuous, somewhat drooping **S. amygdaloides**

26 Leaves lanceolate or somewhat oblanceolate, acuminate; petioles straight; branchlets erect to spreading, not drooping 27

27 Twigs gray-brown to dark brown, closely gray-pubescent the first year and often into the second **S. eriocephala**

27 Twigs yellow or yellowish-gray to yellowish-brown, glabrous **S. lutea**

28 Leaves persistently pubescent, especially on the lower surface (rarely glabrate in age in S. humilis) 29

28 Leaves glabrous or glabrate with age 31

29 Leaves elliptic, narrowly ovate or narrowly obovate, sparsely to densely pubescent beneath; leaf margins flat **S. bebbiana**

29 Leaves linear-oblong to narrowly lanceolate or oblanceolate to narrowly obovate, densely pubescent or white-tomentose beneath (rarely glabrate beneath in *S. humilis*); leaf margins usually revolute 30

30 Leaves linear-oblong to oblong or narrowly lanceolate, white-tomentose beneath; leaf margins revolute; shrubs of cold springs or fens **S. candida**

30 Leaves oblanceolate to narrowly obovate, densely pubescent (rarely glabrate) and greenish beneath; leaf margins flat to slightly revolute; shrubs of drier, often sandy habitats **S. humilis**

31 Leaves entire or nearly so, or with a few scattered inconspicuous teeth, sometimes to crenate-serrate with the teeth distributed unevenly around the margins 32

31 Leaves mostly serrate or finaly serrate, the teeth evenly distributed around the margins 35

32 Small bog shrubs 4–10 dm tall; leaves elliptic-lanceolate to oblanceolate, 2–4.5 cm long, acute to rounded and often apiculate at the tip **S. pedicellaris**

32 Larger shrubs and small trees of various habitats, mostly 2–7 m tall; leaves of various shapes, mostly 3–10 cm long, never apiculate at the tip 33

33 Leaves dull grayish-green above, the lower surface usually rugose, with the veins raised prominently on the lower surface **S. bebbiana**

33 Leaves bright to dark green above, smooth beneath, only the primary veins, if any, raised on the lower surface **34**

34 Leaves entire or nearly so, 3–6 cm long; twigs reddish-brown to nearly black, shiny ***S. planifolia***

34 Leaves, or at least the larger ones, crenate-serrate, 4–10 cm long; twigs yellowish-brown to dark brown, dull ***S. discolor***

35 Stipules persistent, often prominent; leaves ovate to obovate, rounded to cordate at the base ***S. pseudomonticola***

35 Stipules lacking or caducous; leaves generally lanceolate, acute to obtuse at the base **36**

36 Leaves paler below than above, but not glaucous, elliptic-lanceolate to oblanceolate, mostly 2–3.5x longer than wide ***S. maccalliana***

36 Leaves white-glaucous beneath, narrowly lanceolate or narrowly oblanceolate, mostly 3–6x longer than wide ***S. petiolaris***

Salix alba L.

WHITE WILLOW

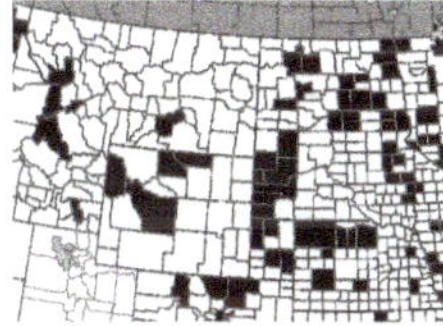

DESCRIPTION Large tree to 20 m tall; **twigs** golden yellow to orange, brittle or sometimes flexible; **branchlets** spreading, golden yellow to dark brown, glabrous with age. **Leaves** dark green and shiny above, white-glaucous beneath, glabrous to sparsely sericeous beneath, lanceolate to narrowly lanceolate, acuminate and often symmetric at the tip, cuneate at the base, mostly 4–10 cm long, 1–2.5 cm wide, serrate, mostly with 7–10 glandular teeth per cm of margin; petioles glandless or with minute vestiges of glands at the summit, 0.5–1.5 cm long; stipules caducous, lanceolate, entire, 2–4 mm long, sericeous. **Catkins** appearing with the leaves; female catkins 3–6 cm long, on leafy branchlets 1–3(5) cm long; bracts yellowish-green to pale yellow, early deciduous, pubescent, ciliate at the tip; stamens 2. **Capsules** ovoid-conic, 3.5–5 mm long, glabrous, nearly sessile or on stipes to 1 mm long. Flowering May, fruiting early June.

HABITAT Introduced from Europe and escaping to wet areas from shelter belts and ornamental plantings throughout the region.

NOTE The golden yellow to orange twigs characteristic of some trees are especially conspicuous during winter and early spring. *S. alba* hybridizes freely with *S. fragilis* and many collections seem to show introgression with that species.

WETLAND STATUS

GP FACW | MIDW FACW | WMTN FACW

Salix amygdaloides Anderss.

PEACHLEAF WILLOW

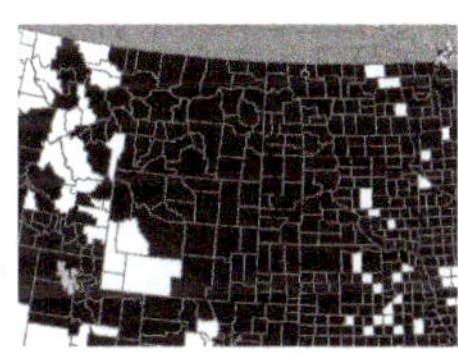

DESCRIPTION Small to medium-sized tree with 1–several trunks, to 12 m tall; **twigs** gray to light yellow, shiny, flexible; **branchlets** spreading to drooping, yellow to dark brown, glabrous. **Leaves** yellowish-green above, pale to white-glaucous beneath, glabrous, lanceolate to ovate-lanceolate, short to mostly long-acuminate with tail-like

Salix alba
WHITE WILLOW

Salix amygdaloides
PEACHLEAF WILLOW

tips, acute to nearly rounded at the base, mostly 3–8 cm long, 1–3 cm wide, occasionally much larger on vigorous shoots, finely serrate; petioles glandless or rarely with vestiges of glands on vigorous shoots, often recurved, 5–20(30) mm long; stipules minute and very early deciduous, occasionally well-developed and persistent on vigorous shoots, reniform, 3–12 mm long, serrate. **Catkins** emerging with the leaves; female catkins 3–8 cm long, on leafy branchlets 1–4 cm long; bracts deciduous, pale yellow, villous on the inside; stamens 4–7. **Capsules** ovoid, 3–5 mm long, glabrous, uncrowded on the axis giving the catkins a loose, open appearance; stipes 1–2 mm long. Flowering May, fruiting June.

HABITAT Floodplains, streambanks, lake and pond borders, moist ravines, ditches and other wet or damp places.

WETLAND STATUS
GP FACW | MIDW FACW | WMTN FACW

Salix bebbiana Sarg.
BEAKED WILLOW

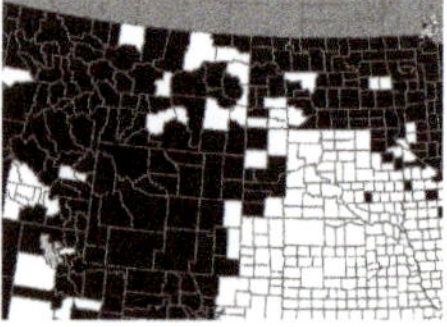

DESCRIPTION Shrub to 4 m tall; **twigs** grayish-brown, closely pubescent to eventually glabrous, gnarled and rough in appearance owing to jutting leaf scars, irregular growth and die back; **branchlets** spreading, yellowish-brown to dark brown, tomentulose, occasionally glabrate toward the base. **Leaves** dull grayish-green and glabrous to pubescent above, pale to gray-pubescent and rugose beneath, with the veins raised prominently on the lower surface, elliptic to narrowly ovate or narrowly obovate, acute to short-acuminate, cuneate at the base, mostly 3–6 cm long, 1–3 cm wide, entire to shallowly toothed; petioles glandless, 5–10(15) mm long; stipules deciduous or persistent on vigorous shoots, ovate to reniform, 2–6 mm long, 1–3 mm wide, shallowly dentate. **Catkins** emerging and maturing with the leaves; female catkins persistent for some time after capsule dehiscence, 2–5 cm long, on short leafy branchlets 0.5–2 cm long, with 2–4 small leaves; bracts persistent, pale with a reddish or darkened tip when young, yellowish to brown with age, villous; stamens 2. **Capsules** ovoid-conic, 5–8 mm long, finely pubescent; stipes 2–5 mm long. Flowering late Apr-May, fruiting late May–June.

HABITAT Wet meadows, streambanks, moist wooded ravines and hillsides, marsh borders and seepage areas.

WETLAND STATUS
GP FACW | MIDW FACW | WMTN FACW

ADDITIONAL SPECIES **Western pussy willow** (*Salix scouleriana* Barr.) is a similar species that occurs on moist slopes at higher elevations in the Black Hills. It differs from *S. bebbiana* in having the leaves arranged in a fanlike fashion at the tips of the branchlets. Also, the leaves have some reddish-brown hairs mixed with silvery ones on one or both surfaces.

Salix bebbiana
BEAKED WILLOW

Salix candida Fluegge
SAGE-LEAVED WILLOW, HOARY WILLOW

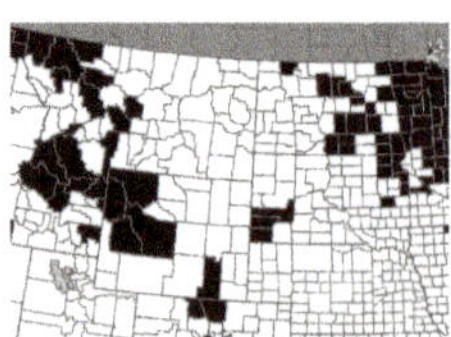

DESCRIPTION Low shrub to 1.5 m tall; **twigs** yellow to reddish-brown or brown, usually with patches of white to-

Salix candida
SAGE-LEAVED WILLOW, HOARY WILLOW

mentum; **branchlets** strongly ascending, yellow to brown, mostly white-tomentose. **Leaves** dark green and glabrate or thinly white-tomentose above, densely white-tomentose beneath, linear-oblong to oblong or narrowly lanceolate, acute at the tip, cuneate at the base, mostly 3–9(11) cm long, 0.5–1.5(2) cm wide, the margin revolute; petioles glandless, 3–10 mm long; stipules persistent, obliquely ovate to lanceolate, 2–10 mm long, tomentose, entire or serrulate. **Catkins** emerging with the leaves; female catkins 1.5–4.5 cm long, on leafy branchlets 4–15 mm long, with 2 or 3 small leaves; bracts persistent, yellow to brown, villous; stamens 2. **Capsules** narrowly ovoid, 4–8 mm long, white-tomentose; stipes 1 mm long. May–June.

HABITAT Cold springs, fens and boggy areas associated with marshes and streams.

WETLAND STATUS
GP OBL | MIDW OBL | WMTN OBL

Salix discolor Muhl.
PUSSY WILLOW

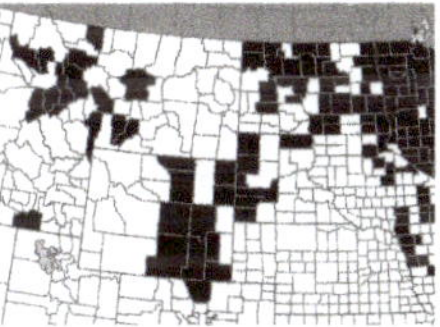

DESCRIPTION Shrub or small tree to 5 m tall; **twigs** reddish-brown to dark brown, dull, glabrous to slightly pubescent, rarely densely pubescent; **branchlets** spreading, yellowish-brown to nearly black, tomentulose, often glabrous with age. **Leaves** bright to dark green above, pale to white-glaucous beneath, glabrous, not rugose beneath, only the primary veins, if any, raised on the lower surface, elliptic to narrowly ovate or narrowly obovate, acute to short-acuminate, cuneate to narrowly rounded at the base, mostly 3–10 cm long, 1–3 cm wide, subentire to more often shallowly and irregularly crenate-serrate; petioles glandless, 5–20 mm long; stipules deciduous, often persistent on vigorous shoots, obliquely ovate to flabellate, 3–10 mm long, about as wide, glabrous, sometimes deeply lobed. **Catkins** emerging and maturing before the leaves; female catkins sessile, sometimes with 2 or 3 minute, bract-like leaves at the base, soon deciduous after capsule dehiscence, 2–6(9) cm long; bracts persistent, black or very dark brown, villous; stamens 2. **Capsules** ovoid with a long neck, 5–10 mm long, finely pubescent; stipes 1.5–4 mm long. Flowering mid Apr–early May, fruiting mid May–early June.

HABITAT Swamps, fens, streambanks, floodplains, marsh borders, ditches and other wet places.

WETLAND STATUS
GP FACW | MIDW FACW | WMTN FACW

Salix discolor
PUSSY WILLOW

Salix eriocephala Michx.
DIAMOND WILLOW, MISSOURI WILLOW

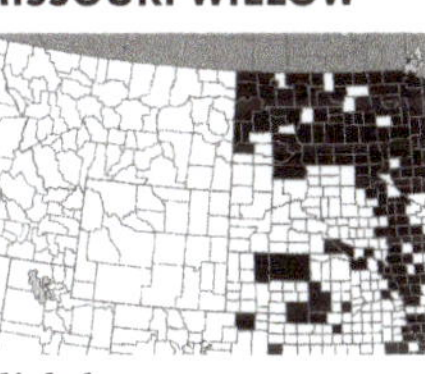

DESCRIPTION Shrub or small tree to 7 m tall; **twigs** gray-brown to dark brown, closely gray-pubescent, the pubescence often patchy; **branchlets** reddish-brown, gray-pubescent. **Leaves** dark green to yellowish-green above, pale to weakly glaucous beneath, glabrous on both sides or pubescent

Salix eriocephala
DIAMOND WILLOW, MISSOURI WILLOW

beneath, lanceolate to somewhat oblanceolate, acuminate at the tip, cuneate, rounded or cordate at the base, 3–8(12) cm long, 1–3(4) cm wide, finely serrate; petioles glandless, 3–15 mm long; stipules persistent on vigorous shoots, semicordate, ovate or reniform, to 12 mm long, glabrous, serrate. **Catkins** emerging with or prior to the leaves; female catkins 2–8 cm long, on short leafy or bracteate branchlets to 1.5 cm long; bracts persistent, brown to nearly black, pubescent; stamens 2. **Capsules** ovoid with a long neck, 4–7 mm long, glabrous; stipes 1–2 mm long. Flowering Apr–early May, fruiting May–early June.

SYNONYMS *Salix rigida* var. *vestita, Salix missouriensis* Bebb.

HABITAT Shores, streambanks, floodplains, ditches and wet meadows, especially along major river courses.

WETLAND STATUS
GP FACW | MIDW FACW | WMTN OBL

Salix interior Rowlee
SANDBAR WILLOW

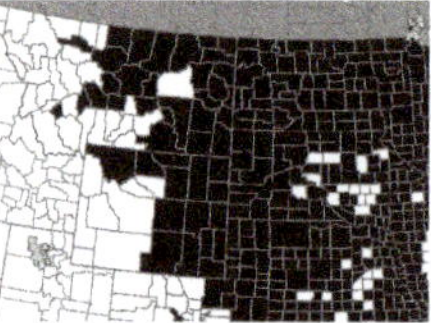

DESCRIPTION Colonial, rhizomatous shrub to 4 m tall, often forming dense thickets; **twigs** light yellow to orange, glabrous; **branchlets** erect, yellow to orange, glabrous. **Leaves** yellowish-green above, the same or paler beneath, initially pubescent and soon glabrous (rarely persistently silvery-pubescent), linear-lanceolate, slowly tapered to an acute tip, acuminate at the base, 4–10 cm long, 2–10 mm wide, remotely and irregularly dentate; petioles glandless, 1–5 mm long; stipules minute or absent. **Catkins** emerging after the leaves, borne on leafy branchlets 0.5–10 cm long, these often branched; female catkins 1.5–8 cm long; bracts deciduous, yellowish; stamens 2. **Capsules** narrowly ovoid, 4–8 mm long, glabrous (although pubescent when immature); stipes 0.5–1 mm long. Flowering May–early June, fruiting June–early July.

HABITAT Shores, streambanks, alluvial bars, ditches and other wet places; often a pioneer species on sand bars and other alluvium; common.

WETLAND STATUS
GP FACW WMTN FACW

NOTE *Salix interior* occurs throughout our area, with leaves usually glabrous at maturity (rarely silvery-pubescent), capsules 5–8 mm long and distinctly stipitate so that the female catkins appear rather loose and elongate. ***Salix exigua*** Nutt. (coyote willow), enters our range from the west, occurring in w SD, w Neb, e Mont and e Wyo. It differs from *S. interior* in having the leaves persistently gray-pubescent (at least beneath); capsules 3–5 mm long, sessile or nearly so, the female catkins mostly dense and short. *S. exigua* is characteristic of w North America.

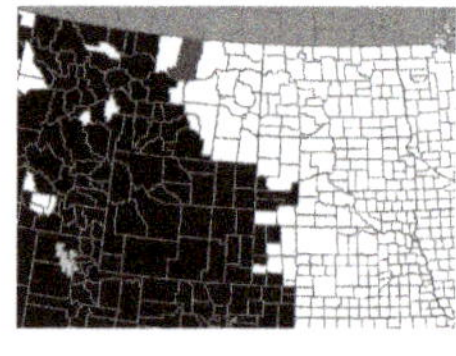

Salix fragilis L.
CRACK-WILLOW, BRITTLE WILLOW

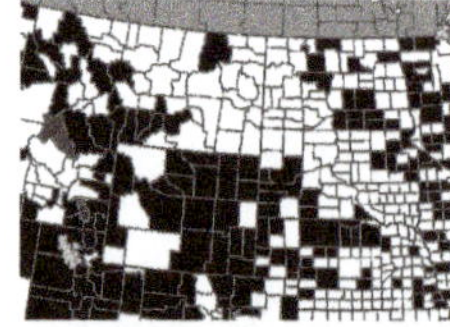

DESCRIPTION Large tree to 20 m tall; **twigs** olive to yellowish-brown, brittle, easily snapping off at the base; **branchlets**

Salix interior
SANDBAR WILLOW

Salix fragilis
CRAK-WILLOW, BRITTLE WILLOW

spreading, green to reddish-brown, eventually glabrous. **Leaves** dark to yellowish-green and shiny above, pale to white-glaucous beneath, glabrous, lanceolate to narrowly lanceolate, acuminate, often asymmetric at the tip, acute at the base, mostly 7–13 cm long, 1.5–3 cm wide, coarsely serrate, mostly with 4–6 glandular teeth per cm of margin; petioles 0.5–1.5 cm long, glandular-viscid at the summit, the glands often stipitate; stipules caducous, narrowly lanceolate, 2–3 mm long when well-developed, pubescent, entire. **Catkins** appearing with the leaves; female catkins 3–6 cm long, on leafy branchlets 1–3(5) cm long; bracts early deciduous, yellowish, pubescent, ciliate at the tip; stamens 2. **Capsules** narrowly conic, 4–5.5 mm long, glabrous, subsessile or on stipes to 1 mm long. Flowering May, fruiting early June.

SYNONYM *Salix* x *fragilis* L.

HABITAT Introduced from Europe and planted as a shade tree, sometimes escaping to wet places throughout the region.

WETLAND STATUS
GP FAC | MIDW FAC | WMTN FAC

NOTE See discussion under *Salix alba.*

Salix humilis Marsh.

PRAIRIE WILLOW

DESCRIPTION Shrub to 3 m tall; **twigs** yellowish-brown to dark brown, gray-pubescent or mostly so; **branchlets** strongly ascending, brown, gray-pubescent. **Leaves** dark green and usually glabrous above, glaucous and densely short-pubescent (rarely glabrate) beneath, with the golden-yellow veins raised prominently on the lower surface, oblanceolate to narrowly obovate, acute, cuneate at the base, mostly (1.5)4–8 cm long, 7–25 mm wide, the margins coarsely and irregularly serrate to subentire, flat to slightly revolute; petioles glandless, 3–10 mm long; stipules commonly persistent on vigorous branchlets, lanceolate to ovate, 3–7 mm long, pubescent, sparsely serrate. **Catkins** emerging and maturing before the leaves; female catkins sessile, 1–3(4) cm long; bracts persistent, dark brown or purplish, villous on the back; stamens 2. **Capsules** ovoid-conic, 4–6(8) mm long, pubescent; stipes 0.5–1.5 mm long. Flowering Apr-May, fruiting May–early June.

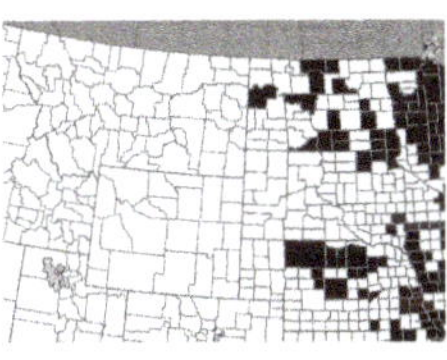

Salix humilis
PRAIRIE WILLOW

SYNONYM *Salix tristis* Ait.

HABITAT Moist or dry places, often in sandhill areas.

WETLAND STATUS
GP FACU | MIDW FACU

Salix lucida Muhl.

SHINING WILLOW

DESCRIPTION Shrub or small tree to 4 m tall; **twigs** gray to yellowish-brown; **branchlets** ascending, yellowish-brown to dark brown, glabrous. **Leaves** yellowish-green to green and semiglossy above, pale beneath, initially reddish-pubescent, soon glabrous, lanceolate to ovate-lanceolate, acuminate to long-acuminate and asymmetric at the tip, broadly cuneate to nearly rounded at the base, 4–8(12) cm long, 1.2–2.5(4) cm wide, finely glandular-serrate; peti-

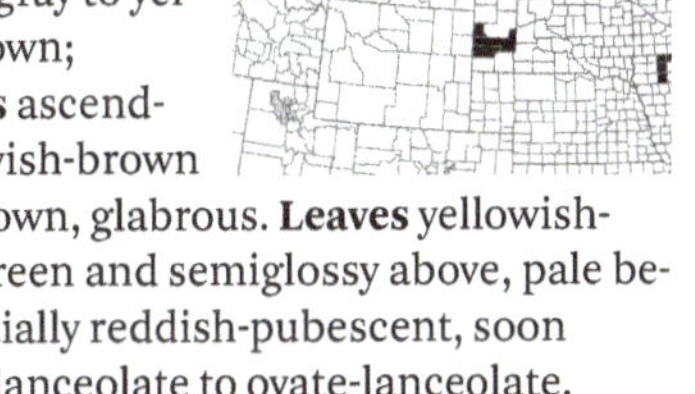

Salix lucida
SHINING WILLOW

oles glandular above, usually with few to several lobate glands, 0.5–1.5(2) cm long; stipules often persistent for some time toward the tips of branchlets, flabellate, well-developed ones 2–3 mm long, 2–4 mm wide, strongly glandular. **Catkins** produced with the leaves; female catkins 1–3 cm long, on leafy branchlets 1–3 cm long; bracts deciduous, yellowish, pubescent; stamens 3–5 or more. **Capsules** ovoid with a long neck, 4–6.5 mm long, glabrous; stipes 0.5–1.5 mm long. Flowering May, fruiting June.

HABITAT Swamps, shores and wet meadows. Fairly common in the Turtle Mts, ND; also SD Black Hills.

WETLAND STATUS
GP FACW | MIDW FACW | WMTN FACW

NOTE Many earlier accounts of *Salix lucida* were based largely on misidentified specimens of *S. amygdaloides*.

Salix lutea Nutt.
YELLOW WILLOW

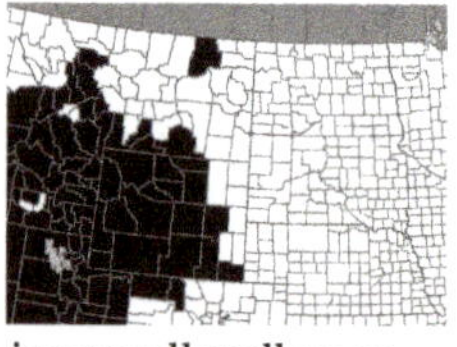

DESCRIPTION Very similar to *Salix eriocephala* and perhaps better treated as a variety of it. More often shrubby, to 5 m tall; twigs usually yellow or yellowish-gray to yellowish-brown, glabrous or nearly so.

SYNONYMS *Salix cordata* Muhl., *Salix rigida* var. *watsonii* (Bebb) Cronq.

HABITAT Shores, streambanks, floodplains, ditches and wet meadows; more common than *S. eriocephala*.

WETLAND STATUS
GP FACW | WMTN OBL

Salix lutea
YELLOW WILLOW

Salix maccalliana Rowlee (*not illustrated*)
MCCALLA'S WILLOW

DESCRIPTION Upright shrub mostly 1–2 m tall; **twigs** reddish-brown to purplish-brown, glabrous; **branchlets** spreading, yellowish to purplish-brown, glabrous. **Leaves** rather firm and leathery, dark green and glossy above, somewhat paler (but not glaucous) and conspicuously reticulate beneath, glabrous, elliptic-lanceolate to oblanceolate, acute to short-acuminate at the tip, acute to obtuse at the base, mostly 4–8 cm long, 12–25 mm wide, the margins glandular-serrate; petioles glandless, 4–10 mm long; stipules lacking. **Catkins** emerging with the leaves; pistillate catkins on short, leafy branchlets; bracts persistent, dark brown to yellowish, pubescent on the back, especially toward the base; stamens 2. **Capsules** elongate-conic, 8–10 mm long, gray-tomentose; stipes 1–2 mm long. Flowering May, fruiting June.

HABITAT Swamps and bogs; rare, with one record from Bottineau Co., ND, more common northward.

WETLAND STATUS
GP OBL | MIDW OBL

Salix pedicellaris Pursh
BOG WILLOW

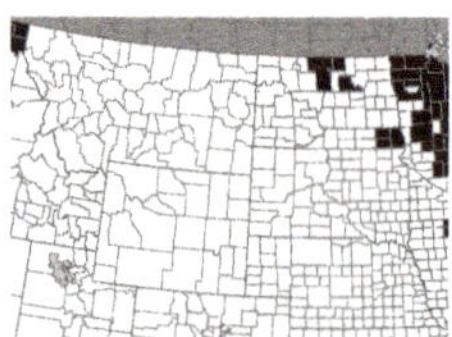

DESCRIPTION Slender shrub 4–10 dm tall; **twigs** grayish-brown, glabrous; **branchlets** erect to spreading, dark brown, glabrous. **Leaves** green above, white-glaucous

Salix pedicellaris
BOG WILLOW

beneath, glabrous, narrowly oblanceolate to oblanceolate or obovate, acute to obtuse and often apiculate at the tip, acute to obtuse at the base, 2–4(6) cm long, 1–1.5(2) cm wide, the margin entire, often slightly revolute; petioles glandless, 2–8 mm long; stipules absent. **Catkins** emerging with the leaves; female catkins 1–3 cm long, on leafy branchlets 1–3 cm long; bracts persistent, yellow to brown, glabrous or pubescent only at the tip; stamens 2. **Capsules** narrowly conic, 5–8 mm long, glabrous; stipes 2–4 mm long. Flowering late May–early June, fruiting June–early July.

HABITAT Sphagnum bogs, swamps and fens; rare.

WETLAND STATUS
GP OBL | MIDW OBL

Salix pentandra L.
LAUREL-LEAVED WILLOW

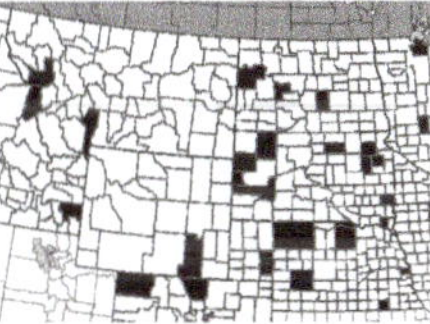

DESCRIPTION Medium-sized tree or shrub 2–8 m tall; **twigs** yellowish-green, shiny; **branchlets** spreading, dark brown and shiny. **Leaves** dark green and glossy above, light green and dull beneath, glabrous, thick and leathery, ovate to ovate-lanceolate, acute to short-acuminate at the tip, obtuse to rounded at the base, mostly 4–10 cm long, 2–3 cm wide, finely glandular-serrate; petioles strongly glandular at the summit, 5–10 mm long; stipules deciduous or persistent for a short time on vigorous shoots, reniform, ca. 1 mm long, 2 mm wide, glandular-dentate. **Catkins** produced after the leaves; female catkins 3–5 cm long, on leafy branchlets 2–3 cm long; bracts early deciduous, yellowish, pubescent; stamens (4)5(–12). **Capsules** ovoid-conic, the 2 halves bulged at the base, 4–5 mm long, glabrous; stipes 0.5–1 mm long. Flowering May–early June, fruiting June–July.

HABITAT Introduced from Europe as an ornamental, occasionally escaping to marsh borders, ditches, streambanks, ravines and other moist places; established as an escape in much of n USA and s Canada.

Salix pentandra
LAUREL-LEAVED WILLOW

Salix petiolaris Sm.
MEADOW WILLOW

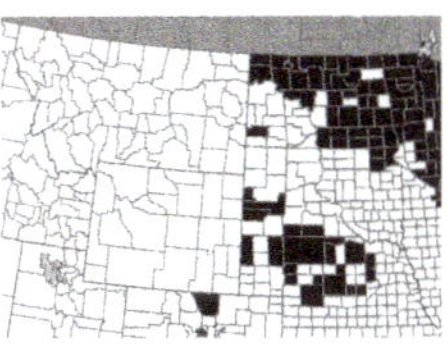

DESCRIPTION
Clumped or few-stemmed shrub to 3 m tall; **twigs** reddish-brown to dark brown or almost black, glabrous; **branchlets** spreading to erect, yellowish-green to dark brown, tomentulose, often glabrous with age. **Leaves** dark green above, white-glaucous beneath, pubescent when young, becoming glabrous with age, narrowly lanceolate to narrowly oblanceolate, acute to abruptly short-acuminate at the tip, acute to slightly rounded at the base, 2.5–8 cm long, 4–15 mm wide, entire to finely serrate; petioles glandless, 3–10 mm long; stipules absent. **Catkins** emerging with the leaves; female catkins 1–3(5) cm long, sessile or on short branchlets to 1.5 cm long, naked or with 2–3 small leaves; bracts persistent, brown, villous; stamens 2. **Capsules** narrowly conic, 5–7 mm long, closely pubescent mostly toward the base; stipes 1–3 mm long. Flowering May, fruiting June.

Salix petiolaris
MEADOW WILLOW

SYNONYM *Salix gracilis* Anderss.

HABITAT Wet meadows, streambanks, shores, ditches and other wet places.

WETLAND STATUS
GP OBL | MIDW OBL | WMTN OBL

Salix planifolia Pursh
PLANELEAF WILLOW

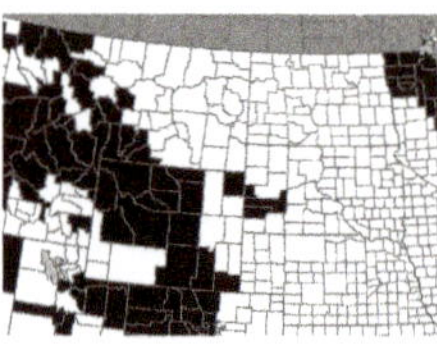

DESCRIPTION Shrub or shrubby tree with clustered trunks, to 3 m tall; **twigs** dark reddish-brown to nearly black, shiny, glabrous; **branchlets** spreading to ascending, brown, glabrous. **Leaves** green above, paler to glaucous beneath, initially short-pubescent but soon glabrous, elliptic to oblanceolate, acute or occasionally obtuse at the tip, rounded to acute at the base, 3–6 cm long, 12–20 mm wide, entire or only sparsely crenulate; petioles glandless, 3–6 mm long; stipules minute, deciduous. **Catkins** emerging slightly before or with the leaves; female catkins 2–4 cm long, sessile or on short branchlets with 1–3 bractlike leaves; bracts persistent, black, villous; stamens 2. **Capsules** ovoid with a long neck, 5–8 mm long, pubescent, nearly sessile or on stipes to 1 mm long. Flowering May, fruiting June.

SYNONYMS *Salix phylicifolia* L. subsp. *planifolia* (Pursh) Hiitonen

Salix planifolia
PLANELEAF WILLOW

HABITAT Streambanks, meadows and moist hillsides at higher elevations in the Black Hills.

WETLAND STATUS
GP OBL | MIDW OBL | WMTN OBL

Salix pseudomonticola Ball
SERVICEBERRY WILLOW

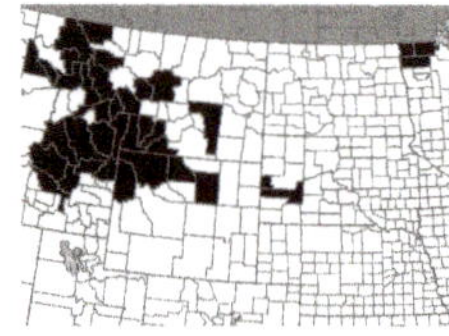

DESCRIPTION Shrub to 3(5) m tall; **twigs** light brown to dark grayish-brown, dull, glabrous; **branchlets** spreading, brown, glabrous. **Leaves** dull green above, paler to glaucous beneath, glabrous, ovate to obovate, acute at the tip, rounded to cordate at the base, 3–8 cm long, 12–35 mm wide, serrate; petioles glandless, 3–15 mm long; stipules persistent, often prominent, broadly ovate, cordate at the base, 5–15 mm long and about as wide, serrate. **Catkins** emerging before the leaves; female catkins 3–7 cm long, sessile or very short-peduncled with 1-few leafy bracts at the base; bracts persistent, dark brown to black, long-villous on the back; stamens 2. **Capsules** ovoid with a narrow neck, 6–8 mm long, glabrous; stipes 1–1.5 mm long. Flowering May, fruiting June.

SYNONYMS *Salix monticola* Bebb ex Coult.

HABITAT Open meadows and streambanks at high elevations in the Black Hills.

WETLAND STATUS
GP FACW | WMTN FACW

Salix pseudomonticola
SERVICEBERRY WILLOW

Salix serissima (Bailey) Fern.

AUTUMN WILLOW

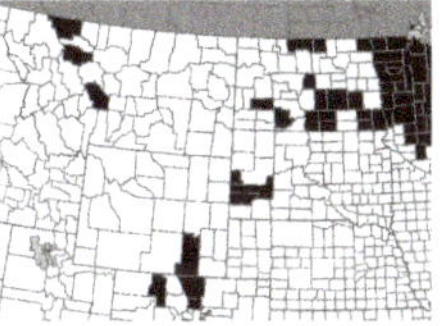

DESCRIPTION Shrub to 3 m tall; **twigs** gray to yellowish-brown, shiny, glabrous; **branchlets** erect to spreading, yellow to dark brown, glabrous. **Leaves** yellowish-green to green and semi-glossy above, white-glaucous beneath, glabrous, elliptic-lanceolate, acute to short-acuminate at the tip, cuneate to narrowly rounded at the base, 4–8 cm long, 1–2.5 cm wide, finely glandular-serrate; petioles glandular at the summit, 0.5–1 cm long; stipules rarely present, flabellate, 1.5 mm long, 2 mm wide, glandular-serrate. **Catkins** emerging after the leaves; female catkins 2–4 cm long, on leafy branchlets 1.5–3 cm long; bracts deciduous, light yellow, pubescent; stamens 3–5 or more. **Capsules** ovoid with a long neck, 7–10 mm long, glabrous; stipes 0.5–2 mm long. Flowering June–early July, fruiting late June–Aug.

HABITAT Swamps, fens and bogs.

WETLAND STATUS
GP OBL | MIDW OBL | WMTN OBL

Salix serissima
AUTUMN WILLOW
(female catkin, leaves)

Saxifragaceae *saxifrage family*

Mitella nuda L.

BISHOP'S CAP

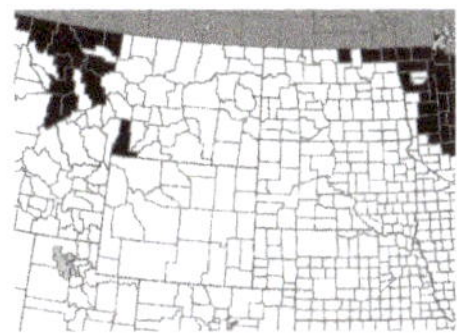

DESCRIPTION Small rhizomatous and often stoloniferous perennial with scapes 0.7- 2.5 dm tall, glandular-pubescent especially upward. **Leaves** all basal or with 1 sessile or short- petioled leaf below the middle on the scape; blades rotundcordate to reniform, 1–3.5 cm across, sparingly hirsute at least on the upper surface, crenate on the margin; petioles mostly 2–9 cm long; stipules brownish, ovate, 2–4 mm long. **Flowers** small, greenish, in racemes of 3–12 flowers; sepals ovate, 1–2 mm long; petals green, pinnately divided into usually 4 pairs of filiform segments, 2–4 mm long; stamens 10; pistil 2-carpellary, stigmas 2, on short divergent styles, ovary ca. 1/2 inferior or less; hypanthium saucer-shaped; pedicels 1–6 mm long. **Capsules** splitting open widely, the seeds rather few, black, shiny, ellipsoid, ca. 1 mm long. June–July.

HABITAT Bogs and swamps, often growing among mosses.

WETLAND STATUS
GP OBL | MIDW FACW

Mitella nuda
BISHOP'S CAP

Tamaricaceae *tamarix family*

Tamarix chinensis Lour.

TAMARISK, SALT CEDAR

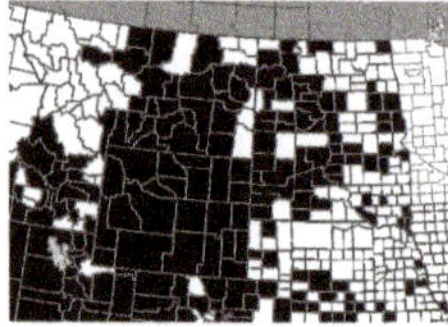

DESCRIPTION Deciduous shrub 1–6 m tall, with reddish-brown bark and glaucous-green foliage. **Leaves** sessile, alternate, small and scalelike, scattered on main shoots but closely overlapping on short, lateral shoots, 1–5 mm long, acute to acuminate, narrowed at base. **Inflorescences** of terminal and lateral bracteate racemes, 1–7 cm long, 3–5 mm thick; bracts similar to the leaves, pink to pale; pedicels shorter than the bracts. **Flowers** perfect, regular, hypogynous, pink or fading to stramineous; sepals 5, spreading, broadly ovate, ca. 0.5 mm long, scarious and erose to denticulate on the margin; petals 5, obovate to elliptic-obovate, 1–1.9 mm long, persistent and pale with age; stamens 5 or seldom more, arising from below and between the lobes of a brown disk, shorter than to surpassing the petals; pistil flask-shaped, usually 3(–5)-carpellary and angled, styles 3(–5), clavate, deciduous, ovary 1-celled. **Fruit** a conic, 3(–5)-valved, many-seeded capsule; seeds tiny, erect, each with a tuft of hairs. June–Sept.

SYNONYMS *Tamarix ramosissima* Ledeb.

HABITAT Streambanks, floodplains, ditches, alkaline or saline flats; sometimes locally common; introduced from Eurasia as an ornamental and now widely established in the s and w USA.

WETLAND STATUS

GP FACW | MIDW FACW | WMTN FAC

Tamarix chinensis
TAMARISK, SALT CEDAR

Urticaceae *nettle family*

Monoecious or dioecious, annual or perennial herbs with watery juice, sometimes beset with stinging hairs, the **leaves** opposite (in those included here), simple, petiolate, usually stipulate. **Flowers** small, greenish and inconspicuous, borne in simple or branched axillary clusters, usually imperfect; perianth consisting of a 3- to 4-parted or toothed calyx, that of the male flowers usually more deeply parted; stamens equal in number to the calyx lobes and opposite them, a vestigial pistil sometimes present; female flowers containing a unicarpellary pistil and sometimes with scalelike rudiments of stamens, style 1, ovary superior, 1-celled, ovule 1. **Fruit** an achene, often enclosed by an accrescent calyx.

1 Plants 1–5 dm tall, glabrous or nearly so, not armed with stinging hairs ... *Pilea*

1 Plants 8–20 dm tall, armed with stiff stinging hairs ... *Urtica*

Pilea CLEARWEED

Monoecious or dioecious, glabrous annuals; **stems** erect to decumbent, simple or branched, rather brittle and watery, translucent; cystoliths appearing as numerous minute, whitish or dark lines on the foliage of dried specimens. **Leaves** simple, opposite; blades thin and translucent, ovate, with 3 major veins arising from the base, broadly cuneate to rounded at the base, serrate, the teeth prominent, obtuse to rounded, the terminal tooth short to elongate; petioles subtending the inconspicuous, connate, membranous stipules. **Flowers** greenish, clustered in axillary cymes; male flowers with 4 sepals and 4 stamens; female flowers with 3 sepals, these often unequal; staminodes minute and scalelike; ovary superior, stigma sessile. **Fruit** a flattened, ovate achene, subtended by the persistent calyx.

1 Achenes dark olivaceous to nearly black with a narrow pale margin ... *P. fontana*

1 Achenes green, often marked with purple ... *P. pumila*

Pilea fontana (Lunell) Rydb.
LESSER CLEARWEED

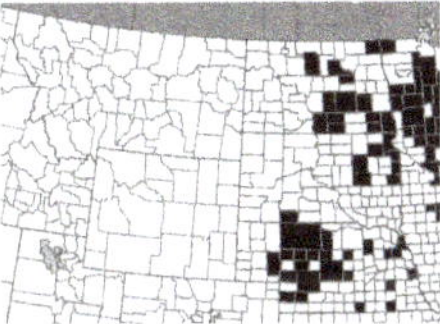

DESCRIPTION Plants 1–4 dm tall, often decumbent. **Leaf blades** mostly 1.5–6 cm long, 1–4 cm wide; petioles mostly 0.5–5 cm long. **Flower clusters** spreading 0.5–5 cm from the stem, the male flowers usually innermost in the clusters when mixed with female flowers. **Achenes** dark olivaceous to nearly black with a narrow pale margin, 1.3–2 mm long, the persistent sepals shorter than to slightly exceeding the achene. Late July–Sept.

HABITAT Cold springs, seeps and boggy places.

WETLAND STATUS
GP OBL | MIDW FACW

Pilea pumila (L.) A. Gray (*not illustrated*)
CANADIAN CLEARWEED

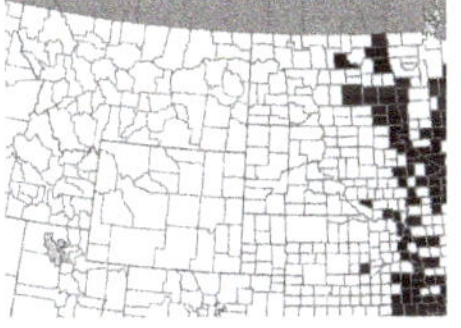

DESCRIPTION Very similar to *Pilea fontana* and differing mainly as follows: Plants sometimes larger, to 5 dm tall. **Leaf blades** to 12 cm long, 8 cm wide, thinner and more translucent than in *P. fontana;* petioles to 8 cm long. **Achenes** green, often marked with purple, 1.3–2 mm long. Late July– Sept.

HABITAT Swampy woods, wooded streambanks.

WETLAND STATUS
GP OBL | MIDW FACW

Pilea fontana
LESSER CLEARWEED

Urtica dioica L.
STINGING NETTLE

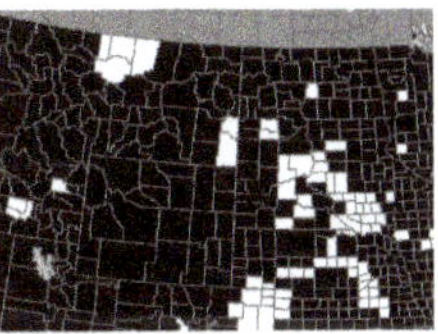

DESCRIPTION Monoecious or dioecious, stout perennial 8–20 dm tall, often forming dense patches by rhizomes, sparsely to moderately clothed with stinging hairs, otherwise glabrous or sparsely to densely puberulent, the **stems** usually simple. **Leaf blades** ovate to lanceolate, often conduplicate, mostly 5–15 cm long, 2–8 cm wide, puncticulate with cystoliths when dried, acute to acuminate, coarsely serrate, cordate to truncate or rounded at the base; petioles mostly 1–6 cm long; stipules linear-lanceolate, 5–15 mm long. **Flower clusters** branched and spreading, usually surpassing the subtending petioles, the clusters all of one sex or some male and some female, the female clusters usually above the male when both are present. **Achenes** tan, ovate, 1–1.2(1.5) mm long, ca. 1/2 as wide, enclosed by the inner pair of sepals which are 1–1.5 mm long, the outer pair ca. 1/2 as long. June–Sept.

SYNONYM *Urtica procera* Muhl.

HABITAT Moist woods, thickets, ditches, shores, streambanks and disturbed areas.

WETLAND STATUS
GP FAC | MIDW FACW | WMTN FAC

ADDITIONAL SPECIES The **wood nettle,** ***Laportea canadensis*** (L.) Wedd., is often abundant in moist woods in e and c parts of our area. This plant also has stinging hairs but differs from stinging nettle in its shorter stature; broader, alternately arranged leaves; and terminal inflorescence.

Urtica dioica
STINGING NETTLE

Verbenaceae *vervain family*

Perennial herbs with 4-angled, erect or mostly prostrate **stems** and simple, opposite, serrate **leaves. Inflorescence** of dense, terminal or axillary, bracteate spikes, these elongate or globose to short-cylindric, elongating as flowering progresses upward from the base. **Flowers** small and numerous, perfect, slightly to obviously irregular; calyx 5-toothed or deeply 2-parted; corolla 5- or 4-lobed, bilabiate or slightly so, blue-violet or white to bluish or purple-tinged; stamens 4, didynamous, epipetalous, included by the corolla or slightly exserted; pistil 2-carpellary, style terminal (not gynobasic as in Lamiaceae), ovary superior, 4- or 2-celled. **Fruit** developing inside the calyx, splitting lengthwise into 4 or 2 nutlets at maturity, each nutlet containing a seed.

1 Flowers white to bluish or purplish, in globose to short-cylindric spikes from the axils **Phyla**

1 Flowers blue-violet, in elongate, terminal spikes **Verbena**

Phyla lanceolata (Michx.) Greene
NORTHERN FOG-FRUIT

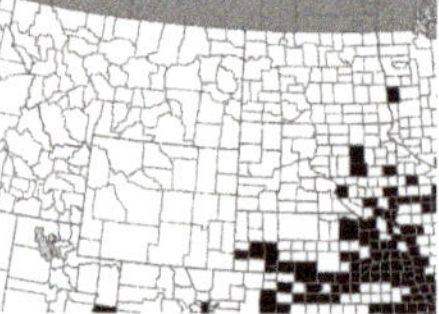

DESCRIPTION Perennial from a branched base, with prostrate to ascending 4-angled **stems**, usually rooting at the nodes, sometimes forming mats, the stem tips and lateral branches ascending to erect; foliage strigose with malpighiaceous hairs. **Leaves** opposite, the blades bright green, elliptic to ovate, oblong, ovate-lanceolate or oblong-lanceolate, 1.5–7.5 cm long, 0.5- 3 cm wide, acute or subacute at the tip, coarsely serrate to below the middle, cuneate at the short-petioled base. **Flowers** tiny and densely crowded in spikes from the axils, the spikes usually single at the nodes, initially globose, elongating to become short-cylindric, 5- 35 mm long, 5–7 mm thick, on slender peduncles 1.5–9 cm long; bractlets ovate to obovate, 2–3 mm long, often rose-pink on the margins. **Calyx** deeply 2-parted and compressed, membranous, about equaling the corolla tube; **corolla** pale blue, purplish or white, marcescent and fading, 3–4 mm long, 4-lobed and bilabiate, the lower lip larger than the upper; stamens included or slightly exsert; ovary 2-celled, each cell containing 1 ovule; stigma thickened, oblique or recurved. **Fruit** included in the calyx, dry, separating into 2 yellowish or olivaceous nutlets 0.9–1.3 mm long. June–Sept.

Phyla lanceolata
NORTHERN FOG-FRUIT

HABITAT Margins of lakes, ponds, streams, ditches and in swales and wet woodlands.

SYNONYM *Lippia lanceolata* Michx.

WETLAND STATUS
GP FACW | MIDW OBL | WMTN OBL

Verbena hastata L.
BLUE VERVAIN

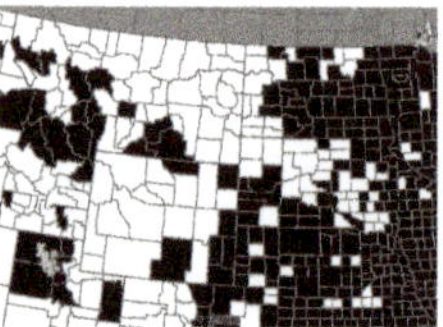

DESCRIPTION Stout, erect, perennial herb 4–12 dm tall from a thick rootstock; **stem** simple from the base, sometimes branched above, 4-angled, hispid to strigose. **Leaves** simple, opposite, short-petioled, the blades lanceolate to ovate-lanceolate or elliptic-lanceolate, 4–12 cm long, 1–5 cm wide, scabrous with stiff appressed hairs, acute to attenuate at the tip, coarsely serrate, occasionally lobed near the base, cuneate to rounded at the base and decurrent on the petiole. **Inflorescence** of (1) few to many, dense, bracteate spikes terminating the main stem and branches; spikes mostly 2–10 cm long, elongating as flowering progresses upward from the base; bracts lance-subulate, mostly 2–3 mm long. **Flowers** numerous, small, dark blue to purple, slightly irregular; **calyx** tubular, unequally 5-toothed, 1.5–3 mm long, stigulose; **corolla** 5-lobed, salverform, weakly bilabiate, surpassing the calyx by 2–5 mm, the limb 2–4 mm across, the tube strigulose; stamens included; style slender, exserted from the calyx after flowering, 2-lobed at the tip, ovary superior, 4-celled, shallowly 4-lobed. **Fruit** oblong, 4-angled, splitting lengthwise into 4 nutlets, these ca. 2 mm long. July–Sept.

HABITAT Wet meadows, shores, streambanks, ditches and springs, where water is fairly fresh.

WETLAND STATUS
GP FACW | MIDW FACW | WMTN FAC

Verbena hastata
BLUE VERVAIN

Violaceae *violet family*

Viola nephrophylla Greene
BOG VIOLET

DESCRIPTION Low, acaulescent perennial from short rhizomes. **Leaf blades** ovate-cordate to reniform, 1–4 cm long, 1.5–4(7.5) cm wide, obtuse to broadly rounded, crenate-serrate; petioles slender, 2–16 cm long. **Flowers** borne singly, nodding on slender peduncles, these about equal to or exceeding the petioles. Flowers violet, irregular, 12–24 mm long including the saccate spur; **sepals** 5, 1/3 to 1/2 the length of the corolla; **petals** 5, bearded toward the base on the inside, or the upper pair not bearded, the lowest petal saccate-spurred; stamens 5, the anthers connivent around the ovary, filaments short and broad, extending beyond the anthers; pistil 3-carpellary, style 1, ovary superior, 1-celled. Fertile cleistogamous flowers often produced after the normal ones on short, prostrate to erect peduncles. **Fruit** a many-seeded, 3-valved, ellipsoid capsule 5–10 mm long. Late May–June; sometimes flowering again late Aug–Sept.

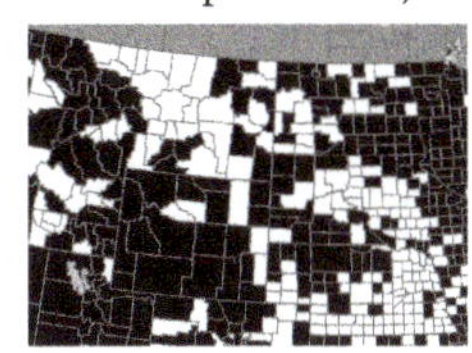

HABITAT Wet meadows, fens, springs, ditches and streambanks.

WETLAND STATUS
GP FACW | MIDW FACW | WMTN FACW

Viola nephrophylla
BOG VIOLET

Class LILIOPSIDA
Monocotyledons

Herbs (those included here) with vascular bundles of the stem usually scattered in the ground tissue, lacking a cambium; leaves often long and linear, sometimes reduced to sheathing at the base of the stem, mostly parallel-veined, the veins rarely forming a network; flower parts usually in multiples of 3; plant embryo usually with 1 cotyledon.

Acoraceae *calamus family*

Acorus calamus L.
SWEET FLAG, CALAMUS

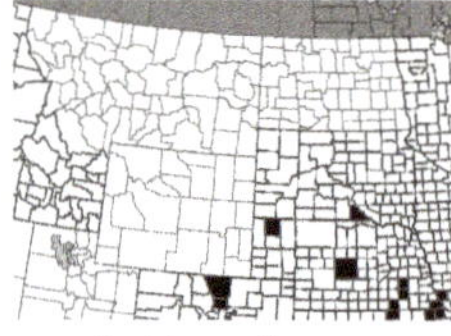

DESCRIPTION Moderately tall, reedlike plant from stout rhizomes; fresh foliage and rhizomes sweetly fragrant when crushed. **Leaves** linear, acute-tipped, 5–15 dm long, 5–20 mm wide, rather tough and leathery, the margins scarious toward the base. **Scape** arising lateral to the leaves, tyigonous; spathe appearing as a leaflike extension of the scape, often as long as the scape itself, reaching about as high as the leaves; spadix protruding laterally from the juncture of the scape and spathe, erect or ascending, long-cylindric, 5–10 cm long, 5–10 mm thick at anthesis, to 20 mm thick in fruit. **Flowers** perfect, yellow or brownish, perianth of 6 chartaceous tepals, deciduous; pistil 2- or 3-carpellary. **Fruit** obpyramidal, hard and dry, 1- to 3-seeded. June–July.

Acorus calamus
SWEET FLAG, CALAMUS

HABITAT Bogs, marshes and stream margins; local in Neb and SD; introduced from Europe and widely established in N America.

NOTE The distribution of *A. calamus* in the northern Great Plains probably reflects patterns of introduction from the east by eastern plains Indian tribes who used the plant for medicinal and religious purposes.

WETLAND STATUS
GP OBL | MIDW OBL | WMTN OBL

ADDITIONAL SPECIES Native plants are considered *Acorus americanus* (Raf.) Raf. and may be distinguished from *A. calamus* by having a raised leaf midvein plus 1–5 other veins raised above the leaf surface. In *Acorus calamus,* only the midvein is prominently raised.

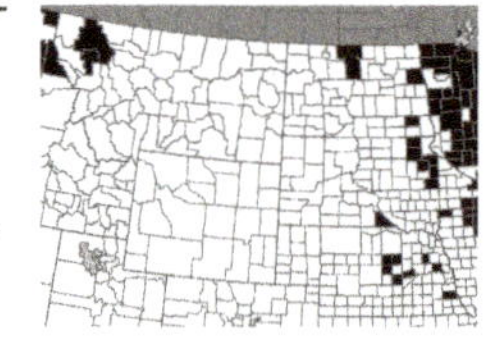

However, the distinctions are not always clear, and populations of both species may be present in the same location. In the n Great Plains, *A. americanus* is uncommon in e and c Neb, ND and e SD, becoming more widespread in the Great Lakes region.

Acorus calamus
rhizome

Alismataceae

water plantain family

Perennial or annual, acaulescent herbs, emergent, stranded or less commonly submersed in habit, often with tuber-bearing rhizomes or cormose at the base when perennial, fibrous-rooted. **Leaves** simple, all basal; blades of floating or emersed leaves oblong-elliptic, lanceolate, ovate or sagittate, often linear when submersed; petioles broadly winged at the plant base, somewhat sheathing. **Inflorescence** a terminal compound panicle with whorled branches, or the flowers mostly in whorls of 3 in a simple or sparsely branched raceme, the whorls of branches or flowers subtended by bracts. **Flowers** perfect, imperfect or both, regular, hypogynous; sepals 3, greenish; petals 3, white or pinkish; stamens 6-many; pistils several to numerous, maturing into laterally flattened or turgid, often beaked achenes.

1 Pistils or achenes in a single whorl on a flat receptacle, mostly fewer than 20 ***Alisma***

1 Pistils or achenes numerous and crowded in a dense cluster on a globose receptacle **2**

2 Flowers mostly imperfect, with the male above the female in the inflorescence; achenes laterally flattened and winged; leaves usually sagittate ***Sagittaria***

2 Flowers all perfect; achenes turgid, strongly ribbed; leaves never sagittate ***Echinodorus***

Alisma WATER PLANTAIN

Submersed to completely emersed perennials, from cormlike rootstocks, fibrous-rooted and lacking rhizomes. **Leaves** emersed or floating, ovate to lance-shaped, never arrowhead-shaped; underwater leaves sometimes ribbonlike (in Alisma gramineum). **Flowers** perfect, in whorled panicles, sepals 3, green; petals 3, white or light pink; stamens 6. **Fruit** a flattened achene in a single whorl on a flat receptacle, style beak small or absent.

NOTE Plants formerly classified as *Alisma plantago-aquatica* L. are now termed either *A. subcordatum* or *A. triviale,* with *A. plantago-gramineum* considered a European species. Old World examples have strongly pink or roseate petals rather than the essentially white petals of our plants.

1 Leaf blades lanceolate to oblong-elliptic or absent when submersed, the leaves then consisting of long, ribbonlike phyllodes; petals usually pinkish, 1–2.5 mm long; mature achenes with a central ridge and 2 lateral ridges down the back ***A. gramineum***

1 Leaves ovate; flower stalks much longer than leaves; petals white **2**

2 Achenes 1.5–2 mm long; larger fruiting heads 3–4 mm in diameter (excluding sepals) ***A. subcordatum***

2 Achenes 2.2–3 mm long; larger fruiting heads 4–7 mm in diameter (excluding sepals) ***A. triviale***

Alisma gramineum Gmel.
NARROW-LEAF WATER-PLANTAIN

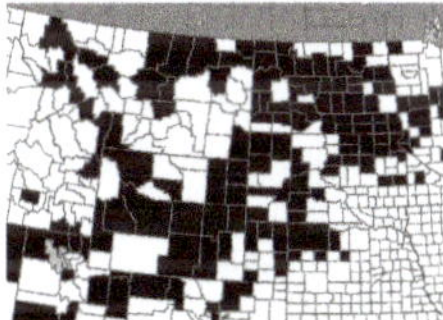

DESCRIPTION Plants upright when submersed or emergent, often low-spreading when stranded, commonly 0.5–2 dm tall, but to 6 dm long in deeper water. **Emersed leaves**, if present, with blades lanceolate to oblong-elliptic, 2–10 cm long, 1–3 cm wide; totally **submersed leaves** phyllodial, long-linear, ribbonlike, to 6 dm long, 2–10 mm wide;

Alisma gramineum
NARROW-LEAF WATER-PLANTAIN

petioles 3–15 cm long and often recurved on stranded plants, to 40 cm long on emergent specimens. Scapes 2–20(40) cm long, commonly short and recurved when stranded. Panicles shorter than to somewhat exceeding the leaves in height; branches spreading to ascending; bracts acute to obtuse, 3–10 mm long. **Flowers** numerous (sometimes few in stunted individuals); sepals obtuse, 2–2.5 mm long; petals usually pinkish, 1–2.5 mm long; pedicels 0.5–3 cm long. **Achenes** 2–2.5 mm long, with a central ridge and 2 lateral ridges down the back. July–Sept.

SYNONYM *Alisma geyeri* Torr.

HABITAT Muddy shores, flats and streambanks or often in shallow water, commonly where the water is brackish.

WETLAND STATUS
GP OBL | MIDW OBL | WMTN OBL

Alisma subcordatum Raf.
AMERICAN WATER-PLANTAIN

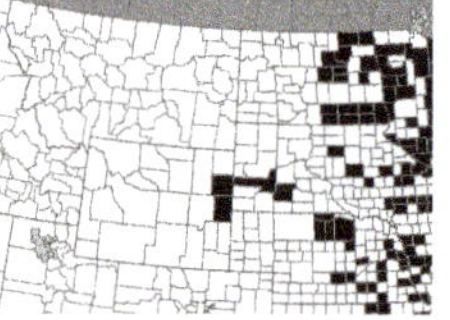

DESCRIPTION **Leaves** ovate to oval, 3–15 cm long and 2– 12 cm wide, rounded to nearly heart-shaped at base; petioles long. **Flowers** clustered on slender stalks 1–10 dm long, in whorls of 3–10; sepals 3; petals white, 3–5 mm long. **Fruit** an achene, 2–3 mm long, with a central groove. July–Sept.

SYNONYM *Alisma plantago-aquatica* var. *parviflorum.*

HABITAT Shallow water marshes, shores, ditches; water usually fresh.

WETLAND STATUS
GP OBL | MIDW OBL

Alisma triviale Pursh
NORTHERN WATER-PLANTAIN

DESCRIPTION **Leaves** usually long-petioled, the blade elliptic to broadly ovate, rounded to subcordate at base, 3–18 cm long. **Flowers** on a scape 1–10 dm long; flower pedicels in whorls of 3–10; sepals obtuse, 2–3 mm long; petals white, about 4 mm long. **Fruit** an achene 2– 3 mm long, usually with a median dorsal groove. June–Sept.

SYNONYMS *Alisma brevipes, Alisma plantago-aquatica* var. *americanum.*

HABITAT Marshes, ponds, and streams; water usually fresh.

WETLAND STATUS
GP OBL | MIDW OBL | WMTN OBL

Alisma subcordatum
AMERICAN WATER-PLANTAIN

Alisma triviale
NORTHERN WATER-PLANTAIN

Echinodorus berteroi (Spreng.) Fassett
UPRIGHT BURHEAD

DESCRIPTION Fibrous-rooted annual or perennial 2–6 dm tall, emersed to emergent or sometimes submersed with floating leaves and then nonflowering. **Leaves** with blades broadly ovate to ovate-lanceolate, more oblong and thin-textured on submersed ones, mostly 3–10 cm long, 1–6 cm wide, rounded at the tip, truncate to cordate or broadly cuneate at the base, gradually tapered and decurrent on the petiole on submersed leaves; petioles trigonous (broadly winged on submersed leaves), mostly longer than the blades. **Inflorescence** oblong, surpassing the leaves, usually with whorls of 3 branches from lower nodes of the scape, the flowers in whorls of 3–9, the whorls of branches and flowers subtended by spreading, lance-subulate bracts to 10 mm long. **Flowers** perfect, usually nonshowy; sepals broadly ovate, 4–5 mm long, reflexed in fruit; petals white, from shorter than the sepals to 9 mm long; stamens usually 12; pistils numerous and crowded, long-styled. **Fruiting heads** globose, 4–6 mm in diameter, appearing burlike due to prominent style beaks; **achenes** somewhat laterally compressed but turgid, 2.5–3.5 mm long including the straight to curved style beak, strongly ribbed lengthwise on the body. July–Sept.

SYNONYM *Echinodorus rostratus* (Nutt.) Engelm.

Echinodorus berteroi
UPRIGHT BURHEAD

HABITAT Shores, mud flats and alluvial bars, sometimes in shallow water.

WETLAND STATUS
GP OBL | MIDW OBL | WMTN OBL

Sagittaria ARROWHEAD, DUCK POTATO

Plants perennial or annual; rhizomes freely produced by perennial species, often bearing starchy tubers in late summer and fall. Emersed and floating **leaf blades** usually sagittate with prominent basal lobes, sometimes blades partly or all ovate to lanceolate or elliptic and lacking basal lobes; a basal rosette of thin, lanceolate to linear submersed leaves characteristic of deeply submersed plants, these often preceding floating and emersed leaves, generally absent on flowering specimens. **Inflorescence** a simple or sparingly branched, bracteate raceme with few to many, mostly 3-flowered whorls, branches sometimes replacing flowers at the lower 1-few nodes of the scape, the inflorescence shorter than to surpassing the leaves. **Flowers** usually male above and female below in the inflorescence, sometimes partly or even mostly perfect or rarely all of one sex, the male flowers soon falling; sepals greenish, persistent, eventually reflexed or sometimes appressed on the fruiting heads; petals white, showy, ephemeral; stamens many, filaments glabrous or pubescent; pistils numerous, densely crowded on a subglobose receptacle. **Achenes** crowded in globose or subglobose heads, laterally flattened and winged, sometimes with a low wing on the faces as well, apically beaked with the persistent style, the beak sometimes obscure.

1 Blades of emersed leaves not sagittate, lacking basal lobes or some occasionally with small caudal lobes — 2

1 Blades of emersed leaves mostly or all sagittate, the basal lobes prominent — 3

2 Female flowers and fruiting heads apparently sessile on the scape — ***S. rigida***

2 Female flowers and fruiting heads obviously pedicellate — ***S. graminea***

3 Sepals appressed to the fruiting heads; pedicels of fruiting heads inflated; plants annual, lacking rhizomes — ***S. calycina***

3 Sepals reflexed on fruiting heads; pedicels of fruiting heads slender; plants perennial, with rhizomes 4

4 Floral bracts acute to obtuse, mostly less than 1 cm long; achene beak projecting horizontally from the apex of the achene ***S. latifolia***

4 Floral bracts acute to acuminate or attenuate, usually more than 1 cm long; achene beak erect or oblique to recurved-ascending 5

5 Achene beak short, erect, 0.1–0.4 mm long; ventral wing of achene usually convex below the beak; basal lobes of leaf mostly shorter than the terminal one ***S. cuneata***

5 Achene beak prominent, oblique to recurved-ascending, 0.5 mm or more long; ventral wing of the achene straight below the beak; basal lobes of leaf usually equaling or longer than the terminal one ***S. brevirostra***

Sagittaria brevirostra Mack. & Bush
SHORT-BEAK ARROWHEAD

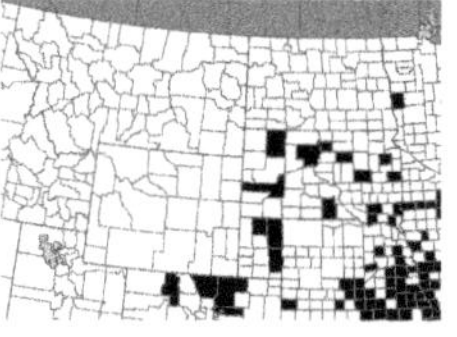

DESCRIPTION Erect, rhizomatous perennial with long-petioled leaves to 6 dm tall. **Leaf blades** sagittate, mostly 10–30 cm long, to 20 cm wide; the basal lobes usually equaling or longer than the terminal lobe. **Inflorescence** usually simple or sometimes branched from the lower 1–3 nodes; bracts 1–5.5 cm long, acuminate or attenuate; pedicels of female flowers ascending, 0.5–2 cm long. **Flowers** male above and female below, often predominantly male or female, seldom all of one sex; sepals reflexed in fruit; petals 1–2 cm long; filaments glabrous, about as long as to longer than the anthers. **Fruiting heads** to 2 cm across, often apically depressed; **achenes** 2–3 mm long, dorsal wing entire to crenulate, truncate apically and not continuous with the beak, separated from the beak by a saddlelike depression, ventral wing straight below the beak; beak obliquely ascending to recurved-ascending (even when immature), 0.5–1.5 mm long at maturity. July–Sept.

SYNONYM *Sagittaria engelmanniana* J. G. Smith.

HABITAT Mud or shallow water of streams and lakes.

WETLAND STATUS
GP OBL | MIDW OBL | WMTN OBL

Sagittaria calycina Engelm.
HOODED ARROWHEAD

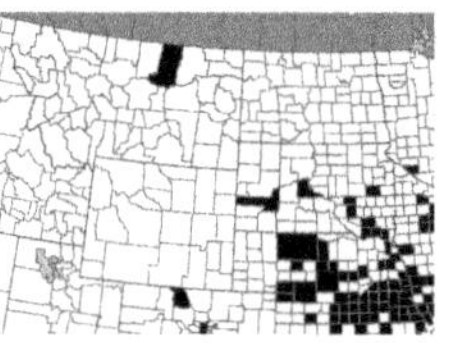

DESCRIPTION Dwarf to large emergent annual 1–10 dm tall, lacking rhizomes and tubers. **Leaves**

Sagittaria brevirostra
SHORT-BEAK ARROWHEAD

with terete, spongy-thickened petioles, erect to spreading; emersed blades sagittate to hastate or earlier ones elliptic-ovate, (1)3–40 cm long, (0.5)2–25 cm wide, the basal lobes much shorter to longer than the terminal lobe. **Inflorescence** ultimately leaning or procumbent, the scape terete and spongy-thickened; bracts membranous, short and broadly rounded at lower nodes, longer and acute to acuminate upward, 3–10 mm long; pedicels of fruiting heads inflated, 0.5–5 cm long, recurved in fruit, those of upper male flowers slender. Lower **flowers** female or perfect (with a single whorl of stamens), upper ones perfect or more often male; sepals broad and obtuse, 5–12 mm long, reflexed in flower but soon appressed to the head of developing fruits and remaining so on mature heads; petals white with a yellow base, usually not much exceeding the sepals; filaments roughened with minute hairs. **Fruiting heads** eventually nodding, to 2 cm across, convex to somewhat flattened apically; **achenes** 2–3 mm long, about equally winged dorsally and ventrally, usually with a prominent resin duct curving over the faces; beak horizontal or oblique, 0.2–0.6 mm long. July–Sept.

SYNONYMS *Sagittaria montevidensis* subsp. *calycinus* (Engelm.) Bogin.

HABITAT Mud or shallow water of ponds, marshes and drying pools.

WETLAND STATUS

GP OBL | MIDW OBL | WMTN OBL

Sagittaria cuneata Sheld.

ARUM-LEAF ARROWHEAD

DESCRIPTION

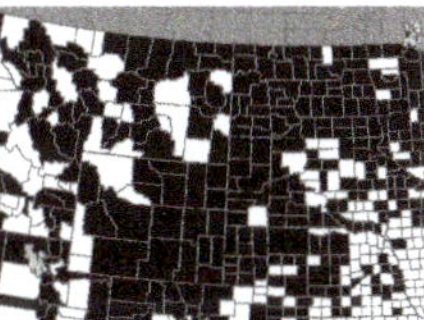

Emersed or emergent perennial with rhizomes and tubers, (0.5)1–6 dm tall, or to 9 dm long when emergent in deep water, often comprised of a basal rosette of thin, subulate or linear-oblong, submersed leaves (phyllodes) during juvenile stages, also commonly producing long-petioled floating leaves with small, ovate to elliptic or sagittate blades when young and submersed, generally not flowering in this condition. **Emersed leaf blades** sagittate (or earliest 1-few sometimes oblong-elliptic), 5–20 cm long, 1–14 cm wide, the basal lobes usually shorter than the terminal one; petioles erect to spreading. **Inflorescences** 1–several, erect to reclining, frequently branched from lower nodes; bracts acute to acuminate or attenuate, (0.5)1–4(5) cm long; pedicels of fruiting heads ascending to spreading, 0.3–2 cm long, those of male flowers longer and more slender. **Flowers** male above and female below or some perfect; sepals ovate, 4–10 mm long, reflexed in flower and fruit; petals white, 7–15 mm long; filaments glabrous, about equaling or longer than the anthers. **Fruiting heads** globose or nearly so, 5–13 mm in diameter; **achenes** 2–3 mm long, dorsal wing wider than the ventral one, rounded above and usually separated from the beak by a concavity, ventral wing broadened upward and usually convex below the beak; beak erect, often obscure, 0.1–0.4 mm long. June–Sept.

Sagittaria calycina
HOODED ARROWHEAD

Sagittaria cuneata
ARUM-LEAF ARROWHEAD

HABITAT Mud or shallow water of marshes, lakes, ponds, streams and ditches.

WETLAND STATUS
GP OBL | MIDW OBL | WMTN OBL

Sagittaria graminea Michx.
GRASS-LEAF ARROWHEAD

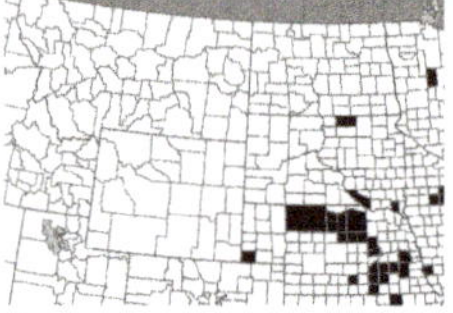

DESCRIPTION Emergent to emersed perennial or perhaps usually annual in our range, producing rhizomes well after flowering in first year plants. **Leaves** variable depending on degree of submergence and age of plant, submersed plants with the earliest leaves comprising a rosette of phyllodes, these tapelike, to 1 cm wide, acute-tipped, often absent by flowering time; emergent leaves or leaves of emersed plants with blades elliptic to linear-oblong, never sagittate, 3–20 cm long, 0.5–3 cm wide, acute or attenuate to the blunt tip. **Inflorescences** 1-few, shorter than to exceeding the leaves, simple, with 2–6(9) whorls of flowers; bracts triangular-ovate, connate in the lower 1/2, 2–8 mm long; pedicels slender to filiform, 1–4 cm long, ascending to spreading. **Flowers** usually male above and female below, rarely all of one sex; sepals ovate, 3–5 mm long, reflexed in fruit; petals white, about equaling to over 2x the sepal length; filaments equaling to longer than the anther, puberulent. **Fruiting heads** globose or subglobose, 5–12 mm across; the receptacle often conspicuously spongy-thickened; **achenes** 1–2 mm long, prominently winged, the dorsal wing rounded over the top of the achene and continuous with the ventral wing; beak minute and projecting laterally or absent. June–Sept.

HABITAT Mud or shallow water of marshes, ponds and ditches; uncommon in Neb and SD; perhaps introduced by waterfowl.

WETLAND STATUS
GP OBL | MIDW OBL | WMTN OBL

Sagittaria latifolia Willd.
DUCK-POTATO

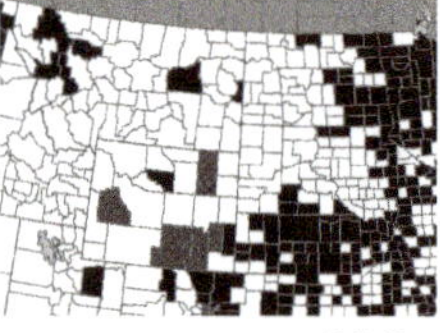

DESCRIPTION Emergent to emersed perennial 2–8 dm tall, or much longer when growing in deep water, rhizomatous and with tubers in autumn. **Leaves** highly variable depending on water depth, emersed leaf blades sagittate, mostly 8–40 cm long, 0.4–15 cm wide, the lobes narrow and linear on plants of deep water to broadly deltoid on emersed specimens, the basal lobes shorter to longer than the terminal one; submersed phyllodes, if produced, apparently gone by flowering time, Inflorescences 1-few, simple or rarely branched from the lowest node; bracts bluntly acute to obtuse, 0.5–1 cm long; pedicels slender, 0.3–3.5 cm long, ascending to spreading. **Flowers** male above and female below or seldom all

Sagittaria graminea
GRASS-LEAF ARROWHEAD

Sagittaria latifolia
DUCK-POTATO

of one sex; sepals ovate, 4–10 mm long, broadly rounded, reflexed in fruit; petals white, 7–20 mm long; filaments glabrous, longer than the anther. **Fruiting heads** globose to somewhat flattened, 1–2.5 cm across; **achenes** 2.5–4 mm long, dorsal wing rounded over the top of the achene, ventral wing gradually widened upward to meet the beak; beak projecting horizontally, tapered, 1–1.5(2.5) mm long. July–Sept.

HABITAT Mud or shallow water of rivers, streams, lakes and marshes.

WETLAND STATUS
GP OBL | MIDW OBL | WMTN OBL

Sagittaria latifolia
DUCK-POTATO (tubers)

Sagittaria rigida Pursh
SESSILE-FRUIT ARROWHEAD

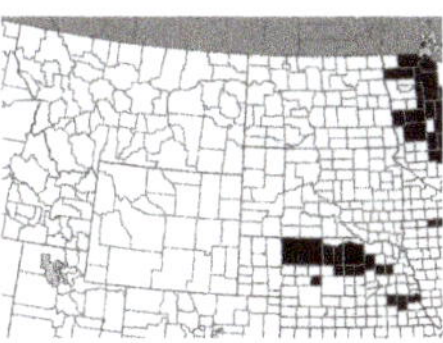

DESCRIPTION Rhizomatous perennial, emersed and erect to submersed and lax, 1–6 dm tall or to 9 dm long in deeper water. **Emersed leaf blades** linear to oblong when emergent in deep water or lanceolate to ovate when growing in shallow water or mud, seldom 1-few of the blades with short, narrow basal lobes, not truly sagittate, mostly 4–15 cm long, to 7 cm wide; petiole sometimes bent near the blade; leaves comprised entirely of lax, linear phyllodes when deeply submersed, and plants generally nonflowering in this condition. **Inflorescence** solitary, simple, erect when emersed or lax in deep water, usually not exceeding the longest leaves, often bent near the lowest node, with 2–8 whorls of flowers; bracts ca. 5 mm **long,** obtuse, connate at the base. Flowers female below, male or some perfect above, the female sessile or nearly so, on pedicels to 3 mm long, the male on slender pedicels 1–3 cm long; sepals ovate, 4–7 mm long, reflexed in fruit; petals white, ca. 2x as long as the sepals; filaments longer than the anther, roughened with minute hairs. **Fruiting heads** subglobose to somewhat flattened, to 1.5 cm across, apparently sessile and appearing bristly due to the rather prominent achene beaks; **achenes** 2.5–4 mm long, narrowly winged; beak inserted laterally and recurved upward, 0.8–1.4 mm long. June–Sept.

HABITAT Mud or shallow water of marshes, ponds, lakes and streams.

WETLAND STATUS
GP OBL | MIDW OBL | WMTN OBL

Sagittaria rigida
SESSILE-FRUIT ARROWHEAD

Araceae *arum family*

Perennial herbs with alternate, simple or compound **leaves**. in our traditional genera of Araceae (*Calla*), **flowers** are small and numerous, mostly single-sexed, staminate flowers usually above pistillate, crowded in a cylindric or rounded spadix subtended by a leaflike spathe; sepals 4–6 or absent; petals absent; stamens mostly 2–6; pistils 1- to 3-chambered. **Fruit** a usually fleshy berry, containing 1 to few seeds, or the entire spadix ripening as a fruit.

Now included in the Araceae are the tiny duckweeds (*Lemna, Spirodela,* and *Wolffia*), aquatic genera formerly treated within Lemnaceae. These are small perennial herbs, floating at or near water surface, single or forming colonies. Plants thallus-like (not differentiated into stems and leaves), the thallus (or frond) flat or thickened; the roots, if present, unbranched, 1 or several from near center of leaf underside; reproducing vegetatively by buds from 1–2 pouches on the sides, the parent and budded plants often joined in small groups. **Flowers** rare, either staminate or pistillate, in tiny reproductive pouches on margins (*Lemna, Spirodela*) or upper surface (*Wolffia*) of the leaves, subtended by a small spathe within the pouch; sepals and petals absent; staminate flowers 1–2, consisting of 1 anther on a short filament; pistillate flower 1 (a single ovary), in same pouch as staminate flowers. **Fruit** a utricle with 1 to several seeds.

1 Plants large, with clearly differentiated normal leaves and with rhizomes **Calla**

1 Plants tiny floating or submerged aquatic species less than 1 mm long, without differentiation into leaves or stems 2

2 Fronds thickened, less than 1.5 mm long, lacking roots **Wolffia**

2 Fronds flat, mostly longer than 1.5 mm long, with 1–several roots trailing from the lower side 3

3 Each frond with 3 or more roots; underside of the frond solid purple **Spirodela**

3 Each frond with a single root; underside of the frond green or purple-tinged or mottled **Lemna**

NOTE The **aquatic liverworts,** *Riccia* and *Ricciocarpos* are occasionally encountered in fresh waters of this region. These nonvascular plants resemble the duckweeds in size and growth habit and, in fact, are sometimes collected with duckweeds. *Riccia fluitans* appears as a very narrow (ca. 1 mm wide), bifurcating, ribbonlike thallus, free-floating and lacking rhizoids. *Ricciocarpos natans* has a broadly lobed rosette form with numerous rhizoids on the underside. When stranded on mud, the thallus is nearly radially symmetric with the rhizoids anchored in the substrate. The free-floating form is more bilaterally symmetric with conspicuous reddish rhizoids trailing beneath the thallus.

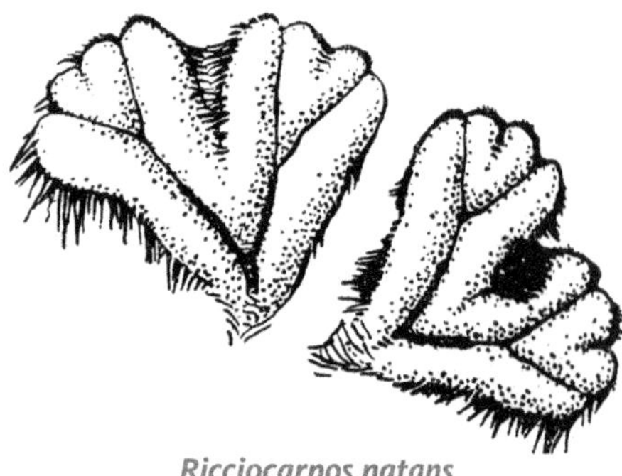

Ricciocarpos natans
RICCIOCARPOS

Calla palustris L.
WATER ARUM

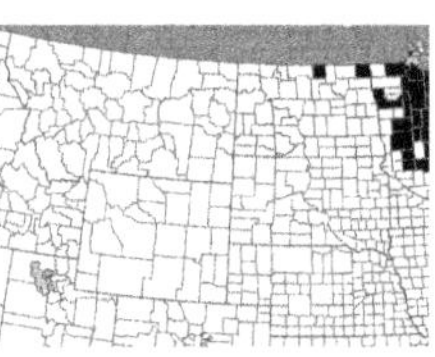

DESCRIPTION Arising from thick sprawling rhizomes, the rhizomes buried in mud or floating under the water surface. **Leaves** petioled, the blades broadly heart-shaped, cordate, mucronate at the tip, 5–15 cm long, about as wide as long; petioles stout, flaccid when submersed, 5–20 cm long, often longer when submersed. **Scape** resembling the petioles; spathe and spadix terminal, the spathe clearly differentiated from the scape, white, ovate, caudate at the tip, exceeding or about equaling the mature spadix; spadix borne on a short stipe, short-cylindric, 3–4 cm long in fruit, 2–3 cm thick. **Flowers** perfect or the uppermost staminate; perianth none; pistil simple, **Fruit** a fleshy berry, ripening red, 8–

12 mm long, containing few seeds. June–July. **HABITAT** Bogs and swamps; n ND, becoming more common north and eastward.

WETLAND STATUS
GP OBL | MIDW OBL | WMTN OBL

Lemna DUCKWEED

Fronds solitary or more often 2-several attached in small thalloid colonies, floating on the water surface or (in *L. trisulca*) submersed, the individual fronds orbicular, ovate, obovate or oblong in shape, sometimes stipitate (*L. trisulca*), all green or often tinged or mottled with red (purple when dried), (1-)3(–5)-nerved, flat to slightly convex on the upper surface, often with one or more papillae (projections) over the midnerve or near the apex, flat or convex to strongly inflated on the underside. **Roots** one per frond (sometimes absent on oldest and youngest fronds), with a short sheath at the base and a usually prominent root cap. Reproductive pouches 2 per frond, one on each lateral margin. **Flowers** rarely produced, consisting of usually 2 stamens (male flowers) and a single pistil (female flower) in each pouch. **Fruits** 1- to several-seeded. Reproduction mostly by budding of new fronds from the reproductive pouches, sometimes producing thickened, rootless turions that sink to the bottom during unfavorable periods.

Since duckweeds can colonize calm or slow-moving water almost anywhere (provided salt concentrations are not excessive), it is not possible to describe specific habitats for them. Also, flowering periods are not given as flowering is so rare and sporadic.

1 Fronds denticulate toward the apex, tapered to a slender stipitate base, the stipe often as long as the main body and commonly attached to the parent frond; colonies star-shaped, usually submersed **L. trisulca**

1 Fronds entire on the margin, nearly rounded and not obviously stipitate at the base, solitary or in tight colonies, these not star-shaped, floating on the water surface or stranded on mud 2

2 Fronds obscurely 1-nerved **L. minuta**

2 Fronds 3- to 5-nerved 3

3 Root sheath winged at the base; root tip sharply pointed; roots not longer than 3 cm; fronds completely green 4

3 Root sheath not winged at the base; root tip mostly rounded; roots often longer than 3 cm; fronds often red-tinged beneath or with red spots on either surface 5

4 Fronds very often with 2–3 papillae in a row on the upper surface above the node (the level at which daughter fronds attach); seeds whitish, with 35–60 faint ribs, not escaping the fruit wall when ripening **L. perpusilla**

4 Fronds with only 1 prominent papilla above the node; seeds brownish, with 8–22 prominent ribs, falling out of the fruit wall when ripening **L. aequinoctialis**

5 Fronds with several about equal sized, small papillae on the upper surface from the midline to the tip (often obscure), very often red-tinged on the lower surface, forming small, obovate to orbicular, rootless, dark green to brown turions under unfavorable conditions, these sinking to the bottom of the water **L. turionifera**

5 Fronds lacking papillae or with one prominent papilla at the apex and another just above the node and with smaller papillae between them, rarely forming turions; if formed, the turionlike fronds have short roots and are slowly forming daughter fronds 6

6 Fronds very often gibbous, strongly convex and obviously inflated beneath, with air spaces often larger than 0.3 mm in diameter, very often 4- to 5-nerved with the nerves all arising from the same point at the node; ovules 1–6; fruit winged **L. gibba**

6 Fronds flat to slightly convex beneath, not inflated, with air spaces less than 0.3 mm in diameter, rarely with more than 3 nerves, but if 4- to 5-nerved, then the outer nerves arising at the base of the inner ones; ovule 1; fruit wingless 7

7 Papilla at apex of the frond very prominent; fronds often red beneath **L. obscura**

7 Papilla at apex of the frond not very prominent; fronds never red beneath **L. minor**

Lemna aequinoctialis Welw.

LESSER DUCKWEED

DESCRIPTION Very similar to *Lemna perpusilla,* except with only 1 prominent papilla above the node on the upper surface. Seeds brownish, with 8–22 prominent ribs, falling from the fruit at maturity.

WETLAND STATUS

GP OBL | MIDW OBL

Lemna gibba L.

INFLATED DUCKWEED

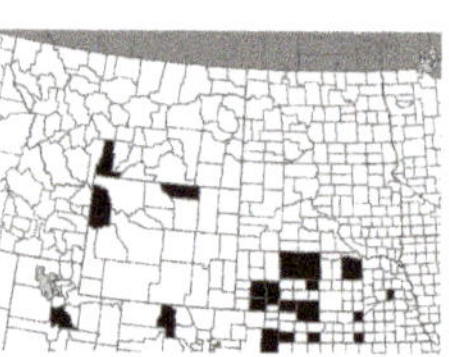

DESCRIPTION **Fronds** orbicular to obovate, 2–5(6) mm long, symmetric or not, green to yellowish-green, often red-tinged or mottled on both surfaces, 3- to 5-nerved, when 4- to 5-nerved the lateral and inner nerves diverging from the same point at the node; upper surface flat to slightly convex, with a prominent central papilla and usually a row of smaller papillae extending to near the apex, lacking a prominent papilla at the apex (except for small fronds); lower surface convex to very often much inflated (gibbous), with air spaces in 2 layers, those of the bottom layer often more than 0.3 mm in diameter. **Turions** not produced, or if turionlike fronds are present, then these have short roots and are slowly growing daughter fronds. Root sheath not winged, root tip mostly rounded. **Fruit** ovoid to ellipsoid, with lateral wings 0.1–0.2 mm wide; seeds (1)2–6, ovoid to ellipsoid.

WETLAND STATUS

GP OBL | MIDW OBL

NOTE Although this species and the following are distinct in fruiting condition, they are often difficult to tell apart vegetatively. This is especially true when dealing with the occasional uninflated forms of *Lemna gibba.*

Lemna minor L.

COMMON DUCKWEED

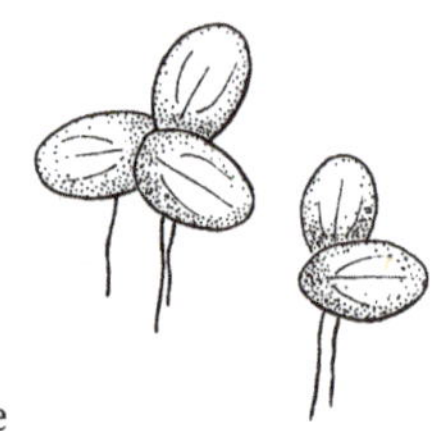

DESCRIPTION **Fronds** nearly orbicular to ellipticobovate, broadest near the middle, 2–4 mm long, symmetric or nearly so, green to yellowish-green, never red-tinged or mottled on either surface, obscurely 3 (–5)-nerved, when 4- to 5-nerved (rare) the lateral nerves arising near the base of the inner ones; both surfaces flat to weakly convex, the upper surface with a low papilla at the apex and often one above the node, usually with a low median ridge or row of smaller papillae between them, the lower surface never inflated, with air spaces in 2 layers but these less than 0.3 mm in diameter. **Turions** not produced. Root sheath not winged, root tip rounded. **Fruit** ovoid to ellipsoid, wingless, 1-seeded.

WETLAND STATUS

GP OBL | MIDW OBL | WMTN OBL

NOTE *Lemna minor* may be limited relative to *L. turionifera* by its inability to form turions. Often, *L. minor* is found in waters kept warm through the winter by active springs.

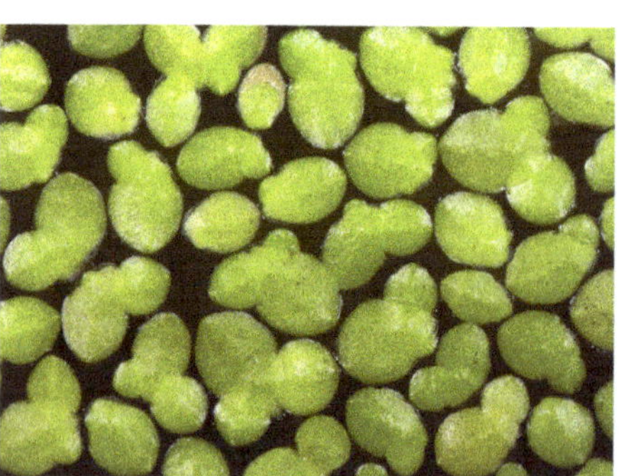

Lemna aequinoctialis
LESSER DUCKWEED

Lemna gibba
INFLATED DUCKWEED

Lemna minuta Kunth

LEAST DUCKWEED

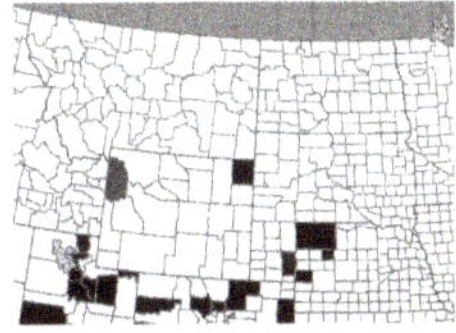

DESCRIPTION **Fronds** oblong to elliptic or somewhat ovoid, 1–2.5 mm long, symmetric or slightly asymmetric at the base, pale green, obscurely 1-nerved (the nerve often nearly obsolete), thin, with only 1 layer of air spaces, hyaline around the margin, flat to slightly convex on both surfaces, sometimes with a few low papillae on the upper surface. **Turions** not produced. Root sheath not winged; root tip pointed. **Fruit** not winged, 1-seeded.

SYNONYMS *Lemna minima* Phil., *Lemna minuscula* Herter

WETLAND STATUS

GP OBL | MIDW OBL

NOTE Local in w Neb Sandhills and ne Wyo. This species is sometimes included in *Lemna valdiviana* Phil., a primarily coastal entity that has been recorded from ne Colorado and Kansas. *L. valdiviana* differs in having longer, more prominently 1-nerved fronds which are strongly asymmetric at the base.

Lemna obscura (Austin) Daubs (*not illus.*)

LITTLE DUCKWEED

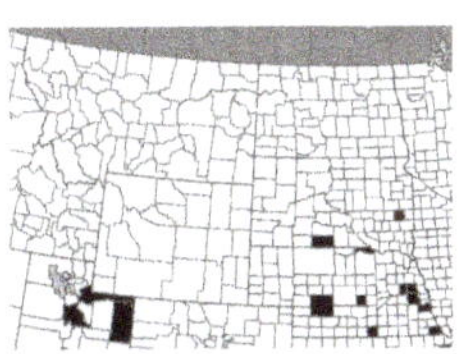

DESCRIPTION Quite similar to *Lemna minor* and differing mainly as follows: **Fronds** obovate to oblong-orbicular, 1.5–3(3.5) mm long, slightly asymmetric, often reddish beneath, obscurely 3-nerved; upper surface flat to slightly convex, with a prominent papilla at the apex, the lower surface convex.

WETLAND STATUS

GP OBL | MIDW OBL | WMTN OBL

Lemna minor **(larger plants)**
Lemna minuta **(smaller plants)**
DUCKWEED

Lemna perpusilla Torr.

MINUTE DUCKWEED

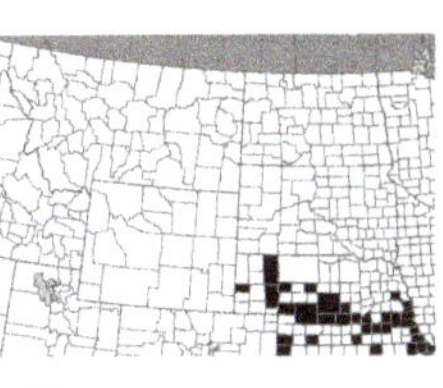

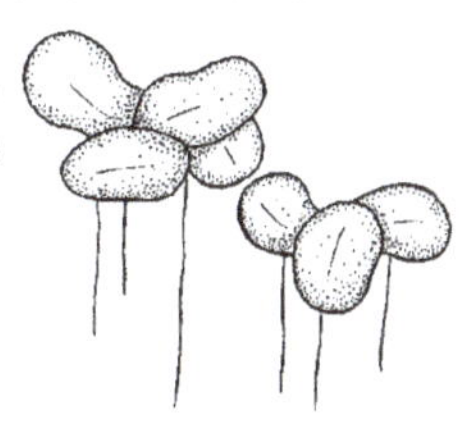

DESCRIPTION **Fronds** obovate to elliptic, 1–3.3 mm long, oblique at the apex and asymmetric at the base, light green, not reddish, obscurely 3-nerved; upper surface slightly convex, with a prominent apical papilla and very often with 1 or 2 others along the midnerve; lower surface flat to slightly convex, with large air spaces in 1 layer. **Turions** not produced. Root sheath with lateral wings; root tip sharply pointed. **Fruit** obliquely attached in the pouch; seeds whitish, with 35–60 faint ribs, not escaping the fruit wall when ripening.

WETLAND STATUS

GP OBL | MIDW OBL

NOTE This species flowers and fruits more freely than other duckweeds.

Lemna trisulca L.

STAR DUCKWEED

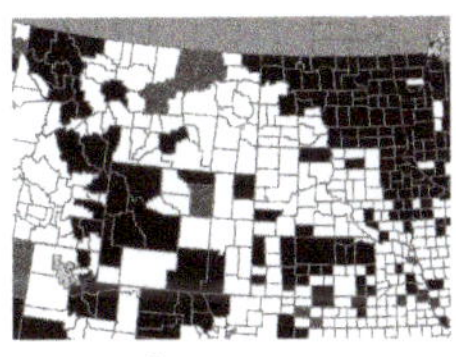

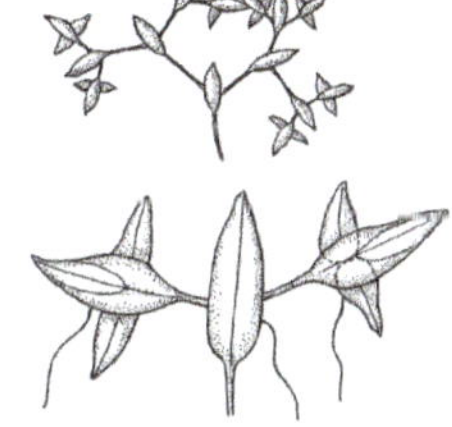

DESCRIPTION **Fronds** usually several to many, attached to form star-shaped colonies, often dense and tangled to form submersed mats, floating at the surface only when flowering, individual fronds oblongelliptic to oblong-lanceolate, denticulate toward the apex, 5–20 mm long including the slender stipitate base, the stipe often as long as the main body and attached to the parent frond, shorter on flowering specimens; frond body symmetrical, dark to pale green and translucent, often reddish or blackish, faintly 3-nerved, flat on both surfaces, the air spaces small and obscure. **Turions** not produced. Root deciduous and thus often lacking on some fronds, the sheath not winged, root tip pointed. **Fruit** 1-seeded.

WETLAND STATUS

GP OBL | MIDW OBL | WMTN OBL

Lemna turionifera Landolt

TURION DUCKWEED

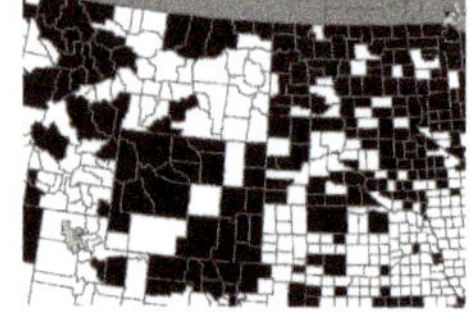

DESCRIPTION Very similar to *L. minor* and often difficult to distinguish from it, differing mainly as follows: **Fronds** often somewhat asymmetric and falcate, broadest toward the apex, often red-tinged beneath and sometimes red-mottled above; upper surface with several papillae of about equal size above the midline, these sometimes obscure. **Turions** produced under unfavorable conditions, appearing as small, obovate to orbicular, thick-textured fronds that lack roots, dark green (often strongly reddened) to brown, eventually breaking free of the parent frond and sinking to the bottom.

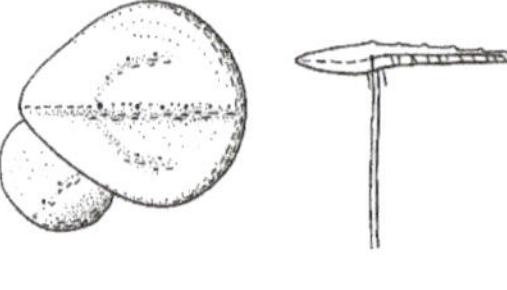

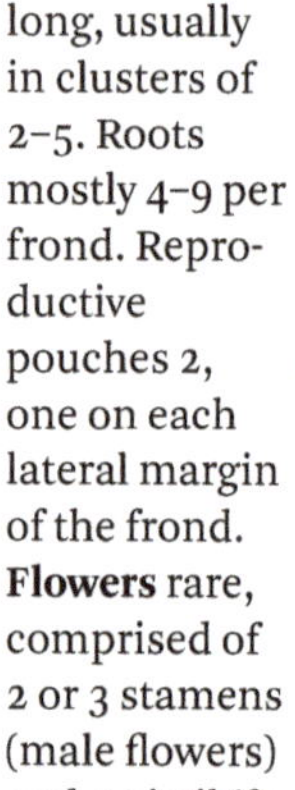

WETLAND STATUS

GP OBL | MIDW OBL | WMTN OBL

Lemna trisulca
STAR DUCKWEED

Spirodela polyrrhiza (L.) Schleid.

DUCK-MEAT

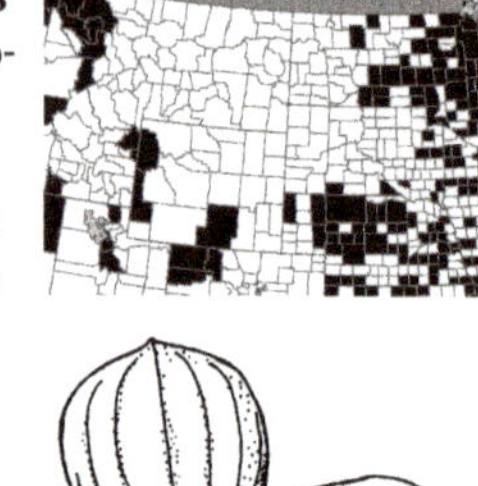

DESCRIPTION **Fronds** flat, orbicular to obovate, green above, dark red or purple beneath, floating on the surface, 3–6 mm long, usually in clusters of 2–5. Roots mostly 4–9 per frond. Reproductive pouches 2, one on each lateral margin of the frond. **Flowers** rare, comprised of 2 or 3 stamens (male flowers) and 1 pistil (female flower) in each pouch. **Fruit** small-winged, 2-seeded. Reproduction mainly by budding of new fronds from the reproductive pouches.

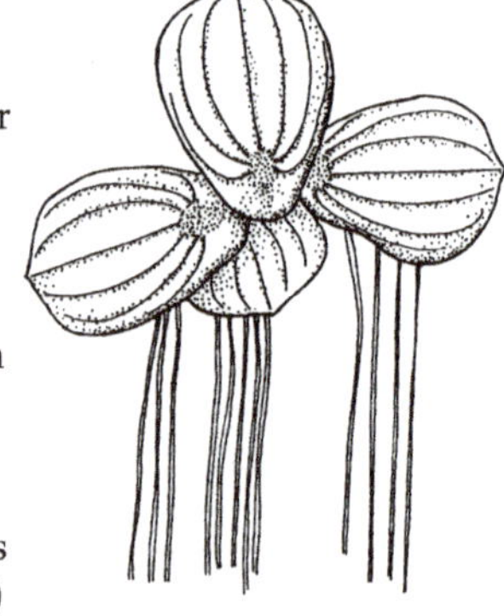

HABITAT Stagnant or slow-moving fresh water.

WETLAND STATUS

GP OBL | MIDW OBL | WMTN OBL

Wolffia WATER-MEAL

Minute, thalloid, rootless plants floating at or just beneath the surface, often abundant and forming a granular scum on the water surface, usually mixed with other Lemnaceae. **Fronds** flat on the upper surface (in those included here), solitary or often paired. Reproductive pouch 1, near the base of the frond. **Flowers** rare and very minute, consisting of 1 stamen (male flower) and 1 pistil (female flower) in the pouch. **Fruit** globose, 1-seeded; reproduction mainly by budding.

These plants are the smallest of all flowering plants and are easily overlooked in mixed culture with larger duckweeds. However, the tiny thalli are fairly easily detected by their granular or mealy feel to the touch and their tendency to adhere to the skin as the hand is withdrawn from the water.

1 Fronds ellipsoid to oblong when viewed from above, 1-1/3–2x longer than wide, with a raised pointed apex, usually brown-punctate when dried ***W. borealis***

1 Fronds orbicular to broadly ovoid when viewed from above, 1–1 1/3x longer than wide, not raised and pointed at the apex, not brown-punctate ***W. columbiana***

Wolffia borealis (Engelm.) Landolt

NORTHERN WATERMEAL

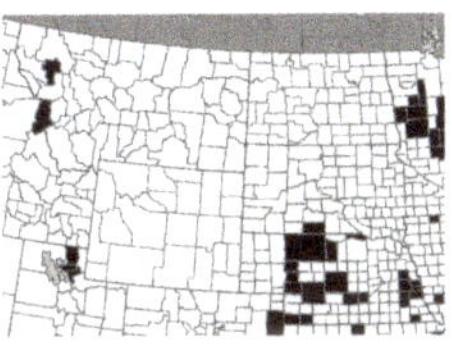

DESCRIPTION **Fronds** ellipsoid to oblong when viewed from the upper side, 0.7–1.2 mm long, shaped somewhat like a deep boat when viewed from the side, with a raised pointed apex, dark green on the flat upper surface, lighter green in the thick lower portion, usually brown-punctate when dried.

SYNONYM *Wolffia punctata* Griseb.

HABITAT Calm, fresh water.

WETLAND STATUS
GP OBL | MIDW OBL | WMTN OBL

Wolffia columbiana Karst. (*not illustrated*)

COLUMBIAN WATERMEAL

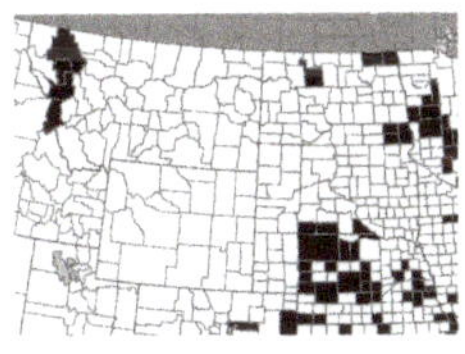

DESCRIPTION **Fronds** orbicular to broadly ovoid (when viewed from above), 0.7–1.4 mm long, with a small, flattened upper surface, otherwise nearly globose or ovoid when viewed from the side, not raised and pointed at tip, green throughout and not brown-punctate.

HABITAT Calm, fresh water.

WETLAND STATUS
GP OBL | MIDW OBL | WMTN OBL

Wolffia borealis
NORTHERN WATERMEAL

Butomaceae
flowering rush family

Butomus umbellatus L.

FLOWERING RUSH

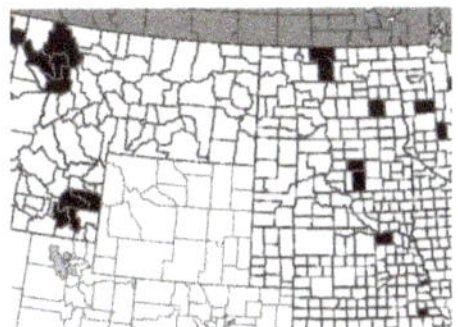

DESCRIPTION Moderately tall, rushlike perennial. **Leaves** basal, from a stout rhizome, stiff and erect when emersed or lax and floating in deep water, long-linear, to 1 m long, 5–10 mm wide. **Inflorescence** a many-flowered umbel borne on a scape 1- 1.5 m tall; bracts subtending the umbel 4, lance-triangular, acute. **Flowers** perfect, regular, 2–3 cm across, pink; sepals 3, petaloid; petals 3; stamens 9-many, anthers elongate, filaments slender; pistils 6, simple, whorled, united at the base, stigmas sessile, apical, ovaries superior; pedicels 5–10 cm long. **Fruit** an indehiscent, many-seeded capsule. June–Aug.

HABITAT Emergent in shallow to moderately deep water of lakes and streams; introduced to N America from Eurasia.

WETLAND STATUS
GP OBL | MIDW OBL | WMTN OBL

Butomus umbellatus
FLOWERING RUSH

Cyperaceae *sedge family*

Mostly perennial, grasslike, rushlike or reedlike plants; stems trigonous or less often terete (or compressed), solid or pithy. **Leaves** in 3 vertical ranks or reduced mainly to sheathing at the base of the stem; leaf blades, when present, grasslike, elongate and parallel-veined, often strongly keeled; sheaths closed around the stem, sometimes splitting with age. **Flowers** much reduced, perfect or unisexual, each subtended by a bract (scale); perianth consisting of 1-many (often 6) small bristles or a single perianth scale, or the perianth lacking; stamens 3 or sometimes 1 or 2; ovary superior, 1-celled, 3-carpellary or less often 2-carpellary, sometimes contained in a saclike covering called the perigynium (*Carex*), ripening into an achene, the style trifid or bifid (stigmas 3 or 21, ovule 1, basally attached. Flowers and their subtending scales arranged in **spikelets** (termed spikes in *Carex*), the spikelets solitary as a single terminal or lateral spike, or the spikelets few to many and arranged in various types of inflorescences, the inflorescence often subtended by 1–several involucral bracts. A large family, of which most members inhabit wet places.

1 Achenes enclosed in a closed sac (perigynium) subtended by a scale, the style protruding through the apex; flowers strictly unisexual (sedges with exclusively staminate flowers should be keyed here) **Carex**

1 Achenes not enclosed in a closed sac, naked beside the subtending scale; at least some flowers bisexual 2

2 Scales of spikelets 2-ranked; spikelets ± flattened in cross-section and always more than one per inflorescence 3

2 Scales of spikelets spirally arranged (or if 2- ranked, the spikelet solitary); spikelets round or several-angled in cross-section, solitary or several to many per inflorescence 4

3 Stems usually ± angled, solid; inflorescences terminal; achenes without subtending bristles **Cyperus**

3 Stems round, hollow; inflorescences in the axils of stem leaves; achenes with subtending bristles **Dulichium**

4 Perianth bristles 6, 3 slender and 3 with an expanded, ± spongy, spoon-like portion at the tip **Fuirena**

4 Perianth bristles absent or 1 to many, all slender 5

5 Spikelet or cluster of spikelets borne on one side of the stem at the base of a single ± erect to somewhat angled or curved involucral bract that appears to be a continuation of the stem 6

5 Spikelet or spikelets terminating the stem or borne both terminally and laterally; if more than one spikelet, the inflorescence with 1 to several spreading to reflexed, leaflike involucral bracts 8

6 Stems less than 0.5 mm thick; plants tiny, less than 10 cm tall **Lipocarpha**

6 Stems thicker than 0.5 mm; plants usually much taller than 10 cm 7

7 Plants colonial from elongate rhizomes, perennial; usually more than 0.8 m tall; anthers 1– 3.5 mm long **Schoenoplectus**

7 Plants tufted, annual, mostly less than 0.8 m tall; anthers 0.3–0.9 mm long **Schoenoplectiella**

8 Spikelet solitary and terminal on the stem (very rarely a few smaller accessory spikelets occur at the base of the terminal spikelet in the bladeless genus *Eleocharis*) 9

8 Spikelets several to many on the stem, terminal or lateral 10

9 Sheaths totally bladeless or at most with an apical tooth up to 1 mm long; achenes usually with an apical tubercle formed by the expanded and persistent base of the style **Eleocharis**

9 Upper sheaths with short green blades 0.3– 12 cm long; achenes blunt at apex, tubercle absent **Eriophorum**

10 Achenes subtended by (12–) 15–50 conspicuous, silky, white or tawny, hair-like bristles many times as long as the achenes **Eriophorum**

10 Achenes subtended by 1–8 bristles, or bristles absent 11

11 Leaves flat or folded; with a definite, ± keeled midrib 12
11 Leaves inrolled and wiry; rounded on the back and without a definite midrib ***Rhynchospora***

12 Achenes with a conspicuous tubercle formed by the expanded, persistent style base ***Rhynchospora***
12 Achenes blunt at apex, without a tubercle; style base, if expanded, not persistent to maturity 13

13 Widest leaves 0.5–3 mm wide; achenes lacking bristles ***Fimbristylis***
13 Widest leaves 4–15 mm wide; achenes subtended by 1–8 bristles 14

14 Spikelets (10–) 15–36 mm long; achenes 3–5 mm long, including apiculus; anthers 4–5 mm long; stems sharply 3-angled nearly or quite to the base; colonial from rhizomes with large corm-like thickenings ***Schoenoplectus***
14 Spikelets 2–10 (–12) mm long, achenes 0.9– 1.2 mm long; anthers 0.5–1.3 mm long; stems terete, obtusely 3-angled, or sharply 3-angled only toward summit; tufted or with rhizomes lacking corm-like enlargements (includes former *S. nevadensis,* now named *Amphiscirpus nevadensis*) ***Scirpus***

Amphiscirpus nevadensis (S.Watson) Oteng-Yeb.

NEVADA BULRUSH

DESCRIPTION

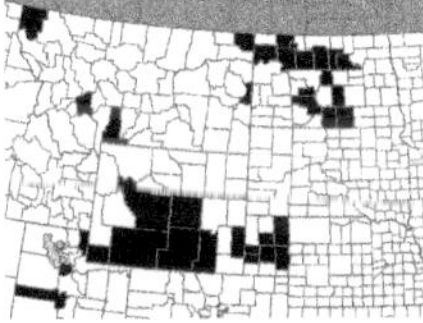

Perennial 1–5 dm tall from deep-seated rhizomes; **culms** erect, terete or nearly so, sheathed by the leaves toward the base. **Leaves** few to several per culm, the blades channeled above with inrolled margins, 0.5–2 mm wide, strongly ascending, seldom overtopping the culm; principal involucral bract erect, resembling a continuation of the culm, sometimes turning brown, 0.7–5(10) cm long; additional involucres 1-few, appearing like enlarged scales subtending the inflorescence, sometimes green. **Inflorescence** appearing lateral, the spikelets directed toward one side of the culm; **spikelets** 1–5(6), sessile, 5–20(25) mm long, 3–6 mm thick; scales brown with a pale midvein, glossy, 4–5 mm long, acute to obtuse; perianth of 1–4 bristles of variable length, shorter than the achene; stamens 3, the filaments straplike, often adherent to the achene; styles bifid. **Achenes** tan to brown, cellular-reticulate, lenticular, plano-convex, 2–2.5 mm long, often nearly as wide, the beak absent or minute. June–Sept.

SYNONYM

Scirpus nevadensis S. Wats.

HABITAT

Alkaline or saline wet meadows, seepage areas and shores.

WETLAND STATUS

GP OBL

NOTE See *Schoenoplectus* key, page xxx.

Carex SEDGE

Monoecious or rarely dioecious, cespitose or rhizomatous, grasslike perennials; **culms** simple, acutely to obtusely trigonous, exceeding or shorter than the leaves. **Leaves** grasslike, the blades flat to involute; sheaths hyaline or cross-rugulose ventrally, hyaline, green-and-white mottled or septate-nodulose dorsally. **Inflorescence** terminal, headlike, dense to open, usually consisting of few to many spikes, or less often of a single spike,

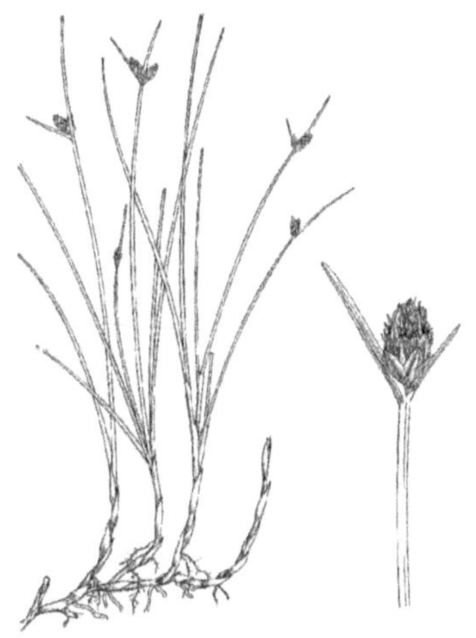

Amphiscirpus nevadensis
NEVADA BULRUSH

the spikes usually subtended by leaflike bracts, the lowermost bracts best developed, sometimes much reduced; spikes unisexual or bisexual, when bisexual, the spikes either *androgynous* (with the staminate flowers above the pistillate) or *gynaecandrous* (with the pistillate flowers above the staminate). **Flowers** unisexual, each subtended by a scale; perianth none; male flowers composed of 3 stamens; female flowers composed of a single ovary contained within a saclike perigynium; perigynia (plural) trigonous to plano-convex, lanceolate to ovate or suborbicular, strongly beaked to beakless, variously nerved, often winged on the margins when plano-convex, sometimes spongy-thickened at the base; beak, when present, entire, obliquely cut or bidentate at the apex; styles bifid or trifid (stigmas 2 or 3), the branches protruding from the beak or tip of the perigynium. **Achenes** developing within the perigynium, lenticular or trigonous, the style jointed with the achene or continuous with it.

Carex is the largest genus of vascular plants in North America; most species are characteristic of wet habitats.

---

1 Spikes only one per culm, terminal, continuous, no lateral spikes present. (Those with 2 or more spikes aggregated in a dense spikelike head will key in the next section, as will those with 1- to few-flowered lateral spikes.) 2

1 Spikes 2 or more per culm, densely crowded in a spikelike head to widely separated, the lateral spikes occasionally only 1-to few-flowered 3

2 Plants tufted; stigmas 3 ***C. leptalea***

2 Plants rhizomatous; stigmas 2 ***C. gynocrates***

3 Spikes mostly unisexual, the terminal ones usually staminate, the lower ones mostly pistillate 4

3 Spikes bisexual 32

4 Stigmas 2; achenes lenticular 5

4 Stigmas 3; achenes trigonous 11

5 Perigynia whitish-pulverulent or eventually turning golden orange at maturity; bract of the lowest pistillate spike obviously sheathing the culm; achenes dark brown at maturity 6

5 Perigynia green, becoming brown at maturity; bract of the lowest pistillate spike not sheathing the culm, or only barely so; achenes tan to brown at maturity 7

6 Terminal spikes staminate (rarely slightly pistillate); scales whitish to tawny; perigynia whitish-pulverulent, golden orange at maturity ***C. aurea***

6 Terminal spikes gynaecandrous or mostly so; scales brown to purplish; perigynia whitish-pulverulent at maturity ***C. garberi***

7 Beaks of the perigynia shallowly bidentate; perigynia 2-ribbed (with a single strong nerve along each edge), conspicuously nerved between the ribs ***C. nebrascensis***

7 Beaks of the perigynia entire; perigynia 2-ribbed and faintly nerved or nerveless between the ribs 8

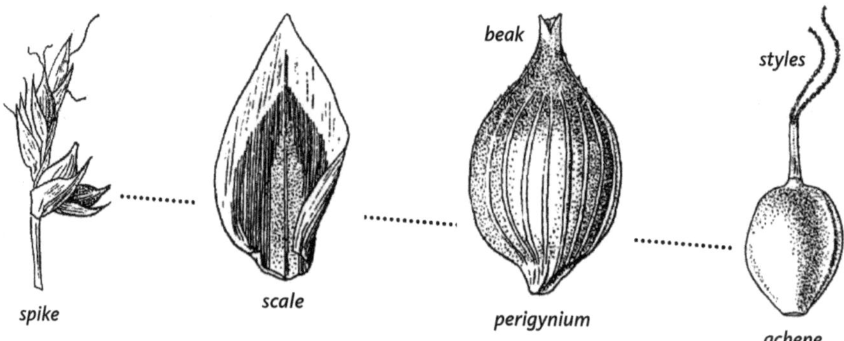

Carex floral parts
(Carex chordorhiza shown)

8 Perigynia usually obovate, broadest near the apex but not inflated; leaves glaucous, the lower ones with well-developed blades **C. aquatilis**

8 Perigynia usually ovate or elliptic, broadest at or below the middle, or if obovate, then inflated at the apex when mature; leaves green, the lower ones bladeless, short and pointed, brown or reddish, sheathing the culm base 9

9 Perigynia inflated at the apex, round-obovate, pale brown, often with darker brown spots **C. haydenii**

9 Perigynia flattened, ovate to elliptic, not inflated, green at the apex, golden to tawny toward the base, sometimes eventually brown 10

10 Basal sheaths disintegrating into ladder-like fibers; ligule at juncture of leaf blade and sheath V-shaped, longer than wide **C. stricta**

10 Basal sheaths not ladder-fibrillose; ligule truncate or low-rounded, wider than long **C. emoryi**

11 Beak of the perigynium entire or obliquely cut (bidentulate) 12

11 Beak of perigynia bidentate 20

12 Terminal spike staminate, pistillate or gynaecandrous, considerably larger than the lateral spikes 13

12 Terminal spike usually staminate throughout and smaller than or equal to the lateral spikes 14

13 Pistillate scales acuminate, awn-tipped; perigynia 2.5–3.5 mm long **C. buxbaumii**

13 Pistillate scales obtuse, acute or mucronate; perigynia 1.5–2.3 mm long **C. hallii**

14 Pistillate spikes drooping on lax, filiform peduncles 15

14 Pistillate spikes not drooping 16

15 Perigynia flattened, the beak less than 0.5 mm long; bract subtending the lowest pistillate spike sheathless or nearly so **C. limosa**

15 Perigynia nearly terete, the beak 0.5 mm or more long; bract subtending the lowest pistillate spike with a well developed sheath **C. capillaris**

16 Perigynia distinctly beaked, the beaks 0.5–1 mm long; pistillate spikes sessile or nearly so, the lower ones sometimes very short-peduncled **C. viridula**

16 Perigynia obscurely beaked, the beaks nearly obsolete; pistillate spikes mostly peduncled (sessile in *C. crawei*) 17

17 Perigynia 2-ribbed (with 2 prominent nerves), otherwise nerveless or with fewer than 10 faint nerves 18

17 Perigynia 2-ribbed, with more than 10 additional strong nerves 19

18 Pistillate spikes closely flowered; perigynia in 6 rows **C. meadii**

18 Pistillate spikes loosely flowered; perigynia in 3 rows **C. tetanica**

19 Plants tufted; staminate spikes short-peduncled or sessile; upper pistillate spikes aggregate **C. granularis**

19 Plants with prolonged rhizomes; staminate spikes long-peduncled; pistillate spikes separate **C. crawei**

20 Spikes on flexuous peduncles, drooping or widely spreading 21

20 Spikes on short erect peduncles or sessile 23

21 Perigynia sessile or nearly so, subterete, thin-textured, rather abruptly tapered to the beak, ascending to widely spreading in the spikes; teeth of the perigynium beak straight or nearly so, to 0.8 mm long **C. hystericina**

21 Perigynia stipitate, trigonous-flattened, coriaceous and firm, gradually tapered to the beak, mostly reflexed in the spikes at maturity; teeth of the perigynium beak straight or recurved, 0.6–2 mm long 22

22 Teeth of the perigynium beak 1.2–2 mm long, ultimately recurved **C. comosa**

22 Teeth of the perigynium beak 0.6–1 mm long, straight or nearly so **C. pseudocyperus**

23 Perigynia pubescent; style jointed with the achene 24

23 Perigynia glabrous; style continuous with the achene 25

24 Leaves flat, 2–5 mm wide **C. lanuginosa**

24 Leaves involute, 2 mm or less wide **C. lasiocarpa**

25 Pistillate spikes 15–30 mm thick; perigynia 10–20 mm long **C. lupulina**

25 Pistillate spikes 8–15 mm thick; perigynia 3.5–11 mm long 26

26 Perigynia strongly 7- to 9-nerved, inflated; style strongly S-curved toward the base 27

26 Perigynia with 10 or more nerves, slightly to strongly inflated; style straight 29

27 Bracts of the pistillate spikes several to many times longer than the inflorescence; perigynia widely spreading, the lower ones reflexed **C. retrorsa**

27 Bracts of the pistillate spikes shorter than to somewhat exceeding the inflorescence; perigynia ascending to spreading 28

28 Perigynia in 8–10 rows, spreading; base of culm spongy-thickened; leaf sheaths strongly nodulose, with conspicuous cross markings between the nerves; widespread in region **C. utriculata**

28 Perigynia in 6–8 rows, ascending; base of culm not spongy-thickened; leaf sheaths not conspicuously nodulose; rare (Holt County, Neb) **C. vesicaria**

29 Teeth of the perigynia less than 1 mm long, erect or slightly curved 30

29 Teeth of the perigynia 1–3 mm long, straight or recurved 31

30 Mature perigynia conspicuously many-nerved, the nerves elevated; plant base purple-tinged, the lowest leaf sheaths bladeless **C. lacustris**

30 Mature perigynia inconspicuously nerved, the nerves impressed or level with the surface; plant base whitish or brownish, the lowest leaf sheaths with blades **C. hyalinolepis**

31 Sheaths pubescent; teeth recurved **C. atherodes**

31 Sheaths glabrous; teeth straight **C. laeviconica**

32 Spikes androgynous 33

32 Spikes gynaecandrous or only the terminal spike gynaecandrous (or rarely staminate in *C. interior*), the lateral ones then pistillate 44

33 Culms arising singly or few together from axils of leaves on older reclining culms; plants of wet sphagnum bogs **C. chordorrhiza**

33 Culms arising from tufts or rhizomes; plants of various habitats, mostly not found in sphagnum bogs 34

34 Beaks of the perigynia entire or obliquely cut 35

34 Beaks of the perigynia distinctly bidentate at the apex 39

35 Perigynia rounded on the margins; spikes few-flowered, with 1–6 perigynia and 1 or 2 staminate flowers **C. disperma**

35 Perigynia sharp-edged, at least between the beak and the body 36

36 Plants forming colonies by long-creeping rhizomes, the culms arising singly or few together 37

36 Plants tufted, forming dense clumps; rhizomes none or very short 39

37 Leaf sheaths green-striate ventrally, prolonged into a conspicuous, hyaline, tubular ligule **C. sartwellii**

37 Leaf sheaths hyaline ventrally, truncate at the summit; ligule inconspicuous **C. praegracilis**

38 Ventral surface of the leaf sheath white-hyaline or only slightly copper-tinged at

the mouth; spikes closely aggregated; perigynia shiny **C. diandra**

38 Ventral surface of the leaf sheath copper-colored, at least at the mouth; lower spikes more or less separate; perigynia dull **C. prairea**

39 Spikes single at each node, usually less than 10 in the head; sheaths tight, not cross-rugulose ventrally 40

39 Spikes 2 or more on a branch at the lower nodes; sheaths usually cross-rugulose ventrally (not cross-rugulose in *C. alopecoidea*) 41

40 Perigynia hidden in the head by scales; scales awn-tipped; leaves 1–2.5 mm wide **C. hookeriana**

40 Perigynia conspicuous in the head; scales acuminate to cuspidate; leaves 3–8 mm wide **C. gravida**

41 Body of the perigynium abruptly contracted into the beak; culms not winged or flattened under pressure, 0.5–1.5 mm wide **C. vulpinoidea**

41 Body of the perigynium gradually tapering into the beak, or if abruptly contracted, then the culms winged and flattened under pressure; culms 1.5–3.5 mm wide when pressed 42

42 Perigynia ovate, rounded at the base, contracted into the beak; sheaths not cross-rugulose ventrally **C. alopecoidea**

42 Perigynia truncate-rounded at the base, tapering into the beak; sheaths cross-rugulose ventrally **C. stipata**

43 Perigynia filled to the margins by the achene; the margins at most sharp-edged, not thin-winged 44

43 Perigynia filled only in the central portion by the achene, the margins thin-winged at least on the upper 1/2 47

44 Lower spikes (when 3 or more) overlapping or nearly so in the head; perigynia widely spreading or the lower ones reflexed at maturity, spongy-thickened at the base, the achene occupying mainly the upper 2/3 of the perigynium body 45

44 Lower spikes usually widely separated in the head; perigynia ascending to spreading-ascending at maturity, not spongy-thickened at the base, the achene essentially filling the body of the perigynium 46

45 Teeth of the perigynium beak obscure, not exceeding 0.25 mm long; scales obtuse; common species **C. interior**

45 Teeth of the perigynium beak sharp, 0.3–0.5 mm long; scales acute to short-cuspidate; rare species **C. sterilis**

46 Perigynia 5–10 per spike, ultimately loosely spreading, with a distinct beak ca. 0.4 mm long or more; leaves 1–2.5 mm wide, green **C. brunnescens**

46 Perigynia 10–30 per spike, ascending, with a minute beak 0.2 mm or less long; leaves 2–4 mm wide, grayish-green to glaucous **C. canescens**

47 Lower bracts of the inflorescence many times longer than the heads 48

47 Lower bracts of the inflorescence little, if at all, longer than the heads, often not evident 49

48 Perigynia ca. 6x longer than wide **C. sychnocephala**

48 Perigynia ca. 3x longer than wide **C. athrostachya**

49 Perigynium beaks slender, subterete, slightly serrulate toward the tip, obliquely cut dorsally; plant of moderate elevations in the Black Hills **C. microptera**

49 Perigynium beaks flattened, winged, serrulate to the tip, bidentate; plants more widespread 50

50 Sheaths of principal leaves green-striate ventrally except for a V-shaped hyaline area at the mouth; leaf blades 3–7 mm wide 51

50 Sheaths of principal leaves with a nerveless, white-hyaline band on the ventral side; leaf blades 0.5–4.5 mm wide 52

51 Spikes globose or subglobose, the tips of the perigynia widely spreading to reflexed at maturity **C. cristatella**

51 Spikes oblong-ovoid, the tips of the perigynia appressed to ascending **C. tribuloides**

52 Perigynia subulate to narrowly ovate-lanceolate, 2.5–4x longer than wide, the marginal wings narrow their entire length **C. scoparia**

52 Perigynia ovate-lanceolate, ovate or orbicular, not more than 2x longer than wide, the marginal wings broad their entire length 53

53 Perigynia ovate-lanceolate or narrowly ovate, 2.1 mm wide or less 54

53 Perigynia suborbicular or orbicular, 2.2 mm wide or wider 55

54 Spikes usually loosely arranged in drooping, moniliform heads, at least the lower spikes usually not overlapping in the head; perigynia stramineous at maturity, the ventral surface nerved **C. tenera**

54 Spikes aggregated into compact heads, strongly overlapping; perigynia brown at maturity, the ventral surface nerveless **C. bebbii**

55 Perigynia (4)5–7 mm long, membranous, thin except where distended by the achene, conspicuously nerved ventrally and dorsally **C. bicknellii**

55 Perigynia 3–4.5 mm long, firm-textured, thickened, plano-convex; nerveless or obscurely few-nerved ventrally, faintly nerved dorsally **C. brevior**

Carex alopecoidea Tuckerm.

FOX-TAIL SEDGE

DESCRIPTION Tufted; **culms** stout, soft, 4–10 dm long, trigonous and sharply winged, flattened when pressed, 1.5–3 mm wide. **Leaves** 3–8 mm wide; sheaths tight, purple-dotted, not cross-rugulose ventrally. **Spikes** bisexual, androgynous, aggregated in heads 1.5–5 cm long; pistillate scales acuminate to cuspidate. **Perigynia** brownish-yellow at maturity, plano-convex, ovate, 3–4.5 mm long, rounded and spongy-thickened at the base, contracted into the beak which is 2/3 to as long as the body; **achenes** lenticular, 1.5–2 mm long; stigmas 2. June–July.

HABITAT Swamps, springs and streambanks.

WETLAND STATUS

GP FACW | MIDW FACW | WMTN OBL

Carex aquatilis Wahl.

WATER SEDGE

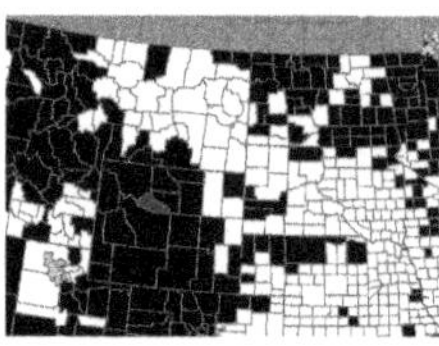

DESCRIPTION Tufted in large or small clumps, with long slender rhizomes; **culms** slender, erect, 2–10 dm long, exceeding the leaves, obtusely trigonous below to sharply trigonous above, usually roughened below the head. **Leaves** glaucous, 2–7 mm wide, the lower ones with well-developed blades; sheaths septate-nodulose dorsally, thin and usually ruptured ventrally, whitish or purplish-dotted. **Spikes** 3–5, the upper staminate, peduncled, the middle and lower ones pistillate or often androgynous, 2–5 cm long; pistillate scales acute to acuminate. **Perigynia** pale green to tawny, often marked or tinged with reddish-brown, usually obovate, broadest near the apex but not inflated, 2–3 mm long; beak minute, entire or oblique; **achenes** lenticular, ca. 1.5 mm long; stigmas 2. June–Aug.

HABITAT Wet meadows, marshes, shores, streambanks, springs, bogs and fens; circumboreal.

WETLAND STATUS

GP OBL | MIDW OBL | WMTN OBL

Carex atherodes Spreng.

SLOUGH SEDGE

DESCRIPTION Loosely tufted from long scaly rhizomes; **culms** erect, trigonous, 5–12 dm long. **Leaves** 3–12 mm wide; sheaths softly pubescent to puberulent dorsally, brown to purple-tinged at the mouth, the lower ones shredding into filaments. **Spikes** unisexual, sessile or short-peduncled, densely flowered; staminate spikes terminal, 2–6; pistillate spikes 2–4, remotely spaced, cylindric, 2–11 cm long; bracts leaflike, exceeding the culm; scales thin, hyaline or pale brown, shorter than the perigynium, the midvein prolonged into a slender awn. **Perigynia** ovoid, long-tapering into a smooth beak, 6–11 mm long, strongly many-nerved, the beak teeth smooth, recurved, 1.2–3 mm long; **achenes** trigonous, 2–2.5 mm long; stigmas 3. June–July.

HABITAT Marshes, wet meadows, ditches, stream and pond margins, usually in shallow water.

WETLAND STATUS

GP OBL | MIDW OBL | WMTN OBL

Carex aquatilis
WATER SEDGE

Carex athrostachya Olney

SLENDER-BEAK SEDGE

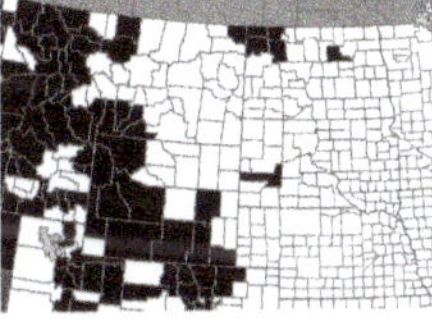

DESCRIPTION Densely tufted, lacking rhizomes; **culms** trigonous, 1.5–10 dm long. **Leaves** shorter than the culms, 1.5–4 mm wide, the lowest leaves bladeless; sheaths hyaline ventrally, brownish-tinged at the base. **Spikes** several, bisexual, gynaecandrous, 4–10 mm long, borne in ovoid to globose-ovoid heads 1–2 cm long; 1–3 of the lowest bracts below the inflorescence prolonged beyond the head, occasionally merely equaling the head, other bracts reduced and inconspicuous; pistillate scales brownish except for the green midrib, longer to shorter than the perigynia, acute to cuspidate. **Perigynia** flattened to slightly plano-convex, narrowly ovate to lanceolate, 2.5–4.5 mm long, ca. 3x longer than wide, faintly several-nerved dorsally, fewer nerved or nerveless ventrally, wing-margined to the base, the beak obliquely cut, serrulate below; **achenes** lenticular, 1.1–1.6 mm long; stigmas 2. July–Aug.

HABITAT Wet meadows and low prairie.

WETLAND STATUS

GP FACW | WMTN FACW

Carex athrostachya
SLENDER-BEAK SEDGE

Carex aurea Nutt.
GOLDEN SEDGE

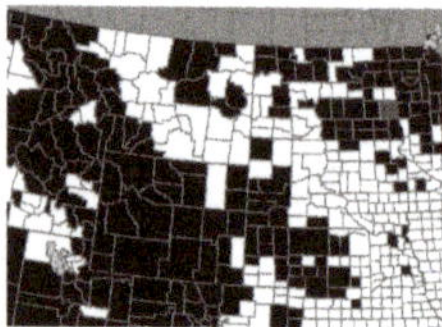

DESCRIPTION Loosely tufted from rhizomes; **culms** erect, trigonous, 0.5–2(3) dm long. **Leaves** 1–4 mm wide; sheaths membranous ventrally, concave at the mouth. **Spikes** 2–5 per culm, ascending, the lower ones peduncled; terminal spikes totally staminate (rarely slightly pistillate at the apex), 3–18 mm long; lateral spikes pistillate, aggregate to widely spaced, 7–15(20) mm long; bract of the lowest spike short-sheathing the culm, surpassing the inflorescence; pistillate scales whitish to tawny, with a green midvein, ovate to round-ovate, acute to nearly rounded, cuspidate, shorter than the perigynia. **Perigynia** whitish-pulverulent, especially at the apex, becoming golden orange at maturity (drying pale), somewhat flattened, elliptic to obovate, beakless or with a very short tubular beak, several-ribbed, (1.5) 2–2.5 (3) mm long; **achenes** dark brown to blackish, lenticular, 1 –1.5 mm long; stigmas 2. June–July.

HABITAT Wet meadows, low prairie, springs, moist woods and along shores, often where sandy.

WETLAND STATUS
GP OBL | MIDW FACW | WMTN FACW

Carex aurea
GOLDEN SEDGE

Carex bebbii Olney ex Fern.
BEBB'S SEDGE

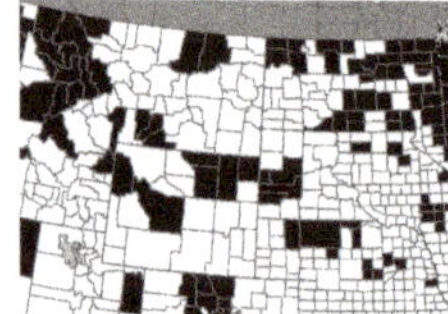

DESCRIPTION Tufted; **culms** sharply trigonous, 1.5–8 dm long. **Leaves** shorter to slightly longer than the culms, 1.5–4.5 mm wide; sheaths white-hyaline ventrally. **Spikes** bisexual, gynaecandrous, 5–10, ovoid to subglobose, 5–8 mm long, closely aggregate in an ovoid head 1.5–2.5 cm long; bracts of the inflorescence reduced and inconspicuous; pistillate scales acute to acuminate, narrower and slightly shorter than the perigynia. **Perigynia** green to brown, plano-convex, ovate, 2.5–3.5 mm long, finely nerved dorsally, nerveless ventrally, narrowly wing-margined to the base; beak ca. 1/3 to 1/2 the entire length of the perigynium, serrulate, shallowly bidentate; **achenes** lenticular, 1–1.5 mm long; stigmas 2. June–Aug.

HABITAT Wet meadows, marshes, streambanks, floodplains, ditches and other wet places.

WETLAND STATUS
GP OBL | MIDW OBL | WMTN OBL

Carex bebbii
BEBB'S SEDGE

Carex bicknellii Britt.

BICKNELL'S SEDGE

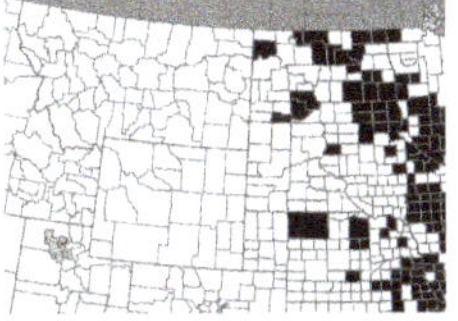

DESCRIPTION Tufted from short rootstocks; **culms** erect, sharply trigonous, 3–12 dm long, overtopping the leaves. **Leaves** 2–4.5 mm wide; sheaths white-hyaline ventrally. **Spikes** bisexual, gynaecandrous, usually 3–7, globose to ovoid above the narrower male portion, 10–18 mm long, aggregate to separate in an ovoid to linear head 1.5–6 cm long; bracts much reduced; pistillate scales obtuse to acute, shorter and much narrower than the perigynia. **Perigynia** stramineous, membranous, thin except where thickened by the achene, broadly ovate, 4–7 mm long, 1/2 to 3/4 as wide, several-nerved on both faces (fewer ventrally), broadly wing-margined to the base, abruptly contracted to the serrulate, bidentate beak which is 1/4 to 1/3 the entire length of the perigynium; **achenes** lenticular, 1.7–2 mm long; stigmas 2. June–July.

HABITAT Wet meadows, low prairie, ditches and shores.

WETLAND STATUS
GP FACW | MIDW FACU

Carex bicknellii
BICKNELL'S SEDGE

Carex brevior (Dewey) Mack. ex Lunell

FESCUE SEDGE

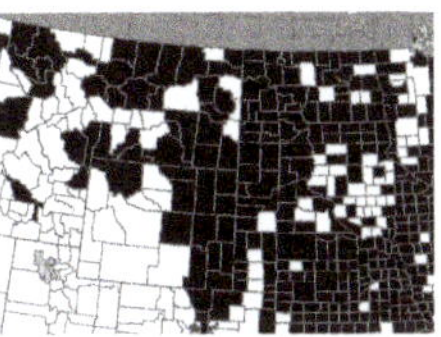

DESCRIPTION Tufted from short rootstocks; **culms** sharply trigonous, 3–10 dm long, exceeding the leaves. **Leaves** 1–4 mm wide; sheaths white-hyaline ventrally, entirely green to green- and-white mottled or white-hyaline between the nerves dorsally. **Spikes** bisexual, gynaecandrous, 2–8, subglobose to ovoid, the lateral ones rounded to clavate at the base, 5–15 mm long, aggregate to somewhat separate in oblong to slender heads 1.5–5 cm long; bracts much reduced, inconspicuous or the lowest seldom exceeding the head; pistillate scales obtuse or acute, shorter and narrower than the perigynia. **Perigynia** green to pale brown, firm-textured, plano-convex, broadly ovate (the body suborbicular), 3–4.5 mm long, 3/5 to 3/4 as wide, several-nerved dorsally, nerveless or obscurely few-nerved ventrally, wing-margined to the base, abruptly contracted or somewhat tapered to the serrulate, bidentate beak which is 1/4 to nearly 1/2 the entire length of the perigynium; **achenes** lenticular, 1.7–2 mm long; stigmas 2. June–July.

HABITAT Wet meadows, low to mesic prairie, ditches, shores and streambanks.

WETLAND STATUS
GP FAC | MIDW FAC | WMTN FAC

Carex brevior
FESCUE SEDGE

Carex brunnescens (Pers.) Poir.

BROWNISH SEDGE

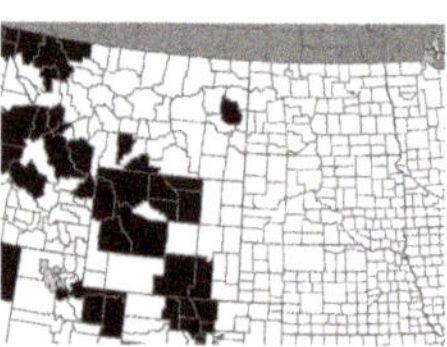

DESCRIPTION Densely tufted from a short, fibrillose rootstock; **culms** slender, sharply trigonous, 0.7–6 dm tall, smooth or slightly roughened below the head, usually surpassing the leaves. **Leaves** deep green, 1–2.5(5) mm wide; sheaths tight, hyaline ventrally. **Spikes** 5–10, all gynaecandrous, ovoid, 4–8 mm long, each with 5–10(15) perigynia, the lower spikes usually widely separated in a head (1.5)3–5 cm long; lowermost bract setaceous, shorter than to exceeding its subtended spike, the remaining bracts shorter, scalelike; pistillate scales ovate, obtuse or acute, somewhat shorter than the perigynia. **Perigynia** filled to the margins by the achene, not winged or sharp-edged, 1.7–2.7 mm long, 1–1.5 mm wide, lightly nerved on both sides, not spongy-thickened at the base, tapered at the apex into a short, minutely bidentate beak 0.4–0.7 mm long, the beak and the upper portion of the perigynium minutely serrulate on the margins and whitish-puncticulate; **achenes** lenticular, 1.2–1.5 mm long, ca. 1 mm wide; stigmas 2. Late May–July.

HABITAT Wet woods and bogs.

WETLAND STATUS
WMTN OBL

Carex brunnescens
BROWNISH SEDGE

Carex buxbaumii Wahl.

BROWN BOG SEDGE

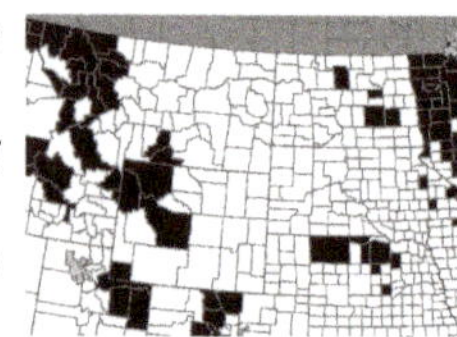

DESCRIPTION Loosely tufted from long rhizomes; **culms** arising singly or few together, lateral to previous year's shoots, trigonous, 3–10 dm long, roughened above, reddish toward the base. **Leaves** 1–3 mm wide, the lowest ones bladeless; lower sheaths shredding into filaments, the upper ones membranous and purple-dotted ventrally. **Spikes** 2–5, close together or rather remote; terminal spike gynaecandrous, slightly to considerably larger than the lateral ones, 1–3 cm long; lateral spikes pistillate, short-cylindric, sessile or nearly so; bracts leaflike, the lowest shorter than to seldom surpassing the head; scales dark brown, acute to acuminate, awn-tipped. **Perigynia** light green, golden toward the base, flattened or distended by the achene, elliptic, 2.5–3.5 mm long, slightly more than 1/2 as wide, 2-ribbed, with 6–8 faint nerves on each face; beak minute, bidentulate; **achenes** trigonous, 1.4–2 mm long; stigmas 3. Late May–June.

HABITAT Wet meadows, springs and fens.

WETLAND STATUS
GP OBL | MIDW OBL

Carex buxbaumii
BROWN BOG SEDGE

Carex canescens L.

HOARY SEDGE

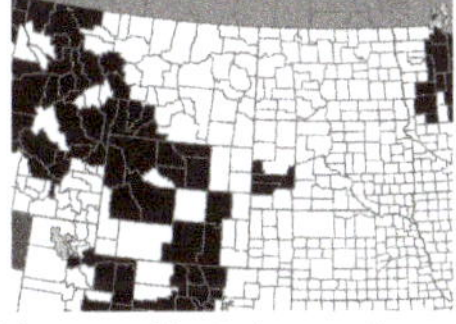

DESCRIPTION Quite similar to *Carex brunnescens,* differing as follows: Foliage glaucous; **leaf blades** 2–4 mm wide. **Spikes** 4–8, ovoid to oblong-cylindric, each containing 10–30 perigynia. **Perigynia** 1.8–3 mm long, 1.3–1.7 mm wide, with a beak 0.4 mm or less long, not noticeably serrulate on the margins. June–July.

HABITAT Wet mountain meadows, streambanks and boggy places in coniferous forest; uncommon in the Black Hills, otherwise more common in mountains farther west; circumboreal.

WETLAND STATUS

GP OBL | MIDW OBL | WMTN OBL

Carex capillaris L.

HAIR-LIKE SEDGE

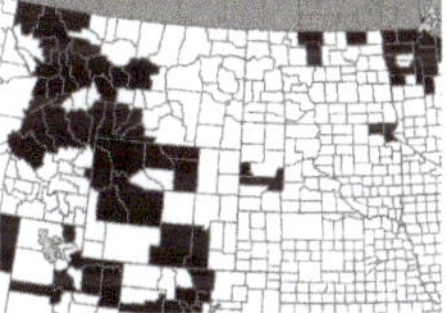

DESCRIPTION Densely tufted; **culms** slender, trigonous, 1.5–4 dm long. **Leaves** mainly basal, much shorter than the culms, 0.7–2.5 mm wide; sheaths tight, truncate at the summit. **Spikes** unisexual; terminal spike staminate, 4–8 mm long; lateral spikes 1–4, remote, loosely flowered, short-cylindric, 5–15 mm long, borne on filiform, spreading to drooping peduncles 5–15 mm long; bract subtending the lowest pistillate spike with a well-developed sheath to 25 mm long; scales white-hyaline marginally, greenish or light brown in the middle, obtuse or acute, shorter but often wider than the perigynium, deciduous. **Perigynia** shiny brown to olive-green, nearly terete, ovoid-ellipsoid, 2.2–3.5 mm long, 2-ribbed, otherwise nerveless, tapered to an obliquely cut, hyaline-tipped beak ca. 0.5 mm or more long; **achenes** trigonous with concave sides, 1.2–1.5 mm long, jointed to the style; stigmas 3. June–July.

HABITAT Boggy places and swamps, usually where wooded.

WETLAND STATUS

GP FACW | MIDW FACW | WMTN FACW

Carex canescens
HOARY SEDGE

Carex capillaris
HAIR-LIKE SEDGE

Carex chordorrhiza Ehrh. ex L.f.
ROPE-ROOT SEDGE

DESCRIPTION **Culms** arising singly or few together from the axils of dried leaves on older, reclining culms, the old, sterile culms elongate, smooth and wiry, fertile culms erect or nearly so, obtusely trigonous, 1–3 dm tall. **Leaves** 1–3 well-developed on each culm, the lower ones tending to be bladeless, blades 1–5 cm long on the fertile culms, to much longer on the sterile ones, 1–2 mm wide; sheaths hyaline ventrally. **Spikes** 3–8, bisexual, androgynous, ovoid, crowded in an oblong to ovoid head 5–15 mm long; bracts none; pistillate scales deep brown, ovate, acuminate, about equaling the perigynia. **Perigynia** brown, compressed-ovoid, 2–3.5 mm long, coriaceous, many-nerved on both faces, obscurely margined; beak short, emarginate; **achenes** lenticular; stigmas 2. June–Aug.

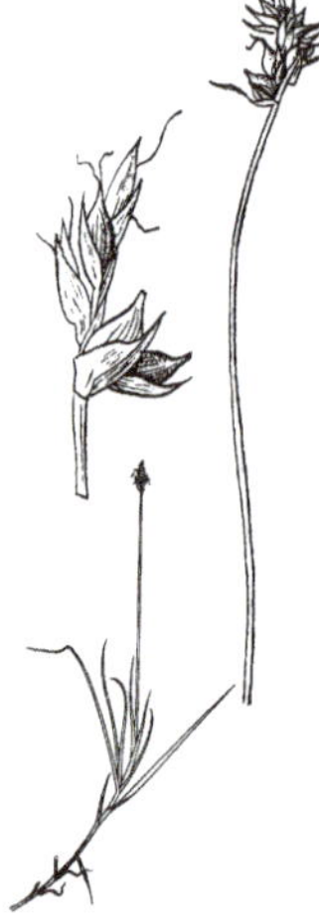

HABITAT Wet sphagnum bogs; circumboreal.

WETLAND STATUS
GP OBL | MIDW OBL

Carex comosa F. Boott
BEARDED SEDGE

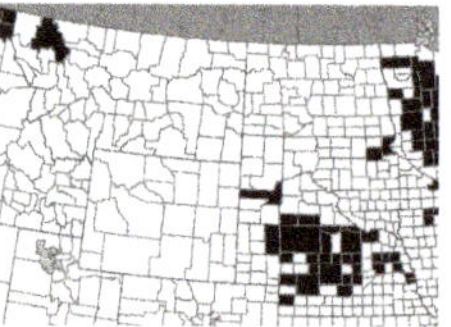

DESCRIPTION Densely tufted, often forming large clumps; **culms** stout, sharply trigonous, 5–15 dm long. **Leaves** 5–12 mm wide; sheaths hyaline ventrally, conspicuously septate-nodulose dorsally. **Spikes** unisexual, the terminal one staminate, 3–7 cm long; lateral spikes pistillate, 3–5, cylindric, 3–8 cm long, 9–12 mm thick, the lower spikes longer-peduncled and eventually drooping; bracts leaflike, much surpassing the inflorescence; pistillate scales with a small hyaline-margined body, tapered into a long, rough awn longer than the body. **Perigynia** numerous, reflexed at maturity, flattened-trigonous, lanceolate with a stipitate base, 5.7–7.7 mm long, firm and shiny, strongly 12- to 17-nerved, tapering to the smooth, slender beak 2–3 mm long, the teeth recurved-spreading, 1.2–2 mm long; **achenes** trigonous, 1.5–2 mm long, continuous with the style; stigmas 3. Late June–Aug.

HABITAT Fresh marshes, swamps and spring-fed streams.

WETLAND STATUS
GP OBL | MIDW OBL | WMTN OBL

Carex crawei Dewey
CRAWE'S SEDGE

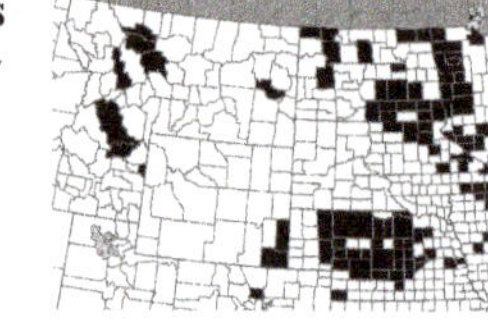

DESCRIPTION **Culms** arising singly or few together from long-creeping rhizomes, weakly trigonous, 0.6–4 dm long. **Leaves** 1–4 mm wide; sheaths tight, hyaline ventrally. **Spikes** unisexual, cylindric, terminal one staminate, lateral ones pistillate, 2–5, remote, the lowest nearly basal, compactly flowered, 1–3 cm long; bracts leaflike with well-developed sheaths, the lowest one best developed, with the sheath 5–15 mm long, the blade not exceeding the terminal spike; scales reddish-brown with a pale or green-

Carex comosa
BEARDED SEDGE

ish midrib, shorter and narrower than the perigynia. **Perigynia** green to brown, ellipsoid to ovoid-ellipsoid, 2.3–3.5 mm long, many-nerved; beak absent or to 0.4 mm long, entire to bidentulate; **achenes** trigonous, 1.3–1.7 mm long; stigmas 3. June–July.

HABITAT Wet meadows, ditches and seepage areas.

WETLAND STATUS

GP FACW | MIDW FACW | WMTN FACW

Carex cristatella Britt.

CRESTED SEDGE

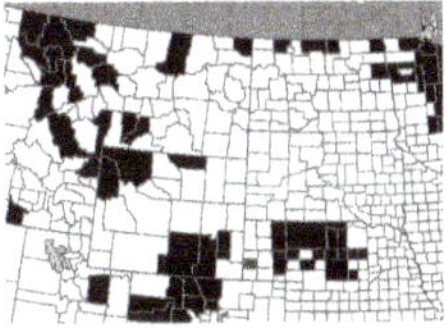

DESCRIPTION Tufted from short rootstocks; **culms** sharply trigonous, 3–10 dm long, slightly shorter to exceeding the leaves. **Leaves** 3–7 mm wide; sheaths loose, green-striate dorsally and ventrally except for the V-shaped hyaline area at the mouth. **Spikes** bisexual, gynaecandrous, 5–12, globose or subglobose, 4–8 mm long, crowded in an ovoid to oblong head 2–3.5 cm long; bracts much reduced, inconspicuous; pistillate scales acute to acuminate, shorter than the perigynia. **Perigynia** widely spreading to reflexed at maturity, green to pale brown, plano-convex, ovate to lanceolate, 2.5–4 mm long, 1/3 to 1/2 as wide, faintly nerved on both sides, strongly winged above the middle, narrower-winged below the middle, tapered to the serrulate, bidentate beak which is 1/3 to 1/2 the total length of the perigynium, often notched or wrinkled at the base; **achenes** lenticular, 1–1.5 mm long; stigmas 2. Late June–Aug.

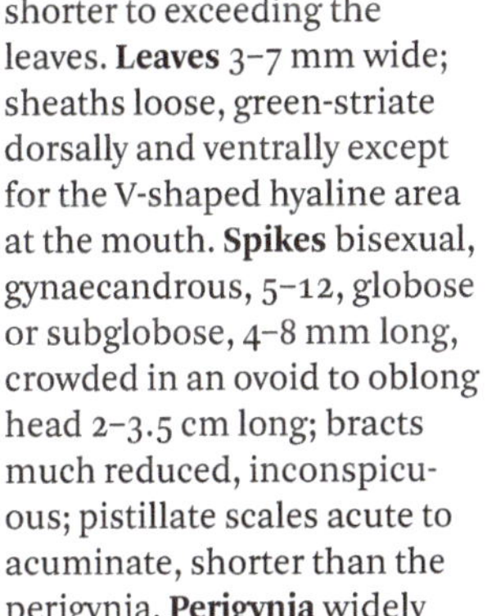

HABITAT Wet meadows, floodplains, shores and streambanks; primarily along rivers and streams.

WETLAND STATUS

GP FACW | MIDW FACW | WMTN FACW

Carex diandra Schrank

LESSER TUSSOCK SEDGE

DESCRIPTION Densely tufted; **culms** sharply trigonous, 3–8 dm long, usually exceeding the leaves. **Leaves** 1–3 mm wide; sheaths pale-striate, white-hyaline ventrally or only slightly copper-tinged at the mouth, prolonged into a ligule 1–3 mm long. **Spikes** bisexual, androgynous, aggregate, continuous or nearly so in ovoid to subcylindric heads 1–4 cm long; bracts reduced, inconspicuous, usually shorter than the spikes; pistillate scales brownish, acute to cuspidate, about equaling the perigynia. **Perigynia** brown, shiny, unequally biconvex, deltoid-ovate, truncate-rounded at the base, 2–3 mm long, ca. 1/2 as wide, few-nerved dorsally, nerveless ventrally, tapering to the serrulate, entire to bidentulate beak which is 1/4 to 1/2 the entire length of the perigynium; **achenes** lenticular, ca. 1 mm long; stigmas 2. June–July.

HABITAT Wet meadows, springs and fens.

WETLAND STATUS

GP OBL | MIDW OBL

Carex diandra
LESSER TUSSOCK SEDGE

Carex disperma Dewey

SOFT-LEAF SEDGE

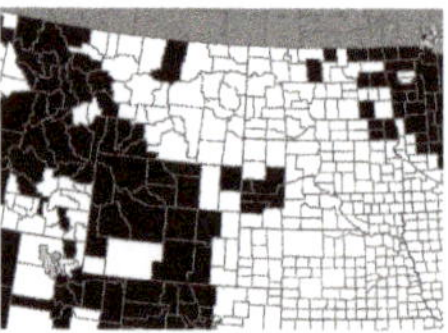

DESCRIPTION Loosely tufted or single from slender rhizomes; **culms** slender, weak, trigonous, 1–4 dm long, shorter than to exceeding the leaves. **Leaves** soft and spreading, 1–2 mm wide, the sheaths tight, hyaline ventrally. **Spikes** bisexual, androgynous, 2–5, sessile, few-flowered and very small, with 1–6 perigynia and 1 or 2 staminate flowers, to 5 mm long, remote or the upper spikes close together in interrupted heads 1.5–2.5 cm long; bracts sheathless, inconspicuous, resembling the pistillate scales or filiform and to 2 cm long; pistillate scales white-hyaline except for the midrib, acuminate or mucronate, mostly 2/3 as long as the perigynium. **Perigynia** unequally biconvex to nearly terete, ellipsoid, 2–3 mm long, strongly nerved and rounded on the margins, otherwise obscurely to plainly many-nerved, the beak minute, oblique; **achenes** lenticular, elliptic, 1.4–1. 7 mm long; stigmas 2. May–June.

HABITAT Swamps, bogs and springs.

WETLAND STATUS
GP FACW | MIDW OBL | WMTN FACW

Carex disperma
SOFT-LEAF SEDGE

Carex emoryi Dewey

EMORY'S SEDGE

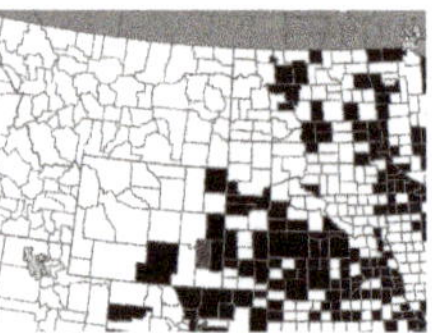

DESCRIPTION Loosely tufted from long scaly rhizomes; **culms** trigonous, 4–10 dm long, exceeding the leaves, smooth to rough. **Leaves** green, 2–5 mm wide, the lower ones bladeless, merely sheathing the base of the culms; leaf sheaths white-hyaline ventrally, white to yellow-tinged dorsally, the lower ones dark red to brown, not shredding into ladderlike fibers; ligule at juncture of blade and sheath truncate to rounded, as wide as or wider than long. **Spikes** 3–7, the terminal 1 or 2 staminate, 2–4.5 cm long, the lateral ones pistillate or androgynous, 2–10 cm long; lowest bract leaflike, sheathless; pistillate scales blunt to acuminate, narrower than the perigynia. **Perigynia** light green, becoming stramineous at maturity, whitish-papillate only at the apex, biconvex, elliptic or ovate-elliptic, 1.6–2.9 mm long, 1/2 to 3/5 as wide; beak to 0.2 mm long, entire; **achenes** lenticular, ca. 1.5 mm long; stigmas 2. Late May–July.

HABITAT Shores, streambanks, wet meadows and seepage areas.

WETLAND STATUS
GP OBL | MIDW OBL | WMTN OBL

NOTE See note under *Carex stricta*.

Carex emoryi
EMORY'S SEDGE

Carex garberi Fern.

ELK SEDGE

DESCRIPTION Much like *Carex aurea.* **Culms** mostly solitary from slender rhizomes, erect, trigonous, 1.5–3 dm long, exceeding the leaves. **Leaves** 2–3 mm wide; sheaths white-hyaline ventrally, concave at the mouth. **Spikes** 2–4 per culm, ascending, the lower ones peduncled; terminal spikes gynaecandrous or some staminate, 8–30 mm long; lateral spikes pistillate, closely spaced or separate, 7–15 mm long; bract subtending the lowest spike short-sheathing the culm and surpassing the inflorescence; pistillate scales brown to purplish with a green midvein, blunt, somewhat shorter than the perigynia. **Perigynia** whitish-pulverulent at maturity, not turning golden orange, obovoid, 2–3 mm long, 2-ribbed, otherwise inconspicuously nerved, beakless or with a minute tubular beak; **achenes** lenticular, ca. 1.5 mm long; stigmas 2. June–July.

HABITAT Wet meadows, seepage areas and boggy places.

WETLAND STATUS
GP FACW | MIDW FACW

Carex garberi
ELK SEDGE

Carex granularis Muhl. ex Willd.

MEADOW SEDGE

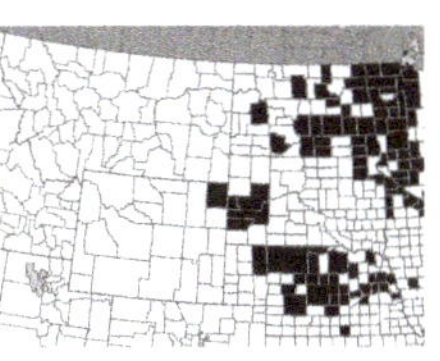

DESCRIPTION Tufted from very short rhizomes; **culms** obtusely trigonous, 1–5 dm long. **Leaves** often exceeding the culms, 3–13 mm wide; sheaths membranous ventrally, septate-nodulose dorsally. Spikes unisexual, the terminal one staminate, sessile or subsessile, the lateral ones pistillate, aggregated around the staminate spike; bracts considerably overtopping the inflorescence, short-sheathing; pistillate scales brownish, acuminate to cuspidate, 1/2 as long as the perigynia. **Perigynia** crowded in several rows, green or olive to brownish, ellipsoid to obovoid, 2–2.5(4) mm long, 1/2 to 2/3 as wide, 2-ribbed with more than 10 additional strong nerves; beak minute or obsolete, entire to slightly bidentulate; **achenes** trigonous, 1.3–1.7 mm long; stigmas 3. June–early July.

HABITAT Wet meadows, streambanks, pond margins, springs and seepage areas.

WETLAND STATUS
GP OBL | MIDW FACW | WMTN OBL

Carex granularis
MEADOW SEDGE

Carex gravida Bailey

HEAVY SEDGE

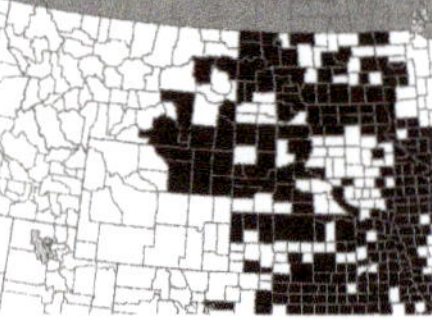

DESCRIPTION Tufted from stout rootstocks; **culms** erect or leaning, sharply trigonous, 3–11 dm long, roughened above. **Leaves** 3–8 mm wide; sheaths whitish and conspicuously septate-nodulose dorsally, white-hyaline ventrally, not cross-rugulose. **Spikes** bisexual, androgynous, single at each node, usually less than 10, in dense, ovoid to oblong heads 1–3.5 cm long; bracts usually inconspicuous, setaceous, seldom exceeding the head; pistillate scales green to brown, acuminate to cuspidate, shorter than the perigynia. **Perigynia** conspicuous in the head, not hidden by the scales, green and pale to yellowish- brown, shiny, plano-convex, ovate, 4–5 mm long, ca. 1/2 as wide, obscurely nerved dorsally, nerveless ventrally, contracted to the serrulate, bidentate beak which is ca. 1/3 the total length of the perigynium; **achenes** lenticular, 1.8–2.2 mm long, about as wide; stigmas 2, the style base enlarged. June–July.

HABITAT Wet meadows, streambanks, floodplains and prairie swales, but more often in moist woods and thickets.

WETLAND STATUS
GP FACW | MIDW FACU | WMTN FACW

Carex gravida
HEAVY SEDGE

Carex gynocrates Wormskj. ex Drej.

NORTHERN BOG SEDGE

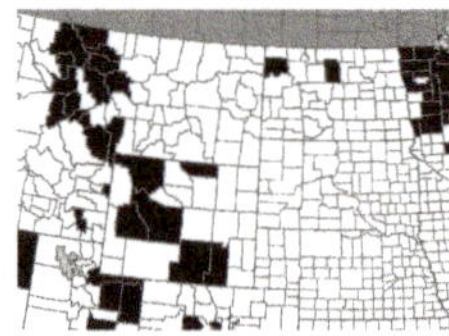

DESCRIPTION **Culms** single or few together from long, very slender rhizomes, slender but stiff, 0.4–3 dm long, smooth, usually exceeding the leaves, brown and sheathed at the base by the previous year's leaves. **Leaves** clustered near the base, the blades narrowly involute and filiform, 0.4–0.9 mm wide. **Spikes** only 1 per culm, all staminate or all pistillate or bisexual and androgynous, 0.5–2 cm long, the staminate spike or portion linear-cylindric, the pistillate spike or portion short-cylindric; bract obsolete; pistillate scales brown or reddish, oblong-ovate, acute or acuminate, shorter than but wider than the perigynium. **Perigynia** 4–10, widely spreading or somewhat reflexed, yellowish to brownish-black and shiny, plump, oblong-ovate, 2.5–4 mm long, 1/2 as wide, spongy at the base, finely many-nerved dorsally, obscurely nerved ventrally, abruptly contracted to the beak; beak nearly entire to obliquely cut, 0.5 mm long; **achenes** lenticular, 1.5 mm long; stigmas 2. June–July.

SYNONYM *Carex dioica* subsp. *gynocrates* (Wormsk.) Hultén

HABITAT Sphagnum bogs and wet, peaty soils.

WETLAND STATUS
GP OBL | MIDW OBL

Carex gynocrates
NORTHERN BOG SEDGE

Carex hallii Olney (*not illustrated*)

DEER SEDGE

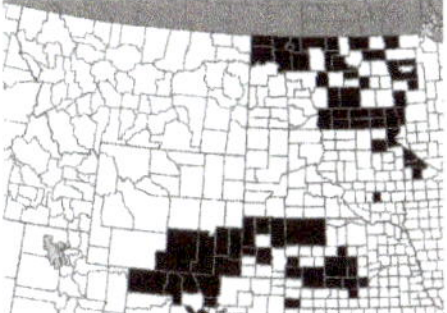

DESCRIPTION Loosely tufted from short rootstocks, also producing long rhizomes; **culms** weakly trigonous below to sharply trigonous above, 1.5–4 dm long. **Leaves** mainly basal, 1/4 to nearly as long as the culms, 1–3 mm wide; previous year's leaves commonly persistent at the base. **Spikes** 1–5, the terminal one staminate, pistillate or gynaecandrous, 1–2.5 cm long; lateral spikes usually present, all pistillate, smaller than the terminal spike; bracts obsolete to setaceous, not exceeding the head; pistillate scales brown with a green midvein and scarious margins, obtuse to cuspidate, equaling to slightly exceeding the perigynia. **Perigynia** green to brown, obovate, 1.5–2.3 mm long, abruptly short-beaked, the beak 0.2–0.4 mm long, entire to bidentulate, papillate and spinulose; **achenes** unequally trigonous, 1.2–1.7 mm long; stigmas 3. June–July.

HABITAT Wet meadows, springs and seepage areas.

WETLAND STATUS
GP FAC | MIDW FACW | WMTN FACW

Carex haydenii Dewey

CLOUD SEDGE

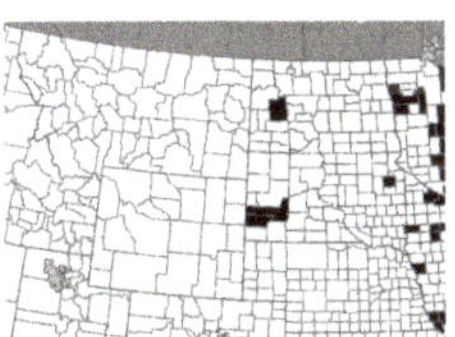

DESCRIPTION Loosely tufted from short rhizomes; **culms** arising from the previous year's tufts of leaves, enveloped at the base by the dried leaves, trigonous, 3–10 dm long, usually surpassing the leaves, roughened above. **Leaves** green, 2–5 mm wide, the lower ones bladeless, merely sheathing the base of the culm; leaf sheaths white to yellowish-hyaline ventrally, green dorsally. **Spikes** 3–6, the upper 1–3 staminate, the terminal one largest, 2–5 cm long, the others much smaller, the lower 2–3 pistillate or androgynous, 1–3 cm long; lowest bract leaflike, sheathless, usually shorter than the head; pistillate scales acuminate, longer than the perigynia. **Perigynia** pale brown at maturity, often with darker brown spots, biconvex, round-obovate, inflated at the apex, 2–2.5 mm long, 3/5 to about as wide; beak minute; **achenes** lenticular, ca. 1 mm long; stigmas 2. Late May–July.

HABITAT Wet meadows, marshes streambanks.

WETLAND STATUS
GP OBL | MIDW OBL | WMTN OBL

Carex hookeriana Dewey

HOOKER'S SEDGE

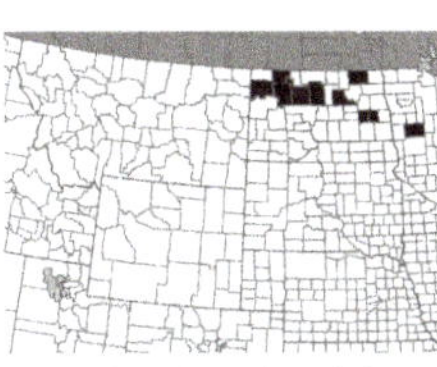

DESCRIPTION Densely tufted from short rootstocks; **culms** slender, rather lax, trigonous, 1.5–6 dm long, exceeding the leaves. **Leaves** 1–2.5 mm wide; sheaths tight, white-hyaline and not cross-rugulose ventrally, green and not septate-nodulose dorsally. **Spikes** bisexual, androgynous, single at each node, usually less than 10, lower ones usually separate, in heads 1.5–5 cm long; bracts obsolete or much reduced and setaceous; pistillate scales green to reddish-brown with broad scarious margins, especially the lower ones awn-tipped, about as long and as wide as the perigynia. **Perigynia** hidden in the head by the scales, brownish, plano-convex, ovate-lanceolate, 2.7–3.5 mm long, membranous-thin and nerveless where distended over the achene, abruptly contracted into a bidentate, serrulate beak which is 1/4 to 1/3 the total length of the perigynium; **achenes** lenticular, 1.5–1.7 mm long; stigmas 2. June–July.

HABITAT Wet meadows, boggy places, moist woods; also in mesic prairie.

WETLAND STATUS
GP UPL | MIDW UPL

Carex haydenii
CLOUD SEDGE

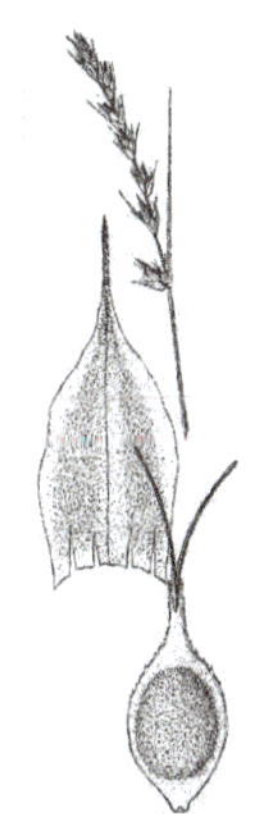

Carex hookeriana
HOOKER'S SEDGE

Carex hyalinolepis Steud. (*not illustrated*)
THINSCALE SEDGE

DESCRIPTION Quite similar to *Carex lacustris* and sometimes treated as a variety of it, differing chiefly as follows: Lower leaf sheaths whitish to pale brown, not red-tinged, rarely breaking into fibers. **Perigynia** 5–8 mm long, with numerous very faint nerves which are slightly elevated or impressed.

HABITAT Wet meadows, marshes and swamps.

WETLAND STATUS
GP OBL | MIDW OBL

Carex hystericina Muhl. ex Willd.
BOTTLEBRUSH SEDGE

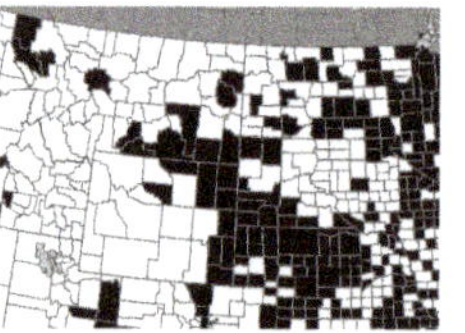

DESCRIPTION Tufted, often in large clumps, with short rhizomes; **culms** erect or leaning, trigonous, 2–10 dm long, usually surpassing the leaves. **Leaves** yellow-green, 3–8 mm wide; sheaths white-hyaline ventrally, green to yellowish or reddish dorsally, the lower sheaths eventually breaking into filaments. **Spikes** unisexual, the terminal one staminate, 1–5 cm long, usually short-peduncled and often subtended by a bract; lateral spikes pistillate or occasionally slightly androgynous, 1–4, short-cylindric, 1–5 cm long, 0.8–1.5 cm thick, separate or aggregate, the lower ones usually nodding on filiform peduncles, the upper ones shorter-peduncled and ascending; pistillate scales inconspicuous, narrow and much shorter than the perigynia, rough-awned. **Perigynia** spreading or ascending, greenish-stramineous, nearly terete at maturity, ovoid, 5–7.5 mm long, strongly 12- to 17-nerved, abruptly contracted to the slender beak which is ca. 1/2 the total length of the perigynium; beak teeth 0.4–1 mm long, erect or nearly so; **achenes** trigonous with concave sides, 1.4–1.7 mm long; stigmas 3. June–July.

HABITAT Shores, streambanks, wet meadows, springs, swamps and fens.

WETLAND STATUS
GP OBL | MIDW OBL | WMTN OBL

Carex hystericina
BOTTLEBRUSH SEDGE

Carex interior Bailey
INLAND SEDGE

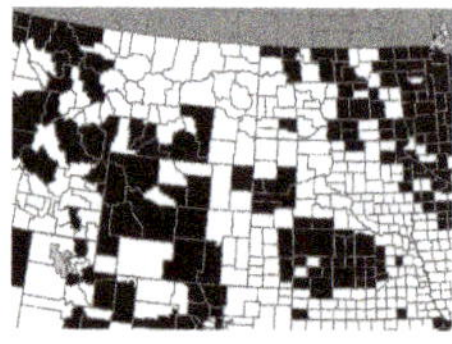

DESCRIPTION Densely tufted; **culms** slender, sharply trigonous, 1–6 dm long, about equaling to exceeding the leaves. **Leaves** 1–2 mm wide; sheaths tight, hyaline ventrally. **Spikes** 2–4, the terminal one gynaecandrous or rarely staminate, the lateral ones pistillate, or seldom gynaecandrous, globose, ca. 5 mm in diameter, overlapping or nearly so in heads 0.8–2.5 cm long, the terminal

Carex interior
INLAND SEDGE

spike often more remote than the lateral ones; bracts much reduced or obsolete; pistillate scales obtuse, much shorter than the perigynium. **Perigynia** brownish-green to brown, concavo-convex, ovate, filled to the margins by the achene, sharp-edged but not wing-margined, 2–3 mm long, 1/2 to 2/3 as wide, several-nerved dorsally, nerveless or few-nerved at the base ventrally, spongy at the base so that the achene fills mainly the upper 2/3 of the perigynium body, truncate-rounded at the base, contracted to a serrulate beak which is 1/4 to 1/3 the entire length of the perigynium; beak teeth obscure, not exceeding 0.25 mm long; **achenes** lenticular, 1.2–1.5 mm long, about as wide; stigmas 2. Late May–July.

HABITAT Wet meadows, shores, streambanks, springs, fens and boggy places.

WETLAND STATUS

GP OBL | MIDW OBL | WMTN OBL

Carex lacustris Willd.

LAKEBANK SEDGE

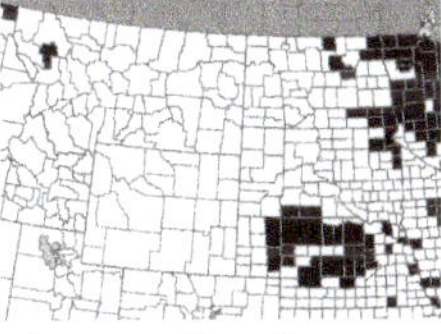

DESCRIPTION Tufted from prolonged scaly rhizomes; **culms** stout, erect, trigonous, 6–13 dm long, roughened. **Leaves** about equaling or slightly exceeding the culm, 6–15 mm wide; sheaths often partly red-tinged, the lower ones disintegrating into a network of fibers. **Spikes** unisexual, the upper 2–4 staminate, sessile, 4–7 cm long, the lower 2–4 pistillate, erect, usually separate, sessile or short-peduncled, cylindric, 3–10 cm long, 3–15 mm thick; bracts leaflike, some or all exceeding the inflorescence; pistillate scales awned or acuminate, the body shorter than the perigynia, hyaline to pale brown on the sides. **Perigynia** olive, flattened-subterete, slenderly ovoid, 5.5–7 mm long, with more than 10 strong, elevated nerves, tapering to a smooth beak ca. 1 mm long, the beak teeth 0.4–0.8 mm long, erect or slightly curved; **achenes** trigonous, 2–2.5 mm long; stigmas 3, the lower portion of the style straight and persistent. June–early July.

Carex lacustris
LAKEBANK SEDGE

HABITAT Swamps, marshes, and springs, usually in shallow water.

WETLAND STATUS

GP OBL | MIDW OBL | WMTN OBL

Carex laeviconica Dewey

SMOOTHCONE SEDGE

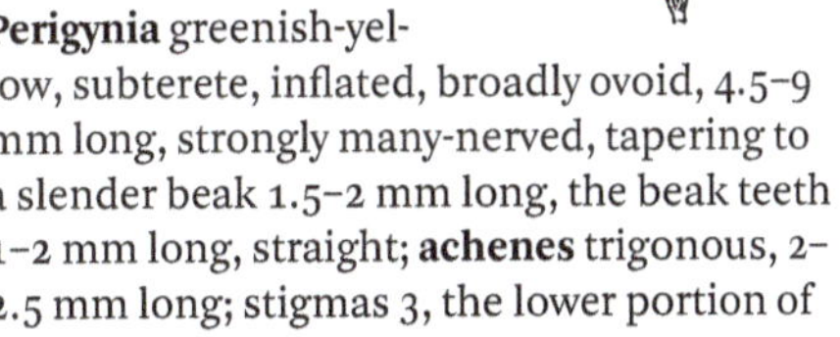

DESCRIPTION Loosely tufted from prolonged scaly rhizomes; **culms** stout, trigonous, 3–12 dm long. **Leaves** shorter than to surpassing the culm, 2–8 mm wide; sheaths glabrous, often purple-tinged below and splitting into a network of fibers. **Spikes** unisexual, the upper 2–6 staminate, 1–4 cm long, the lower 2–4 pistillate, erect, remote, sessile to short-peduncled, cylindric, 3–10 cm long, 6–10 mm thick; bracts leaflike, equaling to surpassing the inflorescence; pistillate scales acute to aristate, the body shorter than the perigynium, hyaline or brown on the sides. **Perigynia** greenish-yellow, subterete, inflated, broadly ovoid, 4.5–9 mm long, strongly many-nerved, tapering to a slender beak 1.5–2 mm long, the beak teeth 1–2 mm long, straight; **achenes** trigonous, 2–2.5 mm long; stigmas 3, the lower portion of

the style straight and persistent on the achene. June–July.

HABITAT Wet meadows, marshes, shores, streambanks, ditches, springs, low wooded areas.

WETLAND STATUS
GP OBL | MIDW OBL | WMTN OBL

Carex lanuginosa Michx.
WOOLLY SEDGE

DESCRIPTION Extensively colonial from scaly rhizomes; **culms** trigonous, 2–10 dm long. **Leaves** 2–5 mm wide; sheaths hyaline ventrally, the lower ones often purple-tinged dorsally, disintegrating and leaving a loose network of fibers. **Spikes** unisexual, the upper 1–3 (usually 2) staminate, 2–6 cm long, the lower 1–3 pistillate, remote, sessile or nearly so, cylindric, 1–4 cm long; bracts leaflike, the lowest one usually surpassing the inflorescence; pistillate scales brown to purplish-brown on the sides, acuminate to awned, shorter to longer than the perigynia. **Perigynia** brownish to yellowish-green to grayish-brown, or sometimes purplish, subterete, oblong-ovoid, 2.5–4(5) mm long, densely pubescent, many-nerved, contracted into the beak which is 1/4 to 1/3 the entire length of the perigynium, the beak teeth 0.3–0.8 mm long, divergent; **achenes** trigonous with concave sides, 1.7–2 mm long; stigmas 3, the style jointed with the achene. June–July.

Carex lanuginosa
WOOLLY SEDGE

SYNONYMS *Carex lasiocarpa* Ehrh., in part; *Carex pellita* Muhl. ex Willd.

HABITAT Wet meadows, marshes, shores, streambanks, springs, ditches and other wet places; common, often abundant.

WETLAND STATUS
GP OBL | MIDW OBL | WMTN OBL

Carex lasiocarpa Ehrh. (*not illustrated*)
WOOLLY-FRUIT SEDGE

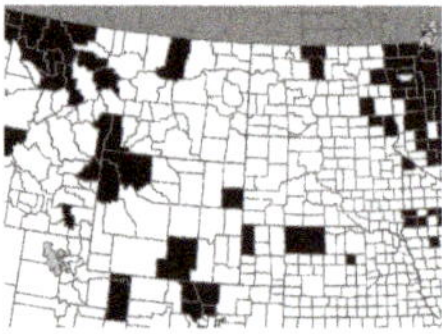

DESCRIPTION Very similar to *Carex lanuginosa* but clearly distinguished from it by the long, attenuate **leaves** which are strongly involute, 2 mm or less wide as folded.

HABITAT Bogs and wet peaty soils; circumboreal.

WETLAND STATUS
GP OBL | MIDW OBL

Carex leptalea Wahl.
BRISTLY-STALK SEDGE

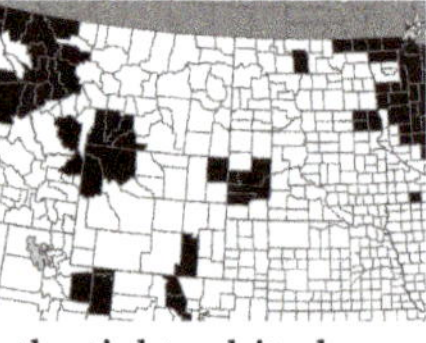

DESCRIPTION Densely tufted; **culms** slender, obtusely trigonous, 1–7 dm long, equaling or exceeding the leaves. **Leaves** 0.5–1.25 mm wide; sheaths tight, white-hyaline ventrally. **Spikes** solitary on the culms, bisexual, androgynous, few-flowered, 5–15 mm long; bracts none; staminate scales obtuse to acute, overlapping around the rachis; pistillate scales obtuse to cuspidate, shorter than the perigynia, or the tip of the lowest scale sometimes prolonged beyond the perigynium. **Perigynia** yellowish-green, nearly terete or somewhat flattened, oblong to narrowly elliptic, 2.5–5 mm long, finely many-nerved, beakless or short-beaked, the orifice entire; **achenes** trigonous-obovoid, 1.5–2 mm long; stigmas 3. June–July.

HABITAT Bogs and wet woods.

WETLAND STATUS
GP OBL | MIDW OBL | WMTN OBL

Carex leptalea
BRISTLY-STALK SEDGE

Carex limosa L.
MUD SEDGE

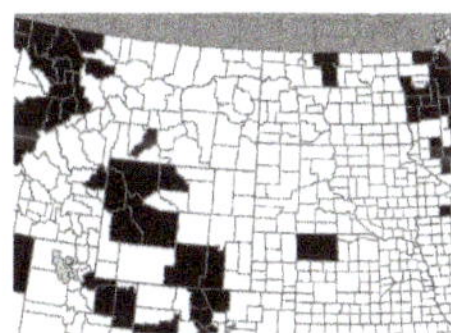

DESCRIPTION Loosely tufted from long scaly rhizomes; **culms** sharply trigonous, usually roughened above, 3–5 dm long, exceeding the leaves. **Leaves** 1–3 mm wide; sheaths hyaline ventrally, shredding into filaments below. **Spikes** unisexual, the terminal one staminate, 1–3 cm long, the lower 1–3 spikes pistillate, drooping on lax, filiform peduncles 1–3 cm long, ovoid to short-cylindric, 1–2 cm long; scales brown, obtuse to cuspidate, about equaling the perigynia in length and width. **Perigynia** glaucous-green, elliptic-ovoid, flattened except where distended by the achene, 2.5–4.2 mm long, 1/2 as wide, strongly 2-ribbed with a few more obscure nerves on each side; beak minute, entire; **achenes** trigonous, ca. 2.2 mm long; stigmas 3. June–July.

HABITAT Bogs and fens; circumboreal.

WETLAND STATUS
GP OBL | MIDW OBL

Carex lupulina Willd.
HOP SEDGE

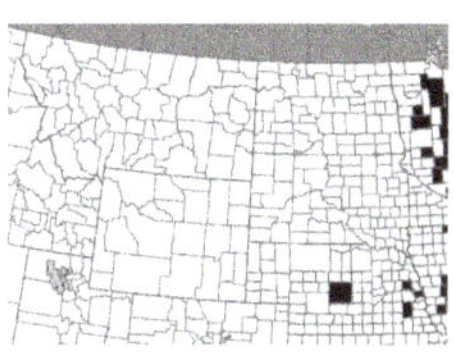

DESCRIPTION Loosely tufted with rhizomes; culms **stout**, trigonous, 3–12 dm long. **Leaves** much surpassing the inflorescence, 4–15 mm wide; sheaths white hyaline ventrally, the lower ones brownish. **Spikes** unisexual, the upper one staminate, short-peduncled, 2–5 cm long; pistillate spikes 2–6, aggregated or at least overlapping, the lowermost sometimes remote, 2.5–6 cm long, 2–3.5 cm thick; bracts leaflike and spreading, sheathing at the base, much surpassing the inflorescence; pistillate scales ovate-lanceolate, acuminate to short-awned, much shorter than the perigynia. **Perigynia** many, ascending to appressed-ascending, greenish-brown, dull, lance-ovoid and inflated, 10–20 mm long, 4–7 mm wide, many-nerved, acuminate to the slender beak which is 1/2 or more of the total length, the teeth 0.7–2 mm long; **achenes** trigonous, 3–4 mm

Carex limosa
MUD SEDGE

long; stigmas 3, the style bent or twisted below the middle and persistent on the achene. June–Aug.

HABITAT Wet woods, swamps, wet meadows, marshes, ditches and shores.

WETLAND STATUS
GP OBL | MIDW OBL

ADDITIONAL SPECIES **Greater bladder sedge** (*Carex intumescens* Rudge) is a similar plant that occurs as a rarity in moist woodland and perhaps along springs and streams in the Black Hills (Pennington and Custer Counties, SD; Crook Co., Wyo). It differs from *C. lupulina* in the relatively few, uncrowded perigynia which are olive-green and glossy, plus the straight to loosely contorted style.

Carex meadii Dewey
MEAD'S SEDGE

DESCRIPTION Tufted from long rhizomes; **culms** stiff, trigonous, 1.5–5 dm long. **Leaves** 2–7 mm wide; sheaths tight, hyaline ventrally, concave at the mouth. **Spikes** unisexual; terminal spike staminate, 1–3 cm long; lateral spikes pistillate, usually remote, at least the lower ones short-peduncled, short-cylindric, usually closely flowered, 16–30 mm long, 4–7 mm thick, with 6 rows of perigynia; bracts sheathing, the blades usually not exceeding the culm; pistillate scales purplish-red or brown on the margins with a prominent green midvein, obtuse or short-awned, as wide but shorter than the perigynia. **Perigynia** yellowish-green or brown, obscurely trigonous, obovoid, 2–3.5 mm long, 2-ribbed, with fewer than 10 additional faint nerves, rounded at the tip, the beak short-tubular and bent, sometimes obsolete; **achenes** trigonous with concave sides, 1.5–2 mm long; stigmas 3. June–July.

SYNONYM *Carex tetanica* var. *meadii* (Dewey) Bailey

HABITAT Moist meadows and prairie.

WETLAND STATUS
GP FAC | MIDW FAC | WMTN OBL

Carex lupulina
HOP SEDGE

Carex microptera Mack.
SMALL-WING SEDGE

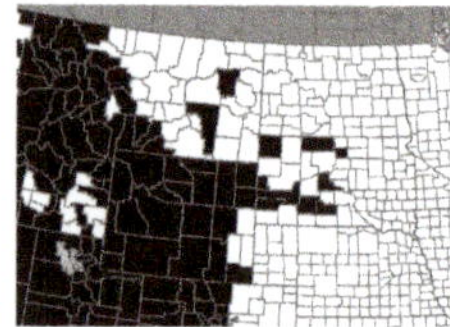

DESCRIPTION Tufted with short rootstocks; **culms** sharply trigonous, 3–10 dm long, surpassing the leaves. **Leaves** 2–6 mm wide; sheaths white hyaline ventrally. **Spikes** bisexual, gynaecandrous, 4–10, closely aggregated in an oblong to ovoid or suborbicular head 10–22 mm long; bracts much reduced, awn-tipped; pistillate scales light to dark brown with a light midrib, acute to obtuse, narrow and shorter than the perigynia. **Perigynia** light green to light brown, membranous, plano-convex, ovate-lanceolate to lanceolate, 3.5–5 mm long, 1–2 mm wide,

Carex microptera
SMALL-WING SEDGE

several-nerved dorsally, few- to several-nerved near the bas.e ventrally, strongly wing-margined to the base, tapering to the dark brown beak; beak terete, serrulate, ca. 1/2 the entire length of the perigynium, obliquely cut dorsally, bidentulate; **achenes** lenticular, 1–1.5 mm long; stigmas 2. June–July.

HABITAT Wet meadows, streambanks and springs.

WETLAND STATUS
GP FAC | WMTN FACU

Carex nebrascensis Dewey
NEBRASKA SEDGE

DESCRIPTION Loosely to densely tufted from long rhizomes; **culms** stout, erect, 2.5–12 dm long, trigonous, shorter than to exceeding the leaves. **Leaves** glaucous, 3–8 mm wide, the lower ones with well-developed blades; sheaths septate-nodulose dorsally, hyaline and often yellow-brown tinged ventrally. **Spikes** unisexual or an occasional 1 or 2 androgynous, the upper 1–2(4) staminate, the terminal one largest, 1.5–4 cm long; lateral spikes pistillate or occasionally 1 or 2 androgynous, 2–5, erect, separate, sessile or the lower ones short-peduncled, 1–5 cm long; pistillate scales brown to nearly black, obtuse to acuminate, mostly about equaling the perigynia. **Perigynia** ascending to spreading, brown or stramineous at maturity, plano-convex or biconvex, oblong-obovate, 3–3.5 mm long, ca. 1/2 as wide, 2-ribbed, conspicuously nerved between the ribs; beak 0.3–0.5 mm long, shallowly bidentate; **achenes** lenticular, 1.5–2 mm long; stigmas 2. June–July.

Carex nebrascensis
NEBRASKA SEDGE

HABITAT Wet meadows, marshes, streams and springs.

WETLAND STATUS
GP OBL | MIDW OBL | WMTN OBL

Carex praegracilis W. Boott
CLUSTERED FIELD SEDGE

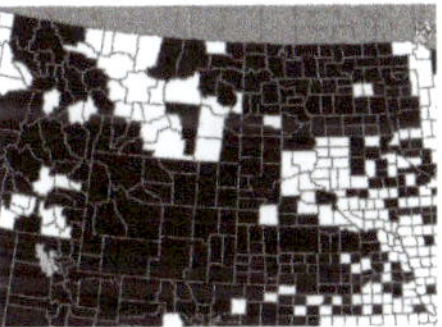

DESCRIPTION Colonial from long black rhizomes; **culms** arising singly or few together, trigonous, 1.5–7 dm long, surpassing the leaves. **Leaves** 2–3 mm wide, coming off the lower portion of the culm; sheaths white-hyaline ventrally, truncate at the summit, ligule inconspicuous. **Spikes** bisexual and androgynous or nearly all staminate or pistillate, 4–8 mm long, the upper ones crowded, the lower ones more separated, in narrowly ovoid to linear-oblong heads 1–4 cm long; bracts obsolete; pistillate scales brown, shiny, shorter than to equaling the perigynia. **Perigynia** greenish-brown to eventually blackish-brown, plano-convex, ovate to

Carex praegracilis
CLUSTERED FIELD SEDGE

ovate-lanceolate, (2.5)3–4 mm long, ca. 1/3 as wide, sharp-edged, nerveless ventrally, obscurely few- to many-nerved dorsally, spongy at the base, tapering into the serrulate beak which is 1/2 or more the length of the body, obliquely cut; **achenes** lenticular, 1.2–1.8 mm long; stigmas 2. June–Aug.

HABITAT Wet meadows, low prairie, shores, streambanks, ditches and other wet or moist places; common and can form dense colonies in open, moist areas.

WETLAND STATUS
GP FACW | MIDW FACW | WMTN FACW

Carex prairea Dewey
PRAIRIE SEDGE

DESCRIPTION Tufted, forming dense clumps from short rootstocks; **culms** sharply trigonous, 5–10 dm long, somewhat exceeding the leaves. **Leaves** 2–3 mm wide; sheaths hyaline ventrally, copper-colored at least at the mouth, prolonged 2–3 mm beyond the base of the blade. **Spikes** bisexual, androgynous, ovoid, 4–7 mm long, the lower ones usually separate, in oblong to linear-oblong heads 3–8 cm long; bracts much reduced, occasionally exceeding the spike; pistillate scales reddish-brown, acuminate, as long as and mostly concealing the perigynia. **Perigynia** brown, dull, plano-convex, lanceolate-ovoid, (2)2.4–3 mm long, 1/2 as wide, few-nerved at the base ventrally, truncate at the base, tapering to a serrulate, obliquely cut beak which is ca. 1/2 the total length of the perigynium; **achenes** lenticular, 1–1.2 mm long; stigmas 2. June–July.

HABITAT Springs, fens, fresh wet meadows and boggy places.

WETLAND STATUS
GP OBL | MIDW OBL | WMTN OBL

Carex pseudocyperus L.
CYPRESS-LIKE SEDGE

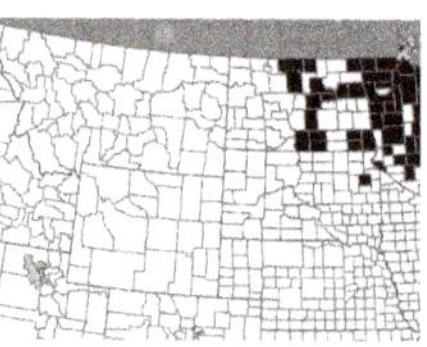

DESCRIPTION Tufted in large clumps from short rootstocks; **culms** stout, trigonous, 3–10 dm long, roughened. **Leaves** 5–15 mm wide; sheaths hyaline ventrally, yellow-tinged dorsally. **Spikes** unisexual, the terminal one staminate, 1.5–7 cm long; lateral spikes pistillate, 2–6, cylindric, 3–8 cm long, ca. 1 cm thick, slender-peduncled, the lower ones drooping; bracts much exceeding the inflorescence; pistillate scales with a very small body, rough awned, the awn shorter to longer than the perigynium. **Perigynia** spreading to reflexed, flattened-trigonous, slenderly ovoid, 4–6.2 mm long, firm and shiny, strongly 12- to 17-nerved, tapering to the smooth beak which is 1–1.5 mm long, the teeth 0.5–1 mm long, straight; **achenes** trigonous, 1.5–1.7 mm long; stigmas 3. July–Aug.

HABITAT Cold springs, bogs and swamps.

WETLAND STATUS
GP OBL | MIDW OBL

Carex retrorsa Schwein.
RETRORSE SEDGE

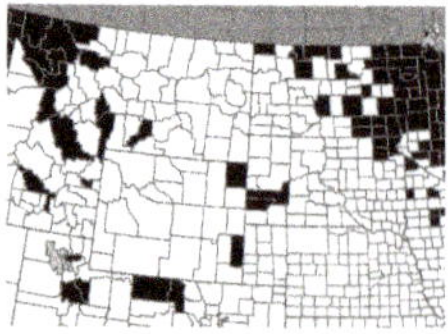

DESCRIPTION Densely clustered; **culms** 4–10 dm long. **Leaves** 3–4 dm long and 4–10 mm wide, flat and soft; sheaths dotted

Carex pseudocyperus
CYPRESS-LIKE SEDGE

with small bumps. **Spikes** either all male or female, or the terminal 1–2 spikes with both male and female flowers, the male above the female, stalkless or lowest spike on a slender stalk; lower spikes 3–8, female, 1.5–5 cm long and 1.5–2 cm wide; pistillate scales conspicuous, shorter and narrower than the perigynia. **Perigynia** crowded in rows, spreading or the lowest perigynia angled downward, smooth and shiny, 6–13-nerved, 7–10 mm long, somewhat inflated, tapered to a long, smooth beak 2–4 mm long, the beak teeth short, to 1 mm long; **achenes** dark brown, 3-angled, 2 mm long, loose in the lower part of the perigynium; stigmas 3. June–Aug.

HABITAT Floodplain forests, swamps, thickets and marshes.

WETLAND STATUS
GP OBL | MIDW OBL | WMTN OBL

Carex sartwellii Dewey
SARTWELL'S SEDGE

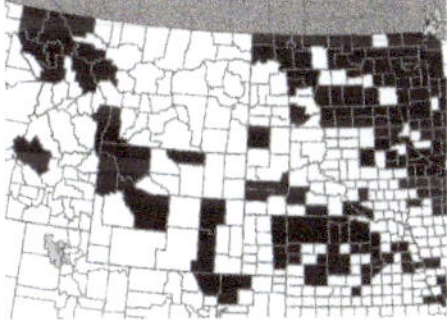

DESCRIPTION Colonial from long, black rhizomes; **culms** arising singly or few together, stiff, sharply trigonous, 3–8 dm long, exceeding the leaves. **Leaves** 2.5–4 mm wide, few per culm, the lowest much reduced, lacking blades; sheaths green-striate ventrally, prolonged into a conspicuous hyaline, tubular ligule. **Spikes** bisexual or the upper ones often staminate, otherwise androgynous, aggregate or the lower ones separate, ovoid, 5–10 mm long, in heads narrowly oblong to conic, 3–6 cm long; bracts reduced, the lower ones sometimes setaceous and exceeding the spike; scales brown with a prominent green midvein, acute to cuspidate, about equaling the perigynia. **Perigynia** tan to brown, plano-convex, ovate, 2.5–3.5 mm long, ca. 1/2 as wide, finely nerved on both sides, sharp-edged, tapered into the serrulate, oblique to minutely bidentate beak which is ca. 1/4 the length of the entire perigynium; **achenes** lenticular, 1–1.5 mm long; stigmas 2. June–July.

HABITAT Wet meadows, marshes, shores, streambanks and ditches.

WETLAND STATUS
GP FACW | MIDW FACW | WMTN OBL

Carex retrorsa
RETRORSE SEDGE

Carex scoparia Schkuhr ex Willd.
BROOM SEDGE

DESCRIPTION Densely tufted, sometimes stoloniferous; **culms** sharply trigonous, usually exceeding the leaves, 2–10 dm long. **Leaves** 1–3 mm wide; sheaths tight, white-hyaline ventrally. **Spikes** bisexual, gynaecandrous, 4–10, ovoid to fusiform, 6–12 mm long, closely aggregate to separate, in an ovoid to linear-oblong head 1–5 cm long; bracts reduced, the lowest often setaceous; pistillate scales acuminate, slightly shorter than the perigynia. **Perigynia** greenish-white, flat, subulate to narrowly lanceolate, (3)4–6.5 mm

Carex sartwellii
SARTWELL'S SEDGE

long, 1/4 to 1/3 as wide, strongly to slightly several-nerved on both sides, narrow-winged for the entire length, gradually tapered to the serrulate, obscurely bidentate beak which is ca. 1/3 the length of the entire perigynium; **achenes** lenticular, 1–1.5 mm long; stigmas 2. June–July.

HABITAT Wet meadows, shores, springs, fens and swamps.

WETLAND STATUS

GP FACW | MIDW FACW | WMTN FACW

Carex sterilis Willd.

DIOECIOUS SEDGE

DESCRIPTION Very similar to *C. interior,* differing mainly as follows: **Leaves** 1–4 mm wide; **spikes** 3–7; pistillate scales acute to cuspidate, about equaling the perigynium body; **perigynium** beak prominent, rough, ca. 1/3 the length of entire perigynium, the teeth sharp, 0.3–0.5 mm long. Late May–July.

SYNONYM

Carex muricata L. var. *sterilis* (Carey)

HABITAT Fens and permanently wet meadows; rare, becoming more common in the Great Lakes region.

WETLAND STATUS

GP OBL | MIDW OBL

Carex stipata Muhl. ex Willd.

SAW-BEAK SEDGE

DESCRIPTION Densely tufted; **culms** concave-trigonous, slightly winged, 2–12 dm long. **Leaves** 4–8 mm wide; sheaths cross-rugulose ventrally, conspicuously septate-nodulose dorsally. **Spikes** bisexual, androgynous, aggregate or the lowest ones often separate, in oblong heads 3–10 cm long; bracts reduced, occasionally setaceous and exceeding the spike; pistillate scales acuminate to cuspidate, 1/2 to 3/4 as long as the perigynia. **Perigynia** yellowish-green to dull brown, plano-convex, ovoid-lanceolate, 3.5–5 mm long, 1/4 to 215 as wide, strongly several-nerved dorsally, strongly few-nerved ventrally, truncate-rounded at the base, tapering to a serrulate, bidentate beak which is 1/2 to 2/3 the length of the entire perigynium; **achenes** lenticular, 1.5–2 mm long; stigmas 2. June–July.

HABITAT Wet meadows, shores, streambanks and swamps.

WETLAND STATUS

GP OBL | MIDW OBL | WMTN OBL

Carex stricta Lam.

TUSSOCK SEDGE

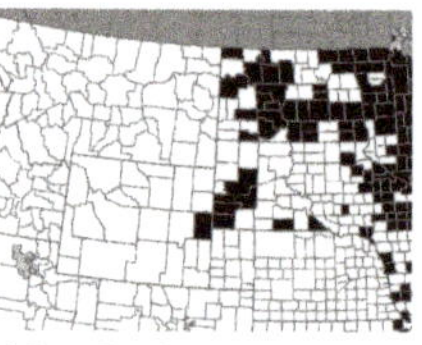

DESCRIPTION Loosely tufted from long scaly rhizomes; **culms** trigonous, 3–10 dm long, exceeding the leaves, rough. **Leaves** green, 2–6 mm wide, the lower ones bladeless, merely sheathing the base of the culm; sheaths white to reddish-brown and hyaline ventrally, green dorsally, the lower ones breaking into ladderlike filaments; ligule at juncture of blade and sheath V-shaped and pointed, longer than wide. **Spikes** unisexual or mostly so, the upper 1–3 staminate, the terminal one 1.5–5 cm long, the lower 2–5 pistillate or some androgynous, 2–8 cm long; lowest bract leaflike, sheathless; pistillate scales obtuse to acuminate, about equaling to exceeding the perigynia in length but narrower. **Perigynia** green at the tip and

Carex stipata
SAW-BEAK SEDGE

margins, golden to tawny toward the middle and base, whitish-papillate from the tip to below the middle, eventually turning brown, biconvex to nearly flat, ovate to elliptic, 1.5–2.7 mm long, 1/2 to 3/5 as wide, 2-ribbed with few weaknerves on both sides; beak tubular, 0.2–0.3 mm long; **achenes** lenticular, 1.3–1.7 mm long; stigmas 2. June–July.

HABITAT Wet meadows, marshes, shores, streambanks, springs and fens.

WETLAND STATUS

GP OBL | MIDW OBL | WMTN OBL

Carex sychnocephala Carey

MANY-HEAD SEDGE

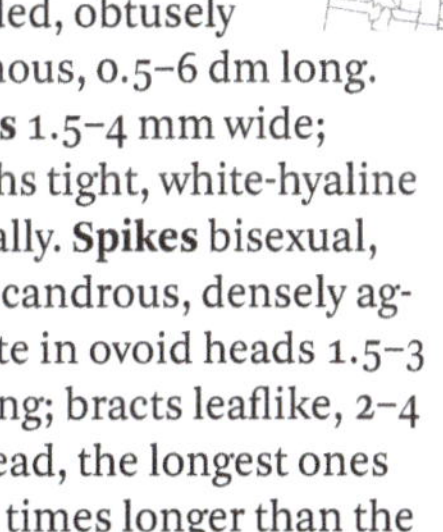

DESCRIPTION Tufted from fibrous roots (often acting as an annual); **culms** often numerous and crowded, obtusely trigonous, 0.5–6 dm long. **Leaves** 1.5–4 mm wide; sheaths tight, white-hyaline ventrally. **Spikes** bisexual, gynaecandrous, densely aggregate in ovoid heads 1.5–3 cm long; bracts leaflike, 2–4 per head, the longest ones many times longer than the heads; pistillate scales hyaline with a green midvein, acuminate or cuspidate, mostly 2/3 as long as the perigynia. **Perigynia** green to stramineous, flat, scale-like, lanceolate, 5–6.5 mm long, 1/6 as wide, slightly nerved dorsally and ventrally, narrowly wing-margined to the base, spongy below the achene at maturity, tapering or contracted to the slender, serrulate beak which is 3–4.5 mm long, slenderly bidentate; **achenes** lenticular, 1–1.5 mm long; stigmas 2. June–Sept.

Carex stricta
TUSSOCK SEDGE

HABITAT Wet meadows, shores, mud flats and streambanks.

WETLAND STATUS

GP FACW | MIDW FACW | WMTN FACW

Carex tenera Dewey

QUILL SEDGE

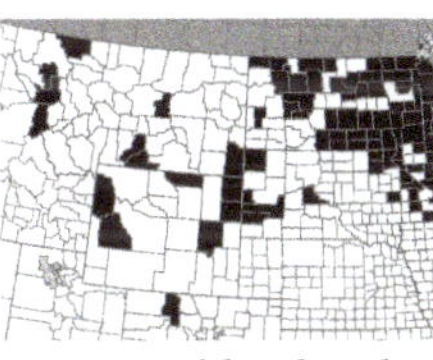

DESCRIPTION Tufted from short rootstocks; **culms** slender, sharply trigonous, 3–8 cm long, exceeding the leaves, rough above. **Leaves** 0.5–3 mm wide; sheaths white-hyaline between the nerves or green-and-white mottled dorsally, white-hyaline ventrally. **Spikes** bisexual, gynaecandrous, 4–8, ovoid to subglobose, 4–10 mm long, usually loosely arranged in moniliform, flexuous or nodding heads 2.5–5 cm long; bracts reduced, occasionally setaceous and exceeding the spike; pistillate scales acute to subacuminate, slightly shorter than the perigynia. **Perigynia** stramineous at maturity, plano-convex, ovate, 2.6–4 mm long, ca. 1/2 as wide, strongly several-nerved dorsally, few-nerved ventrally, wing-margined to the base, tapered

Carex tenera
QUILL SEDGE

or contracted to the serrulate, bidentate beak which is 1/3 to 1/2 the length of the entire perigynium; **achenes** lenticular, 1.2–1.7 mm long; stigmas 2. June–Aug.

HABITAT Wet meadows, springs, streambanks, floodplains, moist woods and thickets.

WETLAND STATUS
GP FACW | MIDW FACW | WMTN FACW

Carex tetanica Schkuhr
RIGID SEDGE

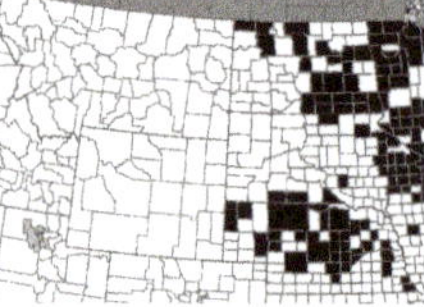

DESCRIPTION Tufted from slender rhizomes; **culms** trigonous, 1–6 dm long, rough above. **Leaves** 1–5 mm wide; sheaths tight, white or yellowish-hyaline ventrally. **Spikes** unisexual, terminal spike staminate, 1–3 cm long; lateral spikes pistillate, usually widely separated, the lower ones peduncled., short-cylindric, loosely flowered, 6–30 mm long, 3–5 mm thick, with perigynia in 3 rows; bracts sheathing, the blades usually not exceeding the inflorescence; pistillate scales purplish-brown on the margins, obtuse to acute or short-awned, as wide but shorter than the perigynia. **Perigynia** green, obscurely trigonous, obovoid-fusiform, 2–4 mm long, ca. 1/2 as wide, 2-ribbed, with fewer than 10 additional faint nerves; beak minute, bent; **achenes** trigonous with concave sides, 2–2.5 mm long; stigmas 3. June–July.

HABITAT Wet meadows, low prairie and boggy or springy places.

WETLAND STATUS
GP FACW | MIDW FACW

Carex tetanica
RIGID SEDGE

Carex tribuloides Wahl.
BLUNT BROOM SEDGE

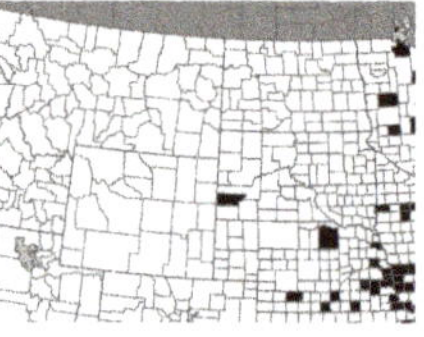

DESCRIPTION Tufted from short rootstocks; **culms** sharply trigonous, 3–9 dm long, exceeding the leaves. **Leaves** rather stiff, mostly 3–7 mm wide; sheaths loose, green-striate and firm ventrally except for the V-shaped hyaline area at the mouth. **Spikes** bisexual, gynaecandrous, 5–15, oblong-ovoid, 6–13 mm long, blunt-tipped, rounded to clavate at the base, densely to loosely aggregated into an ovoid to oblong

Carex tribuloides
BLUNT BROOM SEDGE

head 2–5 cm long; bracts much reduced, inconspicuous; pistillate scales acute to acuminate, shorter than the perigynia. **Perigynia** with tips appressed to ascending in the spikes, light green to pale brown, much flattened except where distended by the achene, lanceolate, 4–6 mm long, 1–2 mm wide, short-stipitate at the base, broadly winged near the middle but nearly wingless around the achene, tapered to the serrulate, bidentate beak which is 1/3 to 1/2 the entire length of the perigynium; **achenes** lenticular, ca. 1.5 mm long; stigmas 2. June–July.

HABITAT Floodplains, wet meadows, shores and ditches.

WETLAND STATUS
GP OBL | MIDW OBL | WMTN FACW

Carex utriculata Boott
BEAKED SEDGE

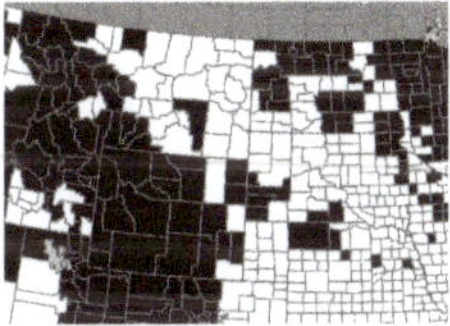

DESCRIPTION Densely tufted from short rootstocks, also with long rhizomes; **culms** bluntly trigonous, 3–12 dm long, spongy-thickened at the base. **Leaves** strongly septate-nodulose, 4–12 mm wide; ligule about as wide as long; sheaths white-hyaline ventrally, conspicuously septate-nodulose dorsally, usually not ladder-fibrillose. **Spikes** unisexual, the upper 2–5 staminate, well above the pistillate spikes, the terminal one 3–6 cm long; lower 2–5 spikes pistillate or occasionally 1 or 2 androgynous, usually remote, cylindric, 1.5–10 cm long, 8–12 mm thick, the upper ones sessile or short-peduncled, lower ones peduncled, erect; bracts shorter than to somewhat exceeding the inflorescence; pistillate scales acute to awned, shorter than, or with awn, to longer than the perigynia. **Perigynia** ascending to ultimately spreading in 8–10 rows in the spike, yellowish-green to brown, shiny, subterete, ovoid, inflated, 3.5–8 mm long, 1/2 to 2/3 as wide, strongly 7- to 9-nerved, contracted to the slender smooth beak 1–2 mm long, the teeth 0.5–0.7 mm long, mostly straight; **achenes** trigonous, 1.7–2 mm long; stigmas 3, the style strongly S-curved toward the base. June–Aug.

HABITAT Wet meadows, marshes, fens, swamps, shores and springs; circumboreal.

WETLAND STATUS
GP OBL | MIDW OBL | WMTN OBL

Carex utriculata
BEAKEDSEDGE

Carex vesicaria L.
LESSER BLADDER SEDGE

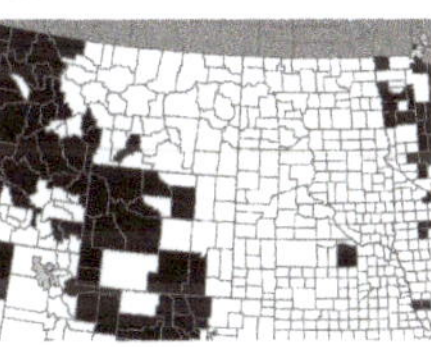

DESCRIPTION Tufted from stout rootstocks, lacking long rhizomes; **culms** sharply trigonous and roughened below the head, 3–10 dm long, not spongy-thickened at the base. **Leaves** not conspicuously septate-nodulose, 2–7 mm wide; ligule much longer than wide; sheaths white-hyaline ventrally, not conspicuously septate-nodulose dorsally, the lowest often breaking into ladderlike fibers. **Spikes** unisexual, the upper 2–4 staminate, well above the pistil-

Carex vesicaria
LESSER BLADDER SEDGE

late, mostly 2–4 cm long; lower 1–3 spikes pistillate, separate, cylindric, 1.5–8 cm long, 4–15 mm thick, sessile or short-peduncled, erect; lowest bract usually exceeding the inflorescence; pistillate scales acute to awned, shorter than to as long as the perigynia. **Perigynia** ascending and imbricate in 6–8 rows in the spike, yellowish-green to brownish, dull, ovoid to globose-ovoid, inflated, 3.5–8 mm long, ca. 3 mm wide, strongly several- to many-nerved, abruptly contracted or tapered to the slender beak 1–2 mm long, the stiff teeth 0.5–1 mm long; **achenes** trigonous, ca. 2.5 mm long; stigmas 3, the style strongly S-curved toward the base. June–Aug.

HABITAT Wet meadows, marshes and shores; circumboreal; in our region, known only from Holt County, Neb, becoming more common e and w of the Great Plains.

WETLAND STATUS
GP OBL | MIDW OBL | WMTN OBL

Carex viridula Michx.
LITTLE GREEN SEDGE

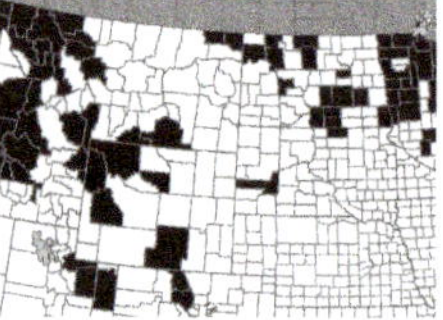

DESCRIPTION Tufted; **culms** stiff, obscurely trigonous, 0.5–4 dm long, exceeding the leaves. **Leaves** 1–3 mm wide; sheaths thin, white-hyaline ventrally. **Spikes** unisexual or mostly so, the terminal one staminate or with a few perigynia at the tip or near the middle, 3–15 mm long, short-peduncled or sessile, surpassing the pistillate spikes or concealed among them; lateral spikes pistillate, 2–6, some often compound, ovoid to short-cylindric, 5–10 mm long, aggregate and sessile or nearly so above, the lower ones often remote and short-peduncled; bracts leaflike, short-sheathing or sheathless, usually erect, much exceeding the heads; pistillate scales brownish on the sides, obtuse to cuspidate, about equaling the perigynia. **Perigynia** yellowish-green to brown, obtusely trigonous, obovoid, 2–3.6 mm long, 2-ribbed, with a few other conspicuous nerves, tapered or contracted to the slightly bidentate beak 0.5–1 mm long; **achenes** trigonous with concave sides, 1–1.2 mm long; stigmas 3. June–July.

HABITAT Calcareous wet meadows, fens, springs, seepage areas and boggy places.

WETLAND STATUS
GP OBL | MIDW OBL | WMTN OBL

Carex viridula
LITTLE GREEN SEDGE

Carex vulpinoidea Michx.
FOX SEDGE

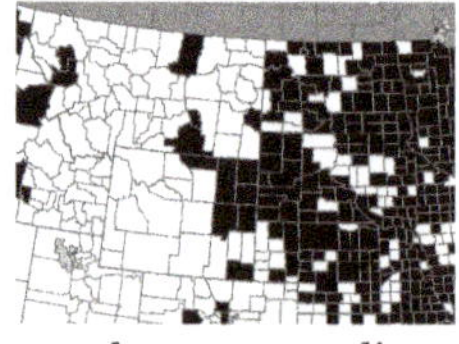

DESCRIPTION Densely tufted from short rootstocks; **culms** stiff, sharply trigonous, not winged nor flattened under pressure, 3–9 dm long, shorter than to exceeding the leaves. **Leaves** 2–4 mm wide; sheaths tight, cross-rugulose and hyaline ventrally, green or green-and-white mottled dorsally. **Spikes** bisexual, androgynous, more than 10,

Carex vulpinoidea
FOX SEDGE

closely aggregate or separate in the lower part, in oblong to cylindric heads 3–9 cm long, with 2-several spikes per branch at the lower nodes; bracts setaceous, slightly to much exceeding the spikes, often 5 cm or more long; pistillate scales aristate, the awns about equaling to much exceeding the perigynia. **Perigynia** yellowish-green, becoming stramineous or brown at maturity, plano-convex, ovate to suborbicular, 2–3 mm long, ca. 1/2 as wide, finely few- to several-nerved dorsally, nerveless ventrally, abruptly contracted to the smooth to serrulate, bidentate beak which is ca. 1/3 the length of the entire perigynium; **achenes** lenticular, 1–1.5 mm long; stigmas 2, the style base swollen above the achene. June–Aug.

HABITAT Wet meadows, marshes, shores, streambanks, ditches, springs and other wet places; common.

WETLAND STATUS
GP FACW | MIDW FACW | WMTN OBL

Cyperus GALINGALE, UMBRELLA SEDGE

Very small to medium-sized, often tufted, grasslike plants, annual or perennial; **culms** simple, sharply trigonous. **Leaves** (excluding the involucres) mostly basal, the blades flat or conduplicate, grasslike, erect or spreading. **Inflorescence** terminal, umbellate, subtended by few to several foliaceous involucral bracts; spikelets numerous, arranged in 1–several hemispheric, subglobose or cylindric spikes, at least one of which (the terminal one) is sessile, the others borne on elongate rays originating from the orifice in the axil of the involucre; scales and their subtended flowers arranged in a distichous manner in the spikelets. **Flowers** perfect; perianth lacking; stamens 1–3; styles trifid or bifid, the **achenes** correspondingly trigonous or lenticular.

---

1 Spikelets overlapping in dense hemispheric to subglobose spikes 2
1 Spikelets loosely arranged in subglobose or cylindric spikes, mostly not overlapping 3

2 Scales with an outwardly curved awn 0.5–1 mm long, conspicuously (5)7- to 9-nerved *C. squarrosus*
2 Scales acuminate, curving outward toward the tip, strongly 3-nerved *C. acuminatus*

3 Achenes lenticular; styles bifid; scales strongly pigmented with purplish-brown; spikes loosely subglobose 4
3 Achenes trigonous; styles trifid; scales pale to stramineous or brown; spikes cylindric, with the spikelets pinnately disposed on a more or less elongate rachis 5

4 Styles exserted ca. 2 mm, persistent, cleft nearly to the base; scales more pigmented near the tip *C. diandrus*
4 Styles exserted less than 1 mm, deciduous, undivided in the lower 1/3; scales more pigmented near the base *C. bipartitus*

5 Scales not overlapping in the spikelets, the tip of each one not reaching the base of the scale directly above it (on the same side of the rachilla) *C. odoratus*
5 Scales overlapping in the spikelets, the tip of each one overlapping with the base of the scale directly above it 6

6 Scales 1–1.5 mm long, 3- to 5-nerved in the green midstripe, the sides nerveless *C. erythrorhizos*
6 Scales mostly 2–4.5 mm long, with 7–13 well distributed nerves, these sometimes faint 7

7 Scales 3–4.5 mm long; achenes less than 1/2 the length of their subtending scales; plants eventually with a hard, cormose base *C. strigosus*
7 Scales 2–2.5 mm long; achenes over 1/2 the length of the scales; plants not cormose at the base 8

8 Plants perennial, producing tubers at the ends of underground scaly stolons (these usually evident even if no tubers are collected); rachilla remaining intact as the scales and achenes drop off, eventually the entire rachilla falling off as a whole *C. esculentus*

8 Plants annual, producing fibrous roots only; rachilla disarticulating between the scales at maturity and falling in segments with the scales and achenes **C. odoratus**

Cyperus acuminatus Torr. & Hook.

TAPER-TIP FLAT SEDGE

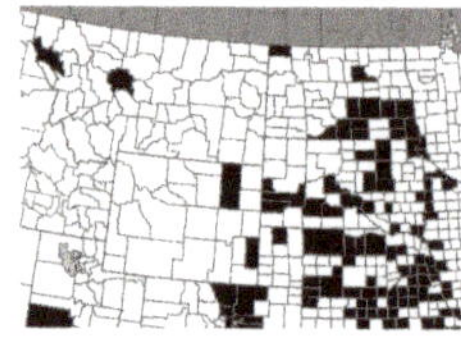

DESCRIPTION Low tufted annual 5–20 cm tall. **Leaves** to as long as the culms or longer, 0.5–2 mm wide; involucral bracts similar, to 3 mm wide, surpassing the inflorescence. **Spikelets** crowded in 1–5 hemispheric to globose spikes, the spikes to 1 cm long, often broader than long, one spike sessile and additional ones borne on rays mostly 0.5–2(3) cm long; spikelets 3–7 mm long; scales 1–2 mm long, acuminate, curving outward toward the tip, pale green, becoming tan at maturity, strongly 3- nerved, the lateral nerves resembling keels; rachilla persistent, the scales falling separately; stamen solitary; styles trifid. **Achenes** tan to pale brown, trigonous, 0.5–0.9 mm long, ca. 1/2 as wide. Late July–Sept.

HABITAT Muddy or sandy streambanks, shores and flats.

WETLAND STATUS
GP OBL | MIDW OBL | WMTN OBL

Cyperus bipartitus Torr.

SHINING CYPERUS, BROOK CYPERUS

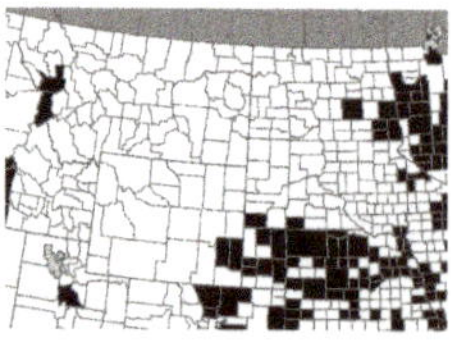

DESCRIPTION Very similar to *Cyperus diandrus,* differing mainly as follows: **Leaves** and involucral bracts 0.5–2 mm wide. Scales of the spikelets closely overlapping, with the achenes usually completely hidden between them, 1.8–2.2 mm long, strongly pigmented (or some lacking pigment or only partially pigmented) with purplish-brown on the sides of the scales, usually darker toward the base, the green keel containing the nerves; styles bifid, exserted less than 1 mm, deciduous, undivided in the lower 1/3. Late July–Sept.

SYNONYM *Cyperus rivularis* Kunth

HABITAT Sandy or muddy shores, streambanks and other wet places.

WETLAND STATUS
GP FACW | MIDW OBL | WMTN OBL

Cyperus diandrus Torr.

LOW CYPERUS

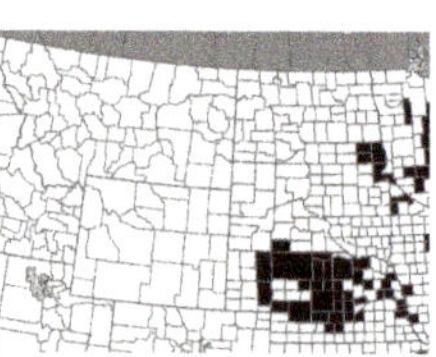

DESCRIPTION Tufted or solitary annual, the **culms** spreading to erect, 5–25 cm tall. **Leaves** shorter than to surpassing the culms, 1–3 mm wide; involucral bracts similar to the leaf blades, overtopping the inflorescence. **Spikelets** loosely arranged in a subdigitate to pinnate manner to form 1–6 rather open, subglobose spikes; rays to 6 cm long; spikelets 4–20 mm long; scales loosely overlapping, the achenes usually visible between them, 2–2.5 mm long, strongly pigmented with purplish-brown along the margins, especially toward the tip, the median portion of the sides of the scale scarious and depressed, the green keel containing the nerves; rachilla persistent, the scales falling separately; stamens 2; styles bifid, exserted ca. 2 mm, persistent, cleft nearly to the base. **Achenes** pale brown, lenticular, ellipsoid to narrowly ovoid, 1–1.2 mm long. Late July–Sept.

Cyperus acuminatus
TAPER-TIP FLAT SEDGE

Cyperus bipartitus
SHINING CYPERUS, BROOK CYPERUS

HABITAT Sandy or muddy shores and streambanks.

WETLAND STATUS
GP FACW | MIDW FACW

Cyperus erythrorhizos Muhl.
RED-ROOTED CYPERUS

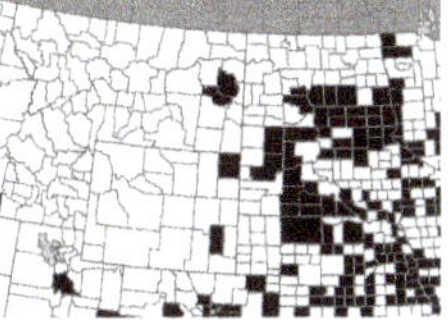

DESCRIPTION Dwarf and tufted to tall and solitary annual 5–60 cm tall, the **culms** stout, erect to spreading; roots blood red when fresh, reddish when dried. **Leaves** shorter than to surpassing the culms, the blades 2–6 mm wide; involucral bracts better developed than the leaves, to 9 mm wide, usually much exceeding the inflorescence. **Spikelets** pinnately disposed on an elongate rachis to form several to many, rather compact, cylindrical spikes, one or a few spikes nearly sessile, with others on rays to 10 cm long; spikelets 2–8 mm long; scales overlapping in the spikelets, 1–1.5 mm long, 3- to 5-nerved in the green midstripe of the keel, the sides nerveless, pale to lustrous copper or auburn; rachilla persistent, bearing deciduous chaffy wings, the scales falling separately; stamens 3; styles trifid. **Achenes** ivory white to pale tan, unequally trigonous, ovoid, 0.6–0.8 mm long. Late July–Sept.

HABITAT Sandy or muddy shores and streambanks.

WETLAND STATUS
GP OBL | MIDW OBL | WMTN OBL

Cyperus diandrus
LOW CYPERUS

Cyperus esculentus L.
YELLOW NUT-SEDGE, CHUFA

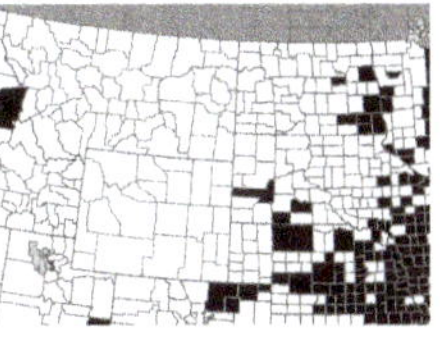

DESCRIPTION Robust tuberiferous perennial 20–60 cm tall; **culms** arising singly, stout, erect; hard nutlike tubers produced at the ends of underground scaly stolons (these evident even if no tubers are collected). **Leaves** shorter than to surpassing the culms, the blades 3–12 mm wide, the involucral bracts similar to the leaves, often greatly exceeding the inflorescence. **Spikelets** pinnately arranged on an elongate rachis to form several to many rather open, cylindrical spikes, a few of the spikes sessile and others extended on rays mostly 2–8(15) cm long; spikelets 4–12 mm long; scales overlapping in the spikelets, (1.8)2–2.5 mm long, stramineous to pale brown, with 7–13 well-distributed, often faint nerves; rachilla persistent, the scales and achenes dropping off separately, eventually the entire rachilla falling as a whole; stamens 3; styles trifid. **Achenes** tan

Cyperus erythrorhizos
RED-ROOTED CYPERUS

Cyperus esculentus
YELLOW NUT-SEDGE, CHUFA

to pale brown, trigonous, ellipsoid to obovoid, 1.2–1.5 mm long. Late July–Sept.

HABITAT Shores, streambanks and other wet places; common as a lawn weed in se USA.

WETLAND STATUS
GP FACW | MIDW FACW | WMTN FAC

Cyperus odoratus L.
COARSE CYPERUS

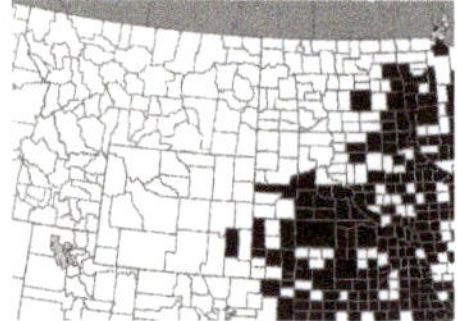

DESCRIPTION Tufted or solitary annual 4–40 cm tall, fibrous-rooted, producing no stolons or tubers, the **culms** stout, erect to spreading. **Leaves** shorter than to surpassing the culms, the blades 2–6 mm wide, the involucral bracts similar, much exceeding the inflorescence. **Spikelets** pinnately disposed on an elongate rachis, forming several to many, rather dense, cylindrical spikes, all of these nearly sessile to form a congested inflorescence, or some of the spikes on rays to 6 cm long; spikelets 4–18 mm long; scales overlapping in the spikelets, 2–2.5 mm long, reddish-brown, with 7–13 well distributed, often faint nerves; rachilla winged, with the margins clasping the achenes, disarticulating between the scales and falling in 1-fruited segments at maturity; stamens 3; styles trifid. **Achenes** brown, trigonous, obovoid-oblong, 1–1.5 mm long. Late July–Sept.

SYNONYMS *Cyperus engelmannii* Steud., *Cyperus ferruginescens* Boeckl., *Cyperus speciosus* Vahl

HABITAT Sandy or muddy shores, streambanks.

WETLAND STATUS
GP FACW | MIDW FACW | WMTN FACW

Cyperus squarrosus L.
AWNED CYPERUS

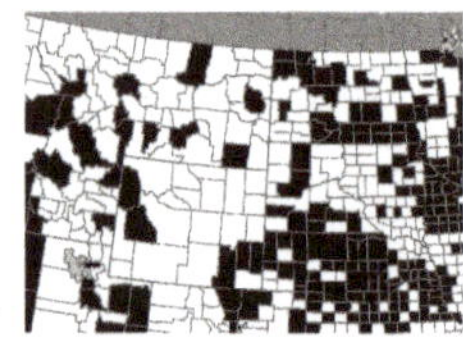

DESCRIPTION Low tufted annual with aromatic foliage, often stunted; **culms** slender, (2)3–15 cm tall. **Leaves** to as long as the culms or longer, 0.5–2 mm wide; involucral bracts similar to the leaves but often somewhat broader, the principal 2 or 3 surpassing the inflorescence. **Spikelets** overlapping in 1–3(4) dense hemispheric to globose spikes, the spikes generally less than 1 cm long, one spike sessile and others, if present, borne on rays to 3 cm long; spikelets 3–7 mm long; scales 1.5–2 mm long including the outwardly curved awn 0.5–1 mm long, pale, turning brown at maturity, conspicuously (5)7- to 9-nerved; rachilla persistent, the scales falling from it at maturity; stamen solitary; styles trifid. **Achenes** pale brown, trigonous, 0.6–1 mm long, 1/3 to 1/2 as wide. Late July–Sept.

SYNONYMS *Cyperus aristatus* Rottb., *Cyperus inflexus* Muhl.

HABITAT Same habitats as *C. odoratus*, but more widespread.

WETLAND STATUS
GP FACW | MIDW FACW | WMTN FACW

Cyperus squarrosus
AWNED CYPERUS

Cyperus strigosus L.

STRAW-COLORED CYPERUS

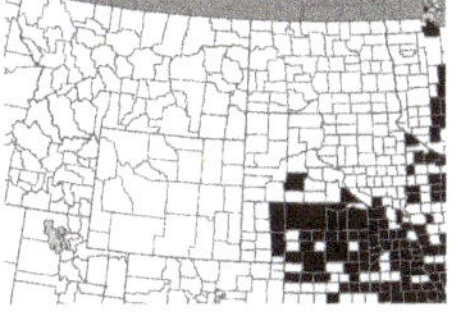

DESCRIPTION Dwarf and tufted to tall and solitary perennial 1–8 dm tall, sometimes flowering the first year, eventually forming a hard, cormlike base; **culms** slender, sharply trigonous. **Leaves** shorter than to surpassing the culms, the blades 2–10(15) mm wide, the involucral bracts leaflike, the largest exceeding the inflorescence. **Spikelets** pinnately arranged and stiffly spreading in several to many cylindrical spikes, often eventually reflexed, the spikes sometimes crowded in dwarf individuals, otherwise mostly on rays 1–12(20) cm long, the longer rays simple (with 1 spike) or often branched above, with few-several spikes in an umbel; spikelets 6–18 mm long, golden-brown, sometimes streaked with red, strongly 7-nerved; rachilla persistent, the scales and achenes falling separately, eventually the rachilla falling as a whole; stamens 3; styles trifid. **Achenes** tan to reddish-brown, trigonous, oblong, 1.5–2.3 mm long. July–Sept.

HABITAT Shores and streambanks, often where sandy.

WETLAND STATUS

GP FACW | MIDW FACW | WMTN FACW

Dulichium arundinaceum (L.) Britt.

THREE-WAY SEDGE

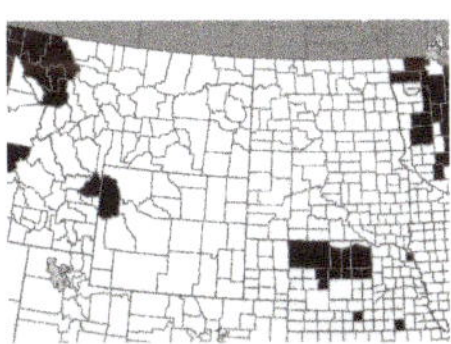

DESCRIPTION

Extensively rhizomatous perennial; **culms** arising singly, simple or seldom with slender branches from lower nodes, 3–10 dm long, hollow, terete or obtusely trigonous. **Leaves** rather evenly distributed along the culm, those in the lower 1/3 to 1/2 of the culm reduced to sheaths only; blades grasslike, rather short, flat, mostly 4–15 cm long, 2.5–8 mm wide; sheaths short, usually overlapping only in the upper portion, convex with a narrow brown band at the mouth. **Inflorescences** axillary, continuously produced from the lower or middle nodes upward, each a short spike of few–10 spikelets, these pinnately arranged, ascending to spreading, 10–25 mm long, mostly 4- to 10-flowered; rachilla winged, disarticulating into 1-fruited segments at maturity; scales (and their enclosed flowers) distichous in the spikelets, green to tan with brownish scarious margins, 5–8 mm long, several-nerved, acute to acuminate. **Flowers** perfect; perianth of 6–9 retrorsely barbellate bristles; stamens 3; styles bifid. **Achenes** tan, oblong, 2.5–3.5 mm long, with a stout, stipitate base, slenderly beaked by the persistent style. Aug–Oct.

HABITAT Marshes, wet meadows, shores and streambanks.

WETLAND STATUS

GP OBL | MIDW OBL

Cyperus strigosus
STRAW-COLORED CYPERUS

Dulichium arundinaceum
THREE-WAY SEDGE

Eleocharis SPIKE-RUSH

Short to intermediate rushlike plants, mostly perennial from rhizomes (tufted and annual in *E. obtusa*), often forming extensive, dense, matlike colonies; **culms** slender and scapose, terete or compressed, sometimes angled and grooved. **Leaves** bladeless, reduced to sheathing at the base of the culm, the sheaths truncate to oblique at the summit. **Inflorescence** consisting of a single spikelet terminating the culm, the spikelet often subtended by 1–2(3) sterile scales (reduced bracts); scales of the spikelets spirally arranged and imbricate. **Flowers** perfect; perianth of usually 6 bristles, but often fewer or more, or the perianth lacking; stamens 3; styles bifid or trifid, the style base swollen and persistent as a tubercle atop the achene (or sometimes confluent with and not always clearly distinct from the achene body). **Achenes** biconvex or trigonous, crowned or beaked by the persistent tubercle.

NOTE The predominant spikerushes of our region's marshes, meadows, shores, and ditches are plants of the ***Eleocharis palustris*** complex. Some authors recognize *E. macrostachya, E. erythropoda, E. smallii* and *E. xyridiformis* as distinct from the principally Eurasian *E. palustris,* but these entities are treated here as part of a broad *E. palustris* taxon.

---

1 Achenes biconvex; styles bifid (often trifid or both trifid and bifid in *E. obtusa*) 2

1 Achenes trigonous to nearly terete; styles trifid 3

2 Tubercles flat, triangular, sharp-edged, tightly fitting to the top of the achene; plants annual, producing fibrous roots only ***E. obtusa***

2 Tubercles thick, deltoid to conic, not sharp on the angles, constricted at the base where attached to the achene; plants perennial, rhizomatous ***E. palustris***

3 Tubercle not obviously differentiated from the achene, nor forming a distinct apical cap 4

3 Tubercle obviously differentiated from the achene, forming a distinct apical cap 6

4 Culms wiry, flattened above, mostly 4–10 dm long, 1–1.4 mm wide, some occasionally arching and rooting at the tip, tufted from a stout erect rootstock, lacking creeping rhizomes; spikelets 6–20 mm long, with 10–20 or more flowers ***E. rostellata***

4 Culms soft, terete, to 3 dm long, less than 1 mm thick, never rooting at the tip, arising in tufts from slender or filiform, creeping rhizomes; spikelets 2–8 mm long, 2- to 9-flowered 5

5 Lowest scale of the spikelet empty; scales mostly 1.5–2(2.5) mm long; achenes 0.9–1.3 mm long ***E. parvula***

5 Lowest scale of the spikelet subtending a flower; scales mostly 2.5–5.5 mm long; achenes 0.9–2.8 mm long ***E. quinqueflora***

6 Achenes 2/3 to fully as wide as long, trigonous, golden to brown, distinctly roughened or pitted ***E. compressa***

6 Achenes twice as long as wide, nearly terete, gray, with several longitudinal ridges and many fine crosslines 7

7 Scales ca. 3 mm long; culms flattened, mostly 0.7–1.5 mm wide ***E. wolfii***

7 Scales ca. 2 mm long; culms scarcely flattened, 0.1–0.5 mm wide ***E. acicularis***

Eleocharis acicularis (L.) Roemer & J.A. Schultes

NEEDLE SPIKE-RUSH

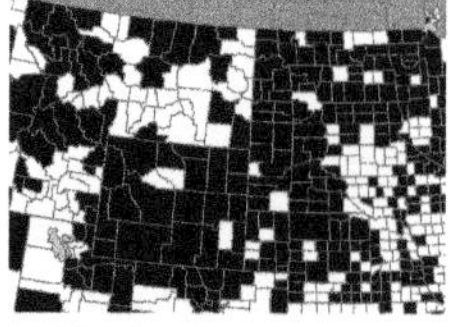

DESCRIPTION Low, tufted, mat-forming perennial from filiform rhizomes; **culms** filiform, 3–10(20) cm long, scarcely flattened, angled by longitudinal grooves, 0.1–0.5 mm thick, rigid when emersed, flexuous and not flowering when submersed; sheaths mostly reddish at the base, membranous, oblique at the summit, often splitting with age. **Spikelets** ovoid, 3–5(8) mm long, 1–1.5 mm thick; scales rather few, all floriferous, pale green, mostly marked with reddish or purplish-brown, ovatelanceolate, with broad scarious margins, ca. 2 mm long; perianth of a few bristles or none; styles trifid. **Achenes** gray, with several longitudinal ridges and numerous fine crosslines, nearly terete, 0.7–0.9 mm long, ca. 1/2 as wide; tubercle turbinate to conic, 1/4 as long as the achene, constricted where attached to the achene. June–early Sept.

HABITAT

Exposed mud or sand, or in shallow water of shores, streambanks, marshes and springs; circumboreal.

WETLAND STATUS

GP OBL | MIDW OBL | WMTN OBL

Eleocharis compressa Sulliv.

FLATSTEM SPIKE-RUSH

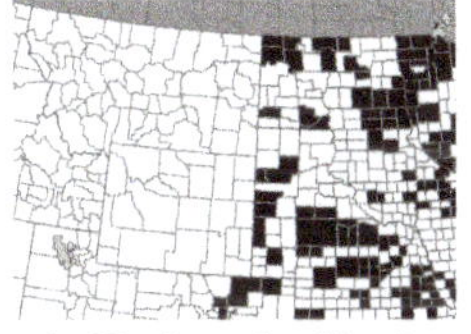

DESCRIPTION Sparsely tufted perennial from stout black rhizomes; **culms** flattened (especially obvious toward the base), 1.5–4 dm tall, angled by longitudinal grooves, 0.5–1 mm wide; sheaths reddish or purplish at the base, firm, slightly oblique at the orifice. **Spikelets** ovoid, blunt, 4–9 mm long, 3–4 mm thick; lowermost scale sterile, encircling the culm; floriferous scales many, reddish-brown on the sides, with a pale midvein, ovate-lanceolate, 2–3 mm long, white-hyaline on the margins, bifid or lacerate at the tip with age; perianth usually lacking or of 1–4 bristles, these reaching to more than 1/2 the length of the achene; styles trifid. **Achenes** golden-yellow to brown, warty or wrinkled on the surface, unequally trigonous, 0.75–1.1 mm long, 2/3 to fully as wide, often persistent after the scales have fallen; tubercle deltoid to depressed-deltoid, to 1/4 the length of the achene, constricted at the attachment to the achene. June–Aug.

HABITAT Low prairie, wet meadows and seepage areas; frequent, but easily overlooked or dismissed for more common species.

WETLAND STATUS

GP FACW | MIDW FACW | WMTN FACW

ADDITIONAL SPECIES **Slender spike-rush** (*Eleocharis tenuis* (Willd.) J.A. Schultes) is reported from c Neb and sc SD. *E. tenuis* has olivaceous achenes with a cellular-reticulate pattern of depressions on the surface. This is in contrast to the golden to brown, warty or wrinkled achenes of *E. compressa*.

Eleocharis obtusa (Willd.) J.A. Schultes

BLUNT SPIKE-RUSH

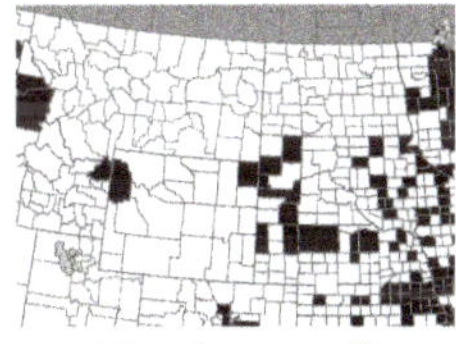

DESCRIPTION Tufted, fibrous-rooted annual 0.5–3 dm tall; **culms** terete, 0.5–1.2 mm thick; sheaths green, firm, oblique at the orifice. **Spikelets** ovoid to lance-cylindric, 4–18 mm long, 2–4 mm thick; sterile basal scales 1 or 2; floriferous scales many, brown on the sides, with a green or pale mid-

Eleocharis acicularis
NEEDLE SPIKE-RUSH

Eleocharis compressa
FLATSTEM SPIKE-RUSH

vein, scarious on the margins, rounded at the tip, 1.7–2.5 mm long; perianth variable, of up to 6(7) brown bristles, shorter than to exceeding the achene, or the perianth lacking; styles bifid or trifid or both in the same spikelet. **Achenes** light green to brown, shiny, biconvex, 0.8–1.3 mm long excluding the tubercle, 0.7–1 mm wide; tubercle flat, triangular, sharp-edged, 0.1–0.3 mm high, tightly fitting to the broad, flat summit of the achene. June–Sept.

SYNONYM

Eleocharis ovata (Roth) Roemer & J.A. Schultes

HABITAT Exposed mud in marshes, ditches, streams and temporary ponds, often where disturbed by cultivation.

WETLAND STATUS

GP OBL | MIDW OBL | WMTN OBL

Eleocharis palustris (L.) Roemer & J.A. Schultes

COMMON SPIKE-RUSH

DESCRIPTION Rhizomatous perennial 1–10 dm tall; **culms** terete, rigid, 0.5–3 mm thick; sheaths reddish or purplish toward the base, usually firm, truncate to oblique at the orifice. **Spikelets** ovoid to lance-cylindric, acuminate to blunt, 5–15(25) mm long, 2–4 mm thick; sterile basal scale 1–2(3), encircling the culm; floriferous scales many, brown, reddish-brown or pale on the sides, green or pale on the midvein, lanceolate to ovate, 2.5–4.5 mm long; perianth variable, usually of 2–6 white to pale brown bristles, shorter than to exceeding the achene; styles bifid. **Achenes** yellow to brown, biconvex, 1.2–1.7 mm long excluding the tubercle, 0.9–1.3 mm wide; tubercle thick, deltoid to conic, 0.25–0.7 mm high, 0.2–0.75 mm wide, constricted at the base. June–Aug.

Eleocharis obtusa
BLUNT SPIKE-RUSH

SYNONYMS *Eleocharis calva* Torr., *Eleocharis* erythropoda Steud., *Eleocharis* macrostachya Britt., *Eleocharis* smallii Britt., *Eleocharis* xyridiformis Fern. & Brack.

HABITAT Marshes (often dominant in shallow marsh zones), wet meadows, ditches, shores and streambanks.

WETLAND STATUS

GP OBL | MIDW OBL | WMTN OBL

Eleocharis parvula (Roemer & J.A. Schultes) Link

LITTLE-HEAD SPIKE-RUSH

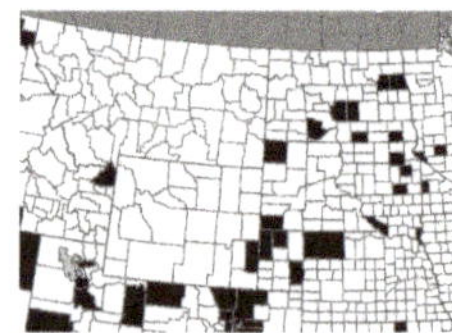

DESCRIPTION Very small, tufted, mat-forming perennial from filiform rhizomes; **culms** filiform, 2–6 cm tall, 0.2–0.3 mm thick; sheaths inconspicuous, membranous. **Spikelets** ovoid, 2–4 mm long, 1–2 mm thick; sterile basal scale 1; floriferous scales 2–9, pale throughout or brown on the sides, ovate, scarious on the margins, 1.5–2(2.5) mm long; perianth none or much reduced; styles trifid. **Achenes** pale brown and cellular-reticulate, trigonous, 0.9–1.3

Eleocharis palustris
COMMON SPIKE-RUSH

mm long including the tubercle, 0.6–0.7 mm wide; tubercle inconspicuous, appearing confluent with the achene, not forming a distinct apical cap. July–early Sept.

HABITAT Wet saline or alkaline flats and shores.

WETLAND STATUS
GP OBL | MIDW OBL | WMTN OBL

NOTE A popular aquarium plant.

Eleocharis quinqueflora (F.X. Hartmann) Schwarz

FEW-FLOWERED SPIKE-RUSH

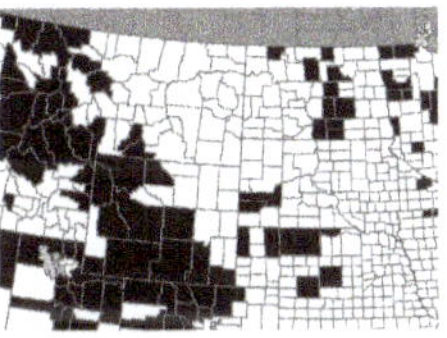

DESCRIPTION Small rhizomatous perennial (0.5)1–3 dm tall; **culms** arising in small tufts or solitary, filiform, longitudinally grooved, less than 1 mm thick; sheaths firm, oblique at the summit. **Spikelets** ovoid, 4–8 mm long, 2–3 mm thick; scales all floriferous, usually 3–9, pale to deep brown, narrowly ovate, scarious on the margins, 2.5–5.5 mm long, the lowest scales longer than the upper ones; perianth shorter than to equaling or exceeding the achene; styles trifid. **Achenes** grayish-brown or brown, trigonous, 0.9–2.8 mm long including the tubercle, ca. 1/2 as wide; tubercle slender, confluent with the achene and appearing like a beak. June–Aug.

SYNONYM *Eleocharis pauciflora* (Lightf.) Link

HABITAT Fens, bogs and springy areas; circumboreal.

WETLAND STATUS
GP OBL | MIDW OBL | WMTN OBL

Eleocharis parvula
LITTLE-HEAD SPIKE-RUSH

Eleocharis rostellata (Torr.) Torr.

BEAKED SPIKE-RUSH

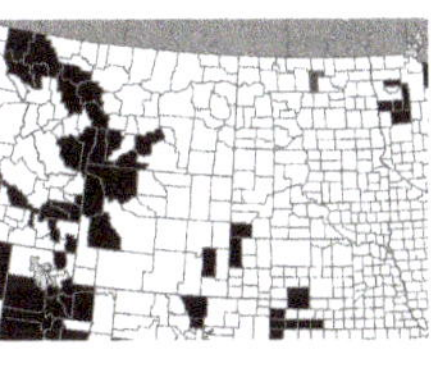

DESCRIPTION Tufted perennial from a stout, erect rootstock, without creeping rhizomes; **culms** wiry, compressed above, mostly 4–10 dm long, 1–1.4 mm wide, the longer ones occasionally arching and rooting at the tip (see photo); sheaths firm, brown, truncate to oblique at the summit. **Spikelets** ovoid to oblong-lanceolate, 6–1 7 mm long, 2.5–4.5 mm thick; sterile basal scale 1; floriferous scales 10–20 or more, pale to dull brown with hyaline margins, 3–5

Eleocharis rostellata
BEAKED SPIKE-RUSH

Eleocharis quinqueflora
FEW-FLOWERED SPIKE-RUSH

mm long; perianth bristles about equaling the achene (including tubercle); styles trifid. **Achenes** olivaceous to brown, trigonous-obovoid, 1.9–3 mm long including the tubercle, ca. 1/2 as wide; tubercle confluent with the achene body and beaklike, light green to brownish, ca. 1/3 the length of the achene. July–Sept.

HABITAT Calcareous wet meadows, seeps and stream margins, often associated with mineral springs.

WETLAND STATUS
GP OBL | MIDW OBL | WMTN OBL

Eleocharis wolfii A. Gray
WOLF'S SPIKE-RUSH

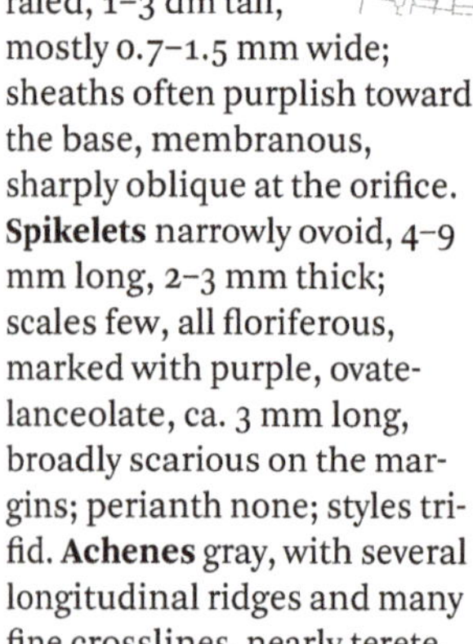

DESCRIPTION Sparsely tufted perennial from slender rhizomes; **culms** flattened, frequently spiraled, 1–3 dm tall, mostly 0.7–1.5 mm wide; sheaths often purplish toward the base, membranous, sharply oblique at the orifice. **Spikelets** narrowly ovoid, 4–9 mm long, 2–3 mm thick; scales few, all floriferous, marked with purple, ovate-lanceolate, ca. 3 mm long, broadly scarious on the margins; perianth none; styles trifid. **Achenes** gray, with several longitudinal ridges and many fine crosslines, nearly terete, 0.7–1 mm long, 1/2 as wide; tubercle conic, 1/4 as long as the achene, constricted at the attachment to the achene. June–July.

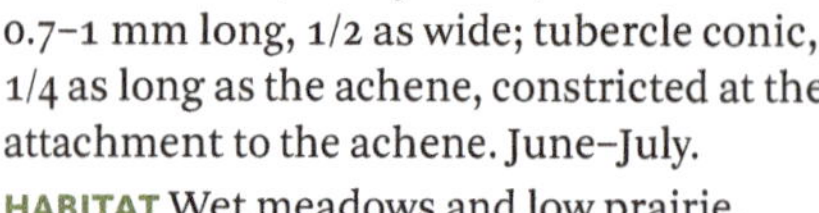

HABITAT Wet meadows and low prairie.

WETLAND STATUS
GP OBL | MIDW OBL

Eriophorum COTTONSEDGE

Solitary or sparsely tufted, grasslike, rhizomatous perennials; **culms** rather stout, terete to weakly trigonous, mostly solid. Leaves mostly basal, sheathing the stems; blades elongate, grasslike, flat or folded or subterete and sometimes channeled on the upper surface. Involucral bracts few, foliaceous. **Inflorescence** terminal, containing few to several spikelets in an umbelliform cyme; **spikelets** having a cotton tuft appearance at maturity due to the elongate, silky perianth bristles; scales many, spirally arranged, scarious. **Flowers** perfect; perianth of numerous, long, persistent, white to rufous bristles, greatly surpassing the scales at maturity; stamens 3; styles trifid. **Achenes** brown, unequally trigonous, sometimes with a short style beak.

1 Involucral bract 1, shorter than the inflorescence ... *E. gracile*
1 Involucral bracts 2 or 3, the longest usually surpassing the inflorescence ... 2

2 Midvein of the scale fading before reaching the very thin tip of the scale ... *E. angustifolium*
2 Midvein of the scale broadening toward the tip of the scale and reaching the tip ... *E. viridicarinatum*

Eriophorum angustifolium Honck.
NARROWLEAF COTTONSEDGE

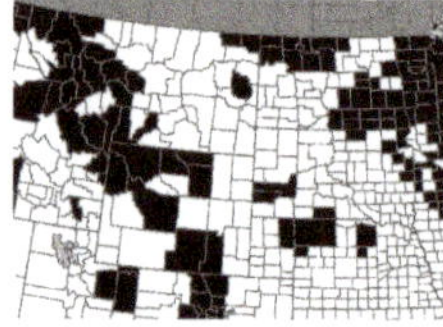

DESCRIPTION **Culms** mostly solitary, weakly trigonous, becoming conspicuously so below the inflorescence, finely ridged, 3–6(9) dm tall, 1.5–3 mm thick. **Leaves** few to several, the uppermost arising in the upper 1/2 of the stem; blades flat or conduplicate, 3–6 mm wide when flat, commonly dying back from the tips; sheaths sometimes reddish, dark-girdled at the tip.

Eriophorum angustifolium
NARROWLEAF COTTONSEDGE

Spikelets 2–8, roughly obconic, 1–3 cm in diameter with the bristles fully expanded; pedicels rather lax, the spikelets drooping; involucral bracts foliaceous, the principal one rather erect and usually surpassing the inflorescence; scales pale throughout or darkened with a pale tip and margins, ovate to lance-subulate, 4–6 mm long, deciduous, the midvein fading before reaching the very thin tip of the scale; perianth bristles white to faintly tawny, sometimes rather short when the achenes fail to develop. **Achenes** dark brown to nearly black, blunt at the tip, 2–3 mm long, ca. 1 mm wide. June–July.

SYNONYM *Eriophorum polystachion* L.

HABITAT Bogs, fens, fresh wet meadows and springs; circumboreal.

WETLAND STATUS

GP OBL | MIDW OBL | WMTN OBL

Eriophorum gracile Koch

SLENDER COTTONSEDGE

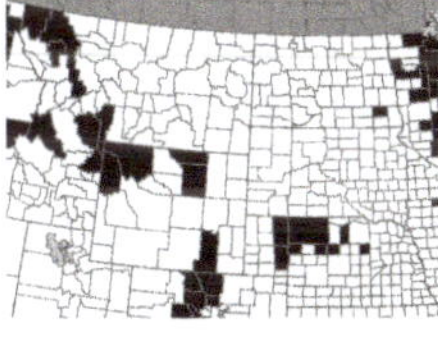

DESCRIPTION **Culms** solitary from rhizomes, slender, weak, subterete, 2–6 dm tall, 1–2 mm thick. **Leaves** few to several, the basal often withered by flowering time, the uppermost arising above or below the middle, with a reduced blade shorter than the sheath; blades channeled on the upper surface, 1–2 mm wide. **Spikelets** 2–5, obovoid to obconic, 1.5–2 cm long with the expanded bristles; pedicels spreading to nodding, to 3 cm long; involucral bracts several but only 1 foliaceous and erect, shorter than the inflorescence; scales pale to blackish-brown, ovate, bluntly acute-tipped; perianth bristles white. **Achenes** pale tan to light brown, narrowly obovate, 2.5–3.5 mm long, 1/3 to 1/4 as wide. June–July.

HABITAT Fens and boggy meadows.

WETLAND STATUS

GP OBL | MIDW OBL

Eriophorum gracile
SLENDER COTTONSEDGE

Eriophorum viridicarinatum (Engelm.) Fern.

DARK-SCALE COTTONSEDGE

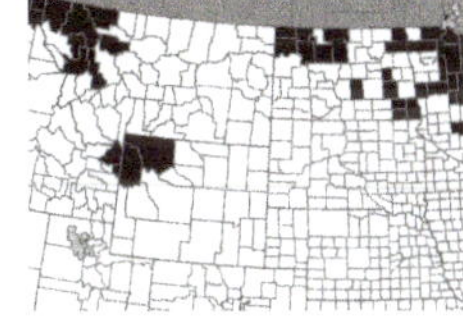

DESCRIPTION Very similar to *Eriophorum angustifolium,* differing mainly as follows: sheaths not dark-girdled at the tip, green throughout; midvein of the scale broadening toward the tip of the scale and reaching the tip.

HABITAT Bogs and swamps.

WETLAND STATUS

GP OBL | MIDW OBL

Eriophorum viridicarinatum
DARK-SCALE COTTONSEDGE

Fimbristylis

Grasslike annuals and perennials; **culms** tufted or solitary, slender. **Leaves** mainly basal, narrowly linear, flat to involute, with or without a ligule of short hairs. **Inflorescence** a terminal umbel or cyme of spike-lets, with 1 or 2 spikelets sessile, subtended by an involucre of 2 or 3 short, leaflike bracts; spikelets several to many, ovoid to oblong-lanceolate, many-flowered; scales spirally arranged and imbricate in the spikelets, all floriferous. **Flowers** perfect; perianth lacking; stamens (1/2–3; styles bifid or trifid, swollen at the base above the attachment to the achene, readily deciduous. **Achenes** lenticular or trigonous, obovoid, smooth to reticulate.

1 Achenes trigonous; styles trifid; plants annual ***F. autumnalis***

1 Achenes lenticular; styles bifid; plants perennial ***F. puberula***

Fimbristylis autumnalis (L.) Roemer & J.A. Schultes (*not illustrated*)

SLENDER FIMBRY

DESCRIPTION Tufted annual with shallow fibrous roots; **culms** 1-many, flattened, 0.5–4 dm tall. Leaves shorter than the culms; blades flat, 1–2 mm wide; ligule a line of short hairs. **Inflorescence** cymose, subtended by 2 or 3 foliaceous bracts, these usually shorter than the inflorescence; pedicels filiform, stiffly spreading to ascending; **spikelets** usually many, solitary or partly in twos or threes on the pedicels, lanceolate to oblong-lanceolate, 3–10 mm long; scales golden-brown or coppery with a prominent green midrib, elliptic-ovate, 1–1.5 mm long, awn-tipped below to mucronate above in the spikelet; stamens 2; styles trifid. **Achenes** ivory to light brown, trigonous, 0.4–0.5 mm long, strongly ribbed on the 3 angles, nearly smooth to cross-reticulate on the faces, sometimes warty toward the base. June–Sept.

HABITAT Shores, streambanks and wet meadows, often where sandy.

WETLAND STATUS

GP OBL | MIDW OBL | WMTN OBL

Fimbristylis puberula (Michx.) Vahl

CHESTNUT FIMBRY

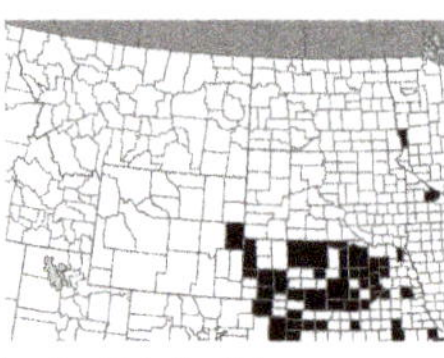

DESCRIPTION Perennial with slender, stiff **culms** 2–7 dm tall, solitary or in small tufts, sometimes bulbous at the base, with short, slender to thick rhizomes. **Leaves** shorter than the culm; blades flat to involute, 0.5–3 mm wide; ligule a line of short hairs or incomplete to absent. **Inflorescence** umbellate or some branches forked, subtended by 2 or 3 foliaceous bracts, the longest of these shorter than to exceeding the inflorescence; pedicels slender, ascending to spreading; **spikelets** few to many, ovoid to ovoid-cylindric, 5–10 mm long; scales pale to dark brown, broadly ovate, mostly 3–3.5 mm long, broadly rounded at the apex, often with a short awn or mucro, scarious and sometimes short-ciliate on the margins, firm, shiny and often finely striate on the back, the lower scales sometimes puberulent; stamens (2)3; styles bifid, fringed with fine hairs below the branches. **Achenes** pale brown, lenticular, 1.2–1.8 mm long, reticulate between fine longitudinal lines on the convex faces, the straplike filaments often adherent to the achene base. June–Aug.

Fimbristylis puberula
CHESTNUT FIMBRY

SYNONYMS *Fimbristylis castanea* (Michx.) Vahl, *Fimbristylis caroliniana* (Lam.) Fern., *Fimbristylis drummondii* Boeckl.

HABITAT Wet meadows, shores and moist to dry prairies, often where sandy.

WETLAND STATUS
GP OBL | MIDW OBL

NOTE We have two varieties: var. *interior* (Britt.) Kral, characterized as follows: culms rarely bulbous at the base, with numerous slender, twisted, orangish rhizomes produced from the rather soft, tufted base; longest involucral bract usually surpassing the inflorescence; spikelet scales glabrous. Var. *puberula*: culms bulbous at the base, arising from a thick rhizome and clothed by the firm, fibrous sheaths of old leaves; longest involucral bract usually much shorter than the inflorescence; spikelet scales short-ciliate.

Fuirena simplex Vahl
UMBRELLA-GRASS

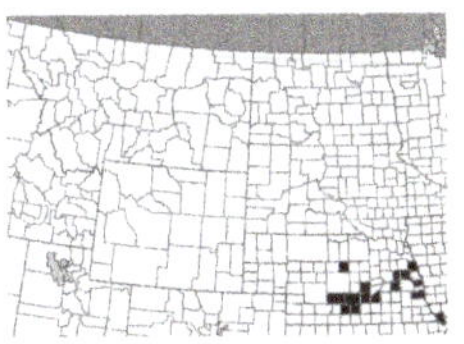

DESCRIPTION Tufted grasslike annual (in this region) with shallow fibrous roots; **culms** slender, weak, mostly 2–6 dm long, sharply trigonous above. **Leaves** grasslike, mostly cauline and extending upward to below the inflorescence; blades soft, flat, 2–5 mm wide, sparingly to moderately pubescent; sheaths glabrous to variously hirsute or hispid, hairy at the collar; ligule membranous with a fringe of short hairs. **Inflorescence** of 1–3 clusters of 2-several spikelets in a terminal head, rarely of a single spikelet (on diminutive culms), each cluster subtended by 2–3 narrow, spreading or reflexed, foliaceous bracts; **spikelets** ovoid to oblong-ovoid, 4–11 mm long, appearing bristly due to recurved awns of scales; scales numerous and imbricate, reddish-brown toward the base, blackish or blackish-brown on the margins, the body obovate, 1.5–2 mm long, short-pubescent, strongly 3-nerved in the middle, the 3 nerves excurrent into a subterminal, recurved awn 1.5–5 mm long. **Flowers** perfect; perianth of 3 stipitate scales alternating with 3 retrorsely barbellate bristles, perianth scales oblong-ovate, 3-nerved, mucronate, persistent to, surpassing and enclosing the achene; stamens 3; styles trifid. **Achenes** light green, sharply trigonous with flat to concave faces, 0.7–1 mm long including the short stipitate base and slender style beak. Aug–Sept.

Fuirena simplex
UMBRELLA-GRASS

HABITAT Stream and pond margins and low areas on floodplains, often where sandy.

WETLAND STATUS
GP OBL | MIDW OBL

Lipocarpha HALFCHAFF SEDGE

Dwarf, tufted, fibrous-rooted annuals 3–15 cm tall; **culms** and leaves filiform; **leaves** 2 per culm, basal, shorter than the culms; involucral bracts 2 or 3, foliaceous, the principal one rather erect, causing the inflorescence to appear lateral. **Inflorescence** of 1–3 sessile, ovoid spikelets, these compact, containing numerous minute, spirally arranged scales; perianth consisting of a single minute inner scale opposite the subtending outer scale, this inner scale often absent; bristles none; stamen solitary; styles bifid. **Achenes** minute, obovate and compressed or cylindrical, black or brown, minutely apiculate.

NOTE *Lipocarpha* now sometimes included within *Cyperus*.

1 Inner scale equaling or exceeding the achene and often cupped around it; mature achenes obovate, compressed, black ***L. drummondii***

1 Inner scale much shorter than the achene or absent; mature achenes cylindrical, brown ***L. micrantha***

Lipocarpha drummondii (Nees) G.C. Tucker
DRUMMOND'S HALFCHAFF SEDGE (*not illus.*)

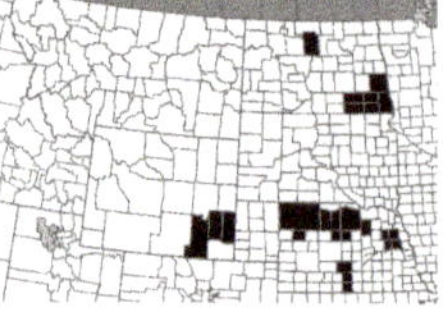

DESCRIPTION Very similar to and often considered a variety of *Lipocarpha micrantha,* differing in having scales with a curved awn ca. 2/3 the length of the scale body; inner perianth scale equaling or exceeding the achene and often cupped around it. Mature achenes black, obovate, compressed. Aug–Sept.

SYNONYMS *Cyperus hemidrummondii* Goetgh., *Hemicarpha drummondii* Nees

HABITAT Sandy shores and streambanks.

WETLAND STATUS
GP FACW | MIDW FACW

Lipocarpha micrantha (Vahl) G.C. Tucker
SMALL-FLOWER HALFCHAFF SEDGE

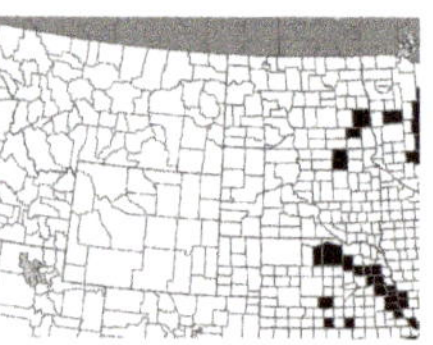

DESCRIPTION **Spikelets** 1–4 mm long; scales with a green midvein, pale or reddish-brown on the sides, obovate to oblanceolate, acute to mucronate, incurved, 0.7–1 mm long; inner scale much shorter than the achene, often bifid, or the inner scale absent. Mature **achenes** brown, cylindrical, 0.5–0.8 mm long. Aug–Sept.

SYNONYMS *Cyperus subsquarrosus* (Muhl.) Bauters, *Hemicarpha micrantha* (Vahl) Britt.

HABITAT Same habitats as *L. drummondii.*

WETLAND STATUS
GP FACW | MIDW OBL

NOTE *Lipocarpha micrantha* is very similar in appearance to *L. drummondii,* which differs by having spikelets with appressed scales (rather than spreading scales), and scales that are more broad in shape (obovate-orbicular). Both of these species differ from bulrushes (*Schoenoplectus*) by having minute internal scales in their spikelets.

Lipocarpha micrantha
SMALL-FLOWER HALFCHAFF SEDGE

Rhynchospora capillacea Torr.
NEEDLE BEAK-RUSH

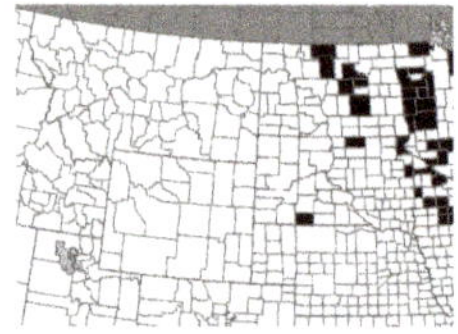

DESCRIPTION Small, tufted, grasslike perennial; **culms** capillary, 0.5–4 dm tall; **leaves** filiform, involute, 0.2–0.4 mm wide. **Inflorescence** of 1 or 2 small, ovoid to oblong clusters of spikelets, sometimes only 1 terminal spikelet present, otherwise the terminal cluster containing 2–10 spikelets; lateral cluster, when present, containing 1–4 spikelets, remote from the terminal cluster, subsessile or short-peduncled, each cluster subtended by 1-few short, setaceous bracts. **Spikelets** ovoid to ellipsoid, 3–7 mm long; scales spirally imbricate, brown, elliptic or elliptic-ovate, the pale midvein prolonged into a cuspidate tip, the lower scales empty, the terminal ones mostly staminate, only 1–2(5) achenes produced from perfect-flowered scales in the middle portion of each spikelet; perianth bristles 6, retrorsely barbellate, exceeding the achene, sometimes some additional shorter bristles present; stamens 3; styles 2. **Achenes** obovate-oblong, biconvex, the body tawny, semiglossy, 1.7–2.1 mm long, truncate above at the juncture with the beak, narrowed below to a stipitate base; beak dull brown, narrowly triangular, 0.7–1.1

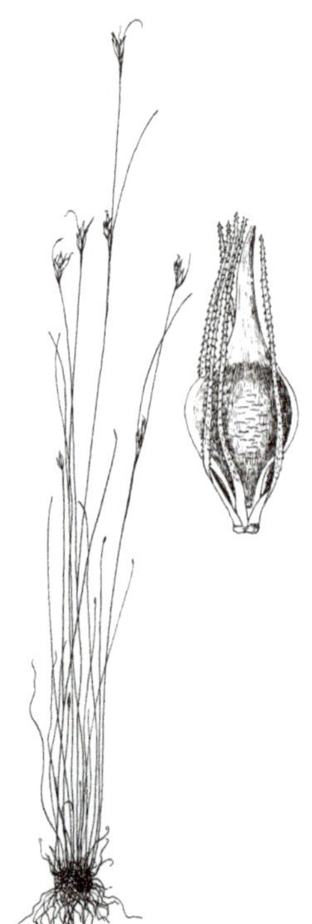

Rhynchospora capillacea
NEEDLE BEAK-RUSH

mm long. July–Aug.

HABITAT Calcareous fens and bogs.

WETLAND STATUS
GP OBL | MIDW OBL

Schoenoplectiella CLUB-RUSH

NOTE See *Schoenoplectus* key below.

Schoenoplectiella saximontana (Fern.) Lye CLUB-RUSH

DESCRIPTION Small, tufted annual 1.5–4 dm tall; **culms** several to many, terete or essentially so, grooved when dried. **Leaves** 1 per culm, mainly basal, sheathing only or with a short blade mostly less than 1(–4) cm long; principal involucral bract prominent, erect, resembling a continuation of the culm, 3–15 cm long; other involucral bracts none or weakly developed, shorter than to surpassing the inflorescence, projecting laterally below the spikelets. **Inflorescence** appearing lateral, comprised of 3–8 spikelets, all essentially sessile or some in clusters of 2–3 at the tips of short pedicels to 1 cm long; **spikelets** ovoid to oblong-cylindric, 4–13 mm long, 2.5–3 mm thick; scales whitish to eventually dull yellowish-brown, with a prominent green midrib, scarious-margined, rotund-ovate, strongly convex dorsally, 2.5–3 mm long, mucronate or the lower scales with an awn to 1 mm long; perianth bristles absent or otherwise variable; stamens 3, the straplike filaments often adherent to the achene; styles trifid. **Achenes** light green to eventually dark brown, strongly trigonous, obovoid, 1.4–1.8 mm long, conspicuously cross-ridged on the surface, short-beaked. Aug–Sept.

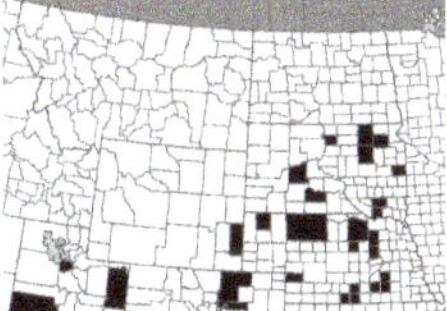

Schoenoplectiella saximontana
CLUB-RUSH

SYNONYMS *Schoenoplectus saximontanus* (Fern.) Raynal, *Scirpus saximontanus* Fern., *Scirpus supinus* L. var. *saximontanus* (Fern.) Koyama

HABITAT Muddy shores and flats, often where previously flooded.

WETLAND STATUS
GP OBL | MIDW OBL | WMTN OBL

NOTE This genus, recently segregated from *Schoenoplectus,* includes our annual, tufted species.

ADDITIONAL SPECIES The similar *Schoenoplectiella hallii* (Gray) Lye is uncommon in nc Neb, but has bifid styles and lenticular (plano-convex rather than strongly trigonous) achenes.

Schoenoplectus BULRUSH

Perennial or annual, tufted or rhizomatous herbs. **Culms** cylindric to strongly 3-angled, smooth, spongy with internal air cavities. **Leaves** basal, rarely 1(–2) on stem; sheaths tubular; ligules membranous; blades well-developed to rudimentary. inflorescences terminal, head-like to openly paniculate; spikelets 1–100 or more; involucral bracts 1–5, leaflike, proximal bract erect to spreading. **Spikelets** terete; scales deciduous, spirally arranged, each subtending a flower, or proximal scale empty, midrib usually prolonged into short awn, margins ciliate. flowers bisexual; perianth of 0–6(–8) spinulose bristles shorter than to somewhat longer than the achene; stamens 3. **Achenes** biconvex to trigonous, with apical beak; rugose or with transverse wavy ridges.

---

1 Principal involucral bract 1, erect or nearly so, resembling a continuation of the culm; inflorescence appearing lateral **2**

1 Principal involucral bracts 2–several, foliaceous and spreading; infloresence terminal **7**

2 Plants annual, tufted; achenes cross-ridged *Schoenoplectiella saximontana*

2 Plants perennial, rhizomatous; achenes smooth or reticulate 3

3 Spikelets 1–6, sessile in a cluster 4
3 Spikelets numerous, pedicelled singly or in clusters, borne in a dense to open panicle 5

4 Stems trigonous; scales aristate, bifid ***S. pungens***
4 Stems terete or nearly so; scales acute to obtuse ***Amphiscirpus nevadensis***

5 Styles trifid; achenes trigonous; perianth of 2–4 bristles; spikelets solitary (rarely paired) on the pedicels ***S. heterochaetus***
5 Styles bifid (rarely trifid); achenes lenticular; perianth of 6 bristles; spikelets mostly in glomerules of (1)2–8(15) on the pedicels 6

6 Spikelets mostly in glomerules of 3–8(15), grayish-brown; culms dark green and resistant to crushing when fresh, fading with drying ***S. acutus***
6 Spikelets mostly single or in pairs, reddish-brown; culms light green and easily crushed when fresh ***S. tabernaemontani***

7 Styles trifid; achenes trigonous; leaf sheaths convex at the mouth, the ventral nerves abruptly divergent at the summit ***S. fluviatilis***
7 Styles bifid; achenes lenticular; leaf sheaths truncate or concave at the mouth, the ventral nerves gradually divergent at the summit ***S. maritimus***

Schoenoplectus acutus (Muhl. ex Bigelow) Á. & D. Löve

HARDSTEM BULRUSH

DESCRIPTION Tall, slender, scapose perennial, densely colonial from extensive, stout rhizomes; **culms** rather stout, terete, 1–2.5(3.5) m long, dark green and resistant to crushing when fresh, fading with drying. **Leaves** consisting of 3–5 basal sheaths, the upper ones with tapering blades to 25 cm long; principal involucral bract erect, appearing as a continuation of the culm, 1.5–10 cm long, eventually turning brown. **Inflorescence** a panicle of up to 60 spikelets, appearing lateral, compact to open, the branches rather stiff; spikelets in glomerules of 2–15 (mostly 3–7) on the pedicels, grayish-brown, 5–15 mm long, 3- 5 mm thick; scales suffused and often spotted with brown or dark red, 2–3.5 mm long, scarious, especially the lower ones often puberulent on the back, acute to slightly cleft at the apex, usually mucronate, the mucro to 0.5 mm long, the margins often finely ciliate; perianth of 6 unequal bristles, usually shorter than the achene body; stamens 3, the flattened filaments often persistent; styles bifid, seldom trifid. **Achenes** light green to dull or dark brown, lenticular, unequally biconvex, the body 1.8–2.2 mm long, 1 .z–1 .9 mm wide, the style beak minute to 0.5 mm long. June–mid Aug.

SYNONYM *Scirpus acutus* Muhl. ex Bigelow

HABITAT Usually emergent in shallow to deep water of marshes, ditches, ponds and lakes, especially where the water is brackish; common, often abundant.

WETLAND STATUS
GP OBL | MIDW OBL | WMTN OBL

Schoenoplectus fluviatilis (Torr.) M.T. Strong

RIVER BULRUSH

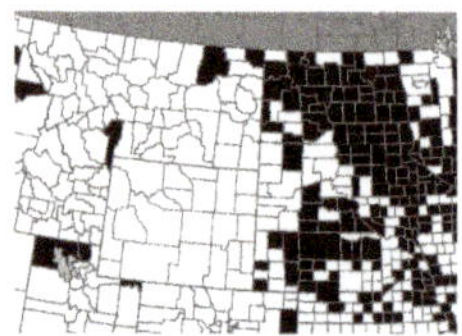

DESCRIPTION Stout perennial 7–15 dm tall, often forming large colonies by the extensive rhizomes; **culms** erect, sharply trigonous, arising from black tuberous thickenings on the rhizomes. **Leaves** sheathing

Schoenoplectus acutus
HARDSTEM BULRUSH

most of the length of the culm; blades ascending, the upper ones commonly surpassing the inflorescence, 6–15 mm wide; sheaths convex at the mouth, the ventral nerves abruptly divergent at the summit; involucral bracts 3–5, foliaceous and spreading, the longest to 30(45) cm long. **Inflorescence** terminal, containing 10–20 spikelets, several of the spikelets nearly sessile in 1 or 2 clusters, others solitary or in glomerules of 2–5 at the ends of ped-icels to 8 cm long; **spikelets** 10–25(35) mm long, 6–12 mm thick; scales golden brown, puberulent on the back, 6–9 mm long, acute to bifid, aristate, the awn 1–3 mm long, curved; perianth of 6 unequal, white to coppery bristles, mostly equaling or exceeding the achene body; stamens 3, strongly exsert at anthesis, with conspicuous elongate yellow anthers 4–6 mm long, the filaments flattened, often persistent to the achene; styles trifid, strongly exsert. **Achenes** tan to grayish-green or grayish-brown, often mottled, trigonous, 3–4.5 mm long, 2–2.8 mm wide, with a beak 0.2–0.5 mm long. June–Aug.

SYNONYMS *Bolboschoenus fluviatilis* (Torr.) Soják, *Scirpus fluviatilis* (Torr.) A. Gray

HABITAT Usually in shallow water of streams, ditches, marshes, lakes and ponds.

WETLAND STATUS

GP OBL | MIDW OBL | WMTN OBL

Schoenoplectus fluviatilis
RIVER BULRUSH

Schoenoplectus heterochaetus (Chase) Soják

PALE BULRUSH

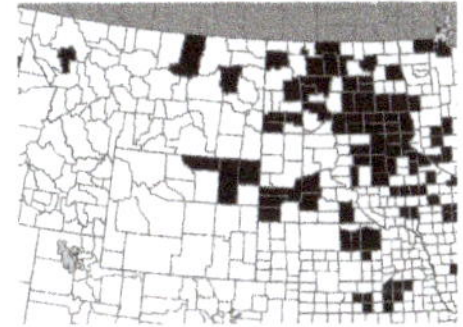

DESCRIPTION Much like *Schoenoplectus acutus* in habit and general appearance; culms more slender, 1–2 (2.3) m long. **Leaves** reduced mainly to 3–4 sheaths at the base of the culm, the upper sheaths with a blade to 6.5 cm long; principal involucral bract erect, 1–15 cm long. **Inflorescence** a rather open, lax panicle of up to 60 spikelets, usually much fewer; **spikelets** solitary (rarely paired) on the pedicels, 5–17 mm long, 3–6 mm thick; scales suffused with light brown to reddish-brown, 2.5–3.5 mm long, scarious, mucronate to aristate, with an awn to 2 mm long; perianth of 2–4 bristles, the 2 bristles on the angles of the achene usually the longest, sometimes exceeding the achene body; stamens 3; styles trifid. **Achenes** light green to dark grayish-brown, unequally trigonous, the ventral face broad and flat, achene body 2–2.5 mm long, 1.5–2 mm wide, with a beak 0.2–0.8 mm long. June–mid Aug.

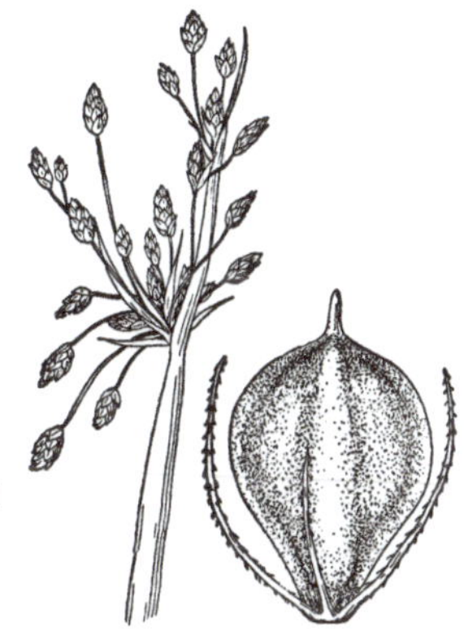

SYNONYM *Scirpus heterochaetus* Chase

HABITAT In same habitats as *S. acutus*, but only dominant where the water is fresh.

WETLAND STATUS

GP OBL | MIDW OBL | WMTN OBL

Schoenoplectus maritimus (L.) Lye

SALTMARSH BULRUSH

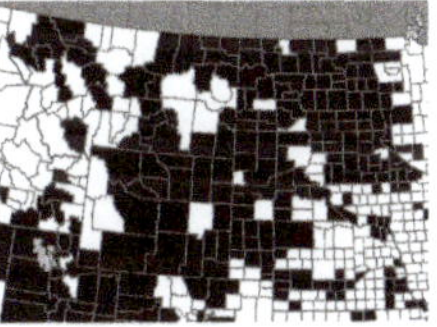

DESCRIPTION Like *Schoenoplectus fluviatilis* in most aspects, differing mainly as follows: Smaller in stature, 3–

9 dm tall. **Leaves** 3–9 mm wide; sheaths truncate or concave at the mouth, the ventral nerves gradually divergent at the summit; involucral bracts 3–5, the longest one sometimes erect, to 30 cm long. **Inflorescence** containing (2)3–20 spikelets, these all sessile in a single cluster, or some solitary or in glomerules of 2–4 spikelets on pedicels to 4 cm long; spikelets 10–25 mm long, 6–9 mm thick; scales 5–7 mm long, usually bifid, aristate, the awn to 2 mm long, mostly straight; perianth of 2–6 coppery bristles, reaching 1/2 the length of the achene body. **Achenes** tan or greenish-brown to brown, lenticular, 2.5–3.7 mm long, 2–2.4(3) mm wide, the beak inconspicuous, to 0.3 mm long. June–Aug.

SYNONYMS *Scirpus maritimus* L., *Scirpus paludosus* A. Nels.

HABITAT In shallow water or mud in the same habitats as *S. fluviatilis*, but especially abundant in brackish or saline situations.

WETLAND STATUS
GP OBL | MIDW OBL | WMTN OBL

Schoenoplectus pungens (Vahl) Palla
THREE-SQUARE

DESCRIPTION Slender rhizomatous perennial 2–10(14) dm tall; **culms** erect to gently curved, trigonous, the sides concave to slightly convex. **Leaves** few, mostly 1–3 per culm, sheathing the base of the culm, the blades typically folded or channeled above, 1–2 mm wide (2–4 mm wide when flat), usually diverging from the culm well within the lower 1/3 of its length and not overtopping it; principal involucral bract erect, resembling a continuation of the culm, 2–8(18) cm long; additional involucral bracts 1–2, appearing as enlarged sterile scales subtending the spikelets. **Inflorescence** appearing lateral, comprised of 1–4(6) sessile spikelets, these 5–20 mm long, 3–5 mm thick; scales light to dark brown or reddish-brown, 3–5 mm long, mostly aristate and bifid, the awn 0.5–2 mm long; perianth of 1–6 bristles, variable in length, shorter than the achene; stamens 3 or sometimes 2, the straplike filaments often remaining attached at the base of the achene; styles trifid or bifid. **Achenes** light green or tan to dark brown, trigonous or lenticular, ventrally flat, the body 2–3 mm long, the beak prominent, 0.3–0.5 mm long. June–Sept.

SYNONYM *Scirpus americanus* Pers., misapplied.

HABITAT Shores, streambanks, wet meadows, ditches, seepage areas and other wet places; common, often abundant.

WETLAND STATUS
GP OBL | MIDW OBL | WMTN OBL

Schoenoplectus maritimus
SALTMARSH BULRUSH

Schoenoplectus pungens
THREE-SQUARE

Schoenoplectus tabernaemontani (K.C. Gmel.) Palla

SOFTSTEM BULRUSH

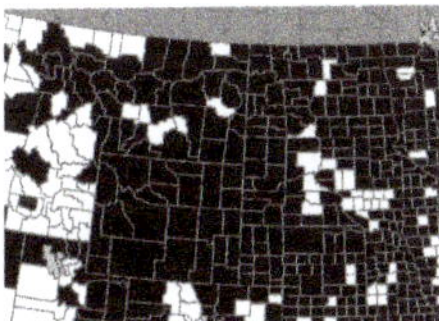

DESCRIPTION Similar to *Schoenoplectus acutus;* **culms** rather stout near the base, terete, 1–3.5 m long, light green, easily crushed when fresh. **Leaves** consisting of 4–5 basal sheaths, the upper ones with a blade to 7 cm long; principal involucral bract erect, 1.5–11 cm long. **Inflorescence** a panicle of up to 235 spikelets, appearing lateral, spreading, rarely congested, lax and drooping; **spikelets** solitary or paired on the pedicels, rarely in glomerules of 3 or more, reddish-brown, 4–13 mm long, 3–4 mm thick; scales suffused with reddish-brown to dark brown, 1.8–3.2 mm long, scarious, often very finely ciliate on the margins, obtuse or rounded, usually mucronate, the mucro to 0.5 mm long; perianth of 6 bristles, mostly equaling or exceeding the achene body; stamens 3; styles bifid. **Achenes** light green to dark brown or black, lenticular, unequally biconvex, the body 1.5–2.1 mm long, 1.3–1.6 mm wide, the beak minute, to 0.2 mm long. June–mid Aug.

SYNONYM *Scirpus validus* Vahl

HABITAT Similar habitats as *S. acutus* but mostly in fresh water and more often in riparian and stream habitats; common, often abundant.

WETLAND STATUS

GP OBL | MIDW OBL | WMTN OBL

Schoenoplectus tabernaemontani
SOFTSTEM BULRUSH

Scirpus BULRUSH

Stout, rushlike perennials, mostly spreading by rhizomes. **Culms** unbranched, 3-angled or round in section, solid or pithy. **Leaves** broad and flat, to narrow and often folded near tip, or reduced to sheaths at base of stems; involucral bracts several and leaflike, or single and appearing like a continuation of the stem. **Spikelets** single, or in panicle-like or umbel-like clusters at ends of stems, or appearing lateral from the stem; the spikelets stalked or stalkless; scales overlapping in a spiral. **Flowers** perfect; sepals and petals reduced to 1–6 smooth or downwardly barbed bristles, or sometimes absent; stamens 2 or 3; styles 2–3-parted. **Achenes** lens-shaped, flat on 1 side and convex on other, or 3-angled, usually tipped with a beak.

1 Perianth bristles much contorted, smooth, not barbellate, equaling or exceeding the scales 2

1 Perianth bristles straight to curved, barbellate, shorter than to equaling the scales 3

2 Perianth bristles conspicuously surpassing the scales, giving the spikelets a woolly appearance ***S. cyperinus***

2 Perianth bristles about equaling the scales, the spikelets not woolly in appearance ***S. pendulus***

3 Leaf sheaths partly red-tinged; styles bifid ***S. microcarpus***

3 Leaf sheaths entirely green; styles trifid ***S. pallidus***

Scirpus cyperinus (L.) Kunth.

WOOL GRASS

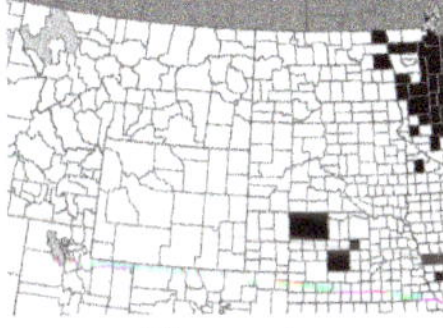

DESCRIPTION Coarse, tufted perennial from short rhizomes; **culms** obtusely trigonous, to 2 m tall. **Leaves** numerous on the culms; blades 3–10 mm wide, scabrous on the margins, involute toward the tips; sheaths brownish, dark brown at the mouth; involucral bracts usually 2–4, foliaceous, spreading, the longest shorter than to surpassing the inflorescence, usually brown or reddish-brown at the base. **Inflorescence** large and spreading, with several to many slender primary branches or rays to 15 cm long, these bearing a secondary set of reduced involucral bracts and branches at their summit, the secondary branches themselves often terminating in a tertiary set of involucral bracts and branches, the ultimate branches spreading to drooping, terminating in glomerules of 2-several spikelets, or some spikelets borne singly, sometimes mostly single; **spikelets** very numerous, ovoid, 3–6 mm long, 2–3 mm wide, appearing woolly due to the long perianth bristles; scales streaked with reddish-brown, sometimes blackish, elliptic-ovate, 1–1.3 mm long, bluntly-acute; perianth bristles 6, smooth, contorted, brownish, much exceeding the scale; styles trifid. **Achenes** whitish to tan, flattened-trigonous, the dorsal angle low, 0.6–0.9 mm long, ca. 1/2 as wide, with a short, slender beak. July–Sept.

HABITAT Wet meadows, marshes and swamps.

WETLAND STATUS
GP OBL | MIDW OBL

NOTE The *Scirpus cyperinus* complex, including this species, *S. atrocinctus* Fern., (and *S. pedicellatus* Fern., found east of our region), is often regarded as a single, highly variable species. Alternately, our two taxa can be separated as follows:

1 Spikelets all or mostly all sessile in clusters of (2–) 3–7 or more **S. cyperinus**

1 Spikelets mostly pediceled, the ultimate branches of the inflorescence typically bearing 1 central, sessile spikelet with 2–3 pediceled ones **S. atrocinctus**

Scirpus atrocinctus flowers and fruits earlier than the other two species, often with inflorescences fully developed by late June, and achenes ripe by late July. *S. atrocinctus* readily hybridizes with *S. cyperinus* to form hybrid swarms.

Scirpus microcarpus J. & K. Presl

REDSTEM BULRUSH

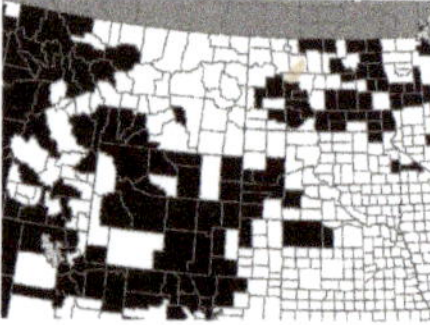

DESCRIPTION Perennial from stout creeping rhizomes, the **culms** arising singly or few together, 5–13 dm tall, weakly trigonous. **Leaves** sheathing most of the culm; blades ascending, the upper ones usually surpassing the inflorescence, 6–11 mm wide; sheaths partly red-tinged; involucral bracts 3–4, foliaceous, the longest to 25 cm long. **Inflorescence** a terminal umbelliform cyme, rather loose and spreading, containing numerous glomerules of 4-many spikelets, the glomerules clustered at the ends of pedicels to 10(15) cm long; **spike-lets** very numerous, 3–5 mm long, 1–1.5 mm thick; scales hyaline except for the green midvein, blackish on the sides, 1.2–1.6 mm long, acute to obtuse, mucronulate; perianth of 4–6, white to tan bristles, exceeding the

Scirpus cyperinus
WOOL GRASS

Scirpus microcarpus
REDSTEM BULRUSH

achene body; stamens 2, the exserted anthers ca. 1 mm long, the filaments often adherent to the achene; styles bifid. **Achenes** light green to pale tan, lenticular, 0.8–1.3 mm long, 0.5–0.7 mm wide, the beak minute. June–July.

SYNONYM *Scirpus rubrotinctus* Fern

HABITAT Streambanks, wet meadows, marshes and springs, where water is fresh.

WETLAND STATUS
GP OBL | MIDW OBL | WMTN OBL

Scirpus pallidus (Britt.) Fern.

PALE BULRUSH

DESCRIPTION Loosely tufted perennial with short rhizomes, 5–17 dm tall, the **culms** trigonous. **Leaves** sheathing most of the culm; blades ascending, usually not surpassing the inflorescence, 5–17 mm wide; sheaths entirely green; involucral bracts 3–4, foliaceous, slightly to much exceeding the inflorescence, the longest to 15 cm long. **Inflorescence** a terminal, rather compact, umbelliform cyme containing many dense glomerules of numerous spikelets, the glomerules densely clustered at the ends of pedicels to 12 cm long; **spikelets** very numerous and crowded, 2–5 mm long, 1–2.5 mm thick; scales hyaline except for the broad, green midvein, blackish on the sides, mostly 2–2.5 mm long including an awn 0.4 mm or more long; perianth mostly of 6 white to tan bristles shorter than or equaling the achene; stamens 3; styles trifid. **Achenes** light green to pale tan, unequally trigonous, the ventral side broadest, the achene body 0.7–1 mm long, 0.5–0.7 mm wide, with a beak to 0.2 mm long. July–Aug.

SYNONYM *Scirpus atrovirens* Willd. var. *pallidus* Britt.

HABITAT Wet meadows, shores, ditches, streambanks, springs and other wet places.

WETLAND STATUS
GP OBL | MIDW OBL | WMTN OBL

ADDITIONAL SPECIES Occasional in the Great Plains is the very similar **dark-green bulrush** (*Scirpus atrovirens* Willd.), of which *S. pallidus* is sometimes treated as a variety. *Scirpus atrovirens* differs from *S. pallidus* chiefly as follows: Scales dark green on the sides, eventually turning brownish or blackish, 1–2 mm long including a mucro 0.1–0.2 mm long.

Scirpus pendulus Muhl.

RUFOUS BULRUSH

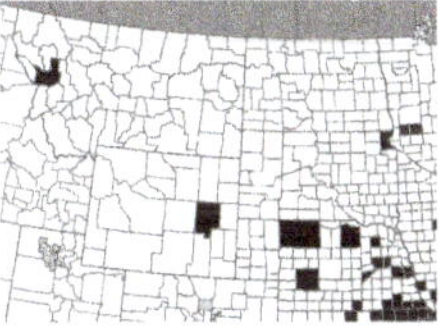

DESCRIPTION Loosely tufted from short, thick rhizomes; **culms** obtusely trigonous, to 1.5 m tall, clothed at the base by old leaf bases. **Leaves** several, scattered on the stem; blades flat, 4–10 mm wide; involucral bracts 3-several, reduced but foliaceous, shorter than the inflorescence, pale brown at the base. **Inflorescence** umbellate, rather open, with several to many primary branches, 1 or 2 lateral inflorescences sometimes produced below the terminal one, ultimate

Scirpus pallidus
PALE BULRUSH

Scirpus pendulus
RUFOUS BULRUSH

branches with 1 sessile and few-several pedicellate spikelets, mostly nodding at the tips; **spikelets** numerous, ovoid to mostly oblong-cylindric, 4–10 mm long, 2–3.5 mm thick, blunt-tipped; scales reddish-brown on the sides with a prominent green midrib, elliptic-ovate, 1.5–2.7 mm long, including the short-awned to cuspidate tip; perianth bristles 6, smooth, much contorted, brownish, longer than the achene but barely, if at all, extended beyond the scale; stamens 3; styles trifid. **Achenes** tan to light brown, trigonous, ellipsoid to obovoid, 1–1.3 mm long, apiculate. June–Aug.

SYNONYM *Scirpus lineatus* Michx.

HABITAT Marshes, wet meadows, streambanks and ditches.

WETLAND STATUS
GP OBL | MIDW OBL | WMTN OBL

Scirpus pendulus
RUFOUS BULRUSH

Scirpus pallidus
PALE BULRUSH

Hydrocharitaceae *frog's-bit family*

Submersed, dioecious, perennial herbs (those included here) of calm or flowing water, densely leafy-stemmed with whorled or decussate, sessile leaves, or plants acaulescent with long, linear, ribbonlike leaves arising in clusters from stolons that creep over the substrate. **Flowers** solitary or numerous, sessile or peduncled, extended or freely floating to the surface at anthesis, arising from a 2-bracted or bifid (trifid) spathe, small and nonshowy, regular; sepals 3; petals (1-)3, seldom absent; male flowers with (1)2(3) or 9 stamens; female flowers with or without staminodes; stigmas 3, often 2-lobed, ovary inferior, ovules few to several, on 3 parietal placentae that extend nearly to the center of the ovary. **Fruit** ovoid or cylindric, few- to several-seeded, rupturing tardily and irregularly, maturing underwater.

1 Leaves very long and ribbon-like (mostly 3–11 mm wide), in a basal rosette ***Vallisneria***

1 Leaves to 6 (–12) cm long, opposite or whorled **2**

2 Leaves whorled, entire ***Elodea***

2 Leaves opposite, minutely denticulate to visibly toothed ***Najas***

Elodea WATERWEED

Stems slender, lax to rather rigid and brittle, sparingly to freely branched, rooting from the lower nodes or sometimes free-floating. **Leaves** sessile, crowded toward the apex, mostly in whorls of 3(4) or opposite and decussate, the margins finely serrulate. **Flowers** minute, delicate, often lost in collection, solitary in the upper leaf axils, subtended by a 2-lobed spathe, usually extended to the water surface by a long, threadlike hypanthium or (in male flowers of *E. nuttallii*) sessile and breaking free to float to the surface; sepals 3; petals 3 or seldom absent; male flowers containing 9 stamens; female flowers with 3 staminodes and 3 stigma lobes protruding at the summit of the hypanthium, the stigmas entire or 2-lobed. **Fruit** ovoid to cylindric, several-seeded.

1 Middle and upper leaves decussate **_E. bifoliata_**

1 Middle and upper leaves in whorls of 3(4) 2

2 Leaves mostly 1.5–3(5) mm wide, obtuse-apiculate at the apex; male flowers extended to the surface by a long, threadlike hypanthium; sepals of female flowers 2–3 mm long **_E. canadensis_**

2 Leaves 0.7–1.5 mm wide, acute or acute-apiculate at the apex; male flowers sessile inside the spathe, breaking free to float to the surface at anthesis; sepals of female flowers ca. 1 mm long **_E. nuttallii_**

Elodea bifoliata St. John (*not illustrated*)

TWO-LEAF WATERWEED

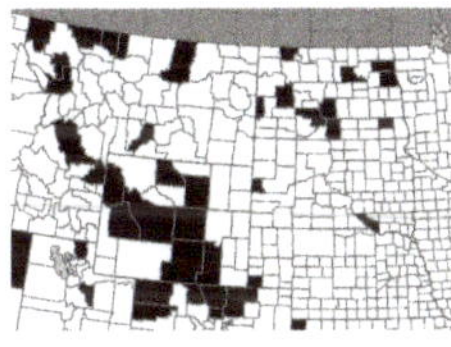

DESCRIPTION **Stems** terete, sparsely branch-ed, 3–10 dm long. **Leaves** all decussate or some lower ones in whorls of 3, linear, 5–20 mm long, 0.5–2 mm wide, acute to obtuse. **Spathes of male flowers** 2–15 cm long, inflated near the summit, to 2.5–4 mm wide; male flowers peduncled by the threadlike hypanthium which is up to 30 cm long; sepals elliptic, 3.5–5 mm long; petals white, linear, 5 mm long; stamens 9, all attached at the summit of the hypanthium. **Pistillate spathes** linear-cylindric, 3–7 cm long; female flowers extended by the elongate, threadlike hypanthium which may attain lengths of several dm; sepals elliptic, 2.8 mm long; petals white, 4 mm long; stigmas entire, 1.5 mm long. **Capsules** ovoid, 10 mm long; seeds cylindric, 6 mm long. July–Aug.

SYNONYMS *Elodea longivaginata* St. John, *Elodea nevadensis* (Planch.) St. John

HABITAT Prairie ponds, lakes and reservoirs.

WETLAND STATUS

GP OBL | MIDW OBL | WMTN OBL

Elodea canadensis Michx.

CANADIAN WATERWEED

DESCRIPTION **Stems** terete, often freely branched, 2–5(10) dm long. Lower **leaves** opposite, reduced, ovate-lanceolate; middle and upper leaves in whorls of 3, linear-lanceolate (on male plants) or oblong-lanceolate to ovate-lanceolate (on female plants), 5–15 mm long, 1–5 mm wide, mostly more than 1.5 mm wide, obtuse-apiculate, more strongly overlapping on pistillate plants. **Winter buds** appearing as short, compact branches in late summer. **Spathes of male flowers** borne in upper axils, ca. 15 mm long, inflated upward to 4 mm wide; male flowers peduncled by the threadlike hypanthium which is 3–20 cm long; sepals elliptic, 3.5–5 mm long; petals white, clawed, linear, 5 mm long; stamens 9, the inner 3 uplifted on a common stalk. **Spathes of female flowers** borne in upper axils, cylindric, 10–20 mm long; female flowers extended to the surface by a threadlike hypanthium 2–28 cm long; sepals oblong-elliptic, 2–2.2 mm long; petals white, oblanceolate, ca. 2.5 mm long; stigmas 2-cleft, 4 mm long. **Fruit** ovoid, ca. 6 mm long; seeds narrowly cylindric, 4.5 mm long, glabrous. June–Aug.

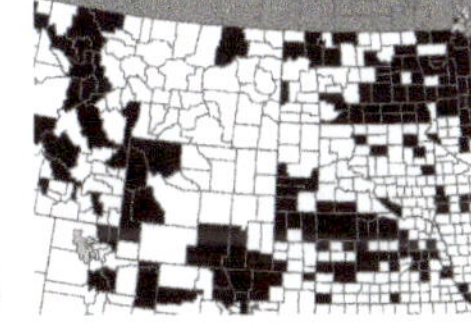

SYNONYM *Anacharis canadensis* (Michx.) Rich.

Elodea canadensis
CANADIAN WATERWEED

HABITAT Quiet water of streams and reservoirs.

WETLAND STATUS
GP OBL | MIDW OBL | WMTN OBL

Elodea nuttallii (Planch.) St. John
WESTERN WATERWEED

DESCRIPTION Similar to *Elodea canadensis* but smaller and more delicate, the **stems** slender, terete, often freely branched. Lower **leaves** opposite and reduced, ovate-lanceolate; middle and upper leaves in whorls of 3 or occasionally 4, linear to linear-lanceolate, 6–1 3 mm long, (0.3)0.7–1.5 mm wide, acute or acute-apiculate at the apex. **Spathes of male flowers** borne in middle axils, sessile, ovoid, ca. 2 mm long, 2-parted but the lobes twisted together so that the spathe appears pointed; male flowers solitary and sessile in the spathe, breaking free and floating to the surface where the flower opens; sepals ovate, ca. 2 mm long, sometimes reddish; petals lacking or to 0.5 mm long, ovate-lanceolate; stamens 9, the inner 3 elevated on a common stalk. **Spathes of female flowers** borne in upper axils, narrowly cylindric but slightly broadened at the base and the bifid tip, 9–25 mm long; female flowers extended to the surface by a threadlike hypanthium to 9 cm long; sepals obovate, ca. 1 mm long; petals white, obovate, longer than the sepals; stigmas slender, shallowly bifid, slightly exceeding the sepals. **Fruit** narrowly ovoid to fusiform, 5–7 mm long; seeds cylindric, 3.5–4.5 mm long, pilose. June–Aug.

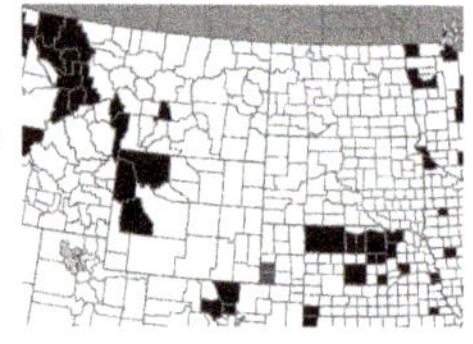

Elodea nuttallii
WESTERN WATERWEED

SYNONYMS *Anacharis occidentalis* (Pursh) Victorin, *Anacharis nuttallii* Planch.

HABITAT Quiet water of streams and lakes.

WETLAND STATUS
GP OBL | MIDW OBL | WMTN OBL

Najas NAIAD, WATERNYMPH

Small, submersed, monoecious or dioecious, perennial herbs; **stems** flexuous, freely branched, anchored by roots. **Leaves** simple, opposite, sessile, abruptly broadened at the base to sheath the stem, the margins faintly to coarsely toothed. **Flowers** minute, imperfect, solitary and sessile in the leaf axils, enclosed by the sheathing leaf bases; male flowers consisting of a single anther enclosed in a membranous envelope, this in turn surrounded by a firmer, entire to 4-lobed, perianth-like structure; female flowers comprised of a single, 1-ovuled pistil, stigmas 2–4, style usually elongate and persistent. **Fruit** a fusiform achene, the pericarp thin and easily removed to expose the single ellipsoid seed.

1 Leaves coarsely spinulose-toothed; seeds 4–5 mm long ... ***N. marina***
1 Leaves very faintly toothed; seeds 2–3 mm long ... **2**

2 Seed coat (under the thin pericarp) shiny, very faintly reticulate, with 30–40 rows of obscure pits across the middle; style beak of achene 1 mm or more long ... ***N. flexilis***
2 Seed coat rather dull, coarsely reticulate, with 10–20 rows of pits across the middle; style beak of mature achene 0.5 mm or less long ... ***N. guadalupensis***

Najas flexilis (Willd.) Rostk. & Schmidt
WAVY WATERNYMPH

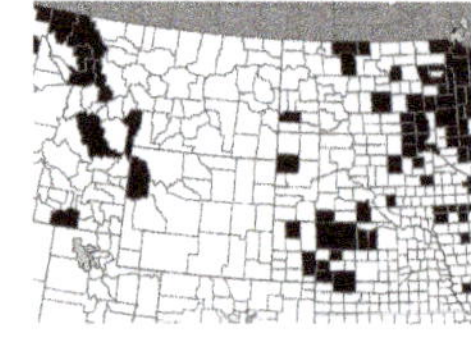

DESCRIPTION Plants monoecious; **stems** branching, 5- 30 cm long. **Leaves** densely clustered at stem tips, linear, attenuate, minutely spinulose on the margins, 1–3 cm long, 0.3–0.7 mm wide. **Achenes** olive-green to reddish, the beak 1 mm or more long; **seeds** 2.5–3 mm long, 1/3 as thick, the seed coat shiny, very faintly reticulate, with 30–40 rows of obscure pits across the middle. July–Sept.

HABITAT Fresh or calcareous water of lakes and ponds.

WETLAND STATUS
GP OBL | MIDW OBL | WMTN OBL

Najas guadalupensis (Spreng.) Magnus.

SOUTHERN NAIAD (*not illustrated*)

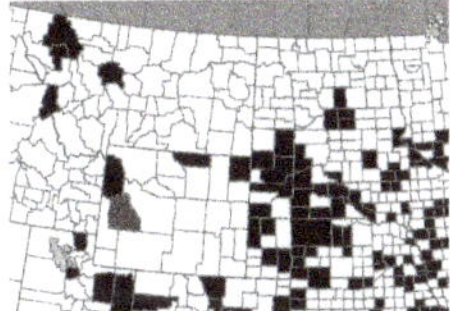

DESCRIPTION Plants monoecious; **stems** branching, 10–30 cm long. **Leaves** linear, acute, minutely spinulose, 1–2 cm long, 0.4–0.8 mm wide; fascicles of smaller leaves appearing in the axils. **Achenes** brown or purplish, the beak 0.5 mm or less long at maturity; **seeds** 2–3 mm long, 1/4 as thick, the seed coat rather dull, coarsely reticulate, with 10–20 rows of pits across the middle. July–Sept.

HABITAT Shallow water of marshes, ditches, lakes and impoundments.

WETLAND STATUS
GP OBL | MIDW OBL | WMTN OBL

Najas marina L.

HOLLY-LEAF WATERNYMPH

DESCRIPTION Plants dioecious; **stems** freely branched, 2–5 dm long, the internodes often spinulose. **Leaves** linear, coarsely spinulose-toothed, sometimes spinulose on the back, 10–20 mm long, 2–4 mm wide (including the teeth), the teeth triangular, projecting 0.5–1 mm, 1–4 mm apart. Achenes olive-green, the beak 0.5–1.5 mm long; **seeds** 4–5 mm long, 1/2 as thick, finely reticulate. July–Sept.

HABITAT Shallow water of brackish lakes and marshes.

WETLAND STATUS
GP OBL | MIDW OBL | WMTN OBL

Najas flexilis
WAVY WATERNYMPH

Vallisneria americana Michx.

EELGRASS, TAPEGRASS, WATER-CELERY

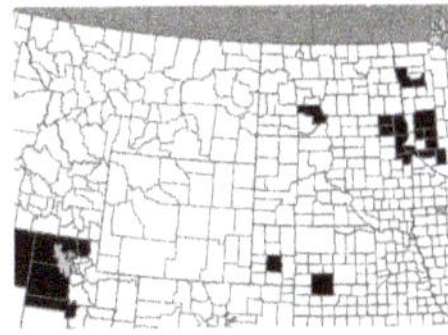

DESCRIPTION Plants acaulescent with long, linear, ribbon-like leaves arising in tufts, fibrous-rooted, producing short stolons from the base, often forming extensive submerged beds. **Leaves** up to 1 m or more long though usually shorter, 3–12 mm wide, with a many-nerved median band bordered by nerveless or sparsely nerved marginal bands, blunt or rounded at the tip. **Male plants** producing a head of numerous flowers within a short-stalked, 2- to 3-parted spathe borne at the base; male flowers minute, with 3 sepals, 1(2) minute petal(s) and (1)2(3) stamens, individually released from the spathe to float freely to the surface where they open. **Female plants** bearing flowers singly on long, slender peduncles that extend the flower to the surface; female flowers with 3 sepals, 3 minute petals and 3 prominent stigmas, sometimes with 1 or more staminodes; ovary enclosed by a hypanthium, linear-cylindric, 2–3 cm long in flower, enclosed at the base by the 2-cleft spathe. **Fruit** cylindric, curved, 4–10 cm long, many-seeded; peduncle becoming loosely coiled to draw the fruit underwater. July–Sept.

HABITAT Uncommon though locally abundant in shallow water of lakes in ne SD; occurring in slow-moving streams e of our range where it is more common.

Najas marina
HOLLY-LEAF WATERNYMPH

WETLAND STATUS
GP OBL | MIDW OBL | WMTN OBL

NOTE Submersed forms of *Sagittaria* spp. are sometimes mistaken for this plant. In sterile material, the distinctive marginal bands of the leaves and the stoloniferous habit are diagnostic for *Vallisneria.*

East of our region, *V. americana* is an important food for waterfowl and other aquatic life. It is apparently limited here by high conductivities, turbidity and unstable water conditions, and is thus not abundant enough to be an important food source.

Vallisneria americana
EELGRASS, TAPEGRASS, WATER-CELERY

Hypoxidaceae
yellow star-grass family

Hypoxis hirsuta (L.) Cov.
YELLOW STARGRASS

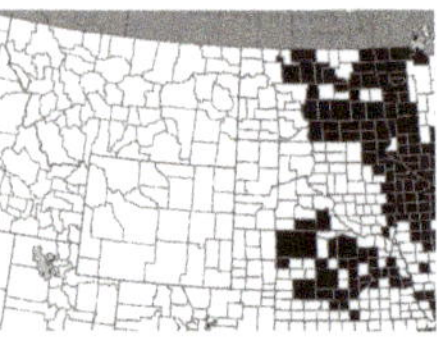

DESCRIPTION Low, acaulescent perennial 6–25 cm tall, from a small shallow corm. **Leaves** grasslike but lacking sheaths, 1–5 mm wide, equaling or surpassing the inflorescence, mostly 5- to 9-nerved. **Scapes** 1-few, slender, lax, sericeous in the upper part. **Flowers** 1–6 (usually 2) per scape, yellow, 1–2.5 cm across, pubescent on the outside of the tepals and ovary; tepals spreading in flower, 5–12 mm long, closing and turning green after flowering, persistent; ovary inferior. **Capsule** ellipsoid, 3–6 mm long; seeds round, black, muricate, 1–1.5 mm in diameter. Late May–July.

HABITAT Wet meadows and low prairie.

WETLAND STATUS
GP FACW | MIDW FAC | WMTN FACW

Hypoxis hirsuta
YELLOW STARGRASS

Iridaceae *iris family*

Perennial herbs with rhizomes, bulbs, or fibrous roots. leaves parallel-veined, narrow, 2-ranked, the margins joined to form an edge facing the stem (equitant). flowers perfect, with 6 petal-like segments, single or in clusters at ends of stem, stamens 3, style 3-parted. fruit a 3-chambered capsule.

1 Flowers more than 2 cm wide; stems not winged; leaves more than 6 mm wide *Iris*

1 Flowers to 2 cm wide; stems winged; leaves to 6 mm wide *Sisyrinchium*

Iris pseudacorus L.

PALE-YELLOW IRIS

Perennial, from thick rhizomes. **Stems** 0.5–1 m long, shorter or equal to the leaves. **Leaves** sword-shaped, stiff and erect, waxy, 1–2 cm wide. **Flowers** several at end of stems, yellow, 7–9 cm wide, sepals spreading, upper portion marked with brown; petals erect, narrowed in middle, 1–2.5 cm long. **Capsules** oblong, 6-angled, 5–9 cm long. may–June.

HABITAT Lakeshores, streambanks, marshes, ditches; introduced from Eurasia and potentially invasive in wetlands and shallow water.

WETLAND STATUS

GP OBL | MIDW OBL | WMTN OBL

Iris pseudacorus
PALE-YELLOW IRIS

Sisyrinchium montanum Greene

BLUE-EYED GRASS

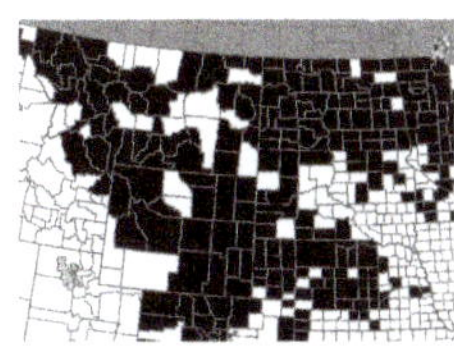

DESCRIPTION Fibrous-rooted perennial 1–4.5 dm tall, often tufted, with mostly basal, grasslike **leaves** 1–3 mm wide and scapose inflorescences, the plants glaucescent, drying pale, sometimes purplish at base; scapes narrowly to broadly winged, 1.5–3 mm wide. **Inflorescence** of 1–several flowers, terminal on the scape, subtended by a 2-bracted spathe, the outer bract (2.5)3.5–6(8.5) cm long, mostly 1.5–3x longer than the inner bract, the margins of the outer bract united for 1–3 mm above the base. **Flowers** blue-violet with a yellow center, perfect, regular; perianth of 6 membranous tepals (5)7–14 mm long, apiculate to emarginate at the tip; stamens 3, united by their filaments around the style; style 3-branched, ovary inferior, 3-celled, globose. **Fruit** a subglobose, loculicidal capsule, pale to stramineous, 4–7 mm in diameter; seeds many, black, globular, shallowly pitted. Late May–July.

HABITAT Wet meadows, ditches, swales and floodplains, also in drier situations.

WETLAND STATUS

GP FAC | MIDW FAC | WMTN FAC

ADDITIONAL SPECIES **White-eyed grass** (*Sisyrinchium campestre*) occurs in e SD and Neb, becoming more common eastward. Although occasionally found in wet meadows, this plant is more characteristic of mesic sites, especially in tallgrass prairie. Distinguished from *S. montanum* by its white to light blue-violet flowers and the fact that the margins of the outer spathe bract are free to the base. The plant is also generally smaller and earlier flowering than *S. montanum.*

Sisyrinchium montanum
BLUE-EYED GRASS

Juncaceae *rush family*

Juncus RUSH

Tufted or rhizomatous grasslike plants, mostly perennial; **culms** elongate, simple, erect; rhizomes often strongly developed. **Leaves** cauline or mainly basal, alternate and usually 3-ranked, reduced to sheathing at the base of the culm in some species; blades, if present, terete and often nodulose-septate, or flat to involute; sheaths broader than the blades, commonly with a pair of auricles at the summit. **Inflorescence** a compact to open cyme of few to many flowers, subtended by 1-few leaflike involucral bracts, the flowers borne singly on pedicels or grouped into glomerules of few to many; each subtended by 1 or 2 bracteoles. **Flowers** perfect, regular; perianth of 6 chaffy tepals, these green to purplish-brown, persistent, scalelike; stamens 6 or 3; carpels 3, united, stigmas 3, ovary superior, 1- or 3-celled, with parietal or axile placentation, ovules numerous. **Fruit** a many-seeded, 3-valved and often 3-angled capsule, 1-locular to completely 3-celled; seeds ovoid to oblong, apiculate or with membranous, taillike appendages at both ends.

---

1 Flowers individually pedicelled, each subtended by a pair of membranous bracteoles; leaves flat to involute or terete and narrowly channeled on the upper side, never septate, or the leaves bladeless, reduced to basal sheaths **2**

1 Flowers grouped into glomerules of few to many, each subtended by a single bracteole; leaves terete or flat, often nodulose-septate **9**

2 Leaves comprised of bladeless basal sheaths; involucral leaf erect, resembling a continuation of the culm, the inflorescence appearing lateral ***J. balticus***

2 Leaves with blades; inflorescence terminal **3**

3 Plants annual; inflorescence making up 1/3 or more of the height of the plant ***J. bufonius***

3 Plants perennial, often rhizomatous; inflorescence proportionately much smaller **4**

4 Leaves, except the involucres, all basal or nearly so, arising from well within the lower 1/4 of the stem **5**

4 Leaves cauline as well as basal, some arising from near or within the upper 1/2 of the stem **8**

5 Leaf blades terete, narrowly channeled on the upper side; capsule exceeding the perianth; seeds ca. 1 mm long, with a membranous appendage at each end ***J. vaseyi***

5 Leaf blades flat or involute; capsule shorter than the perianth; seeds 0.3–0.5 mm long, merely minutely apiculate at each end **6**

6 Auricles flaplike, prolonged 1–5 mm beyond the summit of the sheath ***J. tenuis***

6 Auricles not flaplike, low and rounded, only to 0.5 mm long **7**

7 Auricles cartilaginous, drying yellowish; bracteoles obtuse to subacute; leaf sheaths green, usually not reddish ***J. dudleyi***

7 Auricles membranous, white or spotted with red or brown; bracteoles acute to acuminate or aristate; lower leaf sheaths often reddish ***J. interior***

8 Capsule equaling or barely surpassing the outer tepals; stamens nearly reaching the summit of the perianth ***J. gerardii***

8 Capsule much surpassing the outer tepals; stamens reaching the middle of the perianth ***J. compressus***

9 Leaf blades flat, not hollow, not nodulose-septate or with only incomplete septae **10**

9 Leaf blades terete, hollow, nodulose-septate, the septae complete **12**

10 Leaves equitant, incompletely cross-septate; plants of moderate to high elevations in the Black Hills and westward ***J. ensifolius***

10 Leaves not equitant, not at all cross-septate; plants more widespread in our range **11**

11 Tepals (4)5–6 mm long; stamens 6 *J. longistylis*

11 Tepals 2–3.5 mm long; stamens 3 *J. marginatus*

12 Glomerules obpyramidal to hemispheric, few- to 12-flowered, the flowers projecting upward 13

12 Glomerules spherical, densely many-flowered, the flowers projecting outward in all directions at maturity 16

13 Seeds 0.9–1.8 mm long, with membranous appendages at both ends; stamens 3 14

13 Seeds 0.5 mm or less long, merely pointed at both ends; stamens 6 15

14 Seeds 0.9–1.2 mm long, the body of the seed comprising 3/5 or more of its total length *J. brevicaudatus*

14 Seeds 1.2–1.9 mm long, the body of the seed comprising 1/2 or less of its total length *J. canadensis*

15 Inner tepals shorter than the outer ones, blunt to rounded; capsule rounded at the apex; inflorescence narrow, the branches mostly sharply ascending *J. alpinoarticulatus*

15 Inner tepals longer than or about equaling the outer ones, acute to acuminate; capsule acute at the apex; inflorescence open, the branches spreading *J. articulatus*

16 Stamens 3, opposite the outer tepals 17

16 Stamens 6, opposite both the inner and outer tepals 18

17 Seeds with white, membranous, taillike appendages at both ends; capsule abruptly tapered to an apiculate tip *J. canadensis*

17 Seeds apiculate at both ends, without appendages; capsule gradually tapered to a slender beak ca. 1 mm long *J. scirpoides*

18 Inner and outer tepals equal in length, or nearly so; auricles membranous, yellowish *J. nodosus*

18 Outer tepals distinctly longer than the inner tepals; auricles white-hyaline *J. torreyi*

Juncus alpinoarticulatus Chaix.

ALPINE RUSH

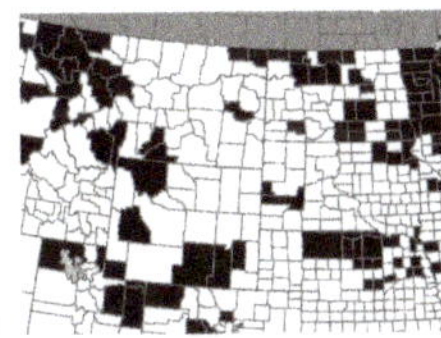

DESCRIPTION Rhizomatous perennial growing in small tufts, 1.5–4 dm tall. **Leaves** mostly basal, 1 or 2 cauline; blades terete, hollow, nodulose-septate, 0.5–1(1.5) mm wide when pressed; sheaths green or often reddish on basal leaves, the margins membranous, shiny, often yellowish, projecting into auricles 0.5–1 mm long at the summit of the sheath. **Inflorescence** slender, 2–15 cm long, sparsely to freely branched, the branches spreading to usually strongly ascending; glomerules often rather few, obpyramidal, 3- to 12-flowered. **Flowers** green to light or dark brown, 2–3 mm long; tepals acute to obtuse or mucronate, the inner ones shorter, broader and tending to be more obtuse than the outer, their margins scarious; stamens 6. **Capsules** oblong, 1-locular, surpassing the perianth, rounded below the apiculate apex; seeds cylindric-ellipsoid to ovoid-ellipsoid, pale brown to brown, ca. 0.5 mm long. July–Sept.

SYNONYM *Juncus alpinus* Vill.

HABITAT Springs, shores, streambanks, fens and boggy places, where water is fresh.

WETLAND STATUS
GP OBL | MIDW OBL | WMTN OBL

Juncus alpinoarticulatus
ALPINE RUSH

Juncus articulatus L.
JOINTED RUSH

DESCRIPTION Usually tufted perennial 2–6 dm tall, with coarse whitish rhizomes. **Leaves** 2–4 per culm, terete, hollow, nodulose-septate, 0.5–2 mm wide when pressed; sheaths green or basal ones sometimes reddish, the margins membranous, white or tawny, prolonged as rounded or acutely rolled auricles 1–2 mm long. **Inflorescence** ovoid to short and depressed in outline, 2–15 cm long, up to 2x longer than broad, the branches mostly divergent to widely ascending; glomerules usually numerous, obpyramidal to nearly hemispheric, 3- to 12-flowered. **Flowers** green to dark brown, mostly 2.5–3 mm long; outer tepals acute to acuminate, often mucronate, about equaling or distinctly shorter than the inner ones; inner tepals acute, scarious-margined; stamens 6. **Capsules** oblong-ovoid, 1-locular, exceeding the tepals, acute or broadly so below the apiculate tip; seeds elliptic-ovoid to obovoid, 0.3–0.5 mm long, apiculate at both ends. July–Sept.

HABITAT Sandy or gravelly shores, streambanks and spring borders.

WETLAND STATUS
GP OBL | MIDW OBL | WMTN OBL

Juncus balticus Willd.
BALTIC RUSH

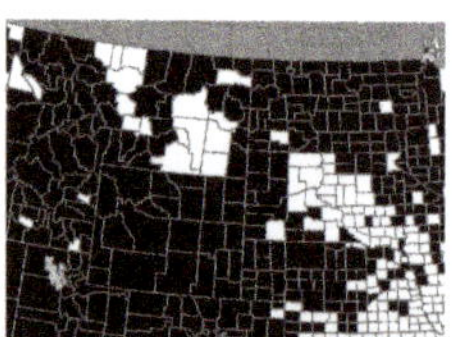

DESCRIPTION Strongly rhizomatous perennial; **culms** slender and tough, dark green, 3–9 dm tall, appearing leafless, arising in rows from thick, brown to black rhizomes. **Leaves**, except for the involucre, comprised of basal sheaths only, these brown to reddish-brown; involucral leaf terminal, erect, terete, resembling a continuation of the culm, **Inflorescence** appearing lateral, subcapitate and dense to spreading and diffuse, extending outward 1–7 cm; bracteoles obtuse to mucronate, 1–2 mm long. **Flowers** borne singly on pedicels, dark brown to purplish-brown, 3.5–5 mm long; tepals acute, hyaline-margined, the inner ones slightly to considerably shorter than the outer; stamens 6. **Capsules** ovoid, 3-locular, shorter than to surpassing the perianth; seeds obliquely ovoid-ellipsoid, grayish-brown, ca. 0.6 mm long, apiculate at both ends. June–Aug.

SYNONYM *Juncus arcticus* Willd.

HABITAT Wet meadows, ditches, seepage areas and shores; common, often abundant.

WETLAND STATUS
GP FACW | MIDW OBL | WMTN OBL

Juncus brevicaudatus (Engelm.) Fern.
NARROW-PANICLE RUSH

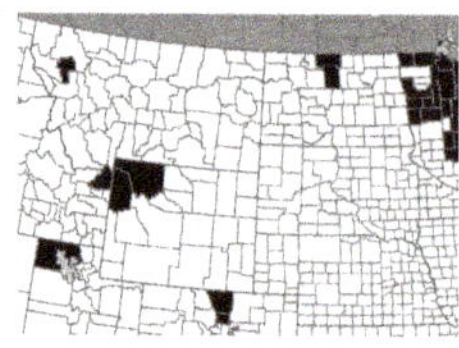

DESCRIPTION Densely tufted perennial 1.5–5 dm tall. **Leaves** both cauline and basal; blades terete, hollow, nodulose-septate, 1–2 mm wide when pressed; sheaths green or sometimes reddish on basal leaves, the margins membranous and prolonged as 2 rounded auricles 1–2 mm long. **Inflorescence** rather narrow, 2–12 cm long, 1/3 to 1/6 as wide, sparsely to freely branched, the branches strongly ascending; glomerules obpyramidal, containing 2–7 flowers. **Flowers** greenish to golden brown or dark brown,

Juncus articulatus
JOINTED RUSH

Juncus balticus
BALTIC RUSH

3–4 mm long; tepals acute to mucronate, with narrow scarious margins; stamens 3. **Capsules** oblong-pyriform to cylindric-ellipsoid, 1-locular, much surpassing the perianth; seeds fusiform, brown, 0.9–1.2 mm long, with membranous, taillike appendages at both ends, the body comprising 3/5 or more of the total length. Aug–Sept.

HABITAT Wet meadows, marshes, fens, shores.

WETLAND STATUS
GP OBL | MIDW OBL

Juncus bufonius L.
TOAD RUSH

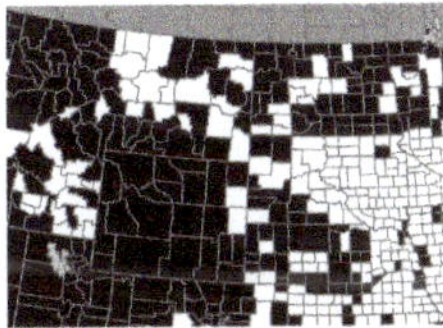

DESCRIPTION Small, tufted annual 5–20 cm tall. **Culms** usually crowded, typically much shorter than the inflorescence, 1–10 cm to the base of the inflorescence. **Leaves** usually shorter than the culm, the blades 0.5–1 mm wide; basal sheaths green to suffused with red or brown, white-hyaline along the margins, rounded to truncate at the summit; auricles lacking. **Inflorescence** 1/3 to nearly the entire length of the plant, freely branched or sparsely so in diminutive specimens; bracteoles obtuse to mucronate, 1–2 mm long. **Flowers** solitary, mostly sessile and secund on the branches, green, 4–6 mm long; tepals with hyaline margins, the outer ones acute, the inner acute to obtuse and shorter than the outer ones; stamens 6. **Capsules** cylindric-ovoid, 1-locular, shorter than the perianth; seeds ovoid, brown, 0.2–0.4 mm long, apiculate at both ends. June–Aug.

HABITAT Shores, mud flats, streambanks and other temporarily flooded places.

WETLAND STATUS
GP OBL | MIDW FACW | WMTN FACW

Juncus canadensis Gay ex Laharpe
CANADIAN RUSH

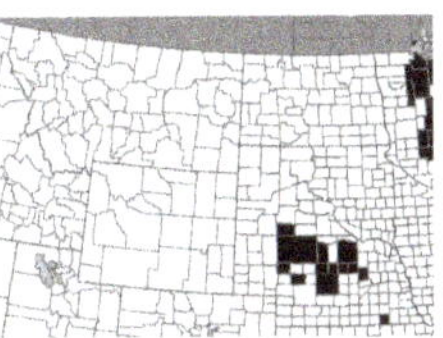

DESCRIPTION Coarse tufted perennial 3–9 dm tall, with stout, rigid culms. **Leaves** 3–4 per culm, terete, hollow, nodulose-septate, 1.5–3 mm wide when pressed; sheaths green or basal ones reddish, with narrow, membranous margins, prolonged as acute to rounded auricles 1–2 mm long. **Inflorescence** open or congested, 2–20 cm long, the branches ascending; glomerules few to many, containing 5–40 or more flowers, obpyramidal to hemispheric when relatively few-flowered, to subglobose when many-flowered. **Flowers** green to brown, (3)4–5 mm long; tepals lance-subulate, the inner slightly to distinctly longer than the outer; stamens 3. **Capsules** ovoid to oblong, 1-locular, equaling to strongly exceeding the inner tepals, abruptly tapered at the apex to an apiculate tip; seeds fusiform, 1.2–1.9 mm long includ-

Juncus brevicaudatus
NARROW-PANICLE RUSH

Juncus bufonius
TOAD RUSH

ing the white, membranous, taillike appendages at both ends, the body 1/2 or less the total length. July–Sept.

HABITAT Sandy shores, marshes and streambanks.

WETLAND STATUS
GP OBL | MIDW OBL | WMTN OBL

Juncus compressus Jacq.
ROUND-FRUIT RUSH

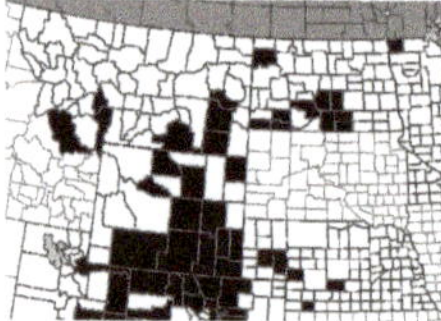

DESCRIPTION Perennial from dark brown to black rhizomes, 2–6 dm tall. **Leaves** cauline as well as basal, some arising near or within the upper 1/2 of the stem; blades flat, to 2 dm long, 1–2 mm wide; sheaths green with broad membranous margins, rounded and entire or cleft into 2 low, rounded auricles at the summit. **Inflorescence** oblong, many-flowered, 2–8 cm long, the branches erect to ascending; involucral leaf seldom surpassing the inflorescence; bracteoles mostly obtuse to rounded, ca. 1 mm long. **Flowers** solitary, mostly on short pedicels, dark brown, 2–3 mm long; tepals blunt and incurved at the tip, the outer ones slightly longer than the inner, the inner ones with a scarious margin toward the tip; stamens 6, reaching the middle of the perianth, the anthers ca. 1 mm or less long, scarcely longer than the filaments. **Capsules** obovoid, 1-locular, much surpassing the perianth; seeds obliquely cylindric-pyriform, brown, 0.5–0.6 mm long. June–Aug.

WETLAND STATUS
GP FACW | MIDW OBL

NOTE Introduced to N America from Europe; now found in Canada and n USA.

Juncus dudleyi Wieg.
DUDLEY'S RUSH

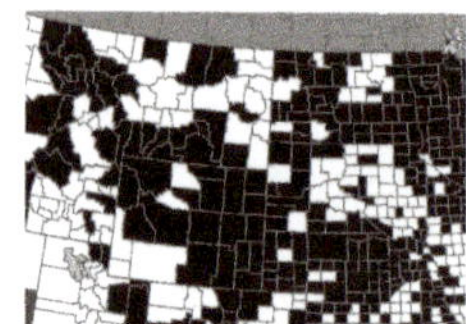

DESCRIPTION Tufted perennial 2–8 dm tall. **Leaves** mostly basal, arising well within the lower 1/4 of the stem; blades typically 1/3 to 1/2 the length of the stem, 0.5–1 mm wide, flat to involute; sheaths green or uncommonly faintly reddish, the margins and auricles firm and cartilaginous, drying yellowish, glossy, auricles low, rounded, to 0.5 mm long. **Inflorescence** compact to spreading, (1)2–5 cm long; involucres 1–3, usually 2, at least one of these exceeding the inflorescence; bracteoles obtuse to subacute, nearly transparent, 1–2 mm long. **Flowers** solitary, sessile to short-pedicelled, green to stramineous or light brown, 4–6 mm long; tepals acuminate, with narrowly hyaline margins, the outer noticeably longer than the inner; stamens 6. **Capsules** ovoid-cylindric to ovoid, 1-locular, shorter than the perianth; seeds obliquely ellipsoid, brown, 0.3–0.4 mm long. June–Aug.

Juncus compressus
ROUND-FRUIT RUSH

Juncus dudleyi
DUDLEY'S RUSH

SYNONYM *Juncus tenuis* var. *dudleyi* (Wieg.) F.J. Herm.

HABITAT Wet meadows, springy or boggy areas, ditches, shores, streambanks.

WETLAND STATUS
GP FACW | MIDW FACW | WMTN FAC

Juncus ensifolius Wikst.

DAGGER-LEAF RUSH

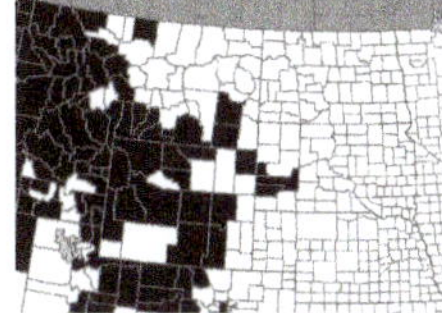

DESCRIPTION Rhizomatous perennial 1.5–6 dm tall; culms single or loosely clustered. **Leaves** 2–4 per culm, the blade folded along the midrib with the margins united, thus laterally flattened and equitant (with one edge toward the culm), 1.5–6 mm wide, incompletely cross-septate; sheaths green or reddish, with broad scarious margins, these often prolonged as low rounded auricles to 0.5 mm long. **Inflorescence** short to oblong, usually few-branched with the branches erect or nearly so; glomerules (1)2-several (seldom more), hemispheric to subglobose, 5- to many-flowered, to ca. 1 cm across. **Flowers** dark brown, 3–4 mm long; tepals lanceolate, the outer tepals acuminate, sharp-pointed, the inner ones shorter, acute, scarious-margined; stamens 6(3). **Capsules** oblong, 1-locular, rounded to the short beak, shorter than to exceeding the tepals; seeds ellipsoid to fusiform, 0.4–0.6 mm long and apiculate at both ends, or sometimes to 1 mm long with a taillike appendage at one or both ends. July–Sept.

HABITAT Margins of springs, streams, ponds and in seepage areas at moderate to high elevations.

WETLAND STATUS
GP FACW | WMTN FACW

Juncus ensifolius
DAGGER-LEAF RUSH

Juncus gerardii Loisel (*not illustrated*)

SALTMEADOW RUSH

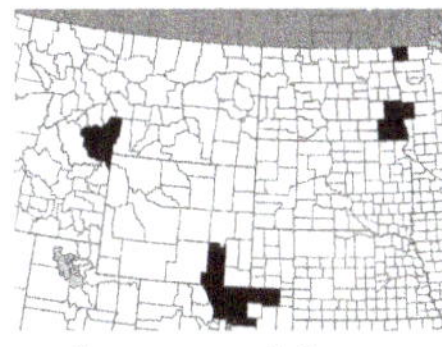

DESCRIPTION Much like *Juncus compressus,* differing mainly as follows: Stamens nearly reaching the summit of the perianth, the anthers 1.5 mm long, much longer than the filaments. **Capsules** ovoid to obovoid, about equaling the perianth. June–Aug.

HABITAT Wet meadows, often where saline.

WETLAND STATUS
gp obl | midw obl

Juncus interior Wieg.

INLAND RUSH

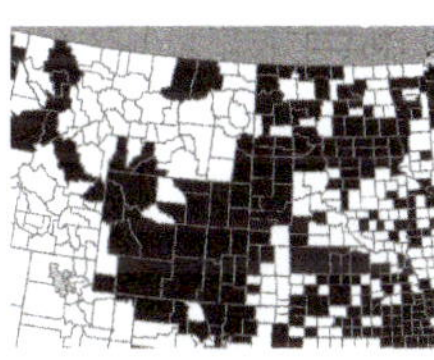

DESCRIPTION Very similar to *Juncus dudleyi,* differing mainly as follows: Leaf sheaths green or often reddish, the auricles and margins membranous, white or suffused faintly with red or brown. Bracteoles

Juncus interior
INLAND RUSH

acute to more often acuminate or aristate. June–Aug.

HABITAT Occurring in the same habitats as *Juncus dudleyi.*

WETLAND STATUS
GP FACW | MIDW FAC | WMTN FAC

Juncus longistylis Torr.
LONG-STYLE RUSH

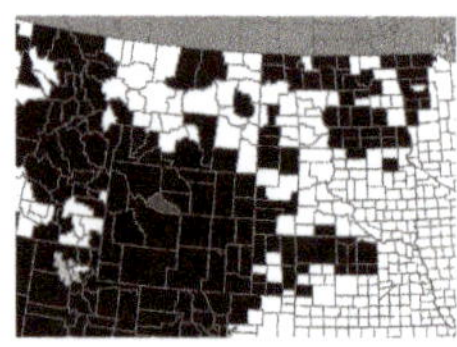

DESCRIPTION Rhizomatous perennial 3–7 dm tall. **Leaves** mostly basal, reduced upward, the upper ones coming off 1/2 to 3/4 the distance up the stem; blades mostly 1/4 to 3/4 the length of the stem, flat and grasslike, 1–3 mm wide; sheaths green, with broad membranous margins extending beyond the sheath to form 2 low, rounded or truncate auricles. **Inflorescence** sparingly branched from the base, comprised of (2)3–5 hemispheric glomerules each containing 3–8 flowers, the glomerules occasionally crowded into one large many-flowered head. **Flowers** brown, 4–6 mm long; tepals acute to mucronate, the outer ones mucronate more often than the inner, the margins hyaline and glistening; stamens 6. **Capsules** cylindric-ellipsoid, 3-locular, shorter than the perianth, rounded to retuse below the beak; seeds cylindric-ellipsoid, 0.3–0.5 mm long, apiculate on the ends. June–Aug.

HABITAT Springs, boggy places and shores, where water is fresh.

WETLAND STATUS
GP FACW | MIDW FACW | WMTN FACW

Juncus marginatus Rostk.
GRASS-LEAF RUSH

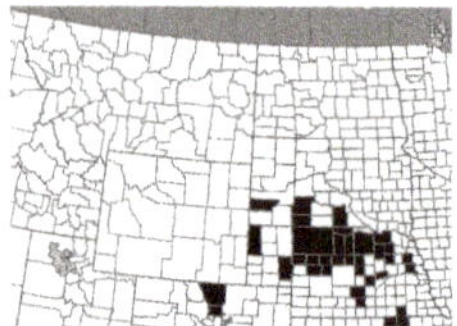

DESCRIPTION Rhizomatous perennial 2–5 dm tall; **culms** solitary or in small tufts, bulbous-thickened at the base. **Leaves** basal and cauline, 2–5 per culm, the blades flat, grasslike, 1–3 mm wide; sheaths green, membranous-margined, extended into 2 low, rounded auricles at the summit. **Inflorescence** open and spreading or sometimes narrow and congested, with 4-many hemispheric to subglobose glomerules containing several to many flowers. **Flowers** pale brown, 2–3.5 mm long; outer tepals broadly lanceolate, acute to acuminate, often awn-tipped; inner tepals elliptic-obovate, broader and longer than the outer ones, with broad, scarious margins, blunt-tipped; stamens 3. **Capsules** obovoid, nearly 3-locular, shorter than to equaling the tepals, rounded to retuse and apiculate at the apex; seeds ellipsoid or asymmetrically so, 0.3–0.5 mm long, apiculate at the ends. June–Sept.

HABITAT Wet meadows, spring borders and streambanks where water is fresh, often where sandy.

WETLAND STATUS
GP FACW | MIDW FACW | WMTN FACW

Juncus longistylis
LONG-STYLE RUSH

Juncus marginatus
GRASS-LEAF RUSH

Juncus nodosus L.

JOINTED RUSH

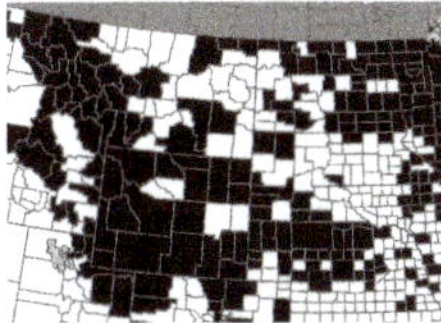

DESCRIPTION Slender rhizomatous perennial 1.5–5 dm tall. **Leaves** cauline and basal, the upper ones and the main involucre usually overtopping the inflorescence; blades terete, hollow, nodulose-septate, 0.5–1.5(2) mm wide when pressed; sheaths green, the margins green, becoming yellowish and membranous toward the summit; auricles membranous, yellowish, 0.5–1.5 mm long. **Inflorescence** usually oblong, 2–6 cm long, sparingly branched, the branches ascending, each branch terminated with a glomerule; glomerules (1)2-several, spherical at maturity, many-flowered, 1 cm or less across. **Flowers** greenish to brown, 3–4 mm long; tepals attenuate, the margins very narrowly hyaline, the outer and inner equal in length or nearly so; stamens 6. **Capsules** subulate, 1-locular, exceeding the perianth, dehiscing from the base and remaining coherent at the tip; seeds cylindric-ellipsoid, pale brown to brown, ca. 0.4 mm long, apiculate on the ends. July–Sept.

HABITAT Wet meadows, springs, fens, shores and streambanks, where water is fresh.

WETLAND STATUS
GP OBL | MIDW OBL | WMTN OBL

NOTE See notes under *J. alpinus* and *J. torreyi.*

Juncus nodosus
JOINTED RUSH

Juncus scirpoides Lam.

NEEDLE-POD RUSH

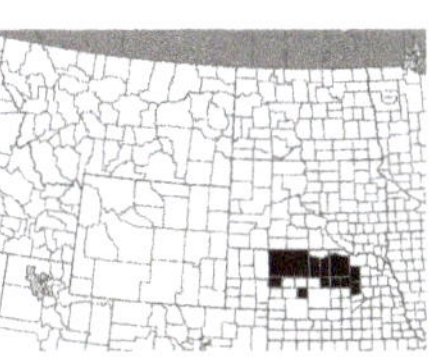

DESCRIPTION Culms solitary or tufted from stout, whitish rhizomes, 2–8 dm tall. **Leaves** mostly 2–3 per culm, the uppermost leaf not surpassing the inflorescence; blades terete, hollow, nodulose-septate, mostly 1–2 mm wide; sheaths green or the lower ones reddish, membranous-margined above, prolonged as membranous auricles (1)2–3 mm long. **Inflorescence** compact and narrow to open with a few spreading branches, (1)2–12 cm long; glomerules (1)2–15, many-flowered, spherical to somewhat lobed, with dense subclusters of flowers, 6–12 mm across. **Flowers** greenish to eventually dull brown, 2.5–4 mm long; tepals attenuate, the outer considerably longer and with sharper, more rigid tips than the inner; stamens 3. **Capsules** subulate, 1-locular, equaling to slightly exceeding the outer tepals, dehiscing from the base and coherent above; seeds ovoid, brown, 0.3–0.5 mm long, apiculate on the ends. June–Sept.

HABITAT Sandy wet meadows, shores and streambanks.

WETLAND STATUS
GP FACW | MIDW FACW

Juncus scirpoides
NEEDLE-POD RUSH

Juncus tenuis Willd.
PATH RUSH

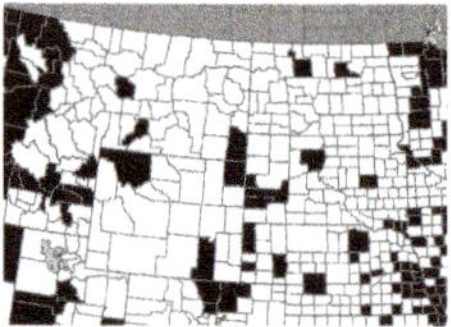

DESCRIPTION Quite similar to *Juncus dudleyi* and *J. interior* which are sometimes treated as variants of this species, differing chiefly as follows: **Foliage** dull to bright green and soft-textured; leaf sheaths with broad membranous margins which are prolonged as flaplike auricles 1–5 mm long. **Inflorescence** variable, from densely flowered and congested to open with sparsely flowered, often secund branches; bracteoles acute to acuminate. June–Aug.

HABITAT Moist to dry habitats, often where disturbed or where soil is compacted.

WETLAND STATUS
GP FAC | MIDW FAC | WMTN FW

Juncus torreyi Cov.
TORREY RUSH

DESCRIPTION Similar to *Juncus nodosus* but more robust, 2–8 dm tall, the stems arising singly from tuberiferous rhizomes. **Leaves** mostly cauline, the upper ones frequently overtopping the inflorescence; blades terete, hollow, nodulose-septate, 1–3 mm wide when pressed; sheaths green, the margins white-hyaline; auricles white-hyaline, 1–3(4) mm long. **Inflorescence** of 1-many glomerules, these spherical, densely flowered, mostly 1 cm or more across. **Flowers** greenish to brown, 3–5 mm long; tepals attenuate, with narrow, white-hyaline margins, the outer tepals distinctly longer than the inner; stamens 6. **Capsules** subulate, 1-locular, about equal to or exceeding the perianth, dehiscing from the base; seeds cylindric-ellipsoid, pale brown to brown, ca, 0.4 mm long. July–Sept.

HABITAT Shores, streambanks, wet meadows, springs and ditches.

WETLAND STATUS
GP FACW | MIDW FACW | WMTN FACW

NOTE A hymenopteran larva is responsible for the formation of bizarre galls which are often seen in this species and sometimes in *J. nodosus*. The galls appear as clusters of overlapping, bractlike leaves, yellow and red in color, typically in the position of the inflorescence on stunted culms.

Juncus tenuis
PATH RUSH

Juncus torreyi
TORREY RUSH

Juncus vaseyi Engelm.
VASEY'S RUSH

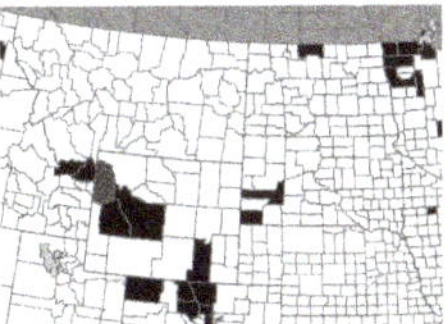

DESCRIPTION Tufted perennial 2–6 dm tall. **Leaves** mostly basal, arising from well within the lower 1/4 of the stem; blades to 3 dm long, usually not surpassing the stem, 0.5–1 mm wide, terete, solid, not nodulose-septate, narrowly channeled on the upper surface; sheaths green or reddish, the margins and auricles membranous, pale green or pale brown. **Inflorescence** rather compact and crowded, 1–4 cm long; main involucre shorter than to greatly surpassing the inflorescence; bracteoles obtuse, scarious, 1–2 mm long. **Flowers** solitary, sessile or on short pedicels, greenish to light brown, 4–5.5 mm long; tepals acute, with narrow hyaline margins, the outer ones averaging only slightly longer than the inner; stamens 6. **Capsules** cylindric, almost completely 3-locular, exceeding the perianth; seeds fusiform, brown, 0.8–1.3 mm long, including the membranous taillike appendages at both ends. July–Aug.

HABITAT Wet meadows and shores.

WETLAND STATUS
GP FACW | MIDW FACW

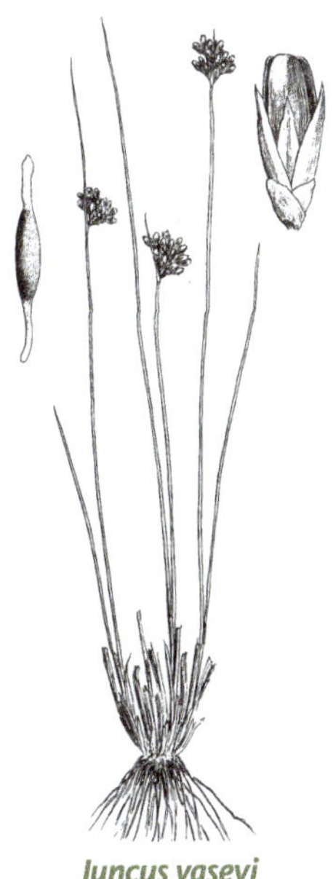

Juncus vaseyi
VASEY'S RUSH

Juncaginaceae
arrow-grass family

Triglochin ARROW-GRASS

Perennial, grasslike herbs, tufted from a creeping rootstock. **Leaves** all basal, slender, terete or somewhat flattened, sheathing at the base, ligulate. **Inflorescence** a terminal, ebracteate, spikelike raceme; pedicels short and decurrent on the scape axis. **Flowers** perfect, regular; tepals 6, in 2 series, deciduous; stamens 6, anthers sessile, nearly as large as the tepals; carpels 3 or 6, each with an apical stigma, ovary superior, eventually splitting into 3 or 6, 1-seeded segments. Plants of wet, often alkaline or saline habitats.

1 Carpels and stigmas 3; fruits linear-clavate, the fruit axis 3-winged ***T. palustris***
1 Carpels and stigmas 6; fruits short-cylindric, the fruit axis terete ***T. maritima***

Triglochin maritima L.
SEASIDE ARROW-GRASS

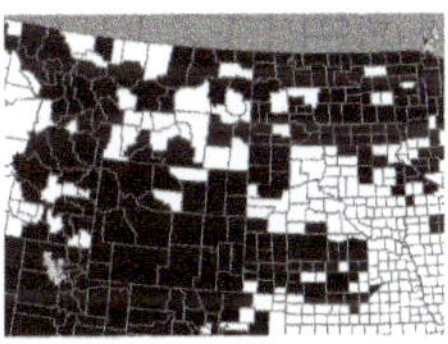

DESCRIPTION Plants (2)3–8 dm tall, tufted from a thick creeping rootstock. **Leaves** erect to curved-spreading, somewhat flattened, especially toward the base, 1/2 to fully the length of the plant, 1.5–3(4) mm wide; ligules entire, acute to rounded, 1–

Triglochin maritima
SEASIDE ARROW-GRASS

5 mm long. **Racemes** densely flowered, (5)10-40 cm long; pedicels 1–6 mm long. **Flowers** 2–3 mm across; tepals 1.5–2 mm long; carpels 6, stigmas 6. **Fruits** short-cylindric, 2–5 mm long, 1–3 mm thick, splitting into 6 oblong segments, the fruit axis terete. June–Aug.

HABITAT Wet meadows, fens, alkaline or saline marshes, ditches and flats, streambanks and other wet places.

WETLAND STATUS
GP OBL | MIDW OBL | WMTN OBL

Triglochin palustris L.
MARSH ARROW-GRASS

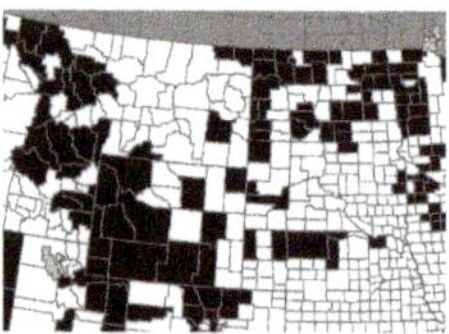

DESCRIPTION Small, slender plants 1.5–4 dm tall. **Leaves** erect, terete, 1/2 to 3/4 the length of the plant, very slender, 0.5–2 mm wide; ligules 2-lobed, 0.5–1.5 mm long. **Racemes** loosely flowered, 10–25 cm long; pedicels 2–5 mm long. **Flowers** small, 1–2 mm across; tepals 1–2 mm long; carpels 3, stigmas 3. **Fruits** linear-clavate, 5–8 mm long, ca. 1 mm thick, splitting upward from the base into 3 elongate segments, the fruit axis 3-winged. Late June–Sept.

HABITAT Springs, fens and seepage areas, often where alkaline or saline.

WETLAND STATUS
GP OBL | MIDW OBL | WMTN OBL

Triglochin palustris
MARSH ARROW-GRASS

Liliaceae *lily family*

Lilium philadelphicum L.
WOOD-LILY

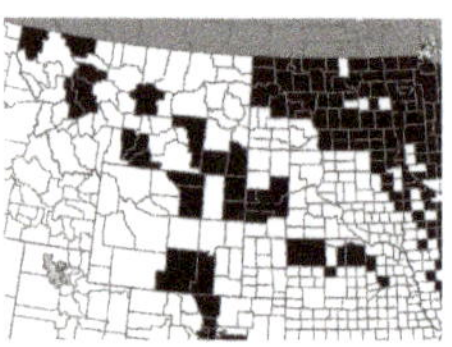

DESCRIPTION Erect, glabrous perennial 2–9 dm tall, from a scaly bulb. **Leaves** cauline, sessile, mostly alternate below to opposite or whorled above, linear to linear-lanceolate, 3–8 cm long, 3–9 mm wide, parallel-veined, blunt-tipped. **Flowers** 1–3, perfect, terminal, erect, large and showy, on peduncles 1–8 cm long; tepals orange-red, yellow and dark-spotted toward the clawed base (very rarely the tepals solid yellow), 4–8 cm long, 0.8–2.8 cm wide; stamens and pistil prominent, abut equaling the perianth; anthers 6–10 mm long; stigma 3-lobed, style elongate, ovary superior. **Capsule** loculicidal, oblong, 2.5–4 cm long; seeds flat, closely packed. Late June–July.

HABITAT Wet meadows, low prairie, boggy places, seepage areas and ditches, also prairie hillsides, woodlands and thickets.

WETLAND STATUS
GP FACU | MIDW FAC | WMTN FAC

Lilium philadelphicum
WOOD-LILY

Orchidaceae *orchid family*

Perennial herbs from rather fleshy, strongly mycorhizal roots. **Leaves** simple, cauline and alternate or mostly basal, sessile and usually sheathing the stem, parallel-veined. **Inflorescence** of 1 or 2 terminal blossoms or of few to many flowers in a terminal bracteate spike or raceme. **Flowers** perfect, strongly irregular, epigynous, very showy in some spp.; perianth comprised of 6 segments, although this is sometimes obscured by fusion; calyx consisting of the uppermost segment and the 2 outermost lateral segments, the sepals sometimes resembling the lateral petals, the lateral sepals free or (in *Cypripedium*) connate to form a single appendage below the lip or (in *Spiranthes*) connivent with the lateral petals to form a hood over the lip; corolla white or otherwise colored, the 2 lateral petals alike, the third petal (the lowermost perianth segment) typically modified and referred to as the lip; stamens 1 or 2, attached to the style and along with the style and stigma forming a short to prominent column, the stigma borne on the lower side near the base of the column, pollen shed in waxy pollinia; ovary inferior, 1-celled, with 3 parietal placentae, or 3-celled, maturing into a many-seeded capsule, the **seeds** very numerous and minute.

The second largest family (or largest by some estimates) of vascular plants, with most representatives occurring in the tropics where many grow as epiphytes. Within the Orchidaceae are a few of our most beautiful and rarest wildflowers, such as the lady's-slippers (*Cypripedium* spp.) and the prairie fringed orchid (*Platanthera praeclara*, listed as a threatened species by the federal government and state of Nebraska).

1 Flowers 1 or 2 per stem, the lip much inflated and pouchlike **_Cypripedium_**
1 Flowers few to many (2–12 in *Liparis*) in a bracteate spike or raceme, the lip not inflated 2

2 Perianth prolonged backward into a spur 3
2 Perianth not spurred 4

3 Spur pouchlike, 2–3 mm long **_Dactylorhiza_**
3 Spur cylindric, 4–50 mm long **_Platanthera_**

4 Flowers in a loose raceme; foliage leaves consisting of a pair of basal leaves **_Liparis_**
4 Flowers in a twisted spike; foliage leaves best developed at the base and reduced upward on the stem **_Spiranthes_**

Cypripedium LADY'S-SLIPPER

Erect plants from coarsely fibrous roots, with rather broad, cauline **leaves** and 1 or 2 large showy flowers on each stem; **stems** simple, usually clumped in groups of few to several. **Flowers** terminal each subtended by a foliaceous bract; lateral sepals resembling the lateral petals in color and texture, connate to form a single, apically bidentate or entire appendage below the lip, upper sepal similar to the lower appendage but with an entire apex; lateral petals free and spreading, lip much inflated and pouchlike, projecting forward; column bent over the orifice of the lip, stamens 2, one on each side of the column, a petaloid staminode projected forward over the column and exceeding it; ovary elongate and curved.

1 Lip yellow **_C. parviflorum_**
1 Lip predominantly white, with pink or purple markings 2

2 Sepals and lateral petals white; lip 3–5 cm long **_C. reginae_**
2 Sepals and lateral petals greenish; lip 1.5–2 cm long **_C. candidum_**

Cypripedium candidum Muhl.
WHITE LADY'S-SLIPPER

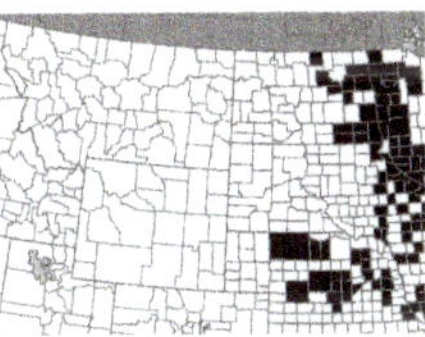

DESCRIPTION Stems 1.5–3 dm tall, glandular-hispid above. **Leaves** appressed-ascending, overlapping and strongly sheathing below, narrowly elliptic, 5–13 cm long, 2–4 cm wide, acute, sparsely glandular-pubescent, reduced to sheathing scales toward the base of the stem. **Flowers** 1 per stem, the subtending leaflike bract erect, 3–6 cm long; sepals and lateral petals greenish, usually purple-striate, sepals 1.5–3 cm long, the upper sepal ovate-lanceolate, acuminate, the lateral sepals connate below the lip with a slightly bidentate tip; lateral petals linear-lanceolate, acute to attenuate, sometimes twisted, 1.5–3.5 cm long; lip white with purple veins, 1.5–2 cm long; ovary densely glandular-pubescent, 1.5–4 cm long. Late May–June.

HABITAT Wet meadows and low prairie, where undisturbed.

WETLAND STATUS
GP OBL | MIDW OBL

STATUS Neb threatened.

Cypripedium parviflorum Salisb.
YELLOW LADY'S-SLIPPER

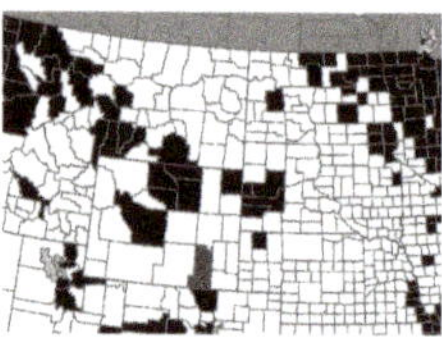

DESCRIPTION Stems 1.5–6 dm tall, pubescent, the hairs often glandular. **Leaves** ascending, sheathing at the base, elliptic, 5–18 cm long, 2–7 cm wide, acute to acuminate, sparsely pubescent. **Flowers** 1 per stem or less often 2, each subtended by an erect leaflike bract 2.5–6 cm long; sepals and lateral petals greenish-yellow to purplish-brown, sepals 2–6 cm long, the upper sepal lanceolate to ovate-lanceolate, acuminate, the lateral sepals connate below the lip, lanceolate, bidentate at the tip; lateral petals lanceolate to linear-lanceolate, acute to attenuate, usually twisted 1–several times, 2–5 cm long; lip yellow, often purple-veined and purplish-dotted around the orifice, 1.5–4 cm long; ovary strongly glandular-pubescent, 1.5–3 cm long. June–July.

SYNONYM *Cypripedium calceolus* L.

HABITAT Wet meadows, bogs, swampy areas and moist forest.

WETLAND STATUS
GP FACW | MIDW FACW | WMTN FACW

NOTE In North America, *C. parviflorum* has been divided into several varieties on the basis of the lip size. Great Plains and e USA species are mostly referable to var. *pubescens* (Willd.) Knight. In the Great Lakes region, var. *makasin* is also present; plants of the n Rocky Mtns are mostly treated as var. *parviflorum.*

Cypripedium reginae Walt.
SHOWY LADY'S-SLIPPER

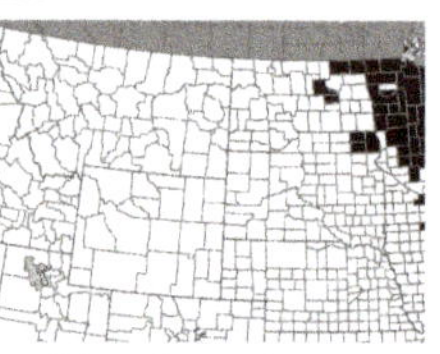

DESCRIPTION Stems 4–10 dm tall, glandular-hirsute. **Leaves** spreading-ascending, sheathing at the base, broadly elliptic, 10–25 cm long, 4–12 cm wide, abruptly acuminate, nearly glabrous to hirsute. **Flowers** 1 or 2 on a stem, the subtending leaflike bract 6–12 cm long; sepals and lateral petals white, sepals 3–4 cm long, the upper sepal broadly elliptic to obovate, obtuse to rounded, the lateral sepals completely fused to form a single, entire appendage under the lip, obtuse at the apex; lateral petals broadly to narrowly lanceolate, obtuse to rounded, about as long as the sepals; lip white, streaked and spotted with pink or purple, 3–5 cm long; ovary glandular-pubescent, 3.5–5 cm long. June–July.

HABITAT Bogs, swamps and springs.

WETLAND STATUS
GP FACW | MIDW FACW

Cypripedium candidum
WHITE LADY'S-SLIPPER

Cypripedium parviflorum
YELLOW LADY'S-SLIPPER

Dactylorhiza viridis (L.) R.M. Bateman, Pridgeon & M.W. Chase

LONG-BRACTED ORCHID

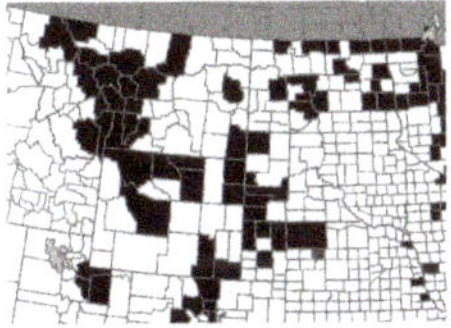

DESCRIPTION Plants 2–5 dm tall, glabrous. **Leaves** elliptic to elliptic-oblanceolate or elliptic-obovate, 4–12 cm long, 1–3(5) cm wide, acute to rounded, reduced in size upward. **Flowers** usually many, ascending, greenish, sometimes tinged with purple, small, not showy; sepals green, 4–6 mm long, the upper sepal broadly ovate and hoodlike, the lateral sepals broadly lanceolate, incurved; petals green, the lateral ones lanceolate, curved upward, shorter than the lateral sepals and nearly hidden by them; lip oblong, 2- or 3-toothed at the tip, the middle tooth shorter than the lateral 2 or obsolete; spur pouchlike, 2–3 mm long, projected forward under the lip; ovary cylindric, 7–11 mm long. June–July.

SYNONYMS *Coeloglossum viride* (L.) Hartman *Habenaria bracteata* (Willd.) R.Br., *Habenaria viridis* (L.) R.Br.

HABITAT Wet meadows, seepage areas, moist or wet woods and thickets.

WETLAND STATUS
GP FACU | MIDW FAC | WMTN FAC

Liparis loeselii (L.) Rich.

FEN-ORCHID

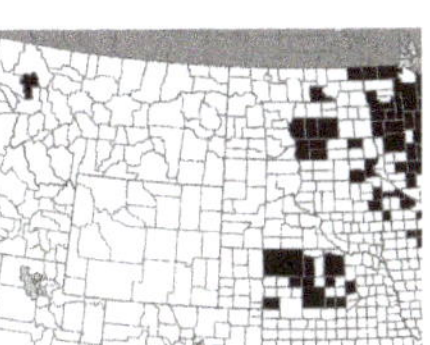

DESCRIPTION Small, erect, glabrous plant 1–2.5 dm tall, arising from a solid bulb. Foliage **leaves** consisting of a pair of basal leaves, ascending, sheathing at the base, shiny, elliptic to lanceolate, 5–15 cm long, 1.5–3 cm wide, obtuse to rounded, subtended below by 1-few basal scale leaves. **Flowers** 2–12, yellowish-green, small, erect, in an open bracteate raceme 2–8 cm long, the bracts minute; sepals narrowly lanceolate and spreading, 4–6 mm long; lateral petals linear, often twisted and bent forward under the lip, 3–5 mm long; lip dilated above the narrow base, decurved, 4–5 mm long; anther 1, terminal on the broad-based, winged column. **Capsules** persistent, short-cylindric, 7–11 mm long. June–July.

HABITAT Swampy woods, fens, boggy places.

WETLAND STATUS
GP OBL | MIDW FACW | WMTN FACW

Liparis loeselii
FEN-ORCHID

Cypripedium reginae
SHOWY LADY'S-SLIPPER

Dactylorhiza viridis
LONG-BRACTED ORCHID

Platanthera BOG ORCHID

Usually stout, erect plants from a cluster of tuberous roots. **Leaves** cauline, ascending to appressed, sheathing and mostly overlapping at the base, reduced upward on the stem, the lowermost often scalelike and sheathing. **Flowers** several to many in a loose to congested, terminal bracteate spike, showy or not, white or greenish; perianth segments free (in those included here); sepals greener and thicker-textured than the white to greenish petals; lip ovate-lanceolate and entire or 3-lobed and fringed, prolonged backward into a spur, the spur commonly curved; stamen 1, the anther attached on top of the short column.

1 Flowers greenish, small, not showy, the lip entire **P. aquilonis**

1 Flowers white, large and showy, the lip 3-lobed and fringed **P. praeclara**

Platanthera aquilonis Sheviak
NORTHERN GREEN ORCHID

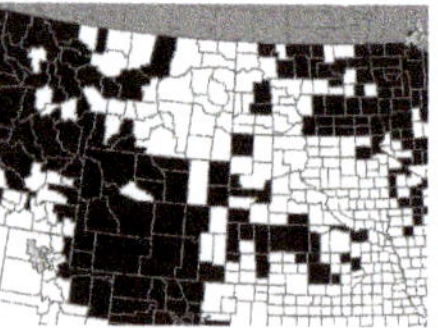

DESCRIPTION Plants 1.5–7 dm tall, glabrous. **Leaves** lanceolate to oblanceolate, 5–20 cm long, 1–3.5(5) cm wide, acute to rounded or mucronate, reduced in size upward. **Spike** elongate, usually dense, 4–20 cm long; bracts lanceolate, acute to blunt, 6- 20 mm long, reduced upward. **Flowers** usually numerous, erect, greenish, small, not showy; sepals green, 2–4 mm long, the upper sepal broadly ovate and hoodlike, the lateral sepals lanceolate to ovate and spreading; petals green to greenish-white, the lateral petals lanceolate, curved upward, 1.5–2.5 mm long; lip ovate-lanceolate, 4–7 mm long, entire; spur inconspicuous, curved forward under the lip, about equaling the lip; ovary cylindric, 6–12 mm long. Mid June–Aug.

Platanthera aquilonis
NORTHERN GREEN ORCHID

SYNONYMS *Platanthera hyperborea* (L.) Lindl. (in part), *Habenaria hyperborea* (L.) R.Br. (in part)

HABITAT Springs, bogs, fens, stream margins, seepage areas, fresh wet meadows and roadside ditches.

WETLAND STATUS
GP OBL | MIDW FACW | WMTN FACW

ADDITIONAL SPECIES *Platanthera huronensis* (Nutt.) Lindl. is similar to *P. aquilonis* and reported from sw SD and e Wyo, and more common in the Rocky Mtn and Great Lakes regions. Both species were formerly included in the *P. hyperborea* complex, that species now considered to be restricted to Greenland.

Platanthera praeclara Sheviak & Bowles
WESTERN PRAIRIE FRINGED ORCHID

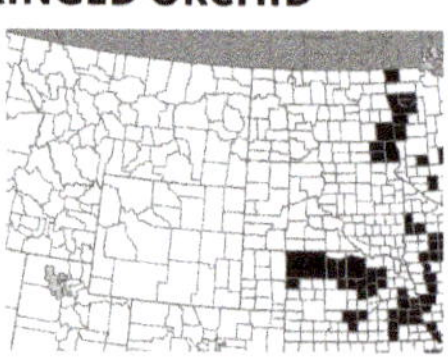

DESCRIPTION Plants 3–8 dm tall, glabrous. **Leaves** lanceolate to broadly linear, 8–20 cm long (including the strongly sheathing base), 1–4 cm wide, blunt-tipped. **Spike** cylindric, 5–15 cm long; bracts lanceolate, blunt, 1–4 cm long. **Flowers** several to many, spreading, white, large and showy; sepals 9–12 mm long, broadly ovate to obovate; lateral petals broadly obovate-cuneate, erose, 10–15 mm long; lip deeply 3-

Platanthera praeclara
WESTERN PRAIRIE FRINGED ORCHID

lobed with fringed margins, 1.5–2.5 cm long and about as wide, the lobes flabelliform, deeply dissected, the spur conspicuous, 2–5 cm long, usually curved; ovary elongate, 2–3 cm long. Late June–July.

SYNONYMS *Habenaria leucophaea* (Nutt.) A. Gray, *Platanthera leucophaea* (Nutt.) Lindl.

HABITAT Wet meadows and low prairie, often where sandy; also in drier prairie.

WETLAND STATUS
GP UPL | MIDW UPL

STATUS Neb threatened, USA threatened.

Spiranthes LADIES'-TRESSES

Slender, erect, mostly glabrous plants from a cluster of tuberous roots. **Leaves** best developed at the base, reduced and bractlike upward, the cauline leaves appressed and sheathing. **Flowers** small, usually many, arranged in usually (1)2–4 spirally twisted rows in a dense, bracteate spike, white or cream-colored; sepals and lateral petals similar, the lateral petals connivent with all 3 sepals or only with the upper one to form a hood over the lip and column; lip oblong or ovate, folded upward near the middle so that the margins embrace the column, curved downward beyond the middle, with a pair of protuberances or thickenings at the base; anther 1, borne on the back of the short column.

1 Lateral petals connivent with all 3 sepals to form a hood, the lateral sepals not free; lip with a distinct constriction at ca. 3/4 of its length resulting in a round-ovate terminal lobe, the base of the lip with a pair of lateral thickenings but these not projecting* ***S. romanzoffiana***

1 Lateral petals connivent with only the upper sepal to form a hood, the lateral sepals free on the sides; lip not constricted or only slightly so near the middle, without a distinct terminal lobe, the base of the lip with a pair of backward projecting protuberances 2

2 Flowers in a single spiral in the spike 3

2 Flowers in 2–4 spirals in the spike 4

3 Pubescence in the inflorescence with knoblike, glandular tips; lip smooth to glandular on the underside ***S. cernua***

3 Pubescence in the inflorescence sharp-tipped; lip prominently papillate on the underside ***S. vernalis***

4 Leaves becoming brown and dried by flowering time; lip not constricted near the middle ***S. magnicamporum***

4 Leaves mostly remaining green at flowering time; lip slightly constricted near the middle ***S. cernua***

*If working with dried material, soak flowers in a wetting solution for several minutes or boiled in water to facilitate dissection.

ADDITIONAL SPECIES **Ute ladies'-tresses** (*Spiranthes diluvialis* Sheviak) is a rare species (federally listed as threatened, state threatened in Neb) of nw Neb and locally westward to Washington and Nevada. In the Great Plains region, plants occur on sand or gravel bars along rivers and streams and in wet meadows that are subirrigated, often seasonally flooded, and remain moist into the summer; soils are usually sandy loams, but also include sands, loams and silt loams.

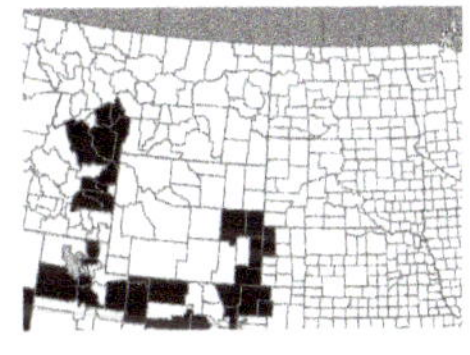

The inflorescence is a loose spike 3–15 cm long of numerous, small white to ivory flowers arranged in a gradual spiral. The lip petal is oval to lance-shaped, narrowed at the middle, and has crispy-wavy margins. Sepals are separate or fused only at the base (not forming a hood-like structure) and are often spreading at their tips. Plants are similar to *Spiranthes romanzoffiana,* but that species has deeply constricted lip petals, sepals fused for at least 1/2 their length into a hood-like tube, and a densely congested inflorescence.

Spiranthes cernua (L.) Rich.
NODDING LADIES'-TRESSES

DESCRIPTION Plants 1–4(6) dm tall, glabrous below, glandular-pubescent in the spike, the hairs with knoblike tips; roots slender but fleshy. **Leaves** mostly green at flowering time, the basal ones linear to

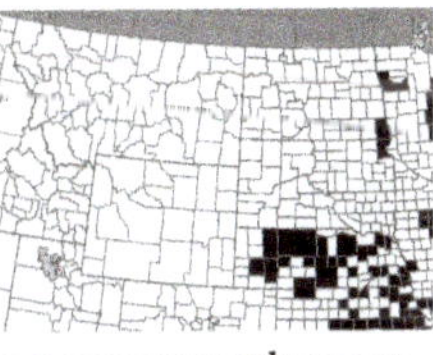

lanceolate, mostly 10–25 cm long, 3–20 mm wide, acute to acuminate, the cauline leaves reduced and becoming bractlike upward. **Spike** with (1)2–4 spiral rows of flowers, 3–18 cm long, the bracts ovate to ovate-lanceolate with a slender, acuminate tip, 8–25 mm long. **Flowers** white, unscented or only weakly scented when fresh; perianth segments sparingly to moderately pubescent on the outside; sepals oblong-lanceolate, 6–10 mm long, the lateral ones free, projected forward and somewhat spreading; lateral petals connivent with the upper sepal to form a hood, linear, acute to obtuse, about equaling the sepals; lip white with a thickened, yellow-green center, oblong to oblong-ovate when flattened, 6–10 mm long, slightly constricted at the middle and often dilated basally curved, downward and abruptly recurved at the rounded tip, crisped to crenulate on the margins toward the tip, the base of the lip with a pair of backward-projecting protuberances. Aug–Oct.

SYNONYM *Spiranthes incurva* (Jennings) M.C. Pace

HABITAT Wet meadows, floodplains and moist prairies.

WETLAND STATUS
GP FACW | MIDW FACW

Spiranthes cernua
NODDINGLADIES'-TRESSES

Spiranthes magnicamporum Sheviak
GREAT PLAINS LADIES'-TRESSES

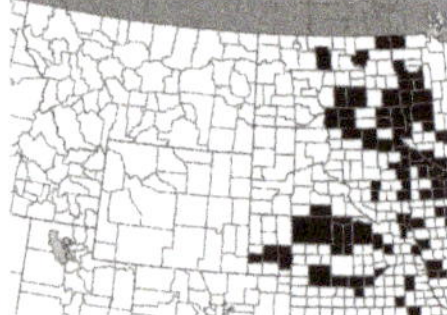

DESCRIPTION Very similar to *Spiranthes cernua,* differing chiefly as follows: Roots more thickened and tuberous. **Leaves** senescent by flowering time, even the cauline ones brown and dried. **Spike** with 2–4 spiral rows of flowers, the bracts to 30 mm long. **Flowers** white to cream-colored, strongly fragrant with the scent of coumarin (similar to vanilla) when fresh; sepals 7–14 mm long; lip 6–1 1 mm long, not constricted near the middle and not dilated at the base, rather evenly curved for its entire length, not abruptly recurved near the tip, the basal protuberances more prominent than in the preceding. Late July–Oct.

HABITAT Wet meadows, moist to dry prairies, ditches and floodplains, often where sandy.

WETLAND STATUS
GP FAC | MIDW FAC | WMTN FAC

Spiranthes magnicamporum
GREAT PLAINS LADIES'-TRESSES

Spiranthes romanzoffiana Cham.

HOODED LADIES'-TRESSES

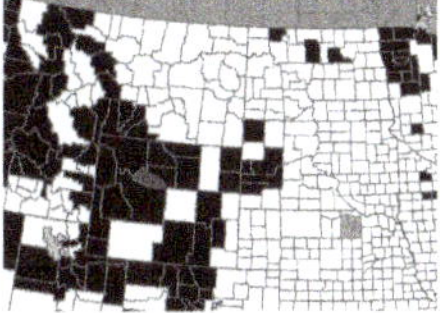

DESCRIPTION Plants 1–3 dm tall, almost entirely glabrous; roots long and fleshy. **Leaves** usually green at flowering time, the basal leaves ascending, linear to narrowly oblong, 8–15 cm long, 2–9 mm wide, blunt, the cauline leaves bractlike. **Spike** with 1–3(4) rows of flowers, 3–8 cm long, the bracts ovate-lanceolate, 10–20 mm long, acuminate. **Flowers** white or cream-colored; perianth segments only slightly glandular-pubescent on the outside, if at all; sepals and the lateral petals alike, ovate-lanceolate, all connivent to form a hood 7–12 mm long, the lateral sepals with only the tips free; lip ovate when flattened, with obscure thickenings at the base which do not project backward; sharply curved downward, recurved at the erose tip, about as long as the hood. Late July–Aug.

HABITAT Springs, fens, bogs, seepage areas.

WETLAND STATUS
GP OBL | MIDW OBL | WMTN FACW

Spiranthes vernalis Engelm. & A. Gray

SPRING LADIES'-TRESSES

DESCRIPTION Plants 2–7 dm tall, finely pubescent above with whitish or reddish-brown, septate hairs, these eglandular, sharp-pointed; roots coarse, fusiform. **Leaves** basal or a few cauline, usually some persistent at flowering time, linear to narrowly lanceolate, 5–30 cm long, to 16 mm wide, acuminate, the basal leaves sheathing the stem, the cauline leaves, when present, much reduced and bractlike upward. **Spike** dense to rather loose with a single twisted spiral of flowers, 3–15 cm long, the bracts ovate to oblong-lanceolate, 7–23 mm long, acuminate, scarious- margined. **Flowers** yellowish to greenish-white, often fragrant; perianth segments pubescent on the outside; sepals linear-lanceolate, 6–8 mm long, the upper acute to obtuse, the lateral ones free, projected forward, acute; lateral petals connivent with the upper sepal to form a hood, linear to linear-elliptic, 5–9 mm long, obtuse; lip broadly ovate to oblong-ovate, 4.5–8 mm long, widest just below the middle, arcuate-recurved, crenulate-wavy on the margins, strongly papillose on the underside, the basal protuberances prominent, projected backward and uncurved, pubescent. Aug–Sept.

HABITAT Wet meadows, moist prairies, ditches, and floodplains, usually where sandy.

WETLAND STATUS
GP FACW | MIDW FAC

Spiranthes romanzoffiana
HOODED LADIES'-TRESSES

Spiranthes vernalis
SPRING LADIES'-TRESSES

Poaceae *grass family*

Perennial or annual, rhizomatous or tufted plants with long, linear, parallel-veined **leaves** arranged alternately in 2 vertical ranks on a terete (rarely compressed), usually hollow stem with swollen, solid nodes; leaf sheaths commonly open longitudinally, the margins overlapping (sheaths tubular with the margins united in *Bromus* and *Glyceria*); a membranous or hairy ligule projecting between the blade and the culm from the summit of the sheath, this sometimes absent. **Flowers** much reduced, perfect or rarely unisexual (the plants then monoecious or dioecious), each subtended by 2 bracts, the larger one (lemma) containing the flower, the smaller one (palea) covering the flower, palea rarely absent, the lemma and palea often adhering to and enclosing the ripe grain, the flower and its surrounding lemma and palea collectively termed the floret; perianth rudimentary, represented by no more than 2 (in taxa of this region) inner, obscure scales (lodicules), these usually undetectable; stamens usually 3 (1–6); ovary superior, 1-celled, never enclosed in a sac; styles bifid, the stigmas plumose; ovule 1, usually parietal. **Florets** arranged in spikelets, these consisting of 1-many florets which are sessile on a shortened axis (rachilla), the lowest 2 bracts of the spikelet empty (the glumes), these rarely absent, often unequal, the lowermost of the glumes usually smaller and called the first glume, the upper one called the second glume, the glumes seldom equal and opposite. **Spikelets** arranged in a variety of inflorescence types, the most common being a panicle in which the spikelets are individually pedicelled in a branching inflorescence, or less often, the spikelets sessile on a rachis in one or more spikes; if solitary, the spike terminal, or if more than one, the spikes 1-sided (the spikelets all attached to one side of the rachis) and arranged in a raceme or panicle; spikelets disarticulating either above or below the glumes at maturity, the glumes remaining attached in the inflorescence if disarticulation occurs above the glumes, or the glumes falling with the florets (and sometimes with other parts of the inflorescence attached) if disarticulation is below the glumes.

Poaceae is the second largest family in our region (after Cyperaceae), with most species characteristic of prairie, but many also occurring in wetland and woodland habitats. The family is also rife with species that occur in a broad array of moisture regimes and many of those are included here.

---

1 Spikelets containing 2 or more functional florets (many may contain only 1 floret in *Catabrosa*) 2

1 Spikelets containing 1 functional floret 16

2 Tall, stout reeds over 2 m tall, with a large plumelike panicle; rachillas hairy ***Phragmites***

2 Plants without the above combination of characters 3

3 Inflorescence a simple, terminal spike, the spikelets sessile on an unbranched rachis ***Elymus***

3 Inflorescence of variously branched types, usually a panicle, if appearing spikelike, then the spikelets individually pedicelled 4

4 Glumes both as long as or exceeding the lowest floret, often as long as the entire spikelet; lemmas awned from the back or awnless 5

4 Glumes, or at least the first one, shorter than the first floret; lemmas awned from the tip or awnless 8

5 Lemmas awned from below the middle of the back ***Deschampsia***

5 Lemmas awnless 6

6 Florets bearded on the callus; spikelets disarticulating above the glumes ***Scolochloa***

6 Florets glabrous on the callus; spikelets disarticulating below the glumes 7

7 Glumes alike, both inflated and broadly boat-shaped; spikelets nearly round in outline ***Beckmannia***

7 Glumes dissimilar, the second much broader than the first; spikelets not round ***Sphenopholis***

8 Spikelets all sessile or very short-pedicelled in an inflorescence of several to many straight branches, much of the inflorescence often enclosed by leaf sheaths **_Leptochloa_**

8 Spikelets, or at least some of them, obviously pedicelled, borne in an open to contracted panicle 9

9 Spikelets unisexual (plants dioecious), in a simple, contracted panicle; low rhizomatous perennial of saline or alkaline soils **_Distichlis_**

9 Spikelets bisexual, in an open to contracted panicle; plants of various habits and habitats 10

10 Lemmas prominently 3-nerved 11

10 Lemmas faintly to prominently 5- to 7-nerved 12

11 Spikelets 2-flowered (many may be 1-flowered); lemmas truncate and erose at the tip **_Catabrosa_**

11 Spikelets several- to many-flowered; lemmas acute-tipped **_Eragrostis_**

12 Callus of the florets bearded with short, stiff hairs **_Scolochloa_**

12 Callus not bearded, although cobwebby hairs may be present at the base of the lemma 13

13 Lemmas awned **_Bromus_**

13 Lemmas awnless 14

14 Lemmas prominently 7-nerved **_Glyceria_**

14 Lemmas faintly 5-nerved 15

15 Lemmas tapered to an acute to blunt tip, often with cobwebby hairs at the base andlor short-pubescent on the back; leaf blades with the margins and midrib converging to a blunt, keeled tip resembling the prow of a boat **_Poa_**

15 Lemmas little tapered to a blunt, often erose tip, glabrous; leaf blades flat or involute and pointed at the tip **_Puccinellia_**

16 Glumes absent 17

16 Glumes present 18

17 Spikelets unisexual, terete, the pistillate above the staminate in a large panicle; leaves smooth **_Zizania_**

17 Spikelets bisexual, strongly compressed; leaves strongly scabrous, abrasive to the touch **_Leersia_**

18 Spikelets containing 1 fertile floret and 1 or 2 sterile or staminate florets below the fertile one (the 2 lower florets represented by a pair of tiny, villous scales appressed to the base of the fertile lemma in *Phalaris,* otherwise the lower floret or florets more lemmalike or glumelike) 19

18 Spikelets containing a single fertile floret only (some reduced sterile spikelets may be present) 22

19 Spikelets disarticulating above the glumes, the 2 lower florets falling with the fertile one as a unit 20

19 Spikelets disarticulating below the glumes, the entire spikelet falling with the fertile floret and the single, glumelike sterile floret intact 21

20 Lower florets sterile, reduced to a pair of villous scales at the base of the hard, shiny fertile one **_Phalaris_**

20 Lower florets staminate, longer than the fertile one and enclosing it, similar to the fertile floret in texture **_Hierochloe_**

21 Spikelets crowded in few to many densely flowered branches; glumes and lemmas with stout, stiff hairs **_Echinochloa_**

21 Spikelets widely spreading in an open panicle; glumes and lemmas glabrous or merely scabrous on the nerves **_Panicum_**

22 Spikelets sessile or essentially so on one or no re rachises, the infloresence a terminal spike or of several to many 1 sided spikes arranged in a raceme or panicle 23

22 Spikelets borne on pedicels in an open or contracted panicle, the panicle sometimes dense and spikelike 25

23 Inflorescence a terminal spike; lemmas with awns several times longer than the body; glumes awnlike **_Hordeum_**

23 Inflorescence of several to many 1-sided spikes arranged in a raceme or panicle; glumes awnless or with awns shorter than the body **24**

24 Glumes unequal, narrow; spikes in a terminal raceme **_Spartina_**

24 Glumes equal, broadly boat-shaped; spikes in a narrow panicle **_Beckmannia_**

25 Spikelets disarticulating below the glumes, falling as an entire unit **26**

25 Spikelets disarticulating above the glumes, the glumes remaining after the florets have fallen **29**

26 Glumes awned **27**

26 Glumes awnless **28**

27 Awns of the glumes several times longer than the body **_Polypogon_**

27 Awns of the glumes shorter than the body **_Phleum_**

28 Panicle dense, cylindric and spikelike **_Alopecurus_**

28 Panicle more open, not cylindric or spikelike **_Cinna_**

29 Lemmas awned from the back, bearded on the callus, 5-nerved **_Calamagrostis_**

29 Lemmas awnless or awned from the tip, glabrous or pilose toward the base, 3-nerved **30**

30 Floret equal to or exceeding one or both glumes (excluding awns, if present); lemmas strongly nerved **_Muhlenbergia_**

30 Floret exceeded by both glumes; lemmas very faintly nerved **_Agrostis_**

Agrostis BENTGRASS

Rhizomatous (sometimes stoloniferous) or tufted perennials with soft, flat leaves and usually open panicles. **Spikelets** 1-flowered, disarticulating above the glumes; glumes subequal, acute to acuminate, scabrous on the keel, 1-nerved; floret exceeded by both glumes; lemma lanceolate, obtuse, more delicate than the glumes, very faintly 3-nerved; palea absent or rudimentary, or to 2/3 as long as the lemma, scarious.

1 Palea present, 1/2 to 2/3 as long as the lemma; plants with rhizomes and/or stolons **_A. stolonifera_**

1 Palea absent or rudimentary; plants usually lacking rhizomes or stolons **2**

2 Panicle open and diffuse, the widely spreading main branches branched and bearing spikelets only above the middle **_A. hyemalis_**

2 Panicle narrow and condensed, the strongly ascending main branches branched and bearing spikelets to near their bases **_A. exarata_**

Agrostis exarata Trin.
SPIKEBENT

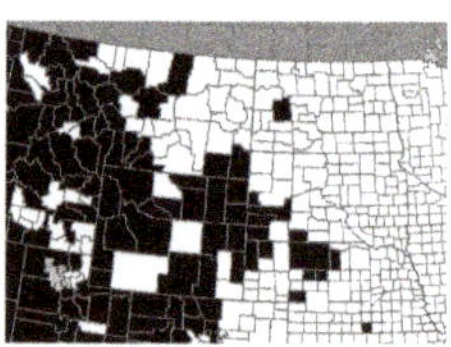

DESCRIPTION Tufted perennial, sometimes with short rhizomes; **culms** erect to sometimes decumbent at the base and rooting at the nodes, 2–10 dm tall. **Leaves** distributed well up the culm, the blades ascending to spreading, flat, 2–10 mm wide, scabrous; sheaths glabrous to scaberu-

Agrostis exarata
SPIKEBENT

lous; **ligule** hyaline, lacerate, 1.5–6 mm long. **Panicle** narrow and condensed, (4)6–25 cm long, continuous or somewhat interrupted mainly in the lower portion, the main branches strongly ascending, bearing short branches and spikelets to near the base. **Spikelets** green or sometimes purplish, 2–3 mm long; glumes lanceolate, acuminate, 1.9–3 mm long, scabrous on the keel; lemma 1.5–2.5 mm long, rarely awn-tipped; palea absent or rudimentary; anthers 0.3–0.5 mm long. **Grain** brownish, oblong, ca. 1 mm long. July–Aug.

HABITAT Moist meadows and streambanks.

WETLAND STATUS
GP FACW | WMTN FACW

Agrostis hyemalis (Walt.) B.S.P.

TICKLEGRASS

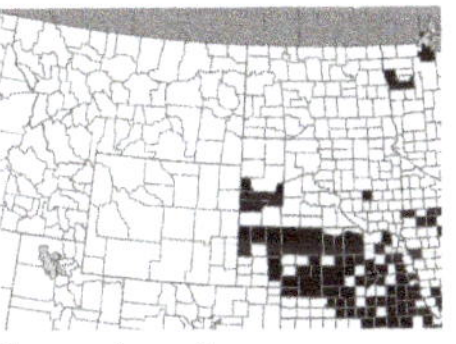

DESCRIPTION Tufted perennial with slender culms, erect to decumbent, 2–6 dm tall. **Leaves** mostly basal, the blades ascending to spreading, flat to involute, 0.5–2 mm wide, glabrous or scaberulous; sheaths glabrous, the **ligule** hyaline, rounded and usually erose, 1–2 mm long. **Panicle** open and diffuse, 8–25(35) cm long, the main branches rather lax, filiform and spreading, themselves branching and bearing spikelets only above the middle. **Spikelets** often purplish, (1.5)2–2.8(3) mm long; glumes lanceolate, acute to acuminate, 1.3–2.8 mm long; lemma 1.2–1.5 mm long; palea absent; anthers 0.2–0.5 mm long. **Grain** brown, narrowly ellipsoid, 1–1.2 mm long. June–Aug.

SYNONYM *Agrostis scabra* Willd.

Agrostis hyemalis
TICKLEGRASS

HABITAT Wet meadows, seepage areas, ditches, streambanks, shores and also in upland situations, often where alkaline.

WETLAND STATUS
GP FACW | MIDW FAC | WMTN FACW

Agrostis stolonifera L.

REDTOP

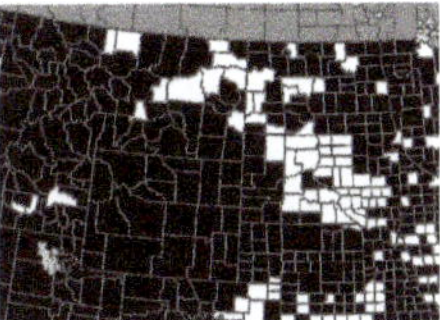

DESCRIPTION Rhizomatous or sometimes stoloniferous perennial, 3–10(14) dm tall; **culms** erect or decumbent at the base. **Leaf blades** ascending, 1–6 mm wide, scabrous; sheaths glabrous, the **ligule** hyaline, usually splitting, 1–5 mm long. **Panicle** open (rarely contracted) but not diffuse, (2)6–20 cm long, the main branches branching and bearing spikelets near the base and toward the tips. **Spikelets** usually purplish, turning pale or whitish after anthesis; glumes ovate-lanceolate to lanceolate, acute to mucronate, 1.5–2.5 mm long; lemma 1.6–2 mm long; palea present, 1/2 to 2/3 as long as the lemma; anthers 0.5–1.3 mm long. **Grain** brown, ellipsoid, 0.8–1.2 mm long. July–Sept.

SYNONYMS *Agrostis alba* L., *Agrostis gigantea* Roth., *Agrostis palustris* Huds.

Agrostis stolonifera
REDTOP

HABITAT Wet meadows, seepage areas, ditches, streambanks and shores; frequent to common; introduced from Europe and cultivated here as a pasture grass, escaped throughout most of Canada and the USA.

WETLAND STATUS
GP FACW | MIDW FACW | WMTN FAC

Alopecurus FOXTAIL

Tufted annuals and tufted or rhizomatous perennials with soft, flat leaves and dense, cylindric, spikelike panicles; **culms** erect or decumbent at the base. **Spikelets** 1-flowered, strongly compressed, densely crowded in the spikelike panicle, disarticulating below the glumes; glumes equal, laterally compressed, usually united toward the base, 3-nerved, often silky hairy on the back, awnless; lemma about as long as the glumes or shorter, laterally compressed with the margins united toward the base, faintly 3- to 5-nerved, awned from the back, the awn not exserted to exserted well beyond the glume tips; palea absent. **Grain** olivaceous to grayish-brown, somewhat obovoid, apiculate at both ends, 1–2 mm long.

1 Anthers more than 1.5 mm long; glumes more than 2.5 mm long; lemmas often more than 2.5 mm long 2

1 Anthers less than 1.5 mm long; glumes usually less than 2.5 mm long; lemmas often less than 2.5 mm long 3

2 Awns elongate, geniculate, 5–10 mm long, the exserted portion 2–6 mm long; plants nonrhizomatous, although some stems tending to root at the lower nodes *A. pratensis*

2 Awns shorter, 0.3–2(6) mm long, seldom exserted as much as 2 mm; plants strongly rhizomatous *A. arundinaceus*

3 Awn exserted just barely to 1.5 mm beyond the glume tips *A. aequalis*

3 Awn exserted 2–4 mm beyond the glume tips 4

4 Plants annual; anthers ca. 0.5 mm long *A. carolinianus*

4 Plants perennial; anthers 0.8–1.5 mm long *A. geniculatus*

Alopecurus aequalis Sobol.
SHORT-AWN FOXTAIL

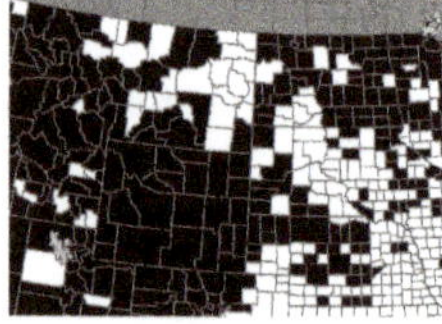

DESCRIPTION Tufted perennial 2–7 dm tall; **culms** erect to decumbent, often rooting at the nodes. **Leaf blades** 1–5 mm wide, finely scabrous above; **ligule** membranous, rounded to elongate, 2–7 mm long. Spikelike **panicle** erect, 3–6(8) cm long, 3–6 mm thick. Glumes 2–2.5 mm long, obtuse-tipped, villous on the keel and lateral nerves; lemma about equaling the glumes; awn from back of lemma exserted just barely to 1.5 mm beyond the glume tips; anthers 0.5–1 mm long. June–Sept.

HABITAT Shallow water or mud of wet meadows, marshes, water-filled ditches, springs, shores and streambanks.

WETLAND STATUS
GP OBL | MIDW OBL | WMTN OBL

Alopecurus arundinaceus Poir.
CREEPING FOXTAIL

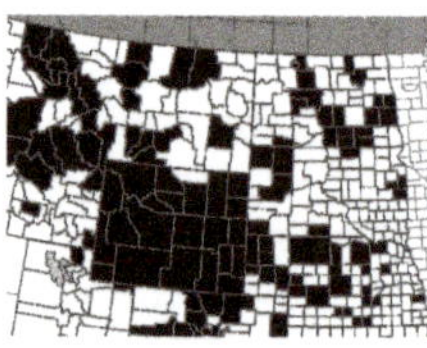

DESCRIPTION Strongly rhizomatous perennial (4)7–11 dm tall, often purplish at the base and in the inflorescence; **culms** stout, erect. **Leaves** mostly toward the base, reduced upward on the culm; blades 3–11 mm wide, smooth or scabrous on the upper surface; **ligule** membranous, 1–3 mm long. Spikelike **panicle** appearing silky due to hairs of the glumes, often grayish-purple, turning gray-brown with age, 3–11 cm long, 6–12 mm thick. Glumes 3.5–5.5 mm long, acute to

Alopecurus aequalis
SHORT-AWN FOXTAIL

acuminate, spreading-villous on the keel, glabrous or with shorter appressed hairs on the sides; lemmas 2–4 mm long; awn of the lemma 0.3–2(6) mm long, straight to slightly geniculate, not exserted or exserted as much as 2 mm; anthers 1.5–2.5 mm long. May–July(Aug).

HABITAT Wet meadows, shores, streambanks, ditches and other wet or moist places; frequently planted in low ground of tame pastures and hay meadows, especially in the n part; rapidly becoming naturalized; introduced from Eurasia and established in parts of s Canada and the northern Great Plains.

WETLAND STATUS
GP FACW | MIDW FACW | WMTN FAC

Alopecurus carolinianus Walt.
CAROLINA FOXTAIL

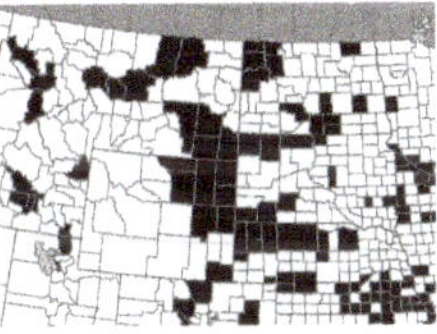

DESCRIPTION Tufted annual (0.5)1–2.5 dm tall; **culms** erect to decumbent. **Leaf blades** 1–3 mm wide, finely scabrous above; **ligule** membranous, rounded to elongate, 1–5 mm long. Spikelike **panicle** erect, 1–5 cm long, 3–5 mm thick. Glumes 1.5–2.8 mm long, obtuse-tipped, villous on the keel and lateral nerves; lemma about equaling the glumes; awn from back of lemma exserted 2–4 mm beyond the glume tips, geniculate; anthers ca. 0.5 mm long. Late May–Aug.

HABITAT Mud flats or wet ground of temporary ponds, wet meadows and low prairie.

WETLAND STATUS
GP FACW | MIDW FACW | WMTN FACW

Alopecurus geniculatus L. (*not illustrated*)
WATER FOXTAIL

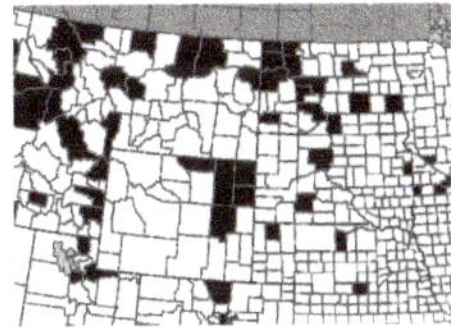

DESCRIPTION Much like *Alopecurus aequalis,* differing mainly as follows: spikelike **panicle** 3–7 cm long, 4–6 mm thick; glumes 2–3 mm long; awn from back of lemma exserted 2–3 mm beyond the glume tips; anthers 0.8–1.5 mm long. June–Sept.

HABITAT Same habitats as *Alopecurus aequalis;* introduced from Eurasia.

WETLAND STATUS
GP OBL | MIDW OBL | WMTN OBL

Alopecurus pratensis L.
MEADOW FOXTAIL

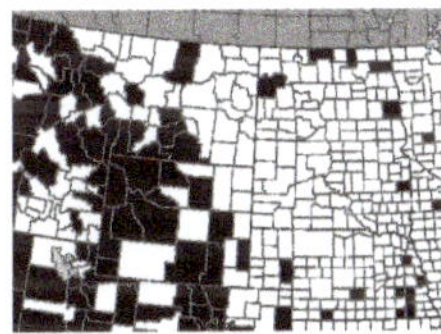

DESCRIPTION Loosely tufted perennial 4–9(12) dm tall; **culms** stout, erect, sometimes rooting at the lower nodes. **Leaves** mostly toward the base; blades 2–7 mm wide, smooth or finely scabrous above; **ligule** membranous, 1–4 mm long. Spikelike **panicle** grayish, turning stramineous with age, 3–12 cm long, 5–11 mm thick; glumes 3.5–6 mm

Alopecurus arundinaceus
CREEPING FOXTAIL

Alopecurus carolinianus
CAROLINA FOXTAIL

long, acute-tipped, villous mainly on the keel and lateral nerves. Lemmas 3.5–5.5 mm long, sometimes slightly surpassing the glumes, awn of the lemma usually geniculate, 5–10 mm long, the exserted portion 2–6 mm long; anthers 1.6–3.5 mm long. May–July.

HABITAT Wet meadows and marshy places; sparingly planted at one time as a forage and hay crop and possibly persisting in low areas; uncommon; introduced from Eurasia and naturalized from Newf. to Alaska and sporadically over the n USA.

WETLAND STATUS
GP FACW | MIDW FACW | WMTN FAC

Beckmannia syzigachne (Steud.) Fern.
WESTERN SLOUGHGRASS

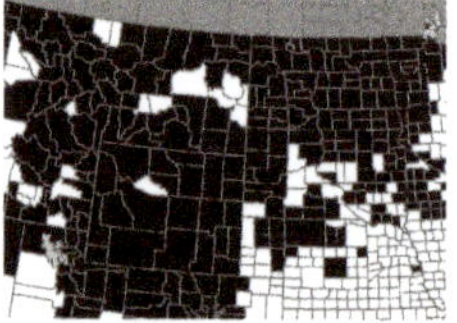

DESCRIPTION Stout annual or shortlived perennial 3–10 dm tall; **culms** solitary or few to several clumped. **Leaf blades** soft, flat, 3–10(13) mm wide, scaberulous; sheaths overlapping, glabrous, the upper one often loosely sheathing the lower portion of the panicle; **ligule** membranous, rounded to acute, 3–6 mm long. **Inflorescence** of many 1-sided spikes arranged in a narrow, continuous or interrupted panicle 8–25(30) cm long, the panicle branches strongly ascending to appressed; spikes bearing several to many spikelets in 2 rows on the rachis, mostly 5–15 mm long. **Spikelets** usually 1-flowered or many often 2-flowered, compressed and overlapping, suborbicular, 2–3 mm long, becoming stramineous at maturity, disarticulating below the glumes; glumes equal, broad, boat-shaped, inflated, apiculate at the tip, 3-nerved, the lateral nerves faint; lemma(s) about as long as the glumes but much narrower, lanceolate, the acuminate tip(s) protruding from between the glume tips, membranous, very obscurely 5-nerved; palea nearly as long as the lemma; anthers 0.5–1.1 mm long. **Grain** golden brown, ellipsoid, 1.5–2 mm long. June–Sept.

HABITAT Wet meadows, marshes, ditches, shores and streambanks.

WETLAND STATUS
GP OBL | MIDW OBL | WMTN OBL

Bromus ciliatus L.
FRINGED BROME

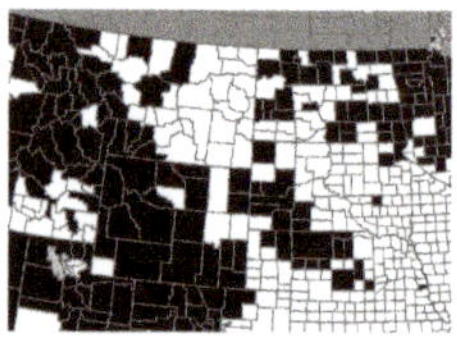

DESCRIPTION Non-rhizomatous perennial 5–12 dm tall; **culms** few together or single, often pubescent at the nodes. **Leaf blades** flat, 3.5–10(12) mm wide, glabrous or pilose mainly on the upper surface; sheaths glabrous to pilose; **ligule** membranous, very short to 1.5 mm long, erose. **Inflorescence** a loose, open panicle 7–20133) cm long, the branches usually drooping; spikelets rather large, 4- to 10-flowered, 14–25(35) mm long, 4–10 mm wide; glumes

Alopecurus pratensis
MEADOW FOXTAIL

Beckmannia syzigachne
WESTERN SLOUGHGRASS

glabrous to scabrous on the nerves, the first glume 4–9.5 mm long, 1-nerved or some rarely 3-nerved, acute, the second glume 6–11(14) mm long, 3-nerved, acute or with a short-awned tip; lemmas mostly 8–15 mm long, reduced upward, 5- to 7-nerved, usually prominently villous along the margins mainly in the lower 1/2 to 3/4, glabrous on the back or short-pubescent toward the base, awned from between the teeth of a minutely bifid apex, the awn 1–6 mm long; palea about equaling the body of the lemma; anthers highly variable in size, 0.7–2(4.6) mm long. **Grain** elongate, about equaling the palea, retained between the lemma and palea. July–Aug (Sept).

HABITAT Wet to moist ground of fresh springs, fens, streambanks and thickets, also in moist woods.

WETLAND STATUS
GP FAC | MIDW FACW | WMTN FAC

NOTE Of the several native species of brome in the region, *B. ciliatus* is the only one found in wet ground with any regularity. Although better known from moist woods in other parts of its range, in our area it demonstrates a clear preference for open, wet to moist places where surface water is fresh.

ADDITIONAL SPECIES **Smooth brome,** ***Bromus inermis*** Leyss., is naturalized throughout our range. Because of its abundance and ubiquity, it is common to find smooth brome in moist meadows and other habitats associated with prairie wetlands. The rhizomatous habit and awnless or very short-awned, glabrous to scabrous lemmas make smooth brome distinctive among our bromes. The weedy nature of the introduced annual bromes accounts for their frequent occurrence in previously flooded areas, e.g., dried shores, floodplains, etc. The most commonly encountered of these are Japanese brome, ***Bromus japonicus*** Thunb. ex Murr. and downy brome, ***Bromus tectorum*** L. Both are more typically upland weeds.

Bromus ciliatus
FRINGED BROME

Calamagrostis REEDGRASS

Perennials of moderate to tall stature, tufted from rhizomes; **leaves** green or glaucous, scabrous; sheaths mostly not overlapping, glabrous; **ligule** prominent, membranous, elongate, usually lacerate, 3–6 mm long. **Panicles** loose and open or dense and contracted, the branches spreading or ascending, scabrous. **Spikelets** 1-flowered, often purplish, disarticulating above the glumes; glumes subequal, the first shorter than the second or occasionally vice versa, lanceolate, acute to acuminate, the first glume 1-nerved, the second 3-nerved; lemma shorter and more delicate than the glumes, lanceolate, cleft or erose at the narrowed tip, 5-nerved, awned from the back, the awn diverging from near or below the middle of the back, nearly reaching to slightly exceeding the tip of the lemma, the base of the lemma (callus) bearded with a tuft of copious hairs, these shorter than to equaling the lemma; palea shorter than the lemma, membranous and appearing nerveless. **Grain** brown, ellipsoid, 1–1.5 mm long.

1 Panicle rather loose and open, the branches ascending to spreading; leaves rather lax, flat, 2–6 mm wide ***C. canadensis***

1 Panicle contracted, dense or interrupted, the branches short, ascending to appressed; leaves stiff, often involute, 1–4 mm wide when flat ***C. stricta***

Calamagrostis canadensis (Michx.) Beauv.
BLUEJOINT REEDGRASS

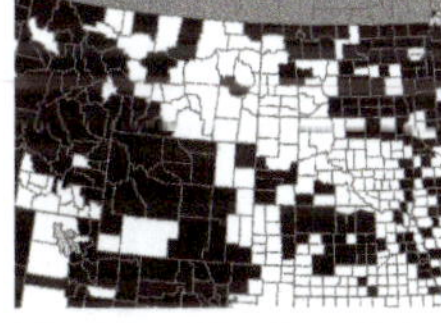

DESCRIPTION **Culms** rather slender, erect, 6–15 dm tall, often rooting from lower nodes when partly submersed; **leaves** green to glaucous, the blades rather lax, flat, 2–6 mm wide. **Panicle** rather loose and open, 8–20 cm long, the branches ascending to spreading. Glumes (1.5)2–4 mm long, glabrous or scaberulous on the back; lemma 1.5–3 mm long, thin, translucent, glabrous to scaberulous; anthers 1–1.7 mm long. Late June–Aug.

HABITAT Wet meadows, shallow marshes, boggy areas, springs and streambanks.

WETLAND STATUS
GP FACW | MIDW OBL | WMTN FACW

Calamagrostis stricta (Timm) Koel.
NORTHERN REEDGRASS

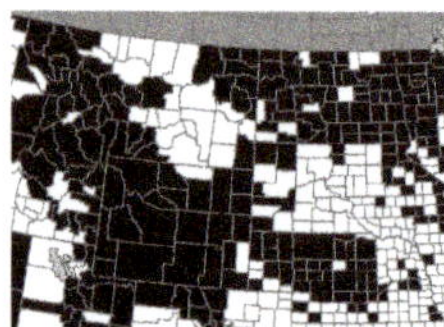

DESCRIPTION Sometimes resembling *Calamagrostis canadensis*, foliage glaucous, the **culms** more stout, erect, 3–12 dm tall; leaf blades stiff, often involute, 1–4 mm wide when flat. **Panicle** contracted, dense or interrupted, 6–15(25) cm long, the branches short, ascending to appressed. Glumes (2.5)3–4(5) mm long, scabrous on the back; lemma 2–3(4) mm long, rather firm, opaque, scabrous; anthers 1.2–2.6 mm long. Late June–Sept.

SYNONYMS *Calamagrostis inexpansa* (Torr.) A. Gray; *Calamagrostis neglecta* (Ehrh.) Gaertn.

HABITAT Wet meadows, shallow marshes, springs, boggy areas, shores, streambanks.

WETLAND STATUS
GP FACW | MIDW FACW | WMTN FACW

Calamagrostis canadensis
BLUEJOINT REEDGRASS

Catabrosa aquatica (L.) Beauv.
BROOKGRASS

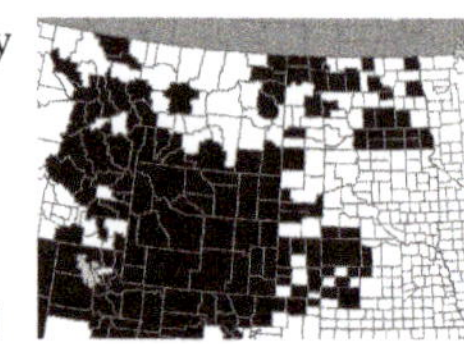

DESCRIPTION Loosely tufted or sprawling perennial 2–4 dm tall; **culms** thick but weak, often prostrate, branching and rooting at the nodes in mud or water. **Leaf blades** soft, flat, mostly less than 15 cm long, 3–8(15) mm wide, glabrous, the tip acuminate, mucronate or obtuse; sheaths glabrous; **ligule** scarious, elongate, 1–4 mm long. **Panicle** pyramidal or oblong, open, 10–20 cm long. **Spikelets** mostly 2-flowered or frequently mostly 1-flowered, golden or brown, 2–4 mm long, disarticulating above the glumes; rachilla joint elongate, the second floret (when present) well above the first; glumes unequal to nearly equal, scarious and nerveless at the tip, 1–2 mm long, the first smaller, ovate, the second broadest toward the tip, truncate and erose; lemmas truncate and erose at the apex, 1.5–2.5 mm long, 3-nerved; palea much like the lemma, 2-nerved; anthers 1–2 mm long. **Grain** brown, ellipsoid, ca. 1.5 mm long. June–early Sept.

HABITAT Springs, streams, seepage areas, where water is fresh.

WETLAND STATUS
GP OBL | MIDW OBL | WMTN OBL

Calamagrostis stricta
NORTHERN REEDGRASS

Cinna WOODREED

Tall, slender, nonrhizomatous to weakly rhizomatous perennials of wet to moist, often shady habitats. **Leaf blades** rather broad, flat and lax; **ligule** prominent, brownish, membranous, lacerate. **Inflorescence** a rather large, closed to open panicle, the branches ascending to spreading or drooping. **Spikelets** small, 1-flowered, laterally compressed, disarticulating below the glumes, the rachilla slightly elongated so that the floret is stipitate at the base, prolonged as a tiny bristle behind the palea; glumes 1- or 3-nerved, strongly keeled, sharply acute, the first a little shorter than the second; lemma similar to the glumes, 3-nerved, more blunt than the glumes, usually with a short awn arising just below the tip; palea 1-nerved, shorter than the lemma. **Grain** golden-brown, ellipsoid, 1.5–2.5 mm long.

1 Second glume 3-nerved, 3.8–6 mm long; anthers 1.1–1.7 mm long; panicle rather condensed, with the branches strongly ascending — ***C. arundinacea***

1 Second glume 1-nerved, 2–3.5 mm long; anthers 0.5–0.8 mm long; panicle usually open, with the branches spreading to drooping — ***C. latifolia***

Catabrosa aquatica
SWEET WOODREED

Cinna arundinacea L. (*not illustrated*)

SWEET WOODREED

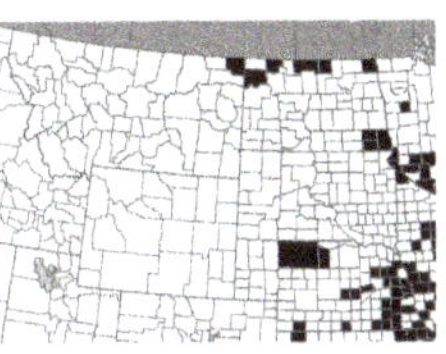

DESCRIPTION Nonrhizomatous to weakly rhizomatous perennial 6–15 dm tall; **culms** solitary or few together, erect, often bulbous at the base. **Leaf blades** 3.5–12(14) mm wide, glabrous to scabrous; sheaths glabrous; **ligule** 2.5–11 mm long. **Panicle** rather condensed, 10–30 dm long, the branches strongly ascending. Glumes narrowly lanceolate, scabrous on the keel, the first glume 1-nerved, 3–4.5 mm long, the second glume 3-nerved, 3.8–6 mm long; lemma 3.5–5 mm long, scaberulous on the back, usually with a short awn 0.2–1.5 mm long, arising just below the tip and often not surpassing it; anthers 1.1–1.7 mm long. Late July–Sept.

HABITAT Wet to moist woods, streambanks.

WETLAND STATUS
GP FACW | MIDW FACW

Cinna latifolia (Trev. ex Goepp.) Griseb.

DROOPING WOODREED

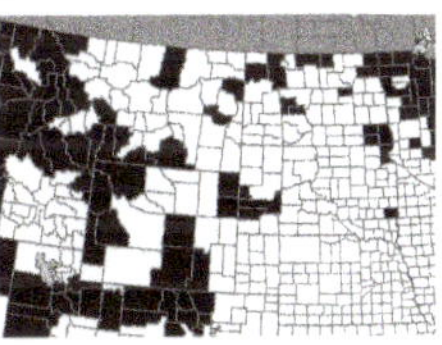

DESCRIPTION Similar to *Cinna arundinacea*, weakly rhizomatous, 6–13 dm tall; **culms** single or few together, erect, not bulbous at the base. **leaf blades** 3–13(17) mm wide, usually scabrous; sheaths smooth

Cinna latifolia
DROOPIN WOODREED

to scaberulous; **ligule** 2–7 mm long. **Panicle** usually loose and open, rarely condensed, 7–35 cm long, the branches lax and usually spreading to drooping. Glumes narrowly lanceolate to lanceolate, scabrous on the keel, both 1-nerved, the first glume 1.6–3(3.5) mm long, the second glume 2–3.5 mm long; lemma 2–3.5 mm long, scaberulous on the back, usually with an awn 0.3–1.5 mm long from just below the tip, usually surpassing the tip; anthers 0.5–0.8 mm long. July–Aug.

HABITAT Wet woods, swamps, springs, where water is fresh.

WETLAND STATUS
GP OBL | MIDW FACW | WMTN FACW

Deschampsia cespitosa (L.) Beauv.
TUFTED HAIRGRASS

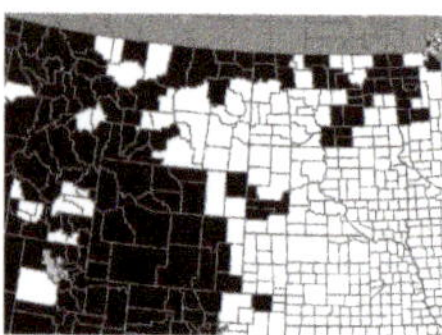

DESCRIPTION Tufted perennial 3–10 dm tall; **culms** stiff, erect, 1–3 mm thick. **Leaves** mostly basal, the blades usually not reaching the panicle, flat or involute, 2–3 mm wide, scabrous on the upper side; sheaths glabrous, the **ligule** white-hyaline, elongate, 2–6 mm long. **Panicle** contracted to open, 1–4 dm long, the filiform branches strongly ascending to loosely spreading, flowered toward the tips. **Spikelets** 2-flowered, purplish and fading to pale with age, 3–4(5) mm long, disarticulating above the glumes, the rachilla short-hairy, prolonged beyond the second floret; glumes acute, 2–4 mm long, subequal, the second somewhat longer than the first, 1-nerved or the second obscurely 3-nerved; lemmas 2- to 4-toothed at the apex, 2–3 mm long, the upper smaller than the lower, scarious, appearing nerveless, awned from near the base on the back, the awn surpassing the apex of the lemma or shorter, the callus short-hairy; palea acute, scarious; anthers 1–1.6 mm long. **Grain** brown, ellipsoid to pyriform, 1–1.5 mm long; late June–Aug.

HABITAT Fresh wet meadows, boggy areas, springs, streambanks.

WETLAND STATUS
GP FACW | MIDW FACW | WMTN FACW

Deschampsia cespitosa
TUFTED HAIRGRASS

Distichlis spicata (L.) Greene
SALTGRASS

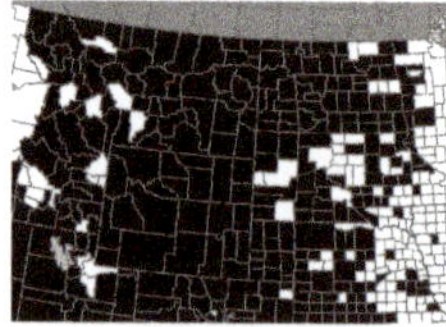

DESCRIPTION Low, extensively rhizomatous, dioecious perennial 1–3 dm tall, the culms wiry, erect. **Leaf blades** strongly ascending, the uppermost reaching or surpassing the panicle, flat to involute, 0.5–3 mm wide, glabrous or with sparse hairs; sheaths glabrous or sparsely hairy, usually long-hairy at the collar; **ligule** inconspicuous. **Panicle** simple, narrow and contracted, rather few-flowered, 3–6 cm long. Spikelets several to many, short-pedicelled, strongly ascending, 8–20 mm long; **staminate spikelets** stramineous, mostly 8- to 15-flowered; **pistillate spikelets** greenish-gray, mostly 7- to 9-flowered, disarticulating above the glumes; glumes unequal, 3- to 7-nerved, the lateral nerves sometimes obscure, the first glume ovate to lanceolate, 1–4 mm long, the second glume lanceolate, 2–5 mm long; lemmas ovate, 3–6 mm long, acute to subacute, keeled, broadly scarious-margined, other-

Distichlis spicata
SALTGRASS

wise firm, many-nerved; palea nearly as long as the lemma, broader than the lemma at the base, coriaceous, enclosing the grain in the pistillate floret; anthers 2–3(4) mm long on male plants. **Grain** dull to coppery brown, the surface rather wrinkled, lance-subulate, 3–4.5 mm long, including the long tapering beak. June–early Sept.

SYNONYM *Distichlis stricta* (Torr.) Rydb.

HABITAT Alkaline or saline flats and shores, also on drier sites.

WETLAND STATUS
GP FACW | MIDW FACW | WMTN FACW

Echinochloa BARNYARD-GRASS

Stout, weedy annuals (1)3–10 dm tall, with flat glabrous **leaf blades** 3–16 mm wide and terminal **panicles** of few to many densely flowered, racemose branches; culms solitary or few to several, erect to decumbent, rather succulent; sheaths glabrous or puberulent at the base; **ligules** absent. **Spikelets** nearly sessile, ovoid, containing one terminal fertile floret and one sterile floret, disarticulating below the glumes; glumes very unequal, the first glume broad with a pointed tip, 1/3 to 1/2 the length of the spikelet (excluding awns, if present), 3-nerved, the second glume broadly ovate, acuminate to apiculate, hispid or papillose-hispid on the back, especially on the nerves, 5-nerved, the outer 2 nerves marginal; sterile lemma resembling the second glume in size and vestiture, awned or awnless; palea of the sterile floret membranous, 1/2 to 3/4 the length of the sterile lemma body; fertile lemma ovate to elliptic, plano-convex, firm, shiny, 2–3 mm long including the short beak; palea of the fertile floret about as long as the lemma body, its margins included by the inrolled margins of the lemma, the tip free. **Grain** retained inside the fertile floret.

NOTE For information on related taxa sometimes encountered in wetland habitats, see the discussion under *Panicum capillare*.

1 Firm, shiny apex of the fertile lemma obtuse or broadly acute, sharply differentiated from the often shriveled beak, the lemma body and the beak separated by a line of minute hairs ***E. crus-galli***

1 Firm, shiny apex of the fertile lemma narrowly acute or acuminate, gradually tapering to the usually stiff beak, the lemma body and beak not separated by a line of minute hairs (the beak itself commonly puberulent) ***E. muricata***

Echinochloa crus-galli (L.) Beauv.
LARGE BARNYARD-GRASS

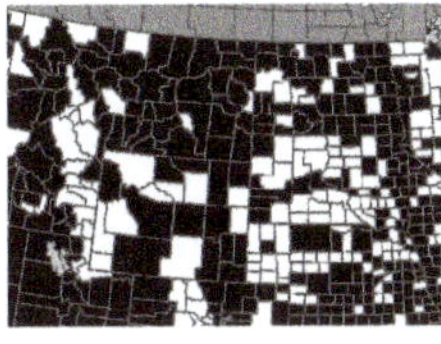

DESCRIPTION **Panicles** green to purplish, seldom strongly purplish; hairs of the panicle branches, or at least some, as long as or longer than the spikelets (excluding awns, if present). **Spikelets** 3–5 mm long (excluding awns); sterile lemma awnless or with an awn to 4 cm or more long, awned and awnless sterile lemmas often occurring in the same panicle; fertile lemmas mostly elliptic, the firm shiny apex obtuse or broadly acute, sharply differentiated from the often shriveled beak, the lemma body and the beak separated by a line of minute hairs. July– Sept.

HABITAT Shores, wet meadows, ditches, streambanks, mud flats and other wet places; mainly in the eastern part but spreading west in the region; introduced from Europe, now occurring throughout much of s Canada and most of the USA, s to Mexico.

WETLAND STATUS
GP FAC | MIDW FACW | WMTN FAC

Echinochloa muricata (Beauv.) Fern.
BARNYARD-GRASS

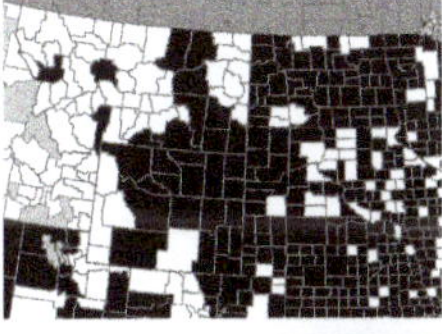

DESCRIPTION **Panicles** green to purplish, sometimes strongly purple, usu-

Echinochloa crus-galli
LARGE BARNYARD-GRASS

ally oblong or pyramidal; hairs of the panicle branches absent or shorter than the spikelets (excluding awns, if present). **Spikelets** 2–3.5 mm long (excluding awns); sterile lemma awnless or with an awn to 6(10) mm long; fertile lemmas mostly ovate, the firm shiny apex narrowly acute or acuminate, gradually tapering to the usually stiff beak, the lemma body and beak not separated by a line of minute hairs (the beak itself commonly puberulent). July–Sept.

SYNONYM *Echinochloa microstachya* (Wieg.) Rydb.

HABITAT Native; same habitats as *Echinochloa crus-galli;* common.

WETLAND STATUS
GP FACW | MIDW OBL | WMTN FACW

Elymus repens (L.) Beauv.
QUACKGRASS

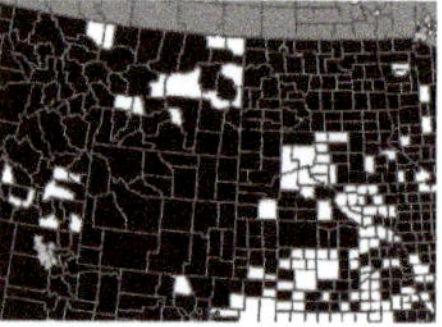

DESCRIPTION Strongly rhizomatous and weedy perennial, the foliage deep green to somewhat glaucous; **culms** erect to decumbent, 5–10 dm tall, smooth and hollow. **Leaf blades** flat to seldom involute, 2–9(14) mm wide, glabrous or pilose on the upper surface; sheaths glabrous or sparsely pilose near the summit, usually with prominent, clawlike auricles at the mouth; **ligule** very short and membranous, to 0.5 mm long, finely erose. **Inflorescence** a balanced, bilateral spike 5–19(26) cm long, continuous or somewhat interrupted in the lower portion; rachis joints flattened, scabrous on the angles. **Spikelets** 1 per node, 3- to 8-flowered, 10–20 mm long; glumes subequal, the first usually slightly shorter than the second, lanceolate, 5–13 mm long, strongly 5- to 7-nerved, firm, glabrous or scaberulous on the midnerve, acute or often awned from a minutely bifid apex, the awn sometimes to 7 mm long; lemmas 7–10(12) mm long, reduced upward, faintly 5-nerved, glabrous, acute or awned like the glumes; palea about equaling the body of the lemma, scaberulous on the margin; anthers prominent, 3–5.5 mm long. **Grain** brown, oblong-cylindric, 4–5 mm long, remaining enclosed by the firm-textured lemma and palea. June–Sept.

SYNONYM *Agropyron repens* (L.) Gould

HABITAT Wet meadows, ditches, streambanks and other wet or moist places; common; introduced from Eurasia.

WETLAND STATUS
GP FACU | MIDW FACU | WMTN FAC

ADDITIONAL SPECIES

A few other members of the wheat tribe (Triticeae) are occasionally encountered in wetland habitats, although not to the extent of *Elymus repens* and *Hordeum jubatum*. Among the wheatgrasses, these include *Elymus smithii* (Rydb.) Gould (synonyms *Agropyron smithii, Pascopyrum smithii*) and *Elymus trachycaulus* (Link) Gould ex Shinners (synonym *Agropyron trachycaulum*)

Elymus smithii, **western wheatgrass** (also classified as *Pascopyrum smithii* (Rydb.) Barkworth & D.R. Dewey), is well known as a dominant species in native mixed grass prairie and is seldom thought of as inhabiting wetlands. However, some ecotypes favor alkaline wet meadows in the central and western parts of the Dakotas, and the species may sometimes occur as an emergent in shallow marshes. *Elymus smithii* resembles *E. repens* in general habit but differs markedly in its strongly glaucous color, narrower, often invo-

Echinochloa muricata
BARNYARD-GRASS

Elymus repens
QUACKGRASS

lute leaves and the glumes and lemmas tapered (not bifid) to an awned tip.

Elymus trachycaulus, **slender wheatgrass**, occasionally occurs on or near shores and streambanks. Unlike the two species already discussed, *E. trachycaulus* is a bunchgrass lacking rhizomes. Occurring in the northern Great Plains are subsp. *trachycaulus*, with lemmas awnless or with awns to 6 mm long, and subsp. *subsecundus*, with lemmas having awns 17–40 mm long.

Elymus trachycaulus and *Hordeum jubatum* are often implicated as parents in the formation of the sterile F1 hybrid *x Elyhordeum macounii* (Vasey) Barkworth & D.R. Dewey (formerly *Elymus macounii* Vasey). Plants have 2 spikelets per node on the rachis, and it is similar to *Hordeum jubatum* in having the rachis disarticulate at maturity; its hybrid nature is plainly evident by the empty florets. The plant is occasional in moist disturbed habitats such as shorelines and streambanks.

Finally, *Elymus canadensis* L., **Canada wildrye**, is a common, tall, tufted perennial of woodlands, moist prairie and road ditches. It is also a frequent inhabitant of streambanks, elevated shorelines and marsh borders. The large, thick, nodding spikes with long, ultimately recurved awns are distinctive for this species in our area.

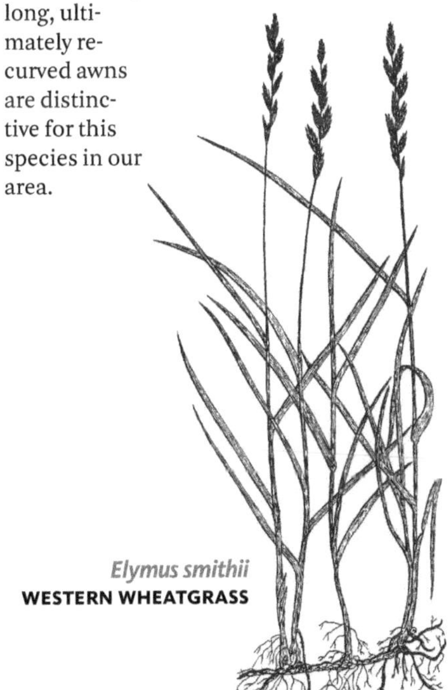

Elymus smithii
WESTERN WHEATGRASS

Eragrostis LOVEGRASS

Low, creeping, mat-forming annuals (those included here), perfect-flowered or dioecious; **culms** mostly spreading and rooting at the nodes. **Leaves** mostly in tufts at the nodes, short-sheathing and with rather short, flat to folded blades. **Inflorescences** usually many, condensed and capitate to somewhat open panicles; spikelets mostly many-flowered, linear to oblong, laterally compressed, the rachilla continuous and remaining intact as the glumes and lemmas fall, the paleas persisting for a short time; glumes unequal, acute to acuminate, (0)1-nerved or the second occasionally 3-nerved; lemmas acute, keeled, prominently 3-nerved; paleas shorter than the lemma, scarious, conspicuously 2-nerved.

ADDITIONAL SPECIES Two weedy species of lovegrass are worth mentioning for their tendency to appear on dry shores and other previously inundated habitats, especially in dry, sandy or gravelly substrates. Both are typical of disturbed upland habitats like roadsides, fields and waste places. **Carolina lovegrass** (*Eragrostis pectinacea* (Michx.) Nees ex Jedw.) is a tufted, nonstoloniferous, introduced grass with open, spreading panicles of narrow, linear, nonglandular spikelets that tend to lie parallel to the panicle branches. **Stinkgrass** (*Eragrostis cilianensis* (All.) Vign. ex Janchen) is another tufted annual with more crowded panicles of broader spikelets and with wartlike glands on the keels of the glumes and lemmas, and also on the pedicels.

1 Plants perfect-flowered; anthers 0.2–0.3 mm long ... *E. hypnoides*

1 Plants dioecious; anthers 1.4–2.3 mm long ... *E. reptans*

Eragrostis hypnoides (Lam.) B.S.P.
TEAL LOVEGRASS

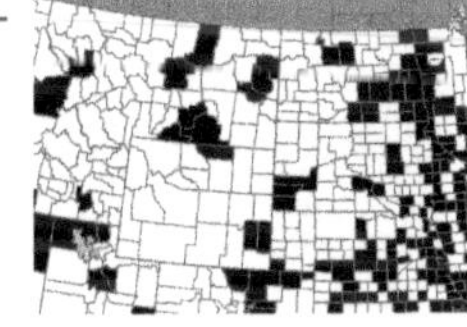

DESCRIPTION Perfect-flowered, mat-forming annual; **culms** mostly spreading and rooting at the nodes, the erect branches (2)5–15 cm tall, glabrous except for short pubescence at the nodes. **Leaf blades**

flat to folded, mostly 1–5 cm long, 1–2(3) mm wide, glabrous or very short-pubescent; sheaths glabrous except for hairs at the summit and sometimes along the margins; **ligule** a tuft of hairs ca. 0.5 mm long. **Panicles** numerous, rather dense to open, oblong, 1.5–6 cm long. **Spikelets** several- to many-flowered, linear, 3–8 mm long; glumes ovate, 1- nerved, the first 0.5–1 mm long, the second 0.5–1.5 mm long; lemmas ovatelanceolate, acute to acuminate, 1.5–2 mm long, hyaline between the 3 nerves; anthers 0.2–0.3 mm long. **Grain** golden brown, ovoid, 0.4–0.6 mm long. Late July–Sept.

HABITAT Wet sandy or muddy streambanks and alluvial bars.

WETLAND STATUS
GP OBL | MIDW OBL | WMTN OBL

Eragrostis reptans (Michx.) Nees (*not illus.*)
CREEPING LOVEGRASS

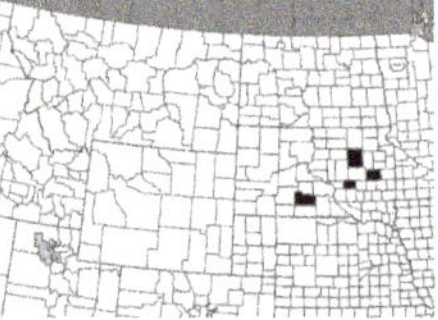

DESCRIPTION Similar to *E. hypnoides* in habit, but plants dioecious. **Culms** freely rooting at the nodes, the erect branches 3–25 cm tall, usually finely puberulent with gland-tipped hairs, seldom glabrate. **Leaves** mostly in tufts at the nodes; blades flat, 1–4 cm long, 1–3 mm wide, the blades and sheaths similarly short glandular-pubescent to seldom glabrate; **ligule** a fringe of hairs ca. 0.5 mm long. **Panicles** numerous, capitate, globose to oblong, mostly 1–3 cm long. **Spikelets** often nearly sessile in the headlike panicles, mostly many-flowered, sometimes purplish, narrowly oblong, (3)5–15 mm long, sometimes 1 or both glumes early deciduous, especially in male plants, lemmas persistent on male plants, eventually deciduous in the female; staminate spikelets with the first glume 1-nerved, 1–3.5 mm long, the second glume 1 (3)-nerved, 2–4 mm long, the lemmas 2–4 mm long, acute; anthers 1.4–2.3 mm long; pistillate spikelets with the first glume (0)1-nerved, 0.2–1.5 mm long, the lemmas 1.5–3.2 mm long, acute to acuminate, sometimes awn-tipped. **Grain** brownish, obovoid, 0.4–0.7 mm long. Late July–Sept.

Eragrostis hypnoides
TEAL LOVEGRASS

SYNONYM *Neeragrostis reptans* (Michx.) Nicora

HABITAT Same habitats as *Eragrostis hypnoides.*

WETLAND STATUS
GP OBL | MIDW OBL

Glyceria MANNAGRASS

Rhizomatous perennials with flat leaf blades and open panicles; sheaths tubular, the margins united for most of their length. **Spikelets** few- to many-flowered, ovoid to linear, subterete or slightly compressed, disarticulating above the glumes and between the florets; glumes unequal, acute or obtuse, scarious, 1-nerved; lemmas rounded on the back, usually obtuse and scarious at the tip, 7-nerved; palea about equaling the lemmas, nerved marginally and winged backward toward the grain; stamens 3 or 2.

1 Spikelets linear, mostly 6- to 13-flowered, 9–15 mm long; panicle branches stiff, erect to ascending **G. borealis**

1 Spikelets ovoid to narrowly ovoid, 3- to 7(9)-flowered, 2.5–7 mm long; panicle branches lax, spreading or drooping at maturity 2

2 First glume 1–2 mm long; lemmas 2.5–3 mm long; panicle 20–35 cm long **G. grandis**

2 First glume 0.5–1 mm long; lemmas 1.5–2 mm long; panicle 6–15(20) cm long **G. striata**

Glyceria borealis (Nash) Batch.

NORTHERN MANNAGRASS

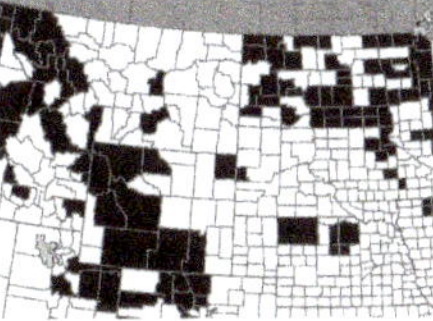

DESCRIPTION Culms 5–13 dm tall, rather weak, sometimes decumbent at the base, often producing adventitious roots from submersed lower nodes. **Leaf blades** 2–6 mm wide, glabrous; sheaths glabrous, the **ligule** scarious, rounded, 3–10 mm long. **Panicle** 15–40 cm long, with stiff, erect to ascending, raceme-like branches. **Spikelets** linear, mostly 6- to 13- flowered, 9–15 mm long; glumes obtuse, the first 1.5–2 mm long, the second 2.5–3.5 mm long; lemmas oblanceolate, 3–4 mm long, obtuse to subacute, scariousmargined, scaberulous on the nerves; anthers 3, 0.5–0.9 mm long. **Grain** dark brown, oblong to ovoid, apiculate at the base, furrowed on the upper face, 1.5 mm long. June–Aug.

HABITAT In shallow water or mud of streams, ditches, ponds and marshes.

WETLAND STATUS
GP OBL | MIDW OBL | WMTN OBL

Glyceria borealis
NORTHERN MANNAGRASS

ADDITIONAL SPECIES Water mannagrass (*Glyceria fluitans* (L.) R. Br.) is a similar species found only in the Black Hills; otherwise it occurs well east and west of the region. It differs from *G. borealis* in the longer, more scabrous lemmas 4–7 mm long. This introduced species is known from several locations in ND and SD.

Glyceria grandis S. Wats.

AMERICAN MANNAGRASS

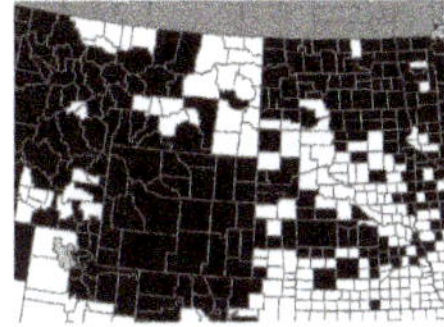

DESCRIPTION Culms 7–14 dm tall, 4–6 mm thick and rather spongy at the base. **Leaves** 4–12 mm wide, glabrous; sheaths glabrous, the **ligule** hyaline, entire or lacerate, 3–7 mm long. **Panicle** 20–35 cm long, oblong to pyramidal, very open with lax, drooping branches at maturity. **Spikelets** purplish, ovate to narrowly ovate, 3- to 6(9)-flowered, 4–7 mm long; glumes ovate to ovate-lanceolate, acute, the first 1–2 mm long, the second 2–2.5 mm long; lemmas purplish and often bronzed at the tip, oblong-lanceolate, 2.5–3 mm long, rounded to truncate; anthers 3, 0.7–1.1 mm long. **Grain** dark brown, ovoid, 1–1.3 mm long. Late June–early Sept.

HABITAT In shallow water or mud of marshes, ditches, streams, lakes and ponds.

WETLAND STATUS
GP OBL | MIDW OBL | WMTN OBL

Glyceria grandis
AMERICAN MANNAGRASS

Glyceria striata (Lam.) Hitchc.

FOWL MANNAGRASS

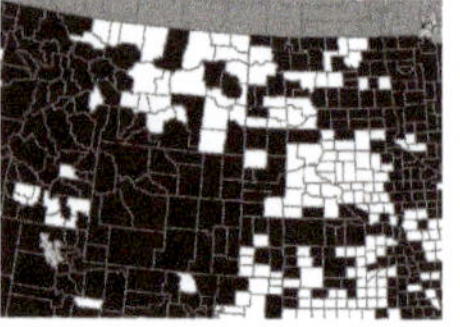

DESCRIPTION **Culms** 2–8 dm tall, slender, clumped or single from rhizomes. **Leaves** 1–4 mm wide, glabrous; sheaths glabrous, the ligule scarious, rounded, 1–2 mm long. **Panicle** 6–15(20) cm long, loosely spreading and open, the branches lax, spreading or drooping. **Spikelets** often purplish, ovate, 3- to 7-flowered, 2.5–4.5 mm long; glumes ovate-elliptic, the first 0.5–1 mm long, the second 0.8–1.5 mm long; lemmas elliptic, obtuse to subacute, 1.5–2 mm long, strongly nerved; anthers usually 2,0.2–0.5 mm long. **Grain** dark brown, shiny, suborbicular, 0.7–1 mm long. June–July, seldom late Aug–Sept.

HABITAT Fresh wet meadows, springs and stream margins.

WETLAND STATUS
GP OBL | MIDW OBL | WMTN OBL

Hierochloe odorata (L.) Beauv.

HOLYGRASS, SWEETGRASS

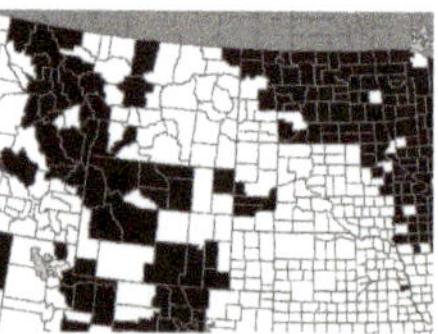

DESCRIPTION Perennial from creeping, often deep-seated rhizomes, sweetly scented (like sweetclover) especially when dried; **culms** erect, (1)2.5–5(7) dm tall, glabrous, sometimes clothed at the base by the previous year's leaves. **Leaf blades** flat (inrolled when young), much shorter (mostly 1–4 cm long) on the culms than on sterile shoots, 2.5–8 mm wide, glabrous or puberulent; sheaths glabrous or puberulent, especially at the collar; **ligules** membranous, 1–4 mm long. **Inflorescence** an ovoid to pyramidal panicle 4–10 cm long, the branches spreading to ascending. **Spikelets** golden-brown or often bicolored when young, with greenish or purplish at the base and golden toward the tips, 3-flowered, the lower 2 florets staminate, the terminal one perfect-flowered, disarticulating above the glumes with the 3 florets falling as a unit; glumes broadly ovate, subequal, 3.5–5.5 mm long, membranous and nearly transparent except at the base, faintly 1- to 3-nerved; lemmas of staminate florets golden-brown, 3–4 mm long, hirsute at the tip, along the margins and at the base, 5-nerved; lemma of the perfect floret 2.5–3.2 mm long, hirsute at the tip, otherwise smooth and shiny, obscurely 3- to 7-nerved; anthers of staminate florets 1.6–2.2 mm long, those of the perfect floret 1–1.6 mm long. **Grain** formed in few spikelets, retained inside the lemma, golden-brown, broadly ellipsoid, 1–1.5 mm long. Late Apr–July.

SYNONYM *Anthoxanthum hirtum* (Schrank) Y. Schouten & Veldkamp

Glyceria striata
FOWL MANNAGRASS

Hierochloe odorata
HOLYGRASS, SWEETGRASS

HABITAT Wet meadows and low prairie, often where sandy.

WETLAND STATUS
GP FACW | MIDW FACW | WMTN FACW

NOTE Dried, braided sweetgrass has long been used in cleansing rituals by Native Americans.

Hordeum jubatum L.

FOXTAIL BARLEY

DESCRIPTION Tufted perennial 2–8 dm tall; **culms** erect or decumbent at the base. **Leaves** usually flat, 1–5 mm wide, glabrous; sheaths glabrous, the **ligule** scarious, truncate, less than 1 mm long. **Inflorescence** a terminal spike, erect to nodding, 3–10 cm long, bristly due to the long, slender, spreading awns of glumes and lemmas; awns (1.512.5–7 cm long. **Spikelets** 1-flowered, 3 at each node of the rachis, the central spikelet fertile, sessile, the lateral 2 reduced, sterile, with awnlike parts, short-pedicelled; central spikelet with the back of the lemma facing away from the rachis; disarticulation above the glumes on the rachilla and also between the nodes on the rachis; lemma lanceolate with a long-awned tip, the lemma body 4–6 mm long, rather firm, obscurely 5-nerved; palea as long as the lemma body; anthers 0.8–1.6 mm long. **Grain** tan to brown, oblong, 3 mm long, the lemma and palea adherent to the grain. June–Sept.

HABITAT Wet meadows, ditches, shores, shallow marshes, seepage areas and other wet or moist places, often where alkaline or saline, also common as a weed of drier disturbed sites; common, often abundant.

Hordeum jubatum
FOXTAIL BARLEY

WETLAND STATUS
GP FACW | MIDW FAC | WMTN FAC

ADDITIONAL SPECIES **Meadow barley** (*Hordeum brachyantherum* Nevski) occurs in w portions of the Great Plains; plants are smaller and with shorter awns (1 cm or less) than in *H. jubatum*. See also the comments under *Elymus repens*.

Leersia oryzoides (L.) Sw.

RICE CUTGRASS

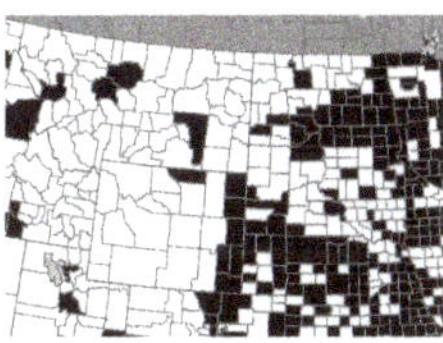

DESCRIPTION Loosely tufted perennial from creeping rhizomes, 3–10 dm tall; **culms** upright to decumbent. **Leaves** flat, 3–11 mm wide, strongly scabrous, abrasive to the touch; sheaths scabrous, usually puberulent at the collar; **ligule** truncate, rather firm, 1 mm long. **Panicle** 10–20 cm long, open with spreading to ascending branches or often partially to completely included by the uppermost leaf sheath, the spikelets then cleistogamous. **Spikelets** strongly compressed, pale

Leersia oryzoides
RICE CUTGRASS

green and becoming stramineous or brown with age, 3.5–5 mm long; glumes absent; lemma boat-shaped, hispid, coarsely so on the keel and margins; palea as long as or slightly exceeding the lemma, coarsely hispid on the keel, the margins tightly included by the lemma; anthers 0.2–2.4 mm long, the smaller sized anthers most typical of cleistogamous spikelets. **Grain** reddish-brown to brown, compressed, asymmetrically pyriform to obovoid, 2.5–3 mm long. Late July–early Sept.

HABITAT Muddy or sandy streambanks and shores.

WETLAND STATUS
GP OBL | MIDW OBL | WMTN OBL

Leptochloa fusca (L.) Kunth
BEARDED SPRANGLETOP

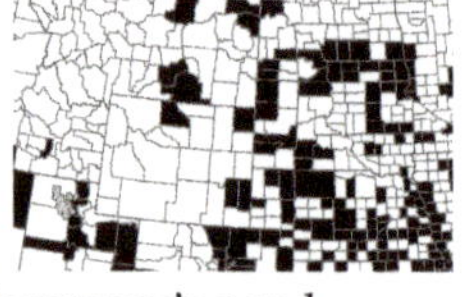

DESCRIPTION Tufted annual 1–4 dm tall, sometimes prostrate, somewhat succulent. **Leaves** strongly sheathing the culms, the upper ones usually surpassing and sheathing the inflorescences, the blades flat to involute, 1–3 mm wide, scaberulous; sheaths often purplish, glabrous; **ligule** elongate, often lacerate, 2–4 mm long. **Inflorescence** of several to many racemose branches, 5–20 cm long, the lower portion usually included by the uppermost sheath, the branches strongly ascending to weakly spreading, scabrous. **Spikelets** sessile or very short-pedicelled on the branches, 6- to 12-flowered, 7–20 mm long, disarticulating above the glumes; rachilla breaking between the florets at maturity into segments of 1 mm; glumes unequal, 1-nerved, the first lanceolate, 2–4 mm long, the second lanceolate to oblong, 4–5 mm long; lemmas acuminate, 4–5 mm long, decreasing upward, minutely notched at the apex and short-awned from between the minute teeth, 3-nerved, pubescent on the nerves toward the base; palea almost as long as the lemma body, nerved along the margins; anthers 0.1–0.4 mm long. **Grain** light brown, ellipsoid, flattened, 1.5–2.5 mm long. July–early Sept.

Leptochloa fusca
BEARDED SPRANGLETOP

SYNONYM *Diplachne fusca* (L.) Beauv., *Leptochloa fascicularis* (Lam.) A. Gray

HABITAT Shores, streambanks, mud or sand flats and other places of temporarily standing water, especially where alkaline or saline.

WETLAND STATUS
GP FACW | MIDW OBL | WMTN FACW

Muhlenbergia MUHLY

Solitary or tufted perennials, many producing creeping, scaly rhizomes; **culms** erect or decumbent at the base, often branching from the nodes; **leaf blades** rather short relative to width, scabrous; sheaths often not overlapping, the internodes exposed; ligules membranous. **Inflorescence** usually a narrow, contracted, spikelike panicle (open and diffuse in *M. asperifolia*), terminal, sometimes axillary panicles also produced. **Spikelets** 1-flowered, disarticulating above the glumes, the glumes remaining after the florets have fallen; glumes unequal to subequal, keeled or rounded on the back, acute to acuminate or awned from the tip, 1-nerved; floret equal to or exceeding one or both glumes (excluding awns, if present); lemma ovate-lanceolate or narrowly lanceolate, acute to acuminate, rarely awned, strongly 3-nerved, pilose at the base in some species; palea nearly as long as the lemma, 2-nerved.

---

1 Panicle open and diffuse, about as broad as long, the filiform branches widely spreading ... ***M. asperifolia***

1 Panicle narrow and contracted, the branches short, ascending to appressed ... **2**

2 Leaf blades usually involute, 0.2–1 mm wide; panicles few-flowered, not glomerulate **M. richardsonis**

2 Leaf blades flat, 2–6(8) mm wide; panicle usually densely flowered and glomerulate 3

3 Glumes acuminate or short-awned, usually not to shortly surpassing the floret; anthers 0.3–0.5 mm long **M. mexicana**

3 Glumes awned, much surpassing the floret; anthers 0.5–1.5 mm long 4

4 Internodes of the culm mostly smooth and shiny; main culms usually branched above; anthers 0.5–0.8 mm long; ligules 0.6–1.5 mm long **M. racemosa**

4 Internodes of the culm puberulent and dull; main culms unbranched or branched from the base; anthers 0.8–1.5 mm long; ligules 0.2–0.6 mm long **M. glomerata**

Muhlenbergia asperifolia (Nees & Meyen) Parodi

ALKALI MUHLY

DESCRIPTION Low perennial with slender scaly rhizomes; **culms** usually decumbent, branching from the lower nodes, 1–5 dm long. **Leaves** strongly ascending, flat, rather short, mostly 2–5 cm long, 1–3 mm wide, pale green; sheaths glabrous, the **ligule** truncate, erose, 0.5–1 mm long. **Panicle** open and diffuse, ovoid to pyramidal, 5–15 cm long, about as wide when fully expanded, the filiform branches widely spreading, scabrous, sparsely flowered. **Spikelets** solitary on filiform pedicels, purplish or dark gray; glumes unequal to subequal, lanceolate to ovate-lanceolate, acuminate, 1/2 to nearly as long as the floret; lemma ovate-lanceolate, narrowed to an acute to rounded tip, 1.2–1.8 mm long; anthers 0.6–1 mm long. **Grain** brown, ellipsoid, 1–1.2 mm long. Mid-July–Sept.

HABITAT Wet meadows, seepage areas, shores and flats, often where alkaline or saline.

WETLAND STATUS
GP FACW | MIDW FACW | WMTN FACW

Muhlenbergia asperifolia
ALKALI MUHLY

Muhlenbergia glomerata (Willd.) Trin.

BRISTLY MUHLY

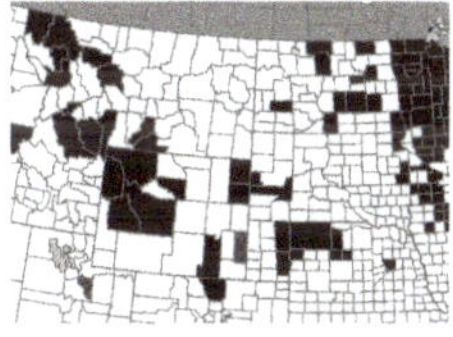

DESCRIPTION Perennial 3–8 dm tall from stout scaly rhizomes; **culms** erect, unbranched or branching from the base, the internodes dull, puberulent. **Leaves** flat, mostly 6–12 cm long, 2–6 mm wide; sheaths glabrous, the **ligule** truncate, erose-ciliate at the tip, 0.2–0.6 mm long. **Panicle** contracted, usually densely flowered, glomerulate, often interrupted, (1)2–11 cm long, 4–10 mm thick.

Muhlenbergia glomerata
BRISTLY MUHLY

Spikelets crowded and appressed in the glomerules, green or purplish; glumes subequal, the first a little shorter than the second, subulate, awned, the bodies 1.5–3 mm long, the awn 1–5 mm long, both glumes (including the awns) much surpassing the floret, 1.3–2x as long as the lemma; lemma lanceolate, acuminate or rarely short-awned, 2–3 mm long, pilose on the callus, pubescent along the margins nearly to the tip; anthers 0.8–1.5 mm long. **Grain** brown to dark brown, oblong, 1–1.2 mm long. Aug–Sept.

HABITAT Wet meadows, springs, fens and boggy areas.

WETLAND STATUS

GP FACW | MIDW FACW | WMTN FACW

NOTE The diploid *M. glomerata* is restricted to permanently wet habitats whereas the tetraploid *M. racemosa* occurs in a variety of upland as well as lowland habitats.

Muhlenbergia mexicana (L.) Trin.

WIRESTEM MUHLY

DESCRIPTION Perennial 2–8 dm tall from stout, scaly rhizomes; **culms** erect or sometimes decumbent at the base, often branching from the nodes; internodes dull and puberulent, especially toward the summit. **Leaves** flat, mostly 5–15 cm long, 2–6 mm wide; sheaths glabrous, the **ligule** entire to erose-ciliolate, 0.5–1 mm long. **Panicle** contracted, densely flowered and glomerulate or sometimes more loosely flowered and very slender, often interrupted, 3–12 cm long, 2–10 mm thick. **Spikelets** crowded, usually appressed in glomerules, green or purplish; glumes subequal, narrowly lanceolate, acuminate or short-awned, shorter than to equaling the floret, sometimes barely surpassing it, 2.5–4 mm long, the awns, if present, to 1.5 mm long; lemma lanceolate, acuminate or rarely short-awned, 2–3 mm long; anthers 0.3–0.5 mm long. **Grain** brown, narrowly ellipsoid, 1.5–2 mm long. Aug–Sept.

Muhlenbergia mexicana
WIRESTEM MUHLY

HABITAT Wet meadows, seepage areas, springs, fens and streambanks.

WETLAND STATUS

GP FACW | MIDW FACW | WMTN FAC

Muhlenbergia racemosa (Trin.) Rydb.

MARSH MUHLY, GREEN MUHLY

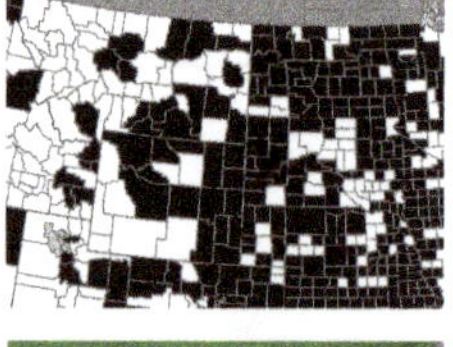

DESCRIPTION Similar to *Muhlenbergia glomerata* but often more robust, 3–10 dm tall; **culms** erect to decumbent, unbranched or more commonly branched from the middle nodes; internodes smooth and shiny, sometimes puberulent-roughened near the summit. **Leaves** flat, 3–17 cm long, 2–5 mm wide; sheaths glabrous, the **ligule** truncate, erose-ciliate, 0.6–1.5 mm long. **Panicle** contracted and often lobed, sometimes interrupted in the lower portion, 3–13 cm long, 3–12 mm thick. **Spikelets** crowded and appressed in oblong to obovoid glomerules, green or purplish; glumes subequal, subulate to an awned tip, the bodies 1.5–3 mm long, the awns 1.5–3.5 mm long, both glumes (including the awns) much surpassing the floret; lemma lanceolate, 2.5–3.8 mm long, tapered to a sharp or minutely awned tip, pilose on the lower 1/2 and on the callus; anthers 0.5–0.8 mm long. **Grain** brown, oblong, 1.5–2 mm long. Late July–Sept (Oct).

HABITAT Wet meadows, shores, streambanks; also in disturbed uplands; often where shady.

WETLAND STATUS

GP FACW | MIDW FACW | WMTN FACW

Muhlenbergia richardsonis (Trin.) Rydb.

MAT MUHLY

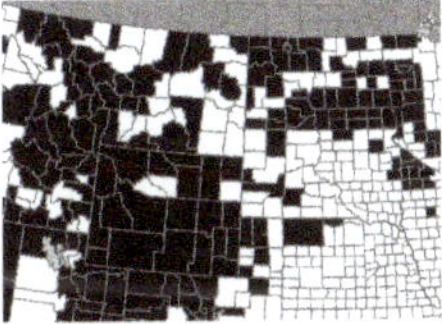

DESCRIPTION Low, loosely tufted perennial from slender scaly rhizomes, (1)2–5 dm tall; **culms** slender, erect or decumbent at the base. **Leaves** usually involute, 1–5 cm long, 0.2–1 mm wide; sheaths glabrous, the ligule elongate, erose, 1–2 mm long. **Panicle** narrow, few-flowered, interrupted or continuous, spikelike, not glomerulate, 2–5(9) cm long. **Spikelets** uncrowded, green or grayish; glumes subequal, ovate to ovate-lanceolate, acute, 1/3 to 1/2 as long as the floret; lemma lanceolate, mucronate, 2–3 mm long; anthers 1–1.5 mm long. **Grain** dark brown, ellipsoid, 1.2–1.5 mm long. Mid July–Sept.

HABITAT Low prairie, wet meadows and seepage areas, often where alkaline or saline.

WETLAND STATUS

GP FAC | MIDW FAC | WMTN FAC

Panicum capillare L.

COMMON WITCHGRASS

DESCRIPTION Weedy, tufted annual 1.5–5(7) dm tall, often purplish at the base and in the panicle; **culms** erect to decumbent and spreading, sometimes sparingly branched below. **Leaf blades** (2)5-i6(2 7) mm wide, hispid on both surfaces or sometimes mainly on the margins toward the base; sheaths prominently papillose-hispid; ligules a fringe of hairs from a membranous base, 0.5–2.2 mm long. **Inflorescence** an open, diffuse panicle obovoid to oblong in outline, usually purplish, mostly 0.9–3 dm long (secondary panicles from axils often smaller when present), eventually breaking off the plant to tumble in the wind, the branches and pedicels widely spreading or the lower ones crowded and partly included in the upper leaf sheath, strongly scabrous. **Spikelets** containing one terminal fertile floret and one sterile floret, disarticulating below the glumes, lanceolate-acuminate in shape; glumes very unequal, the first glume broadly ovate, acute to acuminate, 1–2 mm long, 3- to 5-nerved, sometimes scabrous on the midvein, the second glume ovate-lanceolate, acuminate, 2.5–3.5(4) mm long, 5- to 7-nerved, glabrous or scabrous on the nerves; sterile lemma very similar to the second glume but often a little shorter, lacking a palea; fertile lemma hardened and shiny, elliptic-ovate, 1.5–1.9 mm long, 5- to 7-nerved;

Muhlenbergia richardsonis
MAT MUHLY

Panicum capillare
COMMON WITCHGRASS

palea also hardened, its margins enclosed by the inrolled margins of the lemma; anthers 0.7–1.1 mm long. **Grain** retained inside the fertile floret. July–Sept (Oct).

HABITAT Shores, streambanks, and a variety of disturbed habitats, including roadsides and cropland; very common.

NOTE *Panicum capillare* is an opportunistic weed that ventures into wetland habitats only during drawdown. Otherwise, it is better known as an upland weed.

WETLAND STATUS

GP FAC | MIDW FAC | WMTN FAC

ADDITIONAL SPECIES

Fall panicum, ***Panicum dichotomiflorum*** Michx., is another weedy annual grass of moist, disturbed habitats and sometimes on streambanks, alluvial bars, shores and in wet ditches from s SD to e and c Neb. It differs from *P. capillare* in its more robust habit and essentially glabrous foliage, among other characters. **Switchgrass,** ***Panicum virgatum*** L., is a well known, dominant tallgrass prairie species that often occupies the mesic zone around basins and bordering streams. It differs from *P. capillare* in its taller stature and perennial, rhizomatous habit.

Among grasses of the *Panicum* tribe (Paniceae), the foxtails, *Setaria* spp., sometimes invade previously flooded substrates. Like *Panicums,* these have 2-flowered spikelets that disarticulate below the glumes, with the lower floret empty or staminate and the upper one fertile and hardened. The spikelets are borne in distinctive bristly panicles that are condensed, cylindric and spikelike. Most often encountered in drawdown zones is ***Setaria pumila*** (Poir.) Roemer & J.A. Schultes, **yellow foxtail,** with yellow-green spikelets 2.5–3.5 mm long, and sheaths with glabrous margins.

Phalaris arundinacea L.

REED CANARYGRASS

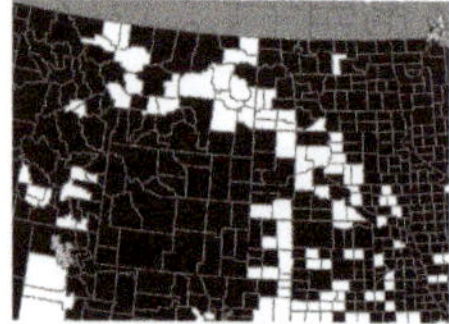

DESCRIPTION Tall, stout, rhizomatous perennial 6–16 dm tall, often densely colonial. **Leaf blades** not surpassing the panicle, flat, 6–16(20) mm wide, glabrous; sheaths glabrous, sometimes purplish; **ligule** scarious, rounded or lacerate, 3–8 mm long. **Panicle** contracted and narrow, dense or interrupted, 6–18 cm long, the spikelets in dense lobelike clusters on short ascending branches. **Spikelets** 3.5–5 mm long, disarticulating above the glumes; glumes subequal, exceeding the fertile floret, ovate-lanceolate, acute to mucronate, whitish-green or purplish and turning stramineous with age, conspicuously 3-nerved; fertile lemma ovate, acute, slightly keeled, 2.5–3.5 mm long, firm and shiny, villous mainly on the margins, faintly 5-nerved; palea as long as the lemma; anthers 1.8–3.4 mm long; sterile florets 2, scalelike and villous, appressed against the base of the fertile one, 1–1.5 mm long. **Grain** brown to dark brown, ovoid, 1.5–2 mm long. June–July.

HABITAT Wet meadows, shallow marshes, wet ditches, shores and streambanks.

WETLAND STATUS

GP FACW | MIDW FACW | WMTN FACW

Phleum pratense L.

TIMOTHY

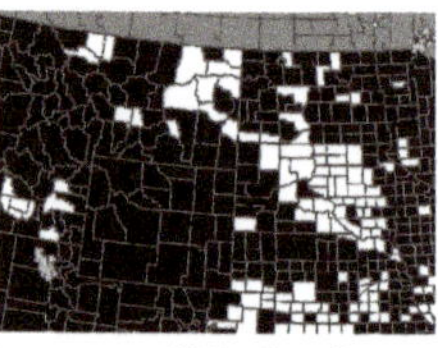

DESCRIPTION Tufted or single-stemmed perennial (4)5–10 dm tall; **culms** erect, smooth, purple or brown-banded at the nodes, bulbous at the base. **Leaf blades** flat, 2–8 mm wide, glabrous or scabrous; sheaths glabrous; **ligules** membranous, 1–4(5) mm long. **Inflorescence** a very condensed, cylindric, spikelike panicle 2–15 cm long, 5–8 mm thick, the branches and pedicels very short, crowded and hidden by the spikelets. **Spikelets** 1-flowered, compressed and U-shaped, crowded and uniformly appressed to

Phalaris arundinacea
REED CANARYGRASS

spreading in the panicle, green or often purple-tipped, turning dull brown in age, disarticulating below the glumes; glumes essentially opposite and equal, strongly compressed and keeled, strongly 3-nerved in the green keeled portion, ciliate on the keel, otherwise membranous and glabrous to puberulent, the body 1.8–3 mm long, rounded to truncate above, the nerves extended into a stout, scabrous awn 0.5–2 mm long; lemma broadly ovate, 1.5–2(2.4) mm long, membranous, glabrous or appressed-puberulent, 5-nerved, the midvein often prolonged into a very short awn; palea membranous, somewhat shorter than the lemma, the 2 nerves sometimes prolonged as very short awns; anthers 1.5–2 mm long. **Grain** dull brown, plump, obovoid, 1.2–1.5 mm long. June–Aug.

HABITAT Wet meadows, low prairie, streambanks, ditches and more upland habitats; a common introduced hay and forage grass now widely established; introduced from Europe, naturalized from Newf. to Alaska, s throughout most of the USA.

WETLAND STATUS
GP FACU | MIDW FACU | WMTN FAC

Phragmites australis (Cav.) Trin. ex. Steud.
COMMON REED

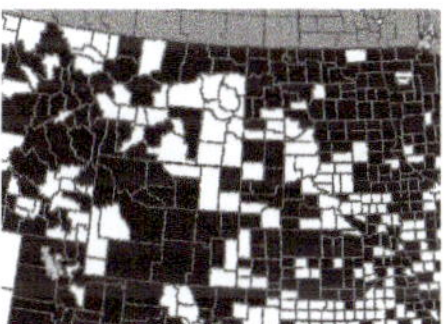

DESCRIPTION Tall, stout reeds 2–4 m tall, extensively colonial from usually deep-seated, scaly rhizomes, these sometimes acting as stolons and creeping over the substrate during drawdown; **culm** hollow, 5–15 mm thick toward the base, the internodes often purplish. **Leaves** broad, 1–3 cm wide; sheaths open, mostly overlapping, the **ligule** white-hyaline, fibrillose, 1 mm long. **Panicle** plumelike, rather densely flowered, purple and turning tawny with age, 15–40 cm long, the branches ascending to curved. **Spikelets** several-flowered, 10–15 mm long, disarticulating above the glumes, the florets decreasing in size upward, the rachilla covered with long silky hairs, these exceeding the florets, exposed after anthesis; glumes 3-nerved (the second rarely 5-nerved), unequal, the first glume ca. 1/2 the length of the second; lemmas long-acuminate, glabrous, 3-nerved; palea much shorter than the lemma, membranous, nerved along the margins. **Grain** seldom produced, dark brown, ellipsoid, 1.2–1.5 mm long. July–Sept.

SYNONYM *Phragmites communis* Trin.

HABITAT Fresh to saline marshes, shores, streams, ditches and seepage areas, in wet ground or shallow water; common, often abundant; an introduced, aggressive invader of wetlands.

WETLAND STATUS
GP FACW | MIDW FACW | WMTN FACW

ADDITIONAL SPECIES *Phragmites australis* subsp. *americanus* (or sometimes considered *P. americanus*) is a native variant of the invasive *P. australis.* the two can usually be separated by the following features: leaves of native *Phragmites* are yellowish, leaves of invasive *Phragmites* have a bluish hue; the stems of invasive *Phragmites* are typically a dull greenish-tan color, native stems are often reddish or purplish; and invasive

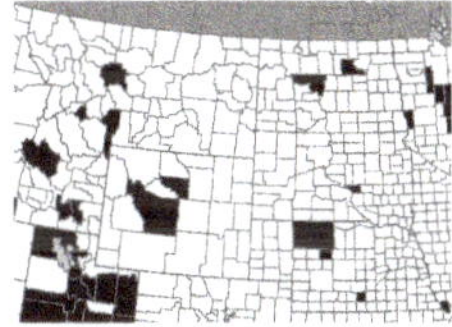

Phleum pratense
TIMOTHY

Phragmites australis
COMMON REED

Phragmites often forms large, tall, dense monocultural stands (native *Phragmites* stands usually appear as stems scattered throughout a diverse wetland community).

Poa BLUEGRASS

Tufted or rhizomatous perennials (those included here) of short to moderate stature, green or glaucous; **culms** erect to decumbent, hollow, sometimes compressed. **Leaves** mostly near the base, the blades flat to folded, the margins and midrib converging to a blunt, keeled leaf tip resembling the prow of a boat; sheaths partly closed, rounded or keeled; **ligules** membranous. **Inflorescence** an open to narrowly contracted panicle of small, mostly 2- to 8-flowered spikelets. **Spikelets** disarticulating above the glumes and between the florets; glumes subequal or the second a little longer than the first, acute; first glume 1- to 3-nerved; second glume 3-nerved; lemmas acute to blunt-tipped, reduced in size upward in the spikelet, faintly 5-nerved or appearing 3-nerved, with the intermediate nerves very obscure or obsolete, keeled or rounded on the back, often with a tuft of cobwebby hairs at the base, this sometimes scant, otherwise glabrous, scabrous or pubescent mainly on the nerves, sometimes pubescent or scabrous between the nerves as well; palea nearly as long as the lemma.

1 Culms strongly compressed, 2-edged; sheaths keeled ***P. compressa***

1 Culms round to slightly compressed, not 2-edged; sheaths rounded **2**

2 Lemmas lacking a tuft of cobwebby hairs at the base, pubescent only on the keel and lateral nerves and sometimes between the nerves on the lower back; foliage glaucous ***P. arida***

2 Lemmas with a tuft of cobwebby hairs at the base, this sometimes scant; foliage green **3**

3 Plants extensively rhizomatous; panicle more compact, mostly 3–13 cm long ***P. pratensis***

3 Plants tufted, sometimes stoloniferous, lacking rhizomes; panicle diffuse and open (contracted when young), mostly 8–30 cm long **4**

4 Lemmas pubescent on the keel and on the marginal nerves toward the base, the intermediate nerves very faint to obsolete ***P. palustris***

4 Lemmas pubescent on the keel only, not on the marginal nerves, the intermediate nerves prominent ***P. trivialis***

Poa arida Vasey
PLAINS BLUEGRASS

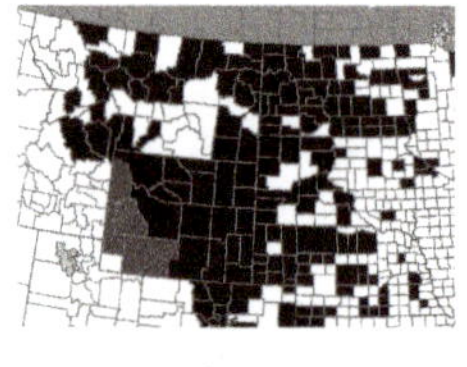

DESCRIPTION Glaucous perennial 3–8 dm tall, with short to long rhizomes, the **culms** arising singly or tufted, erect, terete or somewhat flattened but not 2-edged, smooth. **Leaf blades** flat or involute-folded, 1–4 mm wide; sheaths glabrous, not keeled, closed only near the base; **ligules** 2–5 mm long, acute. **Panicles** usually contracted, sometimes rather open with the branches spreading-ascending, 2.5–1 5 cm long. **Spikelets** 3- to 9-flowered, often attractively tricolored owing to the glaucous glumes and the purple sides and whitish or bronzed tips of the lemmas, pale with age, 4–9 mm long, 1.5–3.5 mm wide; glumes ovate to ovate-lanceolate, obtuse to rounded at the tip, glabrous or scaberulous on the midnerve, somewhat unequal, the first glume 2.3–3.6 mm long, 1- to 3-nerved, the second glume 2.8–4.2 mm long, 3-nerved; lemmas 2.3–4 mm long, weakly keeled to rounded on the back, obtuse to rounded and somewhat erose at the tip, villous on the mid- and lateral nerves, pubescent between the nerves on the lower back, lacking cobwebby hairs at the base; anthers 1.2–2 mm long. **Grain** brown, obovoid, ca. 1 mm long. June–Aug.

HABITAT Alkaline wet meadows, shores and seepage areas, often abundant in such places, also on drier upland sites.

WETLAND STATUS

GP FAC | MIDW FAC | WMTN UPL

Poa compressa L.

CANADA BLUEGRASS

DESCRIPTION Strongly rhizomatous, glaucous perennial 1.5–6(8) dm tall; **culms** erect to decumbent, wiry, compressed and sharply 2-edged. **Leaf blades** flat to folded, 1–3.5(4.5) mm wide; sheaths compressed, strongly keeled, closed only near the base; **ligules** mostly 0.5–2 mm long, truncate. **Panicles** usually narrow and contracted with short branches, sometimes more open with the branches spreading, 1–10 cm long, the lowest branches usually 2 in number. **Spikelets** 2- to 6(7)-flowered, compressed, 2.8–5.5 mm long, 0.8–2 mm wide; glumes broadly lanceolate, subequal, keeled, the first glume 1.2–2.8 mm long, 1- to 3-nerved, the second glume 1.5–3.1 mm long, 3-nerved; lemmas (1.6)2–3 mm long, strongly keeled, usually purple-banded and bronzed at the tip, glabrous on the back or pubescent on the keel and marginal nerves, sometimes with a tuft of cobwebby hairs at the base; anthers 1–1.7 mm long. **Grain** brown, ellipsoid, 1–1.5 mm long. June–Aug.

HABITAT Shores, streambanks and meadows, also moist to dry prairies, woodlands and disturbed areas, often on poor soils; common; introduced from Europe, naturalized from Newf. to Alaska, s throughout most of USA.

WETLAND STATUS

GP FACU | MIDW FACU | WMTN FACU

Poa compressa
CANADA BLUEGRASS

Poa palustris L.

FOWL BLUEGRASS

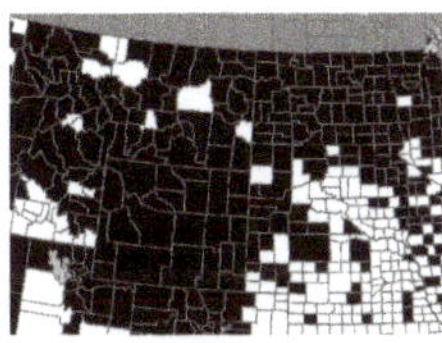

DESCRIPTION Loosely tufted perennial 4–11 dm tall; **culms** terete, usually curved and decumbent at the base, producing roots from lower nodes and thus stolonous (simulating a short-rhizomatous base), usually purplish toward the base. **Leaf blades** ascending to spreading, rather lax, flat, 0.5–5 mm wide, scabrous; sheaths glabrous, rounded to slightly keeled, closed near the base; **ligules** mostly 1.6–5 mm long, rounded or lacerate. **Panicle** diffuse and loosely spreading (contracted when young and emerging from the sheath), 8–30 cm long.

Poa palustris
FOWL BLUEGRASS

Spikelets 2- to 4-flowered, 2–5 mm long, 1.5–2 mm wide; glumes subequal, often purplish, lanceolate, acute, scabrous on the keel, the first glume 1.5–3 mm long, 1- to 3-nerved, the second glume 2–3.1 mm long, 3-nerved; lemmas 2–3 mm long, subacute, often purplish on the sides, bronzed at the tip, pubescent on the marginal nerves and keel toward the base, also bearing a tuft of cobwebby hairs at the base, this sometimes scant, the intermediate nerves very faint to obsolete so that the lemma appears 3-nerved; anthers 0.8–1.2 mm long. **Grain** brown, fusiform to narrowly ellipsoid, ca. 1 mm long. June–early Sept.

HABITAT Wet meadows, marshes, shores, streambanks, ditches and low prairie, also moist woods and hillsides.

WETLAND STATUS

GP FACW | MIDW FACW | WMTN FAC

ADDITIONAL SPECIES **Inland bluegrass,** ***Poa interior*** Rydb., is sometimes encountered on streambanks and in other moist to fairly dry habitats. It is similar to *P. palustris* except for its smaller stature and densely tufted nature. It can be distinguished from *P. palustris* by its smaller panicles, 5–15 cm long, and the shorter ligules, mostly less than 1.6 mm long.

Poa pratensis L.

KENTUCKY BLUEGRASS

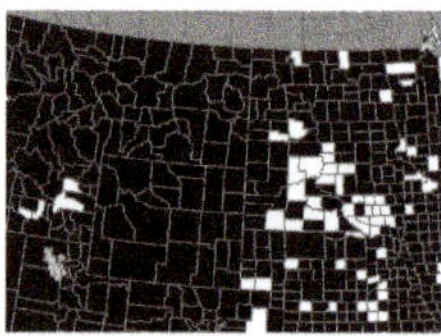

DESCRIPTION Strongly rhizomatous, sod-forming perennial (2)3–8(12) dm tall, the foliage green or slightly glaucous; **culms** erect, terete to slightly flattened, not 2-edged. **Leaf blades** flat to involute-folded, 0.5–4 mm wide, glabrous or seldom lightly pubescent on the upper surface, often slightly scabrous on the margins; sheaths glabrous, rounded to slightly keeled, closed in the lower 1/2; **ligules** 0.5–1.5(2) mm long, truncate. **Panicles** usually open, often pyramidal, 3–13 cm long, the branches spreading to ascending, usually numbering 4–5 at the lowest node, the spikelets crowded on the branches so that the panicle is somewhat condensed. **Spikelets** 2- to 6-flowered, green or purplish, compressed, 3–5 mm long, 1.5–3 mm wide; glumes ovate-lanceolate to lanceolate, scabrous on the keel, mostly unequal, the first glume 1.8–3 mm long, 1- to 3-nerved, acute, the second glume 2.2–3.8 mm long, 3-nerved, acuminate; lemmas 2–3.8 mm long, strongly keeled, acute, sericeous on the keel and marginal nerves toward the base, also with a prominent tuft of cobwebby hairs at the base, glabrous or scabrous above on the keel, often marked with purple on the sides or margins, white or lightly bronzed at the tip, the intermediate nerves usually evident; anthers 1–1.8 mm long. **Grain** brown, ellipsoid, ca. 1.5 mm long, the lemma and palea adherent to it. May–Aug.

HABITAT Wet meadows, shores, streambanks and a great variety of moist to dry habitats; ubiquitous and often abundant, introduced for lawns and pastures but possibly native as well.

WETLAND STATUS

GP FACU | MIDW FAC | WMTN FAC

Poa pratensis
KENTUCKY BLUEGRASS

Poa trivialis L. (*not illustrated*)
ROUGH BLUEGRASS

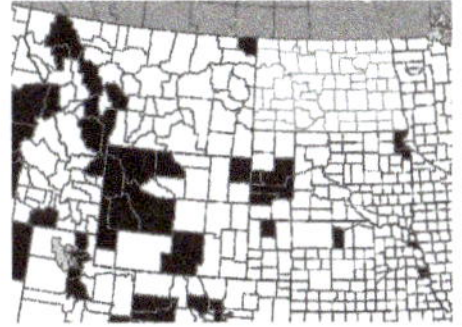

DESCRIPTION Very similar to *Poa palustris,* differing mainly as follows: **Leaves** and sheaths usually more scabrous; **spikelets** 2- to 3-flowered, 3–4 mm long; glumes unequal, narrow and curved, acute to acuminate, the first glume 1.5–2.3 mm long, 1-nerved; second glume 2–3 mm long, 3-nerved; lemmas usually green with whitish tip and margins, sometimes marked with purple and slightly bronzed at the tip, pubescent on the keel only, not on the marginal nerves, the intermediate nerves prominent so that the lemma is plainly 5-nerved. June–July.

HABITAT In wet soil or shallow water around springs or along spring-fed streams; also to be expected in moist, shaded places.

WETLAND STATUS
GP FACW | MIDW FACW | WMTN FAC

NOTE Introduced from Europe and now naturalized over much of e and w N America, probably increasing in our region.

Polypogon monspeliensis (L.) Desf.
RABBITFOOT BEARDGRASS

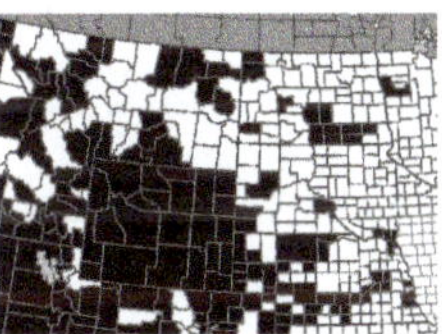

DESCRIPTION Tufted annual 1–5(8) dm tall; **culms** erect or decumbent at the base. **Leaves** flat, 3–7 mm wide, scaberulous; sheaths glabrous, the **ligule** membranous, elongate, 2–7 mm long. **Panicle** ovoid-cylindric to cylindric, rather dense and spike-like, (1)3–8(15) cm long. **Spikelets** 1-flowered, 2 mm long (excluding awns of the glumes), tawny at maturity, disarticulating below the glumes; glumes equal, awned from the 2-lobed apex, hispidulous, 1-nerved, the awn 4–8 mm long; lemma much smaller than the glumes, 0.7–1 mm long, scarious, translucent and shiny, 1-nerved, erose at the apex, with a fragile awn about as long as or longer than the lemma body; palea about as long as the lemma and of the same texture. **Grain** golden-brown, ellipsoid, slightly smaller than the lemma, often retained inside the lemma and palea when shed. July–Sept.

HABITAT Wet sand or mud of streambanks, shores and around springs; introduced from s Europe.

WETLAND STATUS
GP FACW | MIDW OBL | WMTN FACW

Puccinellia ALKALIGRASS

Tufted perennials with flat to involute, mostly basal, pale green leaves and ultimately open panicles, the panicle branches ascending, spreading or reflexed with age, scabrous. **Spikelets** few- to several-flowered, elliptic to linear, subterete, disarticulating above the glumes and between the florets; glumes unequal, acute or obtuse, scarious at the tip, the first 1-nerved, the second 3-nerved; lemmas rounded on the back, acute to obtuse or truncate, scarious and often erose at the tip, often puberulent at the base, faintly 5-nerved; palea about as long as the lemma or shorter, its margins somewhat clasping the grain. **Grain** brown, ellipsoid or narrowly so, 1–1.5 mm long, tomentose at the tip.

1 Lemmas rounded to truncate, not narrowed at the apex, 1.6–2 mm long; panicle branches reflexed at maturity ***P. distans***

1 Lemmas acute to obtuse or rounded, narrowed at the apex, 1.8–3.2 mm long; panicle branches usually ascending to spreading ***P. nuttalliana***

Polypogon monspeliensis
RABBITFOOT BEARDGRASS

Puccinellia distans (L.) Parl.

WEEPING ALKALIGRASS

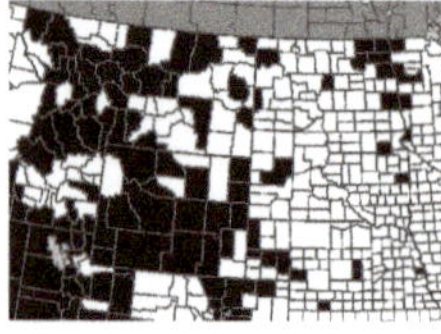

DESCRIPTION **Culms** 1–5 dm tall; **leaves** flat to slightly involute, 1–3 mm wide, glabrous; **ligule** white-hyaline, rounded, 0.5–1 mm long. **Panicle** pyramidal, 5–15 cm long, the branches (at least the lower ones) reflexed at maturity. **Spikelets** 3- to 7-flowered, 3–6 mm long; first glume acute, 0.5–1.5 mm long, the second glume obtuse to rounded, erose; lemmas rounded to truncate, erose, not narrowed at the apex, 1.6–2 mm long; anthers 0.5–0.8 mm long. Late June–Aug.

HABITAT Springs, shores, ditches and wet disturbed places, not necessarily where alkaline or saline.

WETLAND STATUS
GP FACW | MIDW OBL | WMTN FACW

Puccinellia distans
WEEPING ALKALIGRASS

Puccinellia nuttalliana (Schult.) Hitchc.

NUTTALL'S ALKALIGRASS

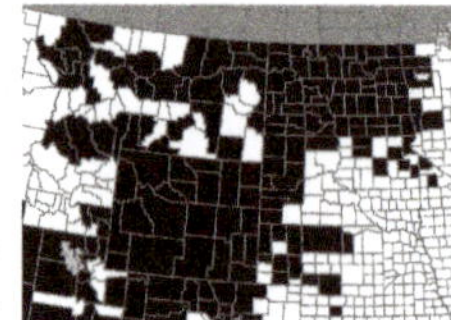

DESCRIPTION **Spikelets** purplish or green and becoming stramineous or purplish-brown at maturity; second glume 1–2 mm long; lemmas 1.8–2.5(3) mm long; anthers 0.4–1 mm long. June–July.

SYNONYMS *Puccinellia airoides* (Nutt.) Wats. & Coult., *Puccinellia cusickii* Weatherby

HABITAT Wet saline or alkaline meadows, shores, streambanks, ditches, flats, seepage areas.

WETLAND STATUS
GP OBL | MIDW OBL | WMTN FACW

Scolochloa festucacea (Willd.) Link

WHITETOP, SPRANGLETOP

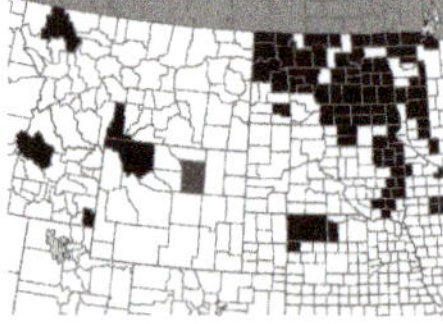

DESCRIPTION Tall rhizomatous perennial, often extensively colonial, 8–15 dm tall; **culms** stout, hollow, 3–5 mm thick

Puccinellia nuttalliana
NUTTALL'S ALKALIGRASS

near the base, usually with a few suckers and adventitious roots at the lower nodes. **Leaves** 3–10 mm wide, long-tapering to the sharp, slender tip, flat or involute along the margins, scabrous above, the sheaths glabrous; **ligule** white-hyaline, delicate, lacerate with age, 4–7 mm long. **Panicle** loose, open, 15–20 cm long, the branches ascending, bearing most of the spikelets above the middle of the panicle. **Spikelets** 3- to 4-flowered, purplish or green, turning stramineous with age, 7–10 mm long, disarticulating above the glumes and between the lemmas, the rachilla extended beyond the last floret; glumes unequal, lanceolate, acute or lacerate at the tip, the first glume 3-nerved, 4–7 mm long, the second glume 5-nerved, 6–9 mm long; lemmas lanceolate, acute or lacerate at the tip, the lower one or two 5–7 mm long, the upper ones smaller, firm, 7-nerved; palea as long as the lemma, firm, finely ciliate along the marginal nerves. **Grain** brown, white-pubescent at the tip, oblong to ellipsoid, furrowed on the upper side, 2–2.5 mm long. June–July.

SYNONYM *Fluminea festucacea* (Willd.) Hitchc.

HABITAT Usually in shallow water of fresh to brackish marshes, lakes and ponds; common and often abundant in e Great Plains.

WETLAND STATUS
GP OBL | MIDW OBL | WMTN OBL

Scolochloa festucacea
WHITETOP, SPRANGLETOP

NOTE This grass is one of few emergent species which can provide a valuable forage for cattle. During dry periods, whitetop is sometimes mowed for hay. Under grazing pressure, *Scolochloa* disappears and is often replaced by undesirable bulrushes.

Spartina CORDGRASS

Coarse perennials spreading by long scaly rhizomes; **culms** stout and erect, tough. **Leaves** flat to involute, tough, scabrous; sheaths glabrous; **ligule** comprised of hairs. **Inflorescence** of several to many 1-sided spikes arranged in a terminal raceme, the spikes ascending to appressed; **spikelets** 1-flowered, strongly compressed and overlapping in 2 rows on one side of the rachis, disarticulating below the glumes; glumes unequal, flattened, hispid or scabrous on the keel, the first glume linear, shorter than the lemma, 1-nerved, the second glume narrowly lanceolate, exceeding the lemma, 1- or 2-nerved along the keel; lemma flattened, lanceolate, weakly hispid or scabrous on the keel, with one strong midvein and 2 obscure lateral ones; palea flattened, often slightly longer than the lemma, very obscurely 2-nerved. **Grain** golden brown, narrowly oblong and compressed, 4–6 mm long.

1 Second glume acute to mucronate; leaves usually involute, 2–5 mm wide when flattened; plants mostly 3–8 dm tall ***S. gracilis***

1 Second glume awned, the awn 2–10 mm long; leaves flat to involute, 3–13 mm wide when flattened; plants mostly 7–18 dm tall ***S. pectinata***

Spartina gracilis Trin. (*not illustrated*)
ALKALI CORDGRASS

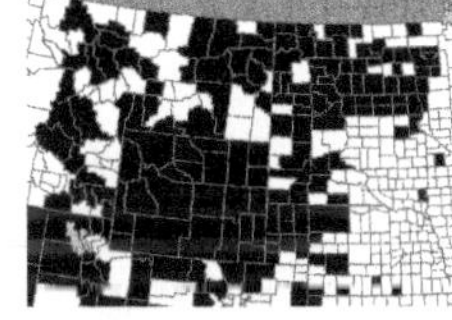

DESCRIPTION Plants mostly 3–8 dm tall; **leaves** usually involute, 2–5 mm wide when flattened. **Inflorescence** of 3–8, 1-sided spikes, the spikes appressed in the raceme, 2–6 cm long. **Spikelets** 7–9 mm long; glumes hispid on the keel, the first glume ca. ½ as long as the second, acuminate, the second glume acute to mucronate; lemma nearly as long as the second glume, 6–8 mm long.

July–Sept.

SYNONYM *Sporobolus hookerianus* P.M. Peterson & Saarela

HABITAT Wet meadows, shores, flats and seepage areas, often where alkaline or saline.

WETLAND STATUS
GP FACW | MIDW FACW | WMTN FACW

Spartina pectinata Link
PRAIRIE CORDGRASS

DESCRIPTION More robust than *Spartina gracilis,* plants mostly 7–18 dm tall; **leaves** flat to involute, 3–13 mm wide when flattened. **Inflorescence** of 4–30, 1-sided spikes, the spikes ascending or appressed, 3–10 cm long. **Spikelets** 8–11 mm long (excluding the awn of the second glurne); glumes strongly scabrous on the keel, the first glume shorter than to equaling the lemma, acuminate or with an awn 1–5 mm long, the second glume with an awn 2–10 mm long; lemma considerably shorter than the second glume, 7–9 mm long. July–Sept.

Spartina pectinata
PRAIRIE CORDGRASS

SYNONYM *Sporobolus michauxianus* (A.S. Hitchc.) P.M. Peterson & Saarela

HABITAT Shallow marshes, wet meadows, ditches, low prairie and other wet or moist places, where water is fresh to brackish; very common, often abundant.

WETLAND STATUS
GP FACW | MIDW FACW | WMTN OBL

Sphenopholis obtusata (Michx.) Scribn.
PRAIRIE WEDGEGRASS

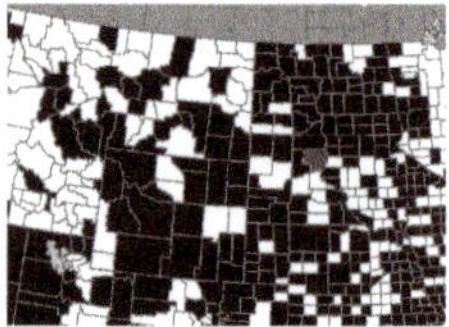

DESCRIPTION Tufted or sometimes solitary perennial 2–8 dm tall; **culms** slender, 0.5–1 mm thick. **Leaves** ascending to spreading, flat, scabrous, 1–7 mm wide; sheaths glabrous, or the basal ones puberulent; **ligule** white-hyaline, lacerate, 1–4 mm long. **Panicle** dense, contracted and spike-like, often interrupted, (1.5)5–12 cm long.

Sphenopholis obtusata
PRAIRIE WEDGEGRASS

Spikelets 2-flowered, 2.5–3.5 inm long, disarticulating below the glumes; glumes 1.5–2.5 mm long, scabrous on the keel, the first glume linear, 1-nerved, the second glume much broader than the first, obovate, 3- to 5-nerved; lemmas acute, 2–3 mm long, scaberulous toward the apex on the keel, 1-nerved; palea exposed, linear, about as long as the lemma, scarious. **Grain** light brown, elongate, flattened, 1.5–2 mm long. Late June–Aug.

HABITAT Low prairie, wet meadows, shores, streambanks and moist woods.

WETLAND STATUS

GP FAC | MIDW FAC | WMTN FAC

Zizania palustris L.

ANNUAL WILDRICE

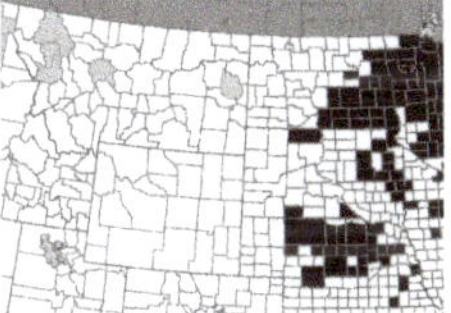

DESCRIPTION Stout annual 5–20 dm tall, with rather fleshy yellowish-orange roots; **culms** solitary or few, robust. **Leaves** flat, 5–25 mm wide, smooth; sheaths puberulent at the collar, otherwise glabrous; ligule hyaline, striate, entire or lacerate, 5–15 mm long. **Panicle** elongate to pyramidal, 20–40 cm long, the lower staminate portion ultimately expanded, the upper pistillate portion remaining contracted. **Spikelets** unisexual, terete, the pistillate above the staminate in the panicle; glumes absent; **staminate florets** purple, 6–12 mm long, the lemmas linear, acuminate or short-awned, strongly 3-nerved, hispid, thin-textured; **pistillate florets** purplish or light green, the lemmas subulate, 10–14(20) mm long and tapering to a slender awn 2–4 cm long, 3-nerved, hispid, chartaceous; palea about as long as the lemma, 3-nerved. **Grain** dark brown to black, slender and elongate, about as long as the body of the lemma, early deciduous. Late July–early Sept.

HABITAT Shallow water or mud of streams, rivers, oxbow swamps and marshes, where water is fresh.

WETLAND STATUS

GP OBL | MIDW OBL | WMTN OBL

NOTE This plant is the source of the commercial wildrice. In the northern Great Plains local populations are generally not large enough to serve this purpose, but wildrice harvesting is practiced in Minnesota where extensive areas in marshes and around lakes may be dominated by the plant. The grain is an excellent food for waterfowl.

Upright pistillate spikelets above the lower spreading staminate spikelets

Zizania palustris
ANNUAL WILDRICE

Pontederiaceae
pickerelweed family

Heteranthera MUD PLANTAIN

Aquatic or amphibious annuals and perennials. **Leaves** alternate, sessile and straplike or differentiated into petiole and expanded blade (membranous-sheathing at the base in the latter). **Flowers** solitary (ours) from the axils, subtended by a sheathing spathe, pale yellow or white to purplish-blue, perfect, regular, hypogynous; perianth of 6 petaloid lobes, united below into a tube, salverform; stamens 3 (in ours), all alike or 1 unlike the other 2, the filaments adnate to the throat of the perianth tube; pistil 3-carpellary, stigma 3-lobed, style 1, ovary incompletely 3-celled. **Fruit** a many-seeded, often indehiscent capsule retained inside the spathe; seeds conspicuously ribbed.

1 Leaves differentiated into petiole and expanded blade, emersed or floating; flowers white to purplish-blue *H. limosa*

1 Leaves linear, straplike, not differentiated into petiole and blade, usually submersed; flowers light yellow *H. dubia*

Heteranthera dubia (Jacq.) MacM.
WATER STARGRASS

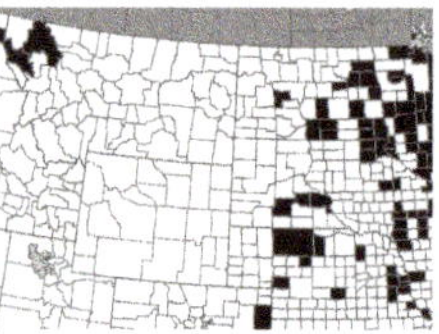

DESCRIPTION Usually submersed or partly floating perennial with lax stems and leaves, or occasionally stranded and forming tiny leafy rosettes on exposed mud or sand. **Stems** slender, elongate, freely branched, often rooting at the lower nodes, to 1 m long. **Leaves** alternate, sessile, stipitate, linear, obtuse to rounded or apiculate at the tip, 2–10(15) cm long, 2–5(7) mm wide, several-nerved, the midvein and others inconspicuous, stipular-winged at the base but not sheathing the stem. **Flowers** solitary in upper leaf axils, the membranous spathe surrounding the perianth tube for much of its length, 10–15 mm long; perianth tube slender, often curved, 15–35(70) mm long, the perianth segments pale yellow, linear, 4–6 mm long; stamens alike. June–Sept.

SYNONYM *Zosterella dubia* (Jacq.) Small

HABITAT Streams, ponds, impoundments.

WETLAND STATUS
GP OBL | MIDW OBL | WMTN OBL

Heteranthera limosa (Sw.) Willd.
MUD PLANTAIN

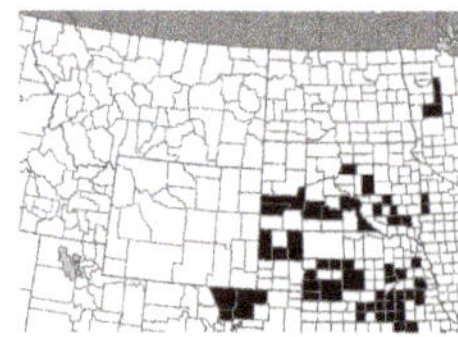

DESCRIPTION Small amphibious annual (in this region), with stems short when emersed (to somewhat elongated and sprawling when in water). **Leaves** differentiated into petiole and blade, with the blades emersed or floating, ovate to elliptic or elliptic-lanceolate, mostly 2–10 cm long, rounded at the tip, subcordate to truncate or sometimes tapered at the base; petioles mostly 4–20 cm long, with a membranous sheath at the base. **Flower** and its elongate spathe borne on a stout peduncle; spathe 2–4 cm long, enclosing the tubular portion of the perianth; perianth segments white to usually purplish-blue, the upper 1–3 yellow at the base, linear-lanceolate, 5–10 mm long, the perianth tube 1–3.5 cm long; stamens dissimilar, 2 with short yellow anthers and the other with a more elongate, blue or yellow anther. June–Sept.

HABITAT Shallow water or mud of ponds and marshes.

WETLAND STATUS
GP OBL | MIDW OBL | WMTN OBL

Heteranthera limosa MUD PLANTAIN

Heteranthera dubia WATER STARGRASS

Potamogetonaceae
pondweed family

Submersed or floating-leaved, rhizomatous aquatics, perennial from the rhizomes or tubers or by budding from the lower nodes, sometimes reproducing and overwintering by free-floating winter buds. **Stems** elongate, flexuous, anchored by roots and rhizomes. **Leaves** alternate, becoming opposite upward in some species, simple; stipules present, fused to each other along one or both margins or adnate to the leaf blade in some linear-leaved species, forming an open or closed sheath around the stem, fibrous or membranous, often rapidly deteriorating. **Submersed leaves** filiform to lanceolate, thin and often flexuous, usually sessile. **Floating leaves** produced in some species, basically elliptic or oblong in outline, petioled, rather leathery with a waxy upper surface. **Winter buds** produced in the leaf axils of some species, these consisting of tightly compressed apices bound by reduced leaves. **Inflorescences** of axillary or terminal spikes bearing few to many minute whorled flowers; peduncles stout to filamentous, usually lifting the spike above the water surface at anthesis, often recurved with age. **Flowers** perfect, regular; perianth of 4 sepaloid bracts (these considered enlargements of the staminal connectives by some authors); stamens 4, each inserted on the claw of a perianth bract; carpels 4, separate, each maturing into a strongly to weakly beaked, nutlet-like fruit.

1 Submersed leaves opposite or whorled, floating leaves absent ***Zannichellia***

1 Submersed leaves alternate, floating leaves (sometimes present) alternate or opposite **2**

2 Flowers 2, at first enclosed in sheathing leaf base, the peduncle elongating and often spiraled or coiled at its base; fruit longstalked; stipular sheath lacking free ligule at summit (the stipule wholly adnate to the leaf blade and merely rounded at the summit); leaf blade terete ***Ruppia***

2 Flowers several to many in a peduncled head or spike; perianth of 4 tepals; fruit ± sessile; stipular sheath absent (stipules entirely free from leaf) or with a short ligule-like extension if stipules fused to the leaf blade **3**

3 Stipules adnate to the leaves for 10–30 mm or more (at least on the larger leaves), adnate for ca. 2/3 of the length of the stipule; leaves all submersed, filiform to narrowly linear (up to 2.5 mm wide) ***Stuckenia***

3 Stipules free from the leaves or adnate for less than half the length of the stipule (adnate for 5 mm or less except in *P. robbinsii*); leaves submersed or floating, filiform to ovate, oblong, or elliptic ***Potamogeton***

Potamogeton PONDWEED

Aquatic perennial herbs, with only underwater leaves or with both underwater and floating leaves, from rhizomes or tubers, sometimes reproducing and over-wintering by free-floating winter buds. **Stems** long, wavy, anchored to bottom by roots and rhizomes. **Leaves** alternate, or becoming opposite upward in some species, simple, with an open or closed sheath at base. **Underwater leaves** usually linear and threadlike, sometimes broader, margins often wavy, usually stalkless. **Floating leaves**, if present, oval or ovate, stalked, with a waxy upper surface. **Flowers** perfect, regular, green to red, in stalked spikes at ends of stems or from leaf axils, usually raised above water surface, the spikes with few to many small flowers; perianth of 4 sepal-like bracts; stamens 4. **Fruit** a 4- parted, beaked achene.

1 Submersed leaves linear, 6 mm or less wide (sometimes wider in *P. epihydrus*), mostly 20x or more longer than wide **2**

1 Submersed leaves linear-lanceolate, lanceolate, oblong or ovate, broader in proportion to the length **9**

2 Stipules adnate to submersed leaf blades for mostly 1–4 mm; embryo coil plainly visible through the papery thin walls of the fruit; small elliptic floating leaves usually present, 5–40 mm long ***P. diversifolius***

2 Stipules free of the leaf blades; fruit walls firm, the embryo coil obscured by the walls of the fruit; floating leaves, if present, mostly larger **3**

3 Leaves dimorphic, both floating and submersed leaves produced 4
3 Leaves all alike, all submersed 5

4 Submersed leaves reduced to phyllodes 1–2 mm wide, these often absent with age; floating leaves rounded to cordate at the base ***P. natans***
4 Submersed leaves ribbonlike, mostly 3–6(10) mm wide, with a cellular-reticulate strip on each side of the midvein forming a conspicuous median band 1–2 mm wide; floating leaves tapered to the petiole ***P. epihydrus***

5 Leaves with many (15–35) fine nerves; mature fruits 4–4.5 mm long ***P. zosteriformis***
5 Leaves 3- to 7(9)-nerved; mature fruits 1.5–3.6 mm long 6

6 Fruits with an undulate to dentate dorsal ridge or keel; glands rarely present at the base of the stipules ***P. foliosus***
6 Fruits dorsally smooth and rounded; glands usually present at the base of the stipules 7

7 Stipules tan to brownish-green, delicate, usually decomposing with age; peduncles filiform to cylindric; winter buds seldom present ***P. pusillus***
7 Stipules whitish, fibrous, the oldest often shredding into fibers; peduncles mostly clavate; indurate winter buds often present 8

8 Leaf tips acute, rarely obtuse; leaves 3- to 5(7)-nerved, 0.6–2 mm wide; peduncles mostly terete; stems mostly terete ***P. strictifolius***
8 Leaf tips rounded to apiculate; leaves 5- to 7(9)-nerved, 1.2–3.2 mm wide; peduncles compressed; stems compressed ***P. friesii***

9 Leaves all submersed, sessile, weakly to strongly clasping at the base, often undulate-crisped 10
9 Floating leaves commonly present by flowering time, occasionally lacking; submersed leaves sessile or petiolate, not clasping the stem; flat to falcate, not undulate-crisped 12

10 Leaf margins finely serrate; fruit beak 2–3 mm long; indurate winter buds commonly produced in upper axils ***P. crispus***
10 Leaf margins entire; fruit beak 1.5 mm or less long; winter buds lacking 11

11 Stems whitish; leaves 10–25 cm long; peduncles over 10 cm long; fruits 4–5 mm long ***P. praelongus***
11 Stems brownish to yellowish-green; leaves less than 10 cm long; peduncles 2–10 cm long; fruits 2.5–3.5 mm long ***P. richardsonii***

12 Upper submersed leaves falcate-folded, 25- to 50-nerved; mature fruits 4–5 mm long ***P. amplifolius***
12 Upper submersed leaves more or less symmetrical and not folded, 3- to 17(19)-nerved; mature fruits 1.7–4 mm long 13

13 Fruits tawny-olive; floating leaves often lacking, thin and delicate, the blade tapering indistinctly into a short petiole; submersed foliage reddish-tinged ***P. alpinus***
13 Fruits brown, reddish-brown or greenish; floating leaves leathery, the blades distinct from the petioles; submersed foliage dark green to brownish-green 14

14 Submersed leaves commonly disintegraticg by fruiting time, tapering to petioles (2)4 cm long or much longer, acute to blunt-tipped; mature fruits brownish to reddish-brown, 3–4 mm long ***P. nodosus***
14 Submersed leaves usually persistent, sessile or tapering to petioles up to 4 cm long, acute to abruptly acuminate or apiculate; mature fruits green or olive, 1.7–3.5 mm long 15

15 Stems usually freely branched, 0.5–1 mm thick; submersed leaves 3-10(15) mm wide, 3- to 7(9)-nerved; floating leaf blades 2–9 cm long, 1–3.5 cm wide; fruiting spikes 1.5–3.5 cm long; fruits 1.7–2.8 mm long, the lateral keels obscure ***P. gramineus***
15 Stems simple or once-branched, 1–5 mm thick; submersed leaves (1)1.5–4 cm wide, 9- to 17(19)-nerved; floating leaf blades 4–

14(19) cm long, 2–7 cm wide; fruiting spikes 2–6 cm long; fruits 2.7–3.5 mm long, the lateral keels strong *P. illinoensis*

Potamogeton alpinus Balbis

REDDISH PONDWEED

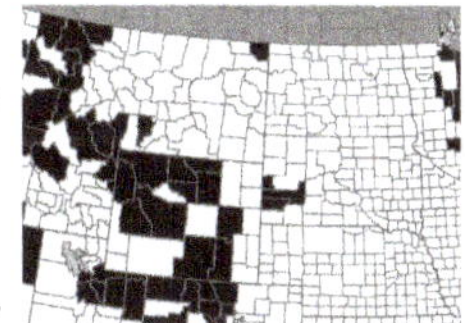

DESCRIPTION **Stems** terete, 1–2 mm thick, simple or rarely branched above, to 1 m long; foliage reddish-tinged, especially the upper leaves and peduncles. **Submersed leaves** linear-lanceolate to linear-oblong, 4–18 cm long, 5–15(20) mm wide, usually 7(-11)-nerved, blunt and obtuse to rarely acutish at the tip, narrowed to a sessile base. **Floating leaves** often lacking, thin and delicate, obovate or oblanceolate to elliptic-oblanceolate, mostly 4–6 cm long, 1–2 cm wide, 7- to 15-nerved, obtuse, tapering indistinctly into a short petiole. Stipules free, membranous, 1–2.5(4) cm long. **Spikes** cylindric, with 5–9 crowded whorls of flowers, peduncles about as thick as the stem, 3–15 cm long. **Fruits** tawny-olive, obliquely obovoid, 2.5–3.5 mm long; dorsal keel usually narrow and prominent; lateral keels absent or low and rounded; beak short, curved backward. July–Sept.

HABITAT Cold streams and lakes.

WETLAND STATUS

GP OBL | MIDW OBL | WMTN OBL

Potamogeton alpinus
REDDISH PONDWEED

Potamogeton amplifolius Tuckerm.

LARGE-LEAF PONDWEED

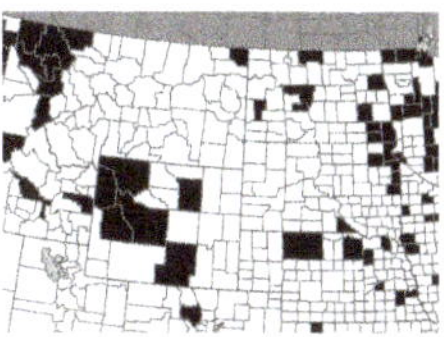

DESCRIPTION **Stems** terete, 2–4 mm thick, simple or occasionally branched above, to 1 m long. **Upper submersed leaves** broadly lanceolate to ovate, falcately folded and often arcuate, 8–20 cm long, 2–7 cm wide, 25- to 50-nerved; **lower submersed leaves** often decayed by fruiting time, lanceolate, often not folded, 19- to 25-nerved; both submersed leaf types obtuse to broadly acute, tapering to petioles 1–6 cm long. **Floating leaves** seldom lacking at flowering time, ovate to elliptic, 5–10 cm long, 3–7 cm wide, 25- to 45-nerved, obtuse or abruptly acute, cuneate or rounded at the base; petioles 5–15 cm long; stipules open and free of the petioles, 5–12 cm long, fibrous and persistent. **Spikes** cylindric, dense, 4–8 cm long in fruit; peduncles broadening upward, 5–20(30) cm long. **Fruits** greenish-brown to brown, obliquely obovoid, 4–5 mm long; dorsal keel prominent, the lateral keels less distinct; beak to 1 mm long. July–Aug.

HABITAT Quiet waters of streams and lakes.

WETLAND STATUS

GP OBL | MIDW OBL | WMTN OBL

Potamogeton crispus L.

CURLY PONDWEED

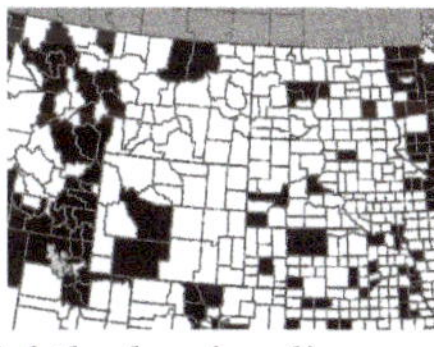

DESCRIPTION **Stems** slightly compressed, mostly 1–2 mm thick, usually branching, mostly 4–8 dm long. **Leaves** all submersed, sessile, slightly clasping, linear-oblong to linear-oblanceolate or oblong to

Potamogeton amplifolius
LARGE-LEAF PONDWEED

oblanceolate, 3–8 cm long, 3–10 mm wide, 3- to 5-nerved, rounded at the tip, narrowed at the base, the margin usually undulate-crisped, finely serrate; stipules slightly adnate at the base, 4–10 mm long, early shredding. Winter buds commonly produced in some of the leaf axils, indurate, ca. 1–2 cm long. **Spikes** dense, short-cylindric, 1–2 cm long; peduncles terete, about as thick as the stem, 2–5(7) cm long. **Fruits** brown, ovoid, the body 2–3 mm long, the beak very prominent, 2–3 mm long; keels low, rounded. Apr–June.

HABITAT Shallow water of lakes, ponds and slow-moving streams, especially recreational waters, where potentially a nuisance; sometimes locally abundant; introduced along both coasts and inland in N America.

WETLAND STATUS
GP OBL | MIDW OBL | WMTN OBL

Potamogeton diversifolius Raf.

WATER-THREAD PONDWEED

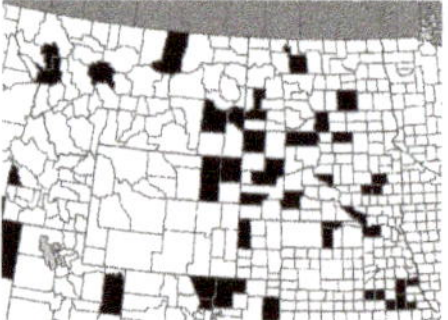

DESCRIPTION **Stems** slender, terete, 0.5–1 mm thick, to 8 dm long, usually with short lateral branches. **Submersed leaves** linear, flat, 1–8 cm long, 0.3–1.5 mm wide, 1(3)-nerved, obtuse to long-acuminate; stipules of submersed leaves 2–18 mm long, membranous, adnate to the leaf blade for mostly 1–4 mm, usually adnate for less than 1/2 the total length. **Floating leaves** sometimes lacking, the blades elliptic to elliptic-lanceolate or elliptic-oblanceolate, 5–40 mm long, 5–20 mm wide, 3- to 17-nerved, acute to rounded at the tip, cuneate to rounded at the base; petioles mostly 5–40 mm long; stipules of floating leaves free of the leaf bases, 2–25 mm long, membranous to weakly fibrous. **Spikes** dimorphic, the lower submersed spikes capitate to ellipsoid, 1.5–6 mm long, few- to several-fruited, the upper spikes ellipsoid to cylindric, 5–30 mm long, usually many-fruited; peduncles slightly clavate, 3–32 mm long. **Fruits** olive to yellowish, round and flattened, with a prominent winged dorsal keel and slightly ridged to winged lateral keels, the keels entire to toothed, the embryo coil plainly visible through the papery thin walls of the fruit, the beak minute. June–Sept.

HABITAT Shallow water of ponds and marshes.

WETLAND STATUS
GP OBL | MIDW OBL | WMTN OBL

ADDITIONAL SPECIES **Spiral pondweed** (*Potamogeton spirillus* Tuck.), similar to *P. diversifolius,* but the underwater leaves typically blunt-tipped, and the floating leaves with a small notch at tip; in *P. diversifolius,* the underwater leaves are generally tapered to a pointed tip, and floating leaves are not notched at tip.

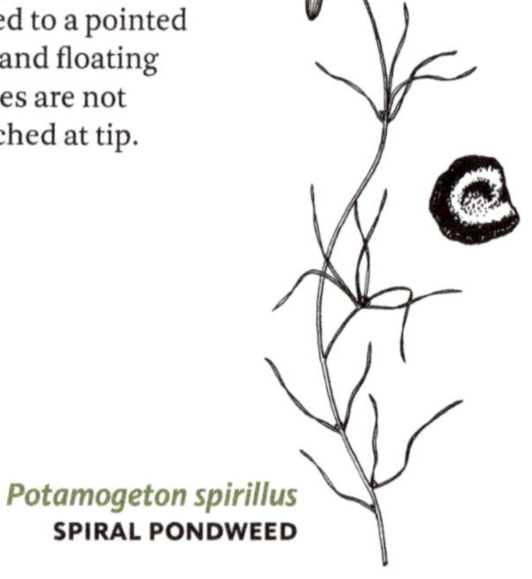

Potamogeton spirillus
SPIRAL PONDWEED

Potamogeton crispus
CURLYPONDWEED

Potamogeton diversifolius
WATER-THREAD PONDWEED

Potamogeton epihydrus Raf.

RIBBONLEAF PONDWEED

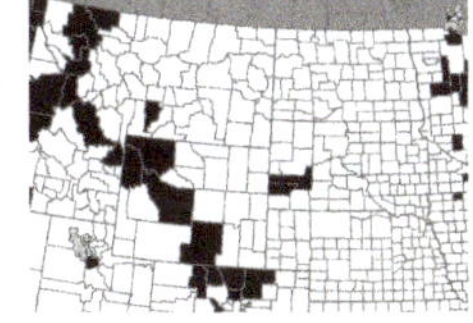

DESCRIPTION **Stems** somewhat flattened, 1–2 mm thick, simple or sparingly branched, to 2 m long. **Submersed leaves** linear, ribbonlike, 5–20 cm long, 3–6(10) mm wide, 5- to 7(13)-nerved, with a cellular-reticulate strip on each side of the midvein forming a conspicuous median band 1–2 mm wide, acute to rather blunt, slightly tapered to the sessile base. **Floating leaves** usually present, elliptic or oblong-elliptic, (2) 3–8 cm long, (5) 10–20 mm wide, mostly 11- to 25-nerved, obtuse to bluntly mucronate at the tip, tapering to flattened petioles which are usually shorter than the blades; stipules free, rather membranous and delicate, 1–3 cm long. **Spikes** dense, cylindric, usually 2–4 cm long; peduncles about as thick as the stem, 3–8 cm long. **Fruits** olivaceous to brown, broadly and obliquely obovate, concave on the sides, 2–3 mm long; dorsal keel prominent, thickly winged; lateral keels low, mostly rounded; beak minute. July–Sept.

HABITAT Stream pools and lakes in the Black Hills.

WETLAND STATUS
GP OBL | MIDW OBL | WMTN OBL

Potamogeton epihydrus
RIBBONLEAF PONDWEED

Potamogeton foliosus Raf.

LEAFY PONDWEED

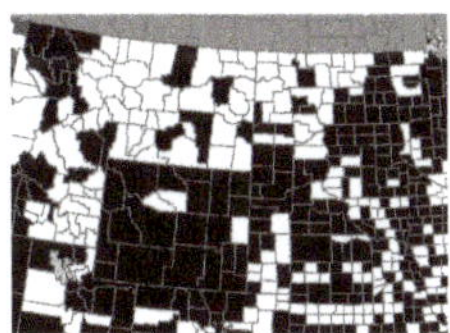

DESCRIPTION **Stems** compressed, mostly 0.5–1 mm wide, freely branched, to 8 dm long. **Leaves** all submersed, linear, 1.3–8.2 cm long, 0.3–2.3 mm wide, 1- to 3(5)-nerved, acute to apiculate at the tip, tapered to the sessile base; stipules free, greenish to brown, mostly 0.5–2 cm long, membranous or fibrous, eventually deteriorating; glands rarely present at the base of the stipules. Winter buds uncommon, lateral, 1–2 cm long. **Spikes** capitate to short-cylindric, 1.5–7 mm long; floral whorls 1 or 2, the whorls 0.6–1.2 mm apart when 2; peduncles usually clavate, recurved, 3–10 mm long. **Fruits** olive to greenish-brown, obliquely obovoid, 1.4–2.7 mm long; dorsal keel ridged or winged, with an undulate to dentate margin 0.2–0.4 mm high; sides rounded to centrally depressed. June–Aug.

HABITAT Shallow water of rivers, streams, lakes and ponds.

WETLAND STATUS
GP OBL | MIDW OBL | WMTN OBL

NOTE The fruit has a characteristic wavy or bumpy keel and a marked beak.

Potamogeton foliosus
LEAFY PONDWEED

Potamogeton friesii Rupr.

FLAT-STALK PONDWEED

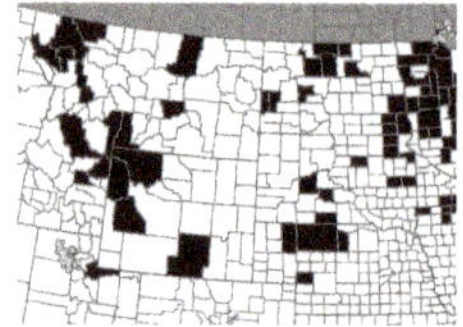

DESCRIPTION Stems compressed, mostly 0.5–1 mm wide, simple or branched, to 1–1.5 m long. **Leaves** all submersed, linear, 2.3–6.5 cm long, 1.2–3.2 mm wide, 5- to 7(9)-nerved, rounded to apiculate at the tip, the margin flat to eventually revolute, tapered to the sessile base; stipules free, white and fibrous, often shredding above, 5–20 mm long; glands present at the base of the stipules. Winter buds common, terminal or lateral, 1.5–5 cm long; inner leaves reduced, arranged into a fan-shaped structure; outer leaves 2–3 per side, apiculate to acute, indurate and corrugated at the base. **Spikes** cylindric, 7–16 mm long; floral whorls 2–5, 1.5–5 mm apart; peduncles slightly clavate, mostly 1.5–4(7) cm long. **Fruits** olive-green to brown, ovoid to obovoid, 1.8–2.5 mm long, rounded on the back. June–Aug.

HABITAT Shallow, fresh to brackish water of lakes and ponds.

WETLAND STATUS

GP OBL | MIDW OBL | WMTN OBL

Potamogeton friesii
FLAT-STALK PONDWEED

Potamogeton gramineus L.

VARIABLE PONDWEED

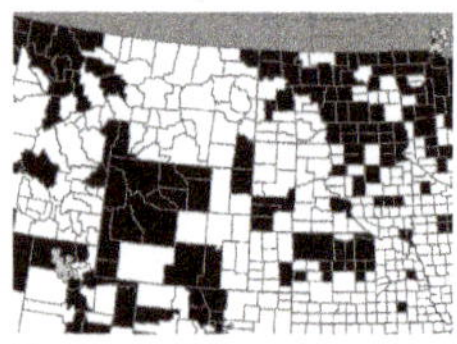

DESCRIPTION Stems subterete, 0.5–1 mm thick, usually freely branched, to 8 dm long. **Submersed leaves** linear to linear-lanceolate or broadly lanceolate, sometimes oblanceolate, 3–9 cm long, 3–15 mm wide, 3- to 7(9)-nerved, acute to acuminate, tapered to the sessile base. **Floating leaves** rarely absent, elliptic or oblong-elliptic, 2–9 cm long, 1–3.5 cm wide, 11- to 19-nerved, obtuse or abruptly acute at the tip, rounded to cuneate at the base; petioles 2–10(15) cm long, shorter than to exceeding the length of the blade; stipules free, persistent, mostly 0.5–4 cm long. **Spikes** dense, cylindric, 1.5–3.5 cm long; peduncles stout, usually broadening upward, 2–10(20) cm long. **Fruits** dull green, obliquely obovoid, 1.7–2.8 mm long, dorsal keel sharp, lateral keels obscure. June–Aug.

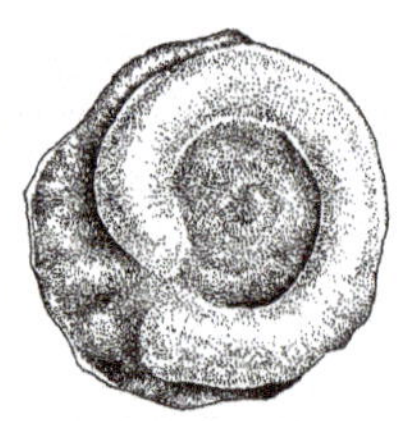

HABITAT Shallow, usually standing water of ponds, lakes, marshes and ditches.

WETLAND STATUS

GP OBL | MIDW OBL | WMTN OBL

NOTE Apparent hybrids between *P. gramineus* and *P. illinoensis* are frequently observed among collections from the Nebraska Sand Hills where they often grow together. The hybrids are typically intermediate between the parents in morphology, but may be mistaken for extremes of either parent. Flowers should be checked for abortive pollen grains to confirm a suspected hybrid.

Potamogeton gramineus
VARIABLE PONDWEED

Potamogeton illinoensis Morong

ILLINOIS PONDWEED

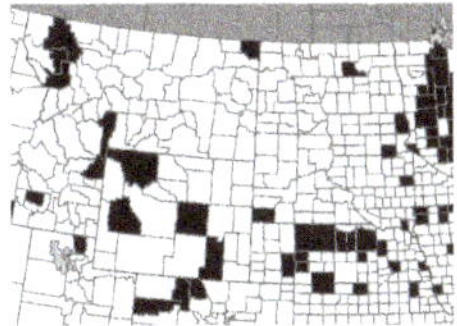

DESCRIPTION Stems subterete, (1)1.5–5 mm thick, simple or branched, to 2 m long. **Submersed leaves** elliptic or oblong-elliptic to lanceolate or linear-lanceolate, sometimes somewhat arcuate, 5–20 cm long, (1)1.5–4 cm wide, mostly 9- to 17(19)-nerved, acute to mucronate, tapered to the subsessile or petioled base, the petioles up to 2–4 cm long. **Floating leaves** often lacking, oblong-elliptic or ovate-elliptic to broadly elliptic, 4–14(19) cm long, 2–7 cm wide, 13- to 29-nerved, obtuse to bluntly mucronate at the tip, rounded to cuneate at the base; petioles 2–9 cm long, shorter than the blades; stipules free, persistent, mostly 2–8 cm long. **Spikes** dense, cylindric, 2–6 cm long; peduncles usually thicker than the stem, 4–20(30) cm long. **Fruits** olive-green or gray-green, obliquely obovoid, 2.7–3.5 mm long, dorsal and lateral keels prominent. June–Sept.

HABITAT Shallow to fairly deep water of sandy lakes and ponds. See discussion under *P. gramineus.*

WETLAND STATUS

GP OBL | MIDW OBL | WMTN OBL

Potamogeton natans L.

FLOATING LEAF PONDWEED

DESCRIPTION Stems slightly compressed, 0.8–2 mm thick, simple or rarely branched, to 2 m long. **Submersed leaves** reduced to linear, bladeless phyllodes, these often disintegrating with age, 10–20 cm long, 1–2 mm wide, tapering to an obtuse tip. **Floating leaves** ovate-lanceolate to ovate-elliptic, 3–10 cm long, 1–5 cm wide, mostly 19- to 35-nerved, acute to obtuse at the tip, rounded to cordate at the base; petioles usually much exceeding the blade in length, usually forming an angle with the blade at their juncture; stipules free, fibrous, persistent or shredding with age, 4–10 cm long. **Spikes** dense, cylindric, 2–5 cm long; peduncles thicker than the stem, 3–10 cm long. **Fruits** greenish-brown to brown, obliquely elliptic-obovoid, 3–5 mm long, often pitted on the sides, the dorsal keel sharp

Potamogeton illinoensis
ILLINOIS PONDWEED

Potamogeton natans
FLOATING LEAF PONDWEED

or rounded with age, the lateral keels obscure. July–Aug.

HABITAT Shallow to rather deep water of lakes and ponds; frequent in the ND Turtle Mts and Neb Sand Hills, otherwise uncommon.

WETLAND STATUS
GP OBL | MIDW OBL | WMTN OBL

Potamogeton nodosus Poir.
LONG-LEAF PONDWEED

DESCRIPTION

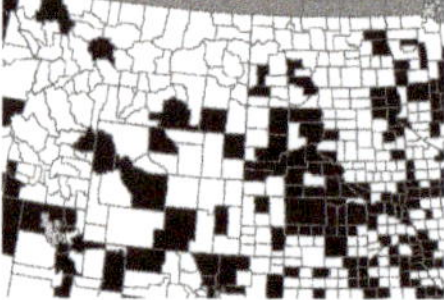

Stems subterete, 1–2 mm thick, simple or seldom branched, to 1.5 m long. **Submersed leaves** commonly deteriorating by fruiting time, linear-lanceolate to elliptic-lanceolate, 10–20(30) cm long, 1–2(3) cm wide, 7- to 15-nerved, acute to blunt-tipped, gradually tapering to petioles mostly (2)4–10 cm long. **Floating leaves** elliptic to oblong-elliptic, 5–13 cm long, (1.5)2–4.5 cm wide, 15- to 25-nerved, acute to nearly rounded at the tip, sometimes obtusely mucronate, acute to somewhat rounded at the base; petioles winged, mostly 2–3 mm wide, 5–20 cm long, usually longer than the blades; stipules free, those of the submersed leaves often decaying early, those of the floating leaves persistent, 3–10 cm long. **Spikes** dense, cylindric, 2–6 cm long; peduncles thicker than the stem, 3–15 cm long. **Fruits** reddish-brown to brown, obovoid, 2.7–4.3 mm long, the dorsal keel sharp, the lateral keels low. July–Aug.

HABITAT Shallow to rather deep water of streams, ponds and reservoirs; frequent in the s part, less common n; cosmopolitan.

WETLAND STATUS
GP OBL | MIDW OBL | WMTN OBL

Potamogeton praelongus Wulfen
WHITESTEM PONDWEED

DESCRIPTION

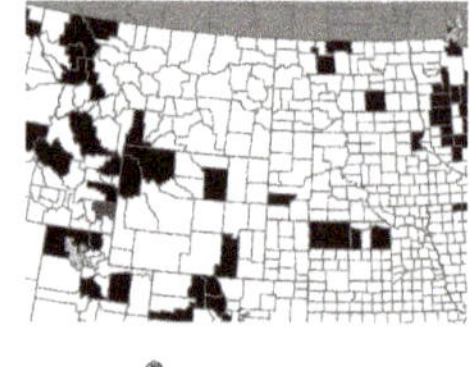

Stems whitish, slightly compressed, 1.5–4 mm thick, sparingly branched, to 2–3 m long, the shorter internodes often zigzag. **Leaves** all submersed, oblong-lanceolate, 10–25(35) cm long, 1–3 cm wide, with 3–5 primary nerves, rounded and cucullate at the tip, the margin entire and somewhat undulate, sessile and weakly to strongly cordate-clasping at the base; stipules free, whitish, 1–3 cm long, fibrous, early shredding. **Spikes** dense, cylindric, 2.5–5 cm long; peduncles elongate and thickening upward, 10–40 cm long. **Fruits** greenish-brown, obovoid, 4–5 mm long, the dorsal keel sharp, the lateral ones obscure. June–July.

HABITAT Deep water of cold, clear lakes.

WETLAND STATUS
GP OBL | MIDW OBL | WMTN OBL

Potamogeton nodosus
LONG-LEAF PONDWEED

Potamogeton praelongus
WHITESTEM PONDWEED

Potamogeton pusillus L.

SMALL PONDWEED

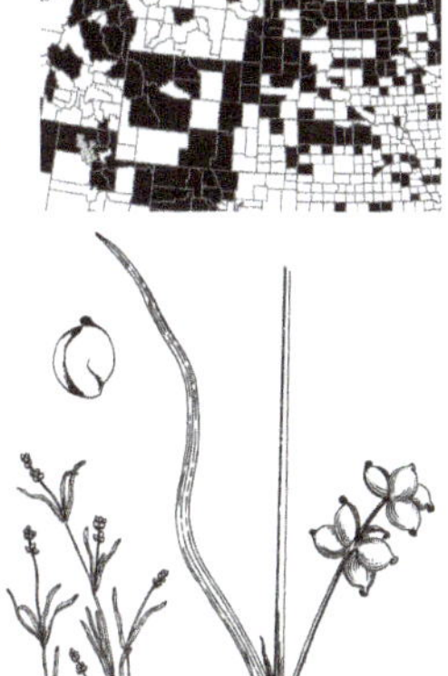

DESCRIPTION Stems terete to subterete, 0.1–0.7 mm thick, simple to freely branching, 2–15 dm long. **Leaves** all submersed, linear, 0.9–6.5 cm long, 0.2–2.5 mm wide, 1- to 3(5)-nerved, acute to obtuse or apiculate at the tip, tapered to the sessile base; stipules free, brownish-green, 3–9 mm long, delicate and nonfibrous, soon decomposing; glands usually present at the base of the stipules. Winter buds sometimes produced, lateral or terminal, 0.9–3.2 cm long; inner leaves rolled into an indurate fusiform structure; outer leaves 1–3 per side, acute to obtuse, without corrugations at the base. **Spikes** short-cylindric to cylindric, 1.5–10 mm long; floral whorls 1–3(4), 1.2–4.7 mm apart; peduncles filiform to cylindric, 0.5–6 cm long. **Fruits** green to brown, obliquely obovoid, 1.5–2.2 mm long, rounded on the back, often concave on the sides. June–Aug.

HABITAT Shallow to deep water of lakes, ponds, marshes, ditches and streams, where water is fresh to brackish; common and often abundant.

WETLAND STATUS

GP OBL | MIDW OBL | WMTN OBL

Potamogeton pusillus
SMALL PONDWEED

ADDITIONAL SPECIES

Two varieties occur in the region:

Var. *tenuissimus* Mert. & Koch (now treated as an accepted species *Potamogeton berchtoldii* Fieb.) and var. *pusillus.*

By far the prevalent form is var. *pusillus,* characterized as follows: leaves with up to 2 rows of lacunae along the midrib, apex acute, rarely apiculate; stipules mostly connate. Spikes usually of 2–4 verticels; peduncles filiform to cylindric, usually 1–3 per plant. Mature fruit widest above the middle, the sides concave, the beak positioned forward.

Potamogeton berchtoldii is of limited occurrence in this region. It differs from var. *pusillus* as follows: leaves with 1–5 rows of lacunae along the midrib, apex acute to obtuse; stipules mostly convolute. Spikes mostly of 1–2 adjacent verticels; peduncles cylindric, usually more than 3 per plant. Mature fruit mostly widest at or below the middle, the sides rounded, the beak positioned centrally.

Potamogeton richardsonii (Benn.) Rydb.

CLASPING-LEAF PONDWEED

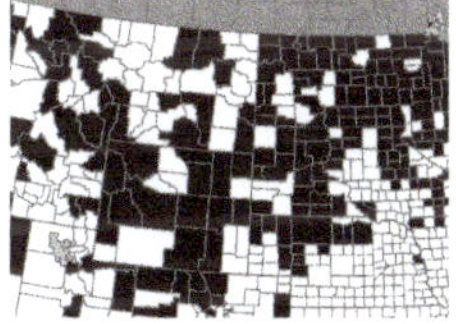

DESCRIPTION Stems brownish to yellowish-green, terete, 1–2.5 mm thick, sparingly to freely branched, mostly 3–

Potamogeton richardsonii
CLASPING-LEAF PONDWEED

10 dm long, the shorter internodes rarely zigzag. **Leaves** all submersed, ovate-lanceolate to lanceolate, 2–10 cm long, 1–2.5 cm wide, with 13–25 prominent nerves, rounded to acute and not cucullate at the tip, the margin entire and undulate-crisped, sessile and strongly cordate-clasping at the base; stipules free, 1–2 cm long, early shredding into whitish fibers. **Spikes** dense, cylindric, 1.5–4 cm long; peduncles strongly recurved in fruit, often thickening upward, 2–10 cm long. **Fruits** green to brown, obliquely obovoid, 2.5–3.5 mm long, rounded to faintly keeled dorsally. June–Aug.

SYNONYM *Potamogeton perfoliatus* L., misapplied.

HABITAT Shallow to moderately deep water of fresh to brackish lakes, ponds, marshes, reservoirs and slow-moving streams.

WETLAND STATUS
GP OBL | MIDW OBL | WMTN OBL

Potamogeton strictifolius Benn.
STRAIGHT-LEAF PONDWEED

DESCRIPTION **Stems** mostly terete, 0.4–0.8 mm thick, simple or branched, to 1 m long. **Leaves** all submersed, linear, 1.2–6.3 cm long, 0.6–2 mm wide, 3- to 5-nerved, acute to attenuate at the tip, the margin often revolute, tapered to the sessile base; stipules free, white, fibrous, shredding at the tip, 6–16 mm long; glands present at the base of the stipules. Winter buds common, terminal or lateral, 2.5–4.8 cm long; inner leaves undifferentiated from the outer ones; outer leaves 3–4 per side, acute, mostly without or rarely with corrugations at the base. **Spikes** cylindric, 6–13 mm long; floral whorls 3–4,1.5–4.2 mm apart; peduncles cylindric, rarely slightly clavate, 1–4.5 cm long. **Fruits** greenish-brown, obovoid, 1.9–2.1 mm long, rounded on the back. July–Aug.

HABITAT Shallow water of ponds, lakes and slow streams.

WETLAND STATUS
GP OBL | MIDW OBL | WMTN OBL

Potamogeton zosteriformis Fern.
FLAT-STEM PONDWEED

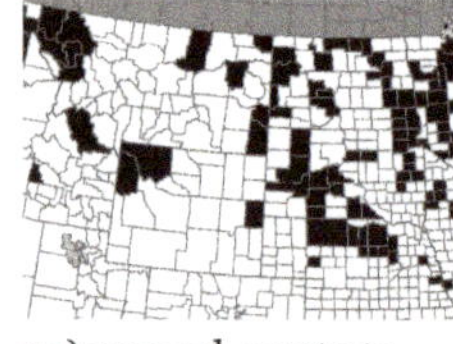

DESCRIPTION **Stems** strongly flattened, freely branched, to 1 m long. **Leaves** all submersed, linear, 5–15(20) cm long, 3–5 mm wide, many (15–35)-nerved, acute to cuspidate at the tip, sessile and slightly narrowed at the base; stipules free, usually whitish, fibrous and shredding with age,, 1–4 cm long. **Spikes** densely flowered, usually un-

Potamogeton strictifolius
STRAIGHT-LEAF PONDWEED

crowded in fruit, cylindric, 1–2.5 cm long; peduncles compressed, 1.5–10 cm long. **Fruits** dark green to brown, obliquely elliptic-ovoid, 4–4.5 mm long, the dorsal keel sharp and somewhat undulate or dentate, the lateral ones obscure. July–Aug.

HABITAT Shallow to deep water of lakes, ponds and marshes; sometimes locally common.

WETLAND STATUS
GP OBL | MIDW OBL | WMTN OBL

Ruppia cirrhosa (Petag.) Grande
DITCH-GRASS, WIDGEON-GRASS

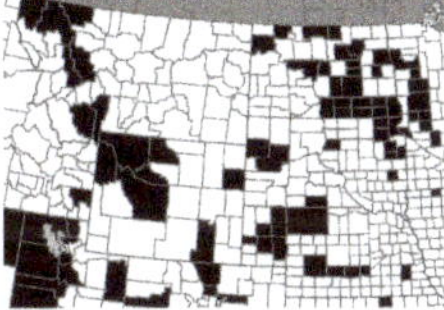

DESCRIPTION Perennial submersed aquatic; **stems** slender and terete, flexuous, anchored by roots, branching at the base and short-branched above, to 6 dm long, the internodes often zigzag. **Leaves** simple, alternate, sessile, the blades filiform, attenuate, mostly 5–15(25) cm long, ca. 0.5 mm wide, stipular-sheathing at the base. **Inflorescence** a reduced axillary spike, included in the leaf sheath at anthesis; peduncles elongating and usually coiling as fruits develop. **Flowers** 2 per spike, perfect; perianth none or minute; stamens 2, briefly adherent during peduncle elongation; carpels 6 or more, separate, stipitate, the stipes gradually elongating from the common base so that the fruits are borne in an umbel. **Fruits** olive-green to black, ovoid, symmetrical to asymmetrical, beaked, 1.5–3 mm long. July–Aug.

Potamogeton zosteriformis
FLAT-STEM PONDWEED

HABITAT Alkaline to saline waters of lakes, ponds and marshes; sometimes locally common; primarily in brackish or saline waters.

NOTE Sometimes placed in own family: Ruppiaceae.

WETLAND STATUS
GP OBL | MIDW OBL | WMTN OBL

Ruppia cirrhosa
DITCH-GRASS, WIDGEON-GRASS

Stuckenia FALSE PONDWEED

Stuckenia is a small genus of perennial aquatic herbs, now segregated from *Potamogeton.* In *Stuckenia,* the stipules are joined to the blade for 2/3 to nearly the entire length of the stipule; in *Potamogeton,* the stipules in most species are free, or if adnate, joined for well less than half the length of the stipule. Also, submersed leaves of *Potamogeton* are translucent, flat, and without grooves or channels; in *Stuckenia,* submersed leaves are opaque, channeled, and turgid.

NOTE *Stuckenia,* although important as a group as waterfowl food (the achenes are eaten), are often difficult to positively identify in the field, the distinguishing features being somewhat hard to see.

---

1 Stipular sheaths of the main stem inflated 2–5x the thickness of the stem; floral whorls 5–12 per spike ***S. vaginata***

1 Stipular sheaths of the main stem about as wide as the stem; floral whorls 2–6 per spike **2**

2 Stems dichotomously branched from the base, mostly unbranched above; fruits oli-

vaceous, 2–3 mm long, the beak flat, inconspicuous ***S. filiformis***

2 Stems dichotomously branched above; fruits yellowish to brown, 3–4 mm long, apiculate, with a beak usually 0.3–0.5 mm long ***S. pectinata***

Stuckenia filiformis (Pers.) Börner

SLENDER PONDWEED

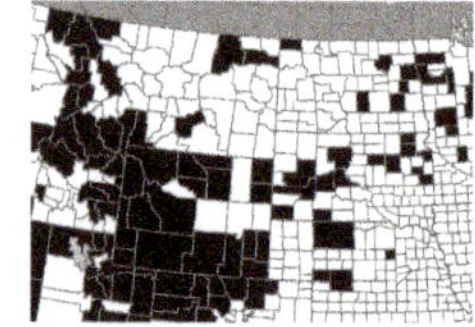

DESCRIPTION **Stems** subterete, to 1 mm wide, branching dichotomously from the base, mostly unbranched above, 1–5 dm long. **Leaves** all submersed, filiform to narrowly linear, 5–12 cm long, 0.2–2 mm wide, 1(3)-nerved, acute to obtuse; stipules adnate to the base of the leaf blade, 1–4 cm long, forming a tight sheath around the stem, the free portion projecting as a ligule 2–10 mm long. **Spikes** elongate, 1–5 cm long, with 2–5 remote to adjacent whorls of flowers; peduncles slender, 2–15 cm long. **Fruits** olivaceous, obovoid, 2–3 mm long, rounded on the back, the beak flat, inconspicuous. July–Aug.

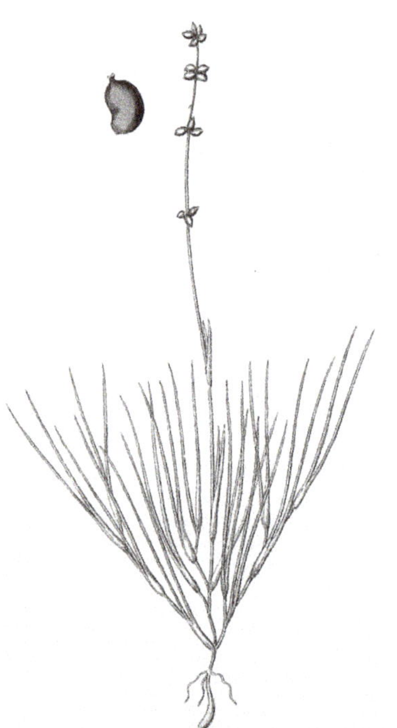

Stuckenia filiformis
SLENDER PONDWEED

SYNONYM *Potamogeton filiformis* Pers.

HABITAT Shallow standing or flowing water.

WETLAND STATUS
GP OBL | MIDW OBL | WMTN OBL

Stuckenia pectinata (L.) Börner

SAGO PONDWEED

DESCRIPTION **Stems** terete, ca. 1 mm thick, or the main stem stouter on deep water forms, sparingly branched at the base, becoming freely dichotomously branched above, 3 –10 dm long. **Leaves** all

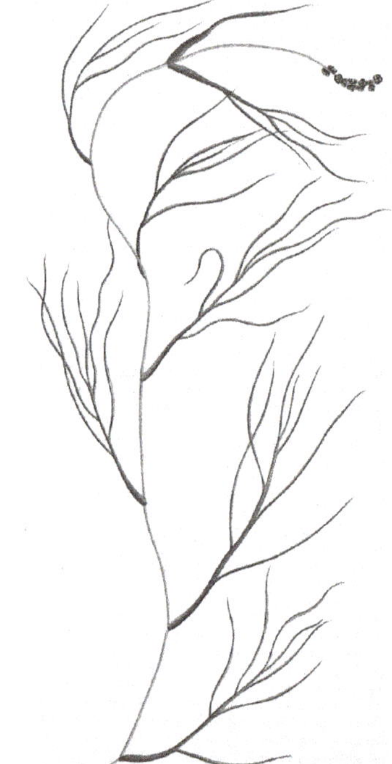

Stuckenia pectinata
SAGO PONDWEED

submersed, filiform to narrowly linear, 3–12 cm long, usually 0.2–1 mm wide, 1- to 3-nerved, acute, sometimes wider with obtuse tips early in the growing season or on plants from running water; stipules adnate to the base of the leaf blade for 1–3 cm, forming a sheath about as wide as the stem, occasionally wider on the main stem, especially in deep water forms. **Spikes** elongate, 1–5 cm long, with 2–5(7) unevenly spaced floral whorls; peduncles lax, filiform, to 15 cm long. **Fruits** yellowish to tawny, drying brown, obliquely obovoid, 2.7–4 mm long, rounded on the back, apiculate due to the style beak which is usually 0.3–0.5 mm long. June–Sept.

SYNONYM *Potamogeton pectinatus* L.

HABITAT Shallow to rather deep, fresh to subsaline water of lakes, ponds, marshes, ditches, rivers and streams; common and often abundant; nearly cosmopolitan.

WETLAND STATUS
GP OBL | MIDW OBL | WMTN OBL

Stuckenia vaginata (Turcz.) Holub
SHEATHED PONDWEED

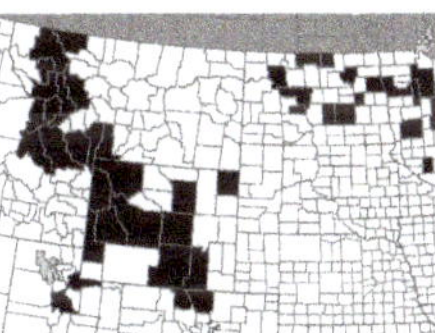

DESCRIPTION **Stems** terete, mostly 1–2 mm thick, freely branched above, to 1.5 m long. **Leaves** all submersed, filiform to narrowly linear, 2–8(30) cm long, 0.5–2 mm wide, 1(3)- nerved, the tip acute to obtuse or sometimes retuse; stipules adnate to the base of the leaf blade for 1–5 cm and sheathing the stem, the sheaths along the main stem inflated 2–5x the thickness of the stem. **Spikes** elongate, 3–6 cm long, with 5–12 evenly spaced floral whorls; peduncles slender and lax, to 10 cm long, often much surpassed by the upper leaves. **Fruits** dark green, obliquely obovoid, ca. 3 mm long, rounded on the back; stigma sessile, forming a low beak. July–Aug.

SYNONYM *Potamogeton vaginatus* Turcz.

HABITAT Deep water of cold, clear lakes.

NOTE Deep water forms of *Stuckenia pectinata,* especially those from wave-swept areas, are often confused with *S. vaginata.* However, in flowering or fruiting condition the two species are easily separated by the number of floral whorls per spike and the shape of the fruit (beaked vs. nearly beakless). Unfortunately, deep water forms of *S. pectinata* often do not flower and they tend to have stout stems sheathed by prominent, somewhat inflated stipules similar to *S. vaginata.* The latter tends to have longer, more filiform leaves (much surpassing the spikes) than deep water forms of *S. pectinata,* and the stipules are more prominently inflated.

WETLAND STATUS
GP OBL | MIDW OBL | WMTN OBL

Stuckenia vaginata
SHEATHED PONDWEED

Zannichellia palustris L.

HORNED PONDWEED

DESCRIPTION Monoecious, perennial, submersed aquatic, often forming extensive mats; stems thin and flexuous, densely leafy, freely branched from the base, anchored by roots, 0.5–5 dm long. **Leaves** simple, opposite, sessile, filiform, 1.5–6 cm long, ca. 0.5 mm wide; stipules membranous and soon disappearing. **Flowers** highly reduced, one staminate and typically 4 (1–5) pistillate flowers at each node; perianth none; staminate flower consisting of a solitary anther raised on a short, slender filament; pistillate flowers located in the same or opposite the axil containing the staminate flower, sessile or short-peduncled as a group, surrounded by a cup-shaped, membranous bract, each flower comprised of a single fusiform carpel with a peltate stigma. **Fruits** often abundantly produced, drupelike, stipitate, mostly 2–4 per node, brown to reddish-brown, crescent-shaped, narrowly undulate to serrate on the keeled edges, 3–5 mm long including the 1–2 mm long persistent style, ca. 1 mm wide. June–Aug.

HABITAT Fresh to brackish water of streams, reservoirs, lakes, ponds, marshes and ditches; common.

WETLAND STATUS

GP OBL | MIDW OBL | WMTN OBL

Zannichellia palustris
HORNED PONDWEED

Scheuchzeriaceae
scheuchzeria family

Scheuchzeria palustris L.

POD-GRASS

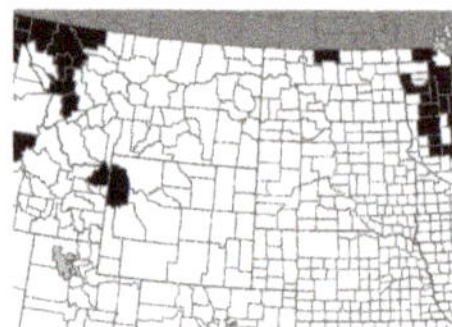

DESCRIPTION Rush-like, bog-dwelling perennial from creeping rhizomes, 2–4 dm tall. **Leaves** alternate, 2-ranked, broadly sheathing at the base with a prominent ligule 2–10 mm long at the juncture of the sheath and blade, basal leaves 1–4 dm long, cauline leaves reduced upward; leaf blades terete or nearly so, 1–3 mm wide, with a small pore at the tip. **Inflorescence** a few- to several-flowered raceme, with a few leaflike bracts in the lower portion; pedicels to 25 mm long. **Flowers** perfect, regular; tepals 6, in 2 series, greenish-white, oblong, ca. 3 mm long, soft, eventually deciduous; stamens 6; carpels 3 (rarely to 6), free or nearly so to the base, stigmas subsessile. **Fruit** a group of 3 (rarely 4, 5 or 6) spreading follicles, these 1- to 2(-several)-seeded, inflated, 5–10 mm long; seeds black, ellipsoid, 4–5 mm long. Flowering late May–June, fruiting July–Aug.

HABITAT Acid bogs, often in sphagnum or sedge mats.

WETLAND STATUS

GP OBL | MIDW OBL | WMTN OBL

Scheuchzeria palustris
POD-GRASS

Typhaceae
cattail family

Family now includes genus *Sparganium* from former family Sparganiaceae.

1 Flowers in a dense, terminal spike of 2 portions, the upper part male and the lower female; plants usually 1 m or more tall; cattails ***Typha***

1 Flowers in few to many dense, globose heads, the upper heads male and the lower female; plants 1 m or less tall; bur-reeds ***Sparganium***

Sparganium BUR-REED

Perennial, reedlike marsh plants, colonial from rhizomes, typically emergent in shallow water. **Stems** simple, terete, stout, usually erect, clothed toward the base by the overlapping leaf bases. **Leaves** broad, long, linear, weakly sheathing, ascending and mostly overtopping the inflorescence. **Flowers** unisexual, the male and female flowers densely crowded in separate globose heads, the **staminate heads** few to many, sessile, borne above the pistillate heads in a simple or sparsely branched inflorescence, the **pistillate heads** 1–several, sessile or peduncled, axillary or supra-axillary in relation to the foliaceous bracts in the lower portion of the inflorescence; perianth (in both male and female flowers) of few to several chaffy, spatulate scales, these appressed to the achenes in the mature pistillate heads; staminate flowers of mostly 3–5 stamens; pistillate flowers each consisting of a simple or 2-carpellary pistil, stigmas 1 or 2. **Fruit** a beaked, nutletlike achene, sessile or short-stipitate in the head.

1 Stigmas 2; achenes obconic to obpyramidal, angled on the sides; staminate heads many ***S. eurycarpum***

1 Stigma 1; achenes fusiform, rounded on the sides; staminate heads usually 2–5 ***S. emersum***

Sparganium emersum Rehm.
NARROW-LEAF BUR-REED

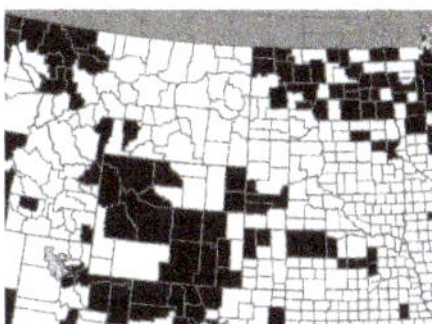

DESCRIPTION Erect or sometimes lax and trailing in the water, at least the upper leaves and inflorescence emergent, 1–6 dm tall (to 10 dm or more long when deeply submersed). **Leaves** yellow-green, flat to keeled, mostly 4–12 mm wide, often scarious-margined toward the base and sometimes strongly dilated. **Inflorescence** 1–2 dm long, unbranched, one or more of the lower pistillate heads usually peduncled, at least one supra-axillary; staminate heads usually 2–5, ca. 1.5 cm in diameter at anthesis; anthers linear, 1–1.4 mm long; pistillate heads 1–4, (1)1.5–2.5 cm in diameter at maturity; stigma 1, linear, 1–1.5(2) mm long. **Achenes** stipitate in the head, on stipes 2–3 mm long, green to brown-olivaceous on the upper 1/2, lighter and usually reddish-brown spotted on the lower 1/2, fusiform, rounded on the sides, slightly constricted at the middle, the body 3–5 mm long, the beak with the stigma 2–4 mm long, curved, the persistent perianth of 4–5 scales reaching to ca. 2/3 the length of the achene. Mid June–Aug.

SYNONYM *Sparganium chlorocarpum* Rydb.

HABITAT Shallow water or mud of marshes, streams, ditches and ponds, where the water is fairly fresh.

WETLAND STATUS
GP OBL | MIDW OBL | WMTN OBL

NOTE Plants are typically lax and sprawling when mostly submersed; when emersed, plants are more rigid and erect.

Sparganium emersum
NARROW-LEAF BUR-REED

Sparganium eurycarpum Engelm.
GIANT BUR-REED

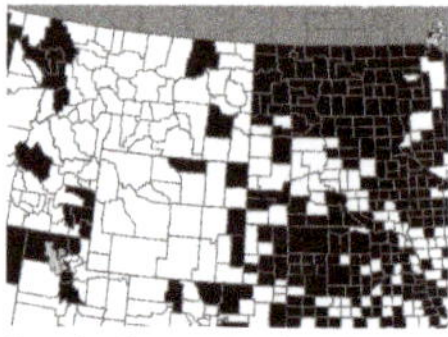

DESCRIPTION Rather robust, typically emergent plants 4–10 dm tall. **Leaves** bright green, 4–11 mm wide, spongy toward the base, sharply keeled on the back, the margins scarious toward the base. **Inflorescence** 1–3 dm long, sparsely branched from the axils of foliaceous bracts, the branches simple, loosely spicate; staminate heads numerous, 1–2 cm in diameter at anthesis; anthers linear, 1–1.8 mm long; pistillate heads 1–4(8), 1.5–2.5 cm in diameter in fruit; stigmas 2, linear, 2–4 mm long. **Achenes** sessile to short-stipitate, brown on the summit, golden-brown on the sides, obconic to obpyramidal, angled on the sides, the body 6–8 mm long, 4–7 mm wide at the summit, the beak (excluding the stigmas) 1–4 mm long, the persistent perianth of 4–8 scales appressed to the achene, the scales about equaling the achene body. Mid June–Aug.

HABITAT Usually in shallow water of marshes, streams, ditches, ponds and lakes, often occurring with cattails; common, often abundant.

WETLAND STATUS
GP OBL | MIDW OBL | WMTN OBL

Typha CATTAIL

Tall, stout, reedlike marsh plants, perennial, extensively colonial from fleshy rhizomes. **Stems** simple, terete, erect, sheathed for most of the length by overlapping leaf sheaths. **Leaves** alternate in 2 ranks, erect, broad, linear, rather spongy, sheathing at the base, the sheaths open, scarious-margined. **Flowers** unisexual, minute, the male flowers above the female in a solitary, terminal, dense, cylindrical spike, the male and female portions of the spike contiguous or separated; perianth absent; **staminate flowers** usually of 3–5 stamens, the anthers linear, filaments often connate, subtended by numerous slender hairs; **pistillate flowers** each comprised of a simple pistil, intermixed with some sterile flowers, the ovary on a short stipe called the gynophore, this with numerous long, slender hairs (the gynophore hairs) near the base, the hairs surpassing the ovary; bracteoles also sometimes present, these intermixed with and about as long as the gynophore hairs, filiform with a broadened, brown tip. **Fruit** a fusiform achene, golden or tawny, 1–1.5 mm long, the style persistent, long and slender with an expanded stigma; mature spike thick, brown and fuzzy in appearance due to the crowded stigmas and gynophore hairs, the upper staminate portion of the spike eventually naked.

Sparganium eurycarpum
GIANT BUR-REED

NOTE Reliable separation of the two common species of *Typha* and their hybrid, *T. x glauca* Godr., is best achieved with floral characters. The reduced size and crowded condition of the flowers in the spike requires that a cluster of female flowers be removed from the spike and observed under magnifications of 20–30x. Higher magnifications may be needed for pollen grains. If the material is dried, the flowers or pollen should be wetted with a wetting solution, e.g., soap-water solution, to restore structures to natural size and shape.

---

1 Pistillate bracteoles absent; stigmas dark brown, lanceolate to ovate-lanceolate; staminate and pistillate portions of the spike usually contiguous ***T. latifolia***

1 Pistillate bracteoles present (these reduced and appearing like gynophore hairs with slightly broadened brown tips in *T. x glauca*); stigmas pale brown, linear to linear-lanceolate; staminate and pistillate portions of the spike usually separated **2**

2 Pistillate bracteoles broader than the linear stigmas; pollen in monads ***T. angustifolia***

2 Pistillate bracteoles narrower than the linear-lanceolate stigmas; pollen usually in a mixture of monads, diads, triads and tetrads ***T. x glauca***

Typha angustifolia L.

NARROW-LEAF CATTAIL

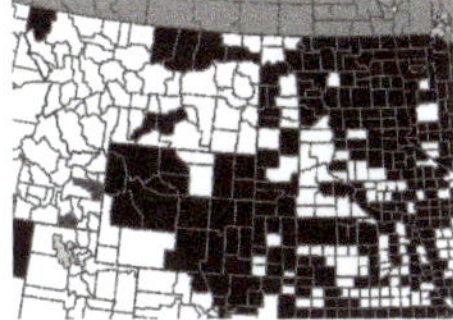

DESCRIPTION Plants mostly 1.5–3 m tall. **Leaves** erect, green, mostly 3–10 mm wide, the auricles of the leaf sheath rounded and surpassing the base of the blade. Staminate and pistillate portions of the **spike** separated by an interval of usually 1–8 cm; **staminate portion** 7–20 cm long, 7–15 mm thick at anthesis, staminate bracteoles brown, anthers 2–3 mm long, pollen released in monads; **pistillate portion** of the spike dark brown, 8–18 cm long, to 2 cm thick at maturity, pistillate bracteoles present, these dark brown at the expanded tip, about equaling the gynophore hairs in length, broader than the linear stigmas, the gynophore hairs brown-pigmented toward the tips; stigmas pale brown, linear, curved with age, 0.8–1.2 mm long. Flowering June, fruiting late July–Sept.

Typha angustifolia
NARROW-LEAF CATTAIL

HABITAT Marshes, shores, streambanks, ditches and margins of lakes and ponds, usually in shallow water; common in the e and c parts, apparently rapidly increasing in the w part, more characteristic of fluctuating water regimes than *T. latifolia,* although often occurring with it; also more tolerant of brackish or saline conditions than *T. latifolia.*

WETLAND STATUS

GP OBL | MIDW OBL | WMTN OBL

ADDITIONAL SPECIES Barely entering our range in the south is the similar **southern cattail** (*Typha domingensis* Pers.) The plant is generally more robust with yellowish to light brown pistillate spikes 2–3 cm thick in fruit. It is most readily distinguishable from *T. angustifolia* by the presence of brown mucilage glands on the upper surface of the leaf blades near their bases. These glands are confined to the inside of the leaf sheaths in *T. angustifolia.* Like *T. angustifolia, T. domingensis* hybridizes with *T. latifolia* to produce intermediate offspring.

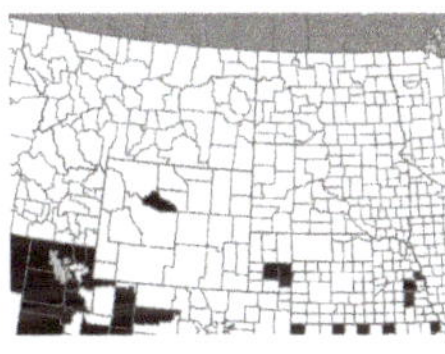

Typha latifolia L.

COMMON CATTAIL

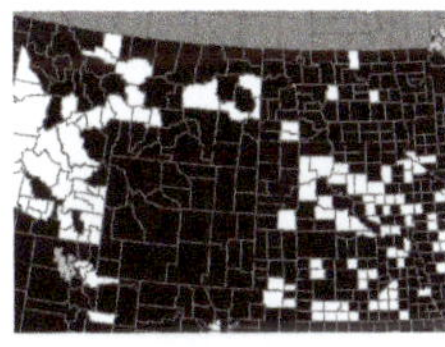

DESCRIPTION Plants mostly 1–2.5 m tall. **Leaves** erect-ascending, glaucous-green when fresh, 5–20 mm wide, the auricles of the sheath round to truncate, not surpassing the base of the blade. Staminate and pistillate portions of the **spike** usually contiguous or separated by an interval of up to 1.5 cm; **staminate portion** 5–15 cm long, 1.5–2 cm thick at anthesis, staminate bracteoles white, anthers 3–4 mm long, pollen released predominantly in tetrads; **pistillate portion** of the spike dark brown, 4–15 cm long, 1.5–3 cm thick at maturity, pistillate bracteoles absent, gynophore hairs white, stigmas dark brown with age, especially toward the tip, lanceolate to ovate-lanceolate, 0.4–0.8 mm long. Flowering June, fruiting late July–Sept.

HABITAT Same habitats as *T. angustifola,* except not found where excessively saline; common, often abundant.

WETLAND STATUS
GP OBL | MIDW OBL | WMTN OBL

Typha x glauca Godr. (*not illustrated*)
HYBRID CATTAIL

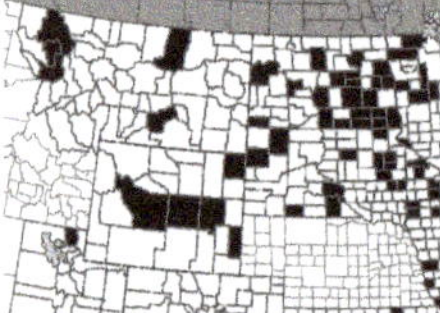

DESCRIPTION *Typha x glauca* is the sterile hybrid between *Typha angustifolia* and *T. latifolia,* usually more robust than the parents, mostly 2–3 m tall, intermediate between the parental spp. in nearly all other details. **Leaves** green to glaucous-green, mostly 5–12 mm wide, the auricles of the leaf rounded, usually surpassing the base of the blade. Staminate and pistillate portions of the **spike** occasionally contiguous or more commonly separated by an interval of up to 4 cm; **staminate portion** 6–18 cm long, 0.8–1.2 cm thick at anthesis, staminate bracteoles pale brown, anthers 2–3 mm long, pollen usually released in a mixture of monads, diads, triads and tetrads; **pistillate portion** of the spike dark brown, 10–20 cm long, 1–2 cm thick at maturity, pistillate bracteoles present, reduced and appearing like gynophore hairs with slightly broadened brown tips, about equaling the gynophore hairs, narrower than the linear-lanceolate stigmas, the gynophore hairs white or slightly brown-pigmented toward the tips, stigmas linear-lanceolate, curved with age, 0.6–1.2 mm long. Flowering June, fruiting late July–Sept.

HABITAT Same habitats as *Typha angustifolia;* common, often abundant, but not reported from Neb; occurs wherever the ranges of *T. angustifolia* and *T. latifolia* are sympatric.

WETLAND STATUS
GP OBL | MIDW OBL | WMTN OBL

NOTE Since *Typha x glauca* is sterile, reproduction is totally vegetative by rhizomes and clone fragmentation. The hybrid cattail is competitively superior to both parents under varying water conditions, and is often viewed as a problem in trying to maintain open water areas in marshes.

Typha domingensis
SOUTHERN CATTAIL

Typha latifolia
COMMON CATTAIL

KEY TO FAMILIES

1 Plants small and flattened, less than 3 cm across, free-floating at or beneath the water surface, often not differentiated into stems and leaves; fine roots often produced on the underside of the plants 2

1 Plants not greatly reduced in size, with stems and leaves, exhibiting various habits but usually anchored in the substrate; roots usually produced 3

2 Plants with tiny, overlapping, scalelike leaves in 2 rows on branched, filiform stems **SALVINIACEAE** (*Azolla*)

2 Plants lacking differentiated leaves and stems **ARACEAE**

3 Plants nonflowering, reproducing by spores (**ferns and fern allies**) 4

3 Plants producing flowers, although these not always conspicuous (**flowering plants**) 12

4 Stems jointed, longitudinally ridged, simple or with whorled branches; leaves scalelike, fused to form a toothed sheath at each joint of the stem; sporangia borne in terminal cones **EQUISETACEAE** (*Equisetum*)

4 Stems not jointed, often subterranean; leaves not scalelike, often with expanded, simple or compound blades; sporangia not borne in cones 5

5 Leaves narrow and grasslike, lacking an expanded blade 6

5 Leaves with an expanded, simple or compound blade 7

6 Leaves arising from a cormlike base; sporangia contained in swollen leaf bases **ISOETACEAE** (*Isoetes*)

6 Leaves arising from filiform rhizomes; sporangia contained in round, hardened sporocarps borne on branches from the rhizomes (also see *Pilularia americana* as described under *Marsilea vestita*) **MARSILEACEAE**

7 Leaves 4-foliate, appearing like a 4-leaved clover **MARSILEACEAE**

7 Leaves simple or pinnately lobed to compound 8

8 Leaf blades simple, entire, subtending a spike of sporangia **OPHIOGLOSSACEAE** (*Ophioglossum*)

8 Leaf blades pinnatifid to pinnately compound; sporangia grouped in sori on modified or unmodified leaves (included here are genera of *Athyrium, Dryopteris, Matteuccia, Onoclea* and *Thelypteris*) 9

9 Fertile fronds brown, structurally unlike the green sterile fronds **ONOCLEACEAE** (*Onoclea, Matteuccia*)

9 Fertile and sterile fronds green, structurally similar 10

10 Indusium attached along its margin to one side of the sorus **ATHYRIACEAE** (*Athyrium*)

10 Indusium attached at its center in the middle of the sorus 11

11 Fronds pubescent, especially on the rachis and midnerves of the pinnae; petioles without scales **THELYPTERIDACEAE** (*Thelypteris*)

11 Fronds glabrous; petioles with membranous scales **DRYOPTERIDACEAE** (*Dryopteris*)

12 Trees and shrubs 13

12 Herbs, woody only at the base (if at all) 19

13 Shrub with pinnately compound leaves **FABACEAE** (*Amorpha*)

13 Trees or shrubs with simple leaves 14

14 Leaves scalelike, 1–5 mm long **TAMARICACEAE** (*Tamarix*)

14 Leaves not scalelike, all or mostly longer than 5 mm 15

15 Trees or shrubs with nonshowy, unisexual flowers in catkins 16

15 Shrubs with greenish, white or yellow flowers, these perfect and not in catkins 17

16 Dioecious trees and shrubs with male and female catkins on separate plants; female catkins bearing capsules, these maturing to release cottony seeds **SALICACEAE**

16 Monoecious shrubs with male and female catkins on the same plant; female catkins bearing winged nutlets **BETULACEAE**

17 Leaves opposite **CORNACEAE** (*Cornus*)

17 Leaves alternate **18**

18 Leaves palmately lobed and veined; fruit a berry **GROSSULARIACEAE** (*Ribes*)

18 Leaves finely toothed, pinnately veined: fruit of 5 (or fewer) follicles **ROSACEAE** (*Spiraea*)

19 Flowers much reduced in size, (2) several to many crowded into involucrate heads and sharing a disklike receptacle; each head resembling a single flower; flowers of the head usually of 2 types, with petaloid ray flowers around the outside and less conspicuous disk flowers in the central portion of the head, or the heads often comprised entirely of ray or disk flowers; involucral bracts herbaceous to membranous, green or sometimes colored, in 1 to many series surrounding the disk, sometimes spiny-tipped and united **ASTERACEAE**

19 Flowers sometimes small in size but not crowded into involucrate heads and not appearing as ray and/or disk flowers **20**

20 Submersed leaves present, finely divided into few to many, filamentous or flat, narrow segments **21**

20 Submersed leaves, if present, entire, toothed or incised but not finely divided into segments **25**

21 Submersed plants with small bladders (1–4 mm in diameter) on filiform segments of finely dissected leaves; flowers irregular and spurred, yellow, in scapose racemes held above the water surface **LENTIBULARIACEAE** (*Utricularia*)

21 Submersed or emergent plants lacking bladders; flowers regular, colored or greenish, not spurred; inflorescence other than a raceme **22**

22 Submersed leaves long-petioled, pinnately compound, with the leaflets variously divided into flat, narrow segments; stems and petioles mostly over 5 mm thick, conspicuously hollow and chambered **APIACEAE** (submersed form of *Sium suave*)

22 Submersed leaves sessile, or if petioled, then palmately divided; stems less than 5 mm thick **23**

23 Leaves alternate; flowers white or yellow **RANUNCULACEAE**

23 Leaves whorled or mostly so; flowers greenish and nonshowy **24**

24 Leaves dichotomously branched, the segments spinulose **CERATOPHYLLACEAE**

24 Leaves pinnately branched, lacking spinulose projections **HALORAGACEAE**

25 Leaves all submersed or some floating also, all basal or borne on stems which are limp when removed from water and usually prostrate when stranded **26**

25 Leaves mostly or all emersed (or mostly large and floating in Nymphaeaceae), borne on erect to procumbent, aerial or partly submersed stems, or the leaves basal; leaves sometimes apparently lacking **37**

26 Leaves all basal **27**

26 Leaves borne on elongate stems **29**

27 Submersed leaves with a median band containing many fine longitudinal veins, the median band bordered by narrow, nerveless or sparsely nerved margins; floating leaf blades none; plants dioecious, producing either oblong female flowers that float on the water surface attached to a long peduncle or producing many tiny male flowers that are released free-floating from a spathe at the plant base **HYDROCHARITACEAE**

27 Submersed leaves lacking a prominent median band of veins, the few to several principal longitudinal veins evenly distributed on both sides of the midvein, often obscure; floating leaves sometimes produced; plants monoecious or hermaphroditic, although flowers often missing in sterile submersed forms **28**

28 Inflorescence a submersed panicle with whorled branches **ALISMATACEAE** (submersed form of *Alisma gramineum*)

28 Inflorescence absent; plants apparently sterile (Several taxa of monocots that are normally emergent will produce sterile submersed forms during juvenile stages when inundated. Some common examples include *Sagittaria* spp. and *Alisma plantago-aquatica,* both of which produce rosettes of brittle, dark green, sword-shaped leaves, often along with long-petioled floating leaves. The floating blades are generally sagittate (*Sagittaria*) or oblong-lanceolate (*Alisma*) in shape (of our Alismataceae only *Alisma gramineum* appears capable of flowering and fruiting underwater and it does so frequently). *Typha* spp., *Sparganium* spp. and some grasses, e.g. *Zizania palustris* and *Glyceria* spp., have sterile submersed forms featuring few to several flexuous, light green, tapelike leaves arising from the base. None of these sterile forms can be confidently identified to species unless one can relate them to more mature examples at the same site or can observe their development through the growing season) **29**

29 Leaves all or mostly alternate, sometimes the uppermost opposite **30**

29 Leaves all opposite or whorled **33**

30 Leaves linear, 2–10(15) cm long, 2–5(7) mm wide, the midvein inconspicuous; flowers pale yellow, solitary in the upper axils; fruit a many-seeded capsule **PONTEDERIACEAE** (*Heteranthera dubia*)

30 Leaves of various shapes and sizes, but if fitting the above dimensions, then the midvein prominent; flowers in various types of inflorescences; fruit hard, one-seeded **31**

31 Flowers pink, in dense, terminal, spike-like racemes **POLYGONACEAE** (*Persicaria amphibia*)

31 Flowers greenish, nonshowy, in terminal or axillary spikes **32**

32 Leaves all submersed, filiform, ca. 0.5 mm wide; flowers enclosed in leaf sheaths at anthesis, the peduncles elongating and often coiling as fruits develop; mature fruits borne in an umbel **POTAMOGETONACEAE** (*Ruppia*)

32 Leaves all submersed or some floating, variously shaped; flowers and fruits borne in elongate to short spikes **POTAMOGETONACEAE**

33 Leaves in whorls of 3 or mostly decussate; flowers axillary and extended to the water surface by a long, threadlike hypanthium **HYDROCHARITACEAE** (*Elodea*)

33 Leaves opposite (most obviously the uncrowded lower leaves); flowers completely contained in the leaf axils **34**

34 Leaf blades abruptly broadened at the base and sheathing the stem; fruits fusiform, only 1 per node **HYDROCHARITACEAE** (*Najas*)

34 Leaf blades of about equal width throughout, only weakly clasping the stem if at all; fruits crescent-shaped or round **35**

35 Fruits mostly 2–4 per node, crescent-shaped with a persistent style **POTAMOGETONACEAE** (*Zannichellia*)

35 Fruits 1 or 2 per node, round; styles deciduous **36**

36 Small, densely branched plants usually sprawling on mud and rooting at the nodes; fruit a capsule containing numerous minute, ridged seeds **ELATINACEAE** (*Elatine*)

36 Irregularly branched plants, normally submersed, the upper leaves sometimes floating; fruit orbicular, cleft down the middle and eventually splitting into 4 nutlets **PLANTAGINACEAE** (*Callitriche*)

37 Plants grasslike, rushlike or reedlike, emergent or terrestrial; leaves long, linear and parallel-veined, often sheathing at the base, or leaf-blades absent, the leaves reduced to sheathing around the base of the stem **38**

37 Plants not grasslike, rushlike or reedlike; leaves of various shapes, net or parallel-veined, seldom sheathing at the base; all or most of the leaves with blades **48**

38 Individual flowers showy, yellow, pink or white to blue-violet **39**

38 Individual flowers not showy, usually greenish to brownish **41**

39 Plants tall, rushlike, with large pink flowers in a terminal umbel **BUTOMACEAE** (Butomus)

39 Plants low, rather grasslike, with yellow or white to blue-violet flowers **40**

40 Flowers usually blue-violet (rarely white), or yellow (Iris), in clusters subtended by a 2-bracted spathe **IRIDACEAE** (*Iris, Sisyrinchium*)

40 Flowers yellow, not subtended by a spathe **HYPOXIDACEAE** (*Hypoxis*)

41 Typically emergent marsh plants with broad (mostly 5 mm or more wide), long, linear leaves; flowers unisexual, arranged in a terminal spike or in globose heads, the male flowers borne above the female **42**

41 Plants of various habits; flowers perfect or imperfect, when imperfect, the male and female flowers contained in small spikes of grasslike plants **43**

42 Flowers in a dense, terminal spike of 2 portions, the upper part male and the lower female; plants usually 1 m or more tall; cattails **TYPHACEAE** (*Typha*)

42 Flowers in few to many dense, globose heads, the upper heads male and the lower female; plants 1 m or less tall; bur-reeds **TYPHACEAE** (*Sparganium*)

43 Inflorescence a dense, cylindric spadix 5–10 cm long, protruding laterally from the base of a leaflike extension of the scape; fresh foliage sweetly fragrant when crushed **ARACEAE** (*Acorus*)

43 Inflorescence of various types, but if spicate, then the flowers borne in 1 or more terminal spikes, or if single and lateral, then the spike much less than 5 cm long; foliage not sweetly fragrant **44**

44 Inflorescence a long, slender, spikelike raceme of many flowers; leaves all basal, terete; fruit splitting lengthwise into 3 or 6, one-seeded segments **JUNCAGINACEAE** (*Triglochin*)

44 Inflorescence of various types; leaves seldom all basal, usually some cauline, or leaves apparently reduced to basal sheaths; leaf blades usually flat **45**

45 Perianth segments evident, consisting of 6 tepals; fruit a capsule or 3 (rarely to 6) follicles **46**

45 Perianth absent or reduced to scales or bristles, the flowers enclosed by 1 or 2 chaffy bracts; fruit an achene or grain **47**

46 Flowers usually numerous in dense clusters or loosely flowered cymes; tepals stiff, chaffy; carpels united, the ovary maturing as a 3-valved, many-seeded capsule; widespread plants **JUNCACEAE** (*Juncus*)

46 Flowers few to several in a bracteate raceme; tepals soft, not chaffy; carpels separate or nearly so to the base, each of the 3(to 6) maturing as a 1 or 2(several) seeded follicle; rare bog plant **SCHEUCHZERIACEAE** (*Scheuchzeria*)

47 Sedges; leaves in 3 vertical ranks on an often trigonous, solid or pithy stem, or leaves reduced to basal sheaths only; leaf sheaths closed around the stem, sometimes splitting with age; flowers each borne in the axil of a scalelike bract and often with several inner subtending scales or bristles; ovary sometimes contained in a saclike covering (perigynium); fruit a beaked achene; styles bifid or trifid **CYPERACEAE**

47 Grasses; leaves in 2 vertical ranks on a terete, usually hollow stem with swollen nodes; leaf sheaths commonly open longitudinally with overlapping margins; flowers each subtended by 2 bracts (lemma and palea) and also by 2 inner, obscure scales (lodicules); ovary never enclosed in a sac; fruit a grain; styles bifid **POACEAE**

48 Broader-leaved monocots; leaves usually curved-parallel veined, sometimes net-veined; perianth parts in multiples of 3 (sometimes obscured by fusion and modification in the Orchidaceae) **49**

48 Dicots; leaves net-veined; perianth segments usually in multiples of 4 or 5 **53**

49 Leaves petiolate **50**

49 Leaves sessile, sometimes sheathing the stem **52**

50 Leaf blades broadly heart-shaped, about as wide as long; inflorescence a short-cylindric spadix subtended by a broadly ovate spathe **ARACEAE** (*Calla*)

50 Leaf blades elliptic, lanceolate, ovate or sagittate; inflorescence other than a spadix 51

51 Flowers in a panicle or branched raceme, perfect or imperfect; ovaries many per flower, superior, maturing as flattened achenes **ALISMATACEAE**

51 Flowers solitary from a narrow spathe, perfect; ovary 1, inferior, maturing as a many-seeded capsule **PONTEDERIACEAE** (*Heteranthera*)

52 Leaves opposite or some whorled above; flowers regular, reddish-orange with dark spots (rarely solid yellow) **LILIACEAE** (*Lilium*)

52 Leaves alternate or mainly basal; flowers irregular, variously colored **ORCHIDACEAE**

53 Leaf blades large and leathery, floating or emergent, oblong to oval or round, with a narrow sinus behind the petiole attachment to the blade, (1)1.5–4 dm long or across; stem a thick, fleshy rhizome buried in the mud; flowers large, 4–15 cm across; water lilies **NYMPHAEACEAE**

53 Leaves smaller and variously shaped; flowers generally smaller 54

54 Small insectivorous bog plant; leaves all basal, the blades rotund, mostly less than 1 cm across, the upper surface covered with reddish, sticky glandular hairs that serve to trap insects **DROSERACEAE** (*Drosera*)

54 Plants not insectivorous; leaves not specialized to trap insects 55

55 Leaves whorled, with 4–12 at each node, the members of each whorl of equal size 56

55 Leaves opposite, alternate or basal, sometimes both opposite and alternate, or if appearing whorled, then the leaves at each node of differing sizes 57

56 Stems weak, ascending to reclining, branched; leaves in whorls of 4–6 **RUBIACEAE** (*Galium*)

56 Stems erect, simple; leaves in whorls of 6–12 **PLANTAGINACEAE** (*Hippuris*)

57 Plants low, succulent, the stems often brittle, green to often strongly red; leaves opposite, small and scalelike; flowers embedded in terminal portions of the fleshy stems **AMARANTHACEAE** (*Salicornia*)

57 Plants without the above combination of characters 58

58 Perianth consisting of a calyx only, the sepals green or often colored and petaloid, sometimes rudimentary; flowers hypogynous 59

58 Perianth consisting of both a calyx and corolla, or if the perianth is of only one series, then the flowers epigynous; flowers hypogynous to epigynous 66

59 Leaves opposite or mostly so 60

59 Leaves alternate or mostly basal (all basal except for leaflike involucres subtending the flowers in *Anemone canadensis*, Ranunculaceae) 62

60 Flowers in terminal inflorescences **CARYOPHYLLACEAE** (*Cerastium*)

60 Flowers solitary in the axils or in axillary clusters 61

61 Flowers white to pinkish, solitary in the axils **PRIMULACEAE** (*Lysimachia maritima*)

61 Flowers greenish, clustered in axillary inflorescences **URTICACEAE**

62 Flowers large and showy, white or yellow, 2–5 cm across **RANUNCULACEAE**

62 Flowers individually small, white, pink or greenish 63

63 Leaves all basal; sepals spurred at the base; pistils many on an elongate receptacle **RANUNCULACEAE** (*Myosurus minimus*)

63 Leaves cauline or partly so; sepals not spurred; pistils 5, 6, 1 or none 64

64 Leaves serrate; pistils 5, fused in a star-shaped ring **PENTHORACEAE** (*Penthorum*)

64 Leaves entire, undulate or lobed; pistil 1 or none 65

65 Stems sheathed at the nodes by a membranous ocrea; perianth petaloid or green to brown and winged **POLYGONACEAE**

65 Stems without ocreae; perianth green, minute and inconspicuous **AMARANTHACEAE**

66 Flowers pouchlike and spurred, yellow to orange-yellow, often reddish-brown spotted **BALSAMINACEAE** (*Impatiens*)

66 Flowers not pouchlike, variously colored 67

67 Inflorescence a simple or compound umbel 68

67 Inflorescence other than an umbel 70

68 Leaves compound; flowers white **APIACEAE**

68 Leaves simple; flowers deeply pink to red or lilac 69

69 Leaves all basal, strongly whitened beneath **PRIMULACEAE** (*Primula*)

69 Leaves opposite, not whitened beneath **APOCYNACEAE** (*Asclepias*)

70 Stamens more than 10, or if as few as 9, then the stamens united below into 3 fascicles 71

70 Stamens 10 or fewer, sometimes united below but not in 3 fascicles 73

71 Leaves opposite **HYPERICACEAE**

71 Leaves alternate or all basal 72

72 Flowers hypogynous, the sepals, petals and stamens attached to the receptacle directly beneath the gynoecium **RANUNCULACEAE** (*Ranunculus*)

72 Flowers perigynous, the sepals, petals and stamens attached around the rim of a saucerlike or disklike hypanthium **ROSACEAE**

73 Leaves pinnately compound, with 3 or more distinct leaflets 74

73 Leaves simple or sometimes pinnately lobed, not divided into distinct leaflets 75

74 Leaves once-pinnate; flowers (or fruits) in simple or branched racemes **FABACEAE**

74 Leaves twice-pinnate; flowers (or fruits) in dense globose clusters **FABACEAE** (*Desmanthus*)

75 Petals separate or lacking 76

75 Petals united, at least toward the base 85

76 Petals present, mainly purple or blue-violet, sometimes white toward the base (actually pink but often drying purple in *Epilobium,* Onagraceae) 77

76 Petals present or absent, white, pink or yellow when present 79

77 Plants acaulescent, the leaves and flowers arising from the base **VIOLACEAE** (*Viola*)

77 Plants caulescent, the leaves and flowers borne on a stem 78

78 Ovary inferior, elongate to linear, appearing like a pedicel of the flower; seeds often with a coma **ONAGRACEAE**

78 Ovary superior, ovoid to globose, enclosed by the tubular calyx; seeds lacking a coma **LYTHRACEAE**

79 Sepals 5; petals 5 80

79 Sepals 4; petals 4 or none 83

80 Leaves all basal, or one leaf sessile above on each flowering scape and the rest basal 81

80 Leaves opposite 82

81 Leaves crenate; flowers greenish, in few- to several-flowered racemes; petals pinnately segmented **SAXIFRAGACEAE** (*Mitella*)

81 Leaves entire; flowers white, solitary on the scapes; petals entire **PARNASSIACEAE** (*Parnassia*)

82 Leaves glandular-serrate **ELATINACEAE** (*Bergia*)

82 Leaves entire **CARYOPHYLLACEAE**

83 Ovary inferior **ONAGRACEAE**

83 Ovary superior, sometimes contained in a calyx cup but free of it 84

84 Leaves alternate, usually shallowly to deeply lobed or compound **BRASSICACEAE**

84 Leaves opposite, entire **LYTHRACEAE**

85 Leaves basal 86

85 Leaves cauline 88

86 Leaves trifoliate **MENYANTHACEAE** (*Menyanthes*)

86 Leaves simple 87

87 Flowers in a spike **PLANTAGINACEAE** (*Plantago*)

87 Flowers solitary on peduncles arising from the plant base **PLANTAGINACEAE** (*Limosella*)

88 Stamens numbering the same as the corolla lobes 89

88 Stamens numbering fewer than the corolla lobes 93

89 Leaves opposite (sometimes appearing whorled because of leaf fascicles in axils) 87

89 Leaves alternate or mostly so 92

90 Flowers yellow **PRIMULACEAE** (*Lysimachia*)

90 Flowers not yellow 91

91 Flowers white or greenish-white, less than 5 mm long; fruit 1 or 2 slender follicles; plants with milky juice **APOCYNACEAE** (*Apocynum*)

91 Flowers purple (drying dark blue), rarely white, 15 mm or more long; fruit a 2-valved capsule; plants with clear juice **GENTIANACEAE**

92 Ovary superior, 4-lobed, splitting into 4 nutlets at maturity; flowers in scorpioid spikes or racemes **BORAGINACEAE**

92 Ovary inferior, or partly so, not lobed, maturing as a capsule; flowers in bracteate racemes or solitary in the upper axils **CAMPANULACEAE**

93 Flowers blue-violet, slightly bilabiate, in dense terminal spikes **VERBENACEAE** (*Verbena*)

93 Flowers of various colors, but when in terminal spikes, the flowers usually pink or yellow, or if blue-violet, then strongly bilabiate 94

94 Flowers in capitate or short-cylindric spikes that are peduncled from the axils **VERBENACEAE** (*Phyla*)

94 Flowers in racemes, spikes, axillary clusters or solitary in the axils 95

95 Ovary 4-lobed, splitting into 4 nutlets at maturity; stems 4-angled **LAMIACEAE**

95 Ovary not lobed, maturing into a 2-valved capsule; stems usually terete 96

96 Leaves all basal, tufted **PLANTAGINACEAE** (*Limosella*)

96 Leaves all or partly cauline 97

97 Leaves pinnatifid 98

97 Leaves entire to serrate or undulate, but not pinnatifid 99

98 Plants annual; flowers axillary **PLANTAGINACEAE** (*Leucospora*)

98 Plants perennial; flowers in terminal spikes **OROBANCHACEAE** (*Pedicularis*)

99 Functional stamens 4 100

99 Functional stamens 2 (a pair of antherless staminodes may also be present) 102

100 Leaves linear, 1-3 mm wide **OROBANCHACEAE** (*Agalinis*)

100 Leaves not linear, wider than 3 mm 101

101 Calyx tubular, the lobes shorter than the tubular portion **PHRYMACEAE** (*Mimulus*)

101 Calyx not tubular, the sepals distinct and imbricate **PLANTAGINACEAE** (*Bacopa*)

102 Flowers in axillary or terminal racemes; calyx lobes 4 **PLANTAGINACEAE** (*Veronica*)

102 Flowers single and pedicelled in leaf axils; calyx lobes 5 103

103 Corolla white to yellow; pair of staminodes absent or very minute; plants glandular-pubescent at least above **PLANTAGINACEAE** (*Gratiola*)

103 Corolla blue-violet; pair of staminodes present; plants glabrous throughout **LINDERNIACEAE** (*Lindernia*)

Wetland regions map (U.S. Army Corps of Engineers). Included in this book are portions of the Western Mountains, Valleys, and Coast (WMTN); Great Plains (GP); and Midwest (MIDW) regions.

Wetland indicator status ratings

Indicator Status	*Abrv.*	*Definition*
Obligate	**OBL**	Almost always occur in wetlands
Facultative Wetland	**FACW**	Usually occur in wetlands, but may occur in non-wetlands
Facultative	**FAC**	Occur in wetlands and nonwetlands
Facultative Upland	**FACU**	Usually occur in non-wetlands, but may occur in wetlands
Upland	**UPL**	Almost never occur in wetlands

GLOSSARY

achene A one-seeded, dry, indehiscent fruit with the seed coat not attached to the mature wall of the ovary.

acid Having more hydrogen ions than hydroxyl (OH) ions; a pH less than 7.

acute Gradually tapered to a tip.

adventive Not native to and not fully established in a new habitat.

alkaline Having more hydroxyl ions than hydrogen ions; pH greater than 7.

alluvial Deposits of rivers and streams.

alternate Borne singly at each node, as in leaves on a stem.

ament Spikelike inflorescence of same-sexed flowers (either male or female); same as catkin.

angiosperm A plant producing flowers and bearing seeds in an ovary.

annual A plant that completes its life cycle in one growing season, then dies.

anther Pollen-bearing part of stamen.

appressed Lying flat to or parallel to a surface.

aquatic Living in water.

areole In leaves, the spaces between small veins.

aromatic Strongly scented.

ascending Angled upward.

asymmetrical Not symmetrical.

auricle An ear-shaped appendage to a leaf or stipule.

awl-shaped Tapering gradually from a broad base to a sharp point.

awn A bristle-like organ.

axil Angle between a stem and the attached leaf.

barb Downward pointing projections.

basal From base of plant.

basic A pH greater than 7.

beard Covering of long or stiff hairs.

berry Fruit with the seeds surrounded by fleshy material.

biennial A plant that completes its life cycle in two growing season, typically flowering and fruiting in the second year, then dying.

blade Expanded, usually flat part of a leaf or petiole.

bog A wet, acidic, nutrient-poor peatland characterized by sphagnum and other mosses, shrubs and sedges. Technically, a type of peatland raised above its surroundings by peat accumulation and receiving nutrients only from precipitation.

boreal Far northern latitudes.

brackish Salty.

bract An accessory structure at the base of some flowers, usually appearing leaflike.

bractlet A secondary bract (*Typha*).

branchlets A small branch.

bristle A stiff hair.

bulblet Small bulb borne above ground, as in a leaf axil.

calcareous fen An uncommon wetland type associated with seepage areas, and which receive groundwater enriched with primarily calcium and magnesium bicarbonates.

calcium-rich Refers to wetlands underlain by limestone or receiving water enriched by calcium compounds.

calyx All the sepals of a flower.

capsule A dry, dehiscent fruit splitting into 3 or more parts.

carpel Fertile leaf of an angiosperm, bearing the ovules. A pistil is made up of one or more carpels.

caryopsis The dry, indehiscent seed of grasses.

catkin Spikelike inflorescence of same-sexed flowers (either male or female); same as ament.

chaff Thin, dry scales; in the Asteraceae, sometimes found as chaffy bracts on the receptacle.

circumboreal Refers to a species distribution pattern which circles the earth's boreal regions.

clasping Leaves that partially encircle the

stem at the base.

cleistogamous Type of flower that remains closed and is self-pollinated.

clumped Having the stems grouped closely together; tufted.

colony-forming A group of plants of the same species, produced either vegetatively or by seed.

column The joined style and filaments in the Orchidaceae.

coma A tuft of fine hairs, especially at the tip of a seed.

composite An inflorescence that is made up of many tiny florets crowded together on a receptacle; members of the Aster Family (Asteraceae).

compound A leaf with two or more leaflets.

concave Curved inward.

cone The dry fruit of conifers composed of overlapping scales.

conifer Cone-bearing woody plants.

convex Curved outward.

corm An enlarged, rounded, underground stem, usually covered with papery scales or modified leaves.

corolla Collectively, all the petals of a flower.

corymb A flat-topped or convex inflorescence.

crisped An irregularly crinkled or curled leaf margin.

crown Persistent base of a plant, especially a grasses.

culm The stem of a grass or grasslike plant, especially a stem with the inflorescence.

cyme A type of inflorescence in which the central flowers open first.

deciduous Not persistent.

dehiscent Splitting open at maturity.

dicots One of two main divisions of the Angiosperms (the other being the Monocots); plants having 2 seed leaves (cotyledons), net-venation, and flower parts in 4s or 5s (or multiples of these numbers).

dioecious Bearing only male or female flowers on a single plant.

disarticulation Spikelets breaking either above or below the glumes when mature, the glumes remaining in the head if disarticulation above the glumes, or the glumes falling with the florets if disarticulation is below the glumes.

discoid In composite flowers (Asteraceae), a head with only disk (tubular) flowers, the ray flowers absent.

disjunct A population of plants widely separated from its main range.

disk In the Asteraceae, the central part of the head, composed of tubular flowers.

dissected Leaves divided into many smaller segments.

disturbed Natural communities altered by human influences.

divided Leaves which are lobed nearly to the midrib.

dolomite A type of limestone consisting of calcium magnesium carbonate.

driftless area Portions of sw Wisconsin, ne Iowa, and se Minnesota that are not covered by glacial drift.

drupe A fleshy fruit with a single large seed such as a cherry.

elliptic Broadest at the middle, gradually tapering to both ends.

emergent Growing out of and above the water surface.

emersed leaf Growing above the water surface or out of water.

endangered A species in danger of extinction throughout all or most of its range if current trends continue.

endemic A species restricted to a particular region.

entire With a smooth margin.

erect Stiffly upright.

escape A cultivated plant which establishes itself outside of cultivation.

evergreen Plant retaining its leaves throughout the year.

exserted Extending beyond the mouth of a structure such as stamens extending out from the mouth of the corolla.

fen An open wetland usually dominated by herbaceous plants, and fed by in-flowing, often calcium- and/or magnesium-rich water; soils vary from peat to clays and silts.

fern Perennial plants with spore-bearing leaves similar to the vegetative leaves and bearing sporangia on their underside, or the spore-bearing leaves much modified.

fibrous A cluster of slender roots, all with the same diameter.

filament The stalk of a stamen which supports the anther.

floating mat A feature of some ponds where plant roots form a carpet over some or all of the water surface.

floodplain That part of a river valley that is occasionally covered by flood waters.

floret A small flower in a dense cluster of flowers; in grasses the flower with its attached lemma and palea.

follicle A dry, dehiscent fruit that splits along one side when mature.

genus The first part of the scientific name for a plant or animal (plural genera).

gland An appendage or depression which produces a sticky or greasy substance.

glaucous Having a bluish appearance.

glumes A pair of small bracts at base of each spikelet the lowermost (or first) glume usually smaller the upper (or second) glume usually longer.

grain The fruit of a grass; the swollen seed-like protuberance on the fruit of some *Rumex*.

gymnosperm Plants in which the seeds are not produced in an ovary, but usually in a cone.

gynophore The central stalk of some flowers, especially in cat-tails (*Typha*).

hardwoods Loosely used to contrast most deciduous trees from conifers.

herb A herbaceous, non-woody plant.

herbaceous Like an herb; also, leaflike in appearance.

hummock A small, raised mound formed by certain species of sphagnum moss.

humus Dark, well-decayed organic matter in soil.

hybrid A cross-breed between two species.

hypanthium A ring, cup, or tube around the ovary; the sepals, petals and stamens are attached to the rim of the hypanthium.

indehiscent Not splitting open at maturity.

indusium In ferns, a membranous covering over the sorus (plural indusia).

inferior The position of the ovary when it is below the point of attachment of the sepals and petals.

inflorescence A cluster of flowers.

insectivorous Refers to the insect trapping and digestion habit of some plants as a nutrition supplement.

interdunal swale Low-lying areas between sand dune ridges.

internode Portion of a stem between two nodes.

introduced A non-native species.

invasive Non-native species causing significant ecological or economic problems.

involucral bract A single member of the involucre; sometimes called phyllary in composite flowers (Asteraceae)

involucre A whorl of bracts, subtending a flower or inflorescence.

irregular flower Not radially symmetric; with similar parts unequal.

joint A node or section of a stem where the branch and leaf meet.

keel A central rib like the keel of a boat.

lance-shaped Broadest near the base, gradually tapering to a narrower tip.

lateral Borne on the sides of a stem or branch.

lax Loose or drooping.

leaf axil The point of the angle between a stem and a leaf.

leaflet One of the leaflike segments of a compound leaf.

lemma In grasses, the lower bract enclosing the flower (the upper, smaller bract is the palea).

lens-shaped Biconvex in shape (like a lentil).

lenticel Blisterlike openings in the epidermis of woody stems, admitting gases to and from the plant, and often appearing as small oval dots on bark.

ligulate Having a ligule; in the Asteraceae, the strap-shaped corolla of a ray floret.

ligule In grasses and grasslike plants, the membranous or hairy ring at top of sheath between the blade and stem.

linear Narrow and flat with parallel sides.

lip Upper or lower part of a 2-lipped corolla; also the lower petal in most orchid flowers.

lobed With lobes; in leaves divisions usually not over halfway to the midrib.

local Occurring sporadically in an area.

low prairie Wet and moist herbaceous plant community, typically dominated by grasses.

margin The outer edge of a leaf.

marl A calcium-rich clay.

marsh Wetland dominated by herbaceous plants, with standing water for part or all the growing season, then often drying at the surface.

megaspore Large, female spores.

microspore Small, male spores.

midrib The prominent vein along the main axis of a leaf.

mixed forest A type of forest composed of both deciduous and conifer trees.

moat The open water area ringing the outer edge of a peatland or floating mat.

monecious Having male and female reproductive parts in separate flowers on the same plant.

monocots One of two main divisions of the Angiosperms (the other being the Dicots); plants with a single seed leaf (cotyledon); typically having narrow leaves with parallel veins, and flower parts in 3s or multiples of 3.

muck An organic soil where the plant remains are decomposed to the point where the type of plants forming the soil cannot be determined.

mucro A sharp point at termination of an organ or other structure.

naked Without a covering; a stalk or stem without leaves.

native An indigenous species.

naturalized An introduced species that is established and persistent in an ecosystem.

needle A slender leaf, as in the Pinaceae.

nerve A leaf vein.

neutral A pH of 7.

node The spot on a stem or branch where leaves originate.

nutlet A small dry fruit that does not split open along a seam.

oblanceolate Reverse lance-shaped; broadest at the apex, gradually tapering to the narrower base.

oblong Broadest at the middle, and tapering to both ends, but broader than elliptic.

obovate Broadly rounded at the apex, becoming narrowed below.

ocrea A tube-shaped stipule or pair of stipules around the stem; characteristic of the Smartweed Family (Polygonaceae).

opposite Leaves or branches which are paired opposite one another on the stem.

organic Soils composed of decaying plant remains.

oval Elliptical.

ovary The lower part of the pistil that produces the seeds.

ovate Broadly rounded at the base, becoming narrowed above; broader than lanceolate.

palea The uppermost of the two inner bracts subtending a grass flower (the lower bract is the lemma).

palmate Divided in a radial fashion, like the fingers of a hand.

panicle An arrangement of flowers consisting of several racemes.

pappus The modified sepals of a composite flower which persist atop the ovary as bristles, scales or awns.

parallel-veined With several veins running from base of leaf to leaf tip, characteristic of most monocots.

peat An organic soil formed of partially decomposed plant remains.

peatland A wetland whose soil is composed primarily of organic matter (mosses, sedges, etc.); a general term for bogs and fens.

pepo A fleshy, many-seeded fruit with a tough rind, as a melon.

perennial Living for 3 or more years.

perfect A flower having both male (stamens) and female (pistils) parts.

perianth Collectively, all the sepals and petals of a flower.

perigynium A sac-like structure enclosing the pistil in *Carex* (plural perigynia).

petal An individual part of the corolla, often white or colored.

petiole The stalk of a leaf.

phyllary An involucral bract subtending the flower head in composite flowers (Asteraceae).

phyllode An expanded petiole.

pinna The primary or first division in a fern frond or leaf (plural pinnae).

pinnate Divided once along an elongated axis into distinct segments.

pinnule The pinnate segment of a pinna.

pistil The seed-producing part of the flower, consisting of an ovary and one or more styles and stigmas.

pith A spongy central part of stems and branches.

pollen The male spores in an anther.

prairie An open plant community dominated by herbaceous species, especially grasses.

prostrate Lying flat on the ground.

raceme A grouping of flowers along an elongated axis where each flower has its own stalk.

rachilla A small stem or axis.

rachis The central axis or stem of a leaf or inflorescence.

radiate heads In composite flowers, heads with both ray and disk flowers (Asteraceae).

ray flower A ligulate or strap-shaped flower in the Asteraceae, where often the outermost series of flowers in the head.

receptacle In the Asteraceae, the enlarged summit of the flower stalk to which the sepals, petals, stamens, and pistils are usually attached.

recurved Curved backward.

regular Flowers with all the similar parts of the same form; radially symmetric.

rhizome An underground, horizontal stem.

rib A pronounced vein or nerve.

rootstock Similar to rhizome but referring to any underground part that spreads the plant.

rosette A crowded, circular clump of leaves.

samara A dry, indehiscent fruit with a well-developed wing.

saprophyte A plant that lives off of dead organic matter.

scale A tiny, leaflike structure; the structure that subtends each flower in a sedge (Cyperaceae).

scape A naked stem (without leaves) bearing the flowers.

section Cross-section.

secund Flowers mostly on 1 side of a stalk or branch.

sedge meadow A community dominated by **sedges** (Cyperaceae) and occurring on wet, saturated soils.

seep A spot where water oozes from the ground.

sepal A segment of the calyx; usually green in color.

sheath Tube-shaped membrane around a stem, especially for part of the leaf in grasses and sedges.

shrub A woody plant with multiple stems.

silicle Short fruit of the Mustard Family (Brassicaceae), normally less than 2x longer as wide.

silique Dry, dehiscent, 2-chambered fruit of the Mustard Family (Brassicaceae), longer than a silicle.

simple An undivided leaf.

sinus The depression between two lobes.

smooth Without teeth or hairs.

sorus Clusters of spore containers (plural sori).

spadix A fleshy axis in which flowers are embedded.

spathe A large bract subtending or enclosing a cluster of flowers.

spatula-shaped Broadest at tip and tapering to the base.

sphagnum moss A type of moss common in peatlands and sometimes forming a continuous carpet across the surface; sometimes forming layers several meters thick; also loosely called peat moss.

spike A group of unstalked flowers along an unbranched stalk.

spikelet A small spike; the flower cluster (inflorescence) of grasses (Poaceae) and sedges (Cyperaceae).

sporangium The spore-producing structure (plural sporangia).

sporophyll A modified, spore-bearing leaf.

spreading Widely angled outward.

spring A place where water flows naturally from the ground.

spur A hollow, pointed projection of a flower.

stamen The male or pollen-producing organ of a flower.

staminode An infertile stamen.

stem The main axis of a plant.

stigma The terminal part of a pistil which receives pollen.

stipe A stalk.

stipule A leaflike outgrowth at the base of a leaf stalk.

stolon A horizontal stem lying on the surface of the soil.

style The stalklike part of the pistil between the ovary and the stigma.

subspecies A subdivision of the species forming a group with shared traits which differ from other members of the species (subsp.).

subtend Attached below and extending upward.

succulent Thick, fleshy and juicy.

superior Referring to the position of the ovary when it is above the point of attachment of sepals, petals, stamens, and pistils.

swale A slight depression.

swamp Wooded wetland dominated by trees or shrubs; soils are typically wet for much of year or sometimes inundated.

talus Fallen rock at the base of a slope or cliff.

taproot A main, downward-pointing root.

tendril A threadlike appendage from a stem or leaf that coils around other objects for support (as in Vitis).

tepal Sepals or petals not differentiated from one another.

terminal Located at the end of a stem or stalk.

thallus A small, flattened plant structure, without distinct stem or leaves.

thicket A dense growth of woody plants.

threatened A species likely to become endangered throughout all or most of its range if current trends continue.

translucent Nearly transparent.

tree A large, single-stemmed woody plant.

tuber An enlarged portion of a root or rhizome.

tubercle Base of style persistent as a swelling atop the achene different in color and texture from achene body .

tundra Treeless plain in arctic regions, having permanently frozen subsoil.

turion A specialized type of shoot or bud that overwinters and resumes growth the following year.

umbel A cluster of flowers in which the flower stalks arise from the same level.

umbelet A small, secondary umbel in an umbel, as in the Apiaceae.

upright Erect or nearly so.

utricle A small, one-seeded fruit with a dry, papery outer covering.

valve A segment of a dehiscent fruit; the wing of the fruit in Rumex.

variety Taxon below subspecies and differing from other varieties within the same subspecies (var.).

vein A vascular bundle, as in a leaf.

velum The membranous flap that partially covers the sporangium in Isoetes.

vine A trailing or climbing plant, dependent on other objects for support.

whorl A group of 3 or more parts from one point on a stem.

wing A thin tissue bordering or surrounding an organ.

woody Xylem tissue (the vascular tissue which conducts water and nutrients).

REFERENCES

NOTE *A sampling of relatively recent treatments of the Great Plains flora are listed below.*

Flora of North America North of Mexico [online and in print]. Flora of North America Editorial Committee, eds. 1993+. 16+ vols. New York and Oxford (*floranorthamerica.org*).

The Flora of Nebraska. 2007. Robert Kaul, David Sutherland, Steven Rolfsmeier.

Flora of the Great Plains. 1986. Ted Barkley, editor. University of Kansas Press.

Floristic Synthesis of North America, Version 1.0. Biota of North America Program (BONAP, in press). 2014. J.T. Kartesz. (*www.bonap.org*)

Grassland Plants of South Dakota and the Northern Great Plains. 1999. Johnson, James R. and Gary E. Larson Ed. Mary Brashier. South Dakota State University College of Atriculture & Biological Science; South Dakota Agricultureal Experiment Station.

Manual of Montana Vascular Plants. 2012. Peter Lesica. Botanical Research Inst. of Texas.

The National Wetland Plant List: 2016 wetland ratings. Lichvar, R.W., D.L. Banks, W.N. Kirchner, and N.C. Melvin. 2016. Phytoneuron 2016-30: 1-17.

Plants of the Black Hills and Bear Lodge Mountains. Gary E Larson and James R Johnson.

Shinner's and Mahler's Illustrated Flora of North Central Texas. 1999. George Diggs, Barney Lipscomb, Robert O'Kennon, and Linny Heagy. Botanical Research Institute of Texas.

Southwestern and Texas Wildflowers (Peterson field guide series) 1984. Theodore F. Niehaus, Charles L. Ripper and Virginia Savage. Houghton Mifflin.

Weeds of Nebraska and the Great Plains. 2nd Edition. 1995. James L Stubbendieck. Nebraska Dept. of Agriculture, Bureau of Plant Industry in cooperation with the University of Nebraska-Lincoln.

Wildflowers & Grasses of Kansas. 2005. Michael John Haddock. University Press of Kansas.

Wildflowers of the Western Plains. 2008. Zoe Merriman Kirkpatrick. University of Texas Press.

NOTE *Selected references from the 1993 USDA Forest Service publication.*

Aiken, S. G. 1981. A conspectus of *Myriophyllum* (Haloragaceae) in North America. Brittonia 33:5 7–69.

Argus, G.W. 1980. The typification and identity of *Salix eriocephala* Michx. (Salicaceae). Brittonia 32:170–177.

Barker, W. T. and J. Hanson. 1976. A record of *Ophioglossum vulgatum* L. for North Dakota. Amer. Fern J. 66:137.

Beal, E. o. 1956. Taxonomic revision of the genus *Nuphar* of North America and Europe. J. Elisha Mitch. Sci. Soc. 72:317–346.

Beetle, A. A. 1943. The North American variations of *Distichlis spicata*. Bull. Torrey Bot.

Beetle, A. A. 1947. *Scirpus*. North Amer. Flora 18:481–504.

Benson, L. 1948. A treatise on the North American Ranunculi. Amer. Midl. Naturalist 40:1–261.

Bogin, C. 1955. Revision of the genus *Sagittaria*. Mem. New York Bot. Gard. 9:179–223.

Cantino, P. D. 1981. Change of status for *Physostegia virginiana* var. *ledinghamii* (Labiatae) and evidence for a hybrid origin. Rhodora 833 11–1 18.

Chinnappa, C. C. and J. K. Morton. 1976. Studies on the *Stellaria longipes* Goldie complex - variation in wild populations. Rhodora 78:488–502.

Clark, H. L. and J. W. Thieret. 1968. The duckweeds of Minnesota. Michigan Bot. 7:67–75.

Clausen, R. T. 1936. Studies in the genus *Najas* in the northern United States. Rhodora 38:333–345.

Cook, C. D. K. 1963. Studies in *Ranunculus* subgenus *Batrachium* (DC.) A. Gray 11. General morphological considerations in the taxonomy of the subgenus. Watsonia 5:294–303.

Cook, C. D. K. 1966. A monographic study of *Ranunculus* subgenus Batrachium (DC.) A. Gray. Mitt. Bot. Staatssamml. Munchen 6:47–237.

Correll, D. and H. Correll. 1975. Aquatic and wetland plants of southwestern United States. Vol. I and 11. Stanford University Press, Stanford, CA.

Cowardin, L. M., V. Carter, F. C. Golet and E. T. LaRoe. 1979. Classification of wetlands and deepwater habitats of the United States. U. S. Fish Wildl. Serv., Biol. Serv. Prog. FWS/OBS–79/31.

Croat, T. B. 1967. The genus *Solidago* of the north central Great Plains (U.S.A.). Unpubl. Ph.D. thesis, Univ. Kans., Lawrence.

Daubs, E. H. 1965. A monograph of Lemnaceae. Illinois Biol. Monogr. 34:1–118.

Dorn, R. D. 1970. The willows of Montana. The Herbarium, Dept. of Botany and Microbiology, Montana State Univ., Bozeman. 18 pp.

Dorn, R. D. 1975. A systematic study of *Salix* section Cordatae in North America. Canad. J. Bat. 53:1491–1522.

Dorn, R. D. 1977. Willows of the Rocky Mountain states. Rhodora 79:390–429.

Duncan, T. 1980. A taxonomic study of the *Ranunculus hispidus* complex in the Western Hemisphere. Univ. of Calif. Bot. Publ. 77:1–125.

Eckenwalder, J. E. 1977. North American cottonwoods (*Populus, Salicaceae*) of sections Abaso and Aigeros. J. Arnold Arbor. 58: 193–207.

Erdman, K. S. 1965. Taxonomy of the genus *Sphenopholis* (Gramineae). Iowa State Col J. Sci. 39:289–336.

Fassett, N. C. 1957. A manual of aquatic plants. With revision appendix by E. C. Ogden. University of Wisconsin Press, Madison.

Fassett, N. C. 1951. *Callitriche* in the New World. Rhodora 53:137–155, 161–182, 185–194, 209–222.

Fassett, N. C. and B. M. Calhoun. 1952. Introgression between *Typha latifolia* and *T. angustifolia*. Evolution 6:367–3 79.

Fernald, M. L. 1905. The North American species of *Eriophorum*. Rhodora 7:81–92, 129–136.

Fernald, M. L. 1932. The linear-leaved North American species of *Potamogeton* section Axillares. Mem. Amer. Acad. Arts 17:1–183.

Fernald, M. L. 1938. *Pilea* in eastern North America. Rhodora 38:169–170.

Fernald, M. L. 1946. North American representatives of *Alisma plantago-aquatica*. Rhodora 48:86–88.

Friedland, S. 1941. The American species of *Hemicarpha*. Amer. J. Bot. 28:855–861.

Froiland, S. G. 1962. The genus *Salix* in the Black Hills of South Dakota. Tech. Bull. U.S.D.A. 1269:1–75.

Godfread, C. and W. T. Barker. 1975. Butomaceae: a new family record for North Dakota. Rhodora 77:160–161.

Godfrey, R. K. and J. W. Wooten. 1979. Aquatic and wetland plants of southeastern United States - Monocotyledons. University of Georgia Press, Athens.

Godfrey, R. K. and J. W. Wooten. 1981. Aquatic and wetland plants of southeastern United States - Dicotyledons. University of Georgia Press, Athens.

Gould, F. W., M. A. Ali and D. E. Fairbrothers. 1972. A revision of *Echinochloa* in the United States. Amer. Midl. Naturalist 87:36–59.

Great Plains Flora Association. 1977. Atlas of the flora of the Great Plains. Iowa State University Press, Ames.

Great Plains Flora Association. 1986. Flora of the Great Plains. University Press of Kansas, Lawrence.

Harms, L. J. 1968. Cytotaxonomic studies in *Eleocharis* subser. Palustres: central United States taxa. Amer. J. Bot. 55:966–974.

Hartog, C. den and F. van der Plas. 1970. A synopsis of the Lemnaceae. Blumea 18:355–368.

Hauke, R. L. 1965. Preliminary reports on the flora of Wisconsin. No. 54. Equisetaceae - Horsetail Family. Trans. Wisconsin Acad. Sci. 54:331–346.

Haynes, R. R. 1974. A revision of North American *Potamogeton* subsection Pusilli (Potamogetonaceae). Rhodora 76:564–649.

Heiser, C. B. 1969. The North American sunflowers (*Helianthus*). Mem. Torrey Bot. Club 22:1–218.

Henderson, N. C. 1962. A taxonomic revision of the genus *Lycopus*. Amer. Midl. Naturalist 68:95–138.

Hendricks, A. J. 1957. A revision of the genus *Alisma* (Dill.) L. Amer. Midl. Naturalist 58:470–493.

Hermann, F. J. 1970. Manual of carices of the Rocky Mountains and Colorado Basin. U.S.D.A. Forest Serv. Handbook No. 374, Washington, D.C.

Hitchcock, A. S. 1971. Manual of the grasses of the United States. Vol. 1 and 2. Revised by Agnes Chase. Dover Publications, Inc., New York.

Hoch, P. C. and P. H. Raven. 1977. New combinations in *Epilobium* (Onagraceae). Ann. Missouri Bot. Gard. 64:136.

Hotchkiss, N. and H. L. Dozier. 1949. Taxonomy and distribution of North American cattails. Amer. Midl. Naturalist 41:237–254.

Howell, J. T. 1947. Remarks on *Triglochin concinna*. Leafl. W. Bot. 5:13–19.

Iltis, H. H. 1965. The genus *Gentianopsis* (Gentianaceae). Transfers and phytogeographic comments. Sida 2:129–154.

Kolstad, O. A. 1966. The genus *Carex* of the High Plains, Prairie Plains and associated woodlands in Kansas, Nebraska, South and North Dakota. Unpubl. Ph.D. thesis, Univ. Kans., Lawrence.

Kral, R. 1971. A treatment of *Abildgaardia*, *Bulbostylis* and *Fimbristylis* (Cyperaceae) for North America. Sida 4:57–227.

Kral, R. 1978. A synopsis of *Fuirena* (Cyperaceae) for the Americas north of South America. Sida 7:309–354.

Landolt, E. 1980. Biosystematic investigation of the family of duckweeds (Lemnaceae), vol. 1. Veroff. Geobot. Inst. ETH, Stiftung Rubel, Zurich, 70:1–247.

Larson, G. E. 1976. The Potamogetonaceae in North Dakota. Prairie Natur. 8:1–18.

Lee, D. W. 1975. Population variation and introgression in North American *Typha*. Taxon 24:633–641.

Love, A. 1961. Some notes on *Myriophyllum spicatum*. Rhodora 63:139–145.

Love, D. and F. J. Bernard. 1958. *Rumex stenophyllus* in N. America. Rhodora 60:54–57.

Luer, C. A. 1975. The native orchids of the United States and Canada excluding Florida. The New York Botanical Garden.

Lundell, C. L. 1959. Studies in *Physostegia*. I. New species and observations on others. Wrightia 2:4–12.

Magrath, L. K. 1973. The native orchids of the prairies and plains region of North America. Unpubl. Ph.D. thesis, Univ. Kans., Lawrence.

Marcks, B. G. 1974. Preliminary reports on the flora of Wisconsin. No. 66. Cyperaceae I1 - Sedge Family 11. The genus *Cyperus* - the umbrella sedges. Trans. Wisconsin Acad. Sci. 62:261–284.

Martens, T. R. and P. H. Raven. 1965. Taxonomy of *Polygonum*, section Polygonum (Avicularia) in North America. Madrono 18:85–92.

Mathias, M. E. and L. Constance. 1942. A synopsis of the American species of *Cicuta*. Madrono 6:145–151.

Muenscher, W. C. 1944. Aquatic plants of the United States. Comstock Publishing Co., Inc., Ithaca, NY.

Mulligan, G. A. 1980. The genus *Cicuta* in North America. Can. J. Bot. 58:1755–1767.

Mulligan, G. A. and A. E. Porsild. 1966. *Rorippa calycina* in the Northwest Territories. Canad. J. Bot. 44:1105–1106.

Munz, P. A. 1965. *Epilobium*. In North Amer. Flora Ser. 11. Part 5. New York Bot. Gard. 198–225.

Nichols, S. A. 1975. Identification and management of Eurasian water milfoil in Wisconsin. Trans. Wisconsin Acad. Sci. 63:116–128.

Ogden, E. C. 1943. The broad-leaved species of *Potamogeton* of North America north of Mexico. Rhodora 45:57–105, 119–163, 171–214.

Ownbey, G.B. and T. Morley. 1991. Vascular plants of Minnesota: A checklist and Atlas. University of Minnesota Press, Minneapolis.

Patten, B. C., Jr. 1954. The status of some American species of *Myriophyllum* as revealed by the discovery of intergrade material between *M. exalbescens* Fern. and *M. spicatum* L. in New Jersey. Rhodora 56:213–225.

Pelvit, J. and W. T. Barker. 1975. North Dakota ferns and fern allies. Tri-College Center for Environmental Studies Publ. Ser. 1:1–24.

Pennell, F. W. 1921. *Veronica* in North and South America. Rhodora 23:1–22, 29–41.

Petrick-Ott, A. J. 1974. A county checklist of the ferns and fern allies of Kansas, Nebraska, South Dakota and North Dakota. Rhodora 77:478–511.

Pogan, E. 1963. Taxonomical value of *Alisma triviale* Pursh and *A. subcordatum* Raf. Can. J. Bot. 41:1011–1013.

Pohl, R. W. 1969. *Muhlenbergia* subgenus *Muhlenbergia* (Gramineae) in North America. Amer. Midl. Naturalist 82:512–542.

Pohl, R. W. 1978. How to know the grasses, third edition. Wm. C. Brown Publishers, Dubuque, Iowa.

Puff, C. 1975. The *Galium trifidum* group (Galium sect. Aparinoides, Rubiaceae). Canad. J. Bot. 54:1911–1945

Puff, C. 1977. The *Galium obtusum* group (Galium sect. Aparinoides, Rubiaceae). Bull. Torrey Bot. Club 104:202–208.

Raup, H. M. 1959. The willows of boreal western America. Contr. Gray Herb. 185:1–95.

Ray, J. D., Jr. 1956. The genus *Lysimachia* in the New World. Illinois Biol. Monogr. 24:1–160.

Rechinger, K. H., Jr. 1937. The North American species of *Rumex*. Publ. Field Mus. Nat. Hist., Bot. Ser. 17:1–151.

Reznicek, A. A. and R. S. W. Bobbette. 1976. The taxonomy of *Potamogeton* subsection Hybridi in North America. Rhodora 78:650–673.

Rossbach, C. B. 1939. Aquatic *Utricularias*. Rhodora 41:113–128.

Rossbach, R. P. 1940. *Spergularia* in North America and South America. Rhodora 42:57–83, 105–143, 158–193, 203–213.

Schuyler, A. E. 1967. A taxonomic revision of North American leafy species of *Scirpus*. Proc. Acad. Nat. Sci. Philadelphia 119: 295–323.

Schuyler, A. E. 1974. Typification and application of the names *Scirpus americanus* Pers., *S. olneyi* Gray and *S. pungens* Vahl. Rhodora 76:51–52.

Sherff, E. E. 1937. The genus *Bidens*. Publ. Field Mus. Nat. Hist., Bot. Ser. 16:1–709.

Sheviak, C. J. 1973. A new *Spiranthes* from the grasslands of central North America. Bot. Mus. Leafl. 23:285–297.

Sheviak, C. J. and M. L. Bowles. 1986. The prairie fringed orchids: a pollinator isolated species pair. Rhodora 88:267–290.

Shinners, L. H. 1953. Synopsis of the United States species of *Lythrum* (Lythraceae). Field and Lab. 21:80–89.

Smith, S. G. 1967. Experimental and natural hybrids in North American *Typha* (Typhaceae). Amer. Midl. Naturalist 78125 7–287.

St. John, H. 1965. Monograph of the genus *Elodea*. Part 4: The species of the eastern and central North America. Rhodora 6711–35, 155–180.

Stanford, E. E. 1925. The amphibious group of *Polygonum*, subgenus Persicaria. Rhodora 27:109–112, 125–130, 146–152, 156–166.

Stason, M. 1926. The marsileas of the western United States. Bull. Torrey Bot. Club 53:473–478.

Stewart, R. E. and H. A. Kantrud. 1971. Classification of natural ponds and lakes in the glaciated prairie region. U. S. Fish Wildl. Serv. Resource Publ. 92.

Stewart, R. E. and H. A. Kantrud. 1972. Vegetation of prairie potholes, North Dakota, in relation to quality of water and other environmental factors. U. S. Geol. Surv. Prof. Paper 585-D.

Stuckey, R. L. 1968. Distributional history of *Butomus umbellatus* in the western Lake Erie and Lake St. Claire region. Michigan Bot. 7:134–142.

Stuckey, R. L. 1972. Taxonomy and distribution of the genus *Rorippa* in North America. Sida 4:279–430.

Svenson, H. K. 1957. *Eleocharis*. North Amer. Flora 18:509–540.

Tryon, R. M. 1954. The ferns and fern allies of Minnesota. Univ. of Minnesota Press, Minneapolis.

Wahl, H. A. 1952–1953 (1 954). A preliminary study of the genus *Chenopodium* in North America. Bartonia 27:1–46.

Ward, R. L. 1973. A cytotaxonomic study of the *Scirpus lacustris* L. complex in the northern Great Plains. Unpubl. Ph.D. thesis, N. Dak. State Univ., Fargo.

Weigand, K. M. 1900. *Juncus tenuis* Willd. and some of its North American allies. Bull. Torrey Bot. Club 27:511–527.

Wheeler, G. 1981. A study of the genus *Carex* in Minnesota (Volumes I and 11). Unpubl. Ph.D. thesis, Univ. Minn., St. Paul.

Wooten, J. W. 1973. Taxonomy of seven species of *Sagittaria* from eastern North America. Brittonia 25:64–74.

INDEX

Synonyms are listed in italics.

www.ingramcontent.com/pod-product-compliance
Lightning Source LLC
Chambersburg PA
CBHW051242020426
42333CB00025B/3023

* 9 7 8 1 9 5 1 6 8 2 1 5 6 *